Comparative Archaeology and Paleoclimatology

Socio-cultural responses to a changing world

Edited by

Maximilian O. Baldia
Timothy K. Perttula
Douglas S. Frink

BAR International Series 2456
2013

Published in 2016 by
BAR Publishing, Oxford

BAR International Series 2456

Comparative Archaeology and Paleoclimatology

ISBN 978 1 4073 1064 0

BAR Publishing is the trading name of British Archaeological Reports (Oxford) Ltd. British Archaeological Reports was first incorporated in 1974 to publish the BAR Series, International and British. In 1992 Hadrian Books Ltd became part of the BAR group. This volume was originally published by Archaeopress in conjunction with British Archaeological Reports (Oxford) Ltd / Hadrian Books Ltd, the Series principal publisher, in 2013. This present volume is published by BAR Publishing, 2016.

Printed in England

BAR
PUBLISHING

BAR titles are available from:

BAR Publishing
122 Banbury Rd, Oxford, OX2 7BP, UK
EMAIL info@barpublishing.com
PHONE +44 (0)1865 310431
FAX +44 (0)1865 316916
www.barpublishing.com

Content

List of Figures

List of Tables

List of Authors

Maximilian O. BALDIA, Institute for the Study of Earth and Man, Southern Methodist University. Mailing address: The Comparative Archeology WEB, 816 / 814 Hillside Avenue, Plainfield, NJ 07060-3105, USA,
Email: baldia.1@osu.edu

Matthew BOULANGER, Archaeometry Laboratory, University of Missouri Research Reactor, 1513 Research Park Drive, Columbia, MO 65211, USA,
Email: boulangerm@missouri.edu

David CALADO, Instituto Português do Património Arquitectónico (IPPAR), State Department for Cultural Affairs, Portugal,
Email: calado.david@gmail.com

Maria Dolores CÀMALICH, University of La Laguna, Spain

William J. FOLAN, Centro de Investigaciones Historicas y Sociales, Autonomous University of Capuche, Mexico

Douglas S. FRINK, Physical and Earth Sciences, Worcester State College, 486 Chandler Street, Worcester, MA 01602-2597, USA,
Email: Douglas.Frink@worcester.edu, dsfrink@aol.com

Sándor GULYÁS, University of Szeged, Department of Geology and Paleontology, H-6722 Szeged, Egyetem u. 2-6, Hungary. Email: gulyas-sandor@t-online.hu

Joel GUNN, New South Associates, Inc., Mebane, North Carolina, USA,
Email: jdgunn3@mindspring.com, jdgunn@uncg.edu

Joseph M. HERBERT, Colorado State University, Archaeologist, Cultural Resources Program, Fort Bragg. Mailing address: Department of the Army, Directorate of Public Works (Imse Brg Pw-Herbert), Hq. Fort Bragg Garrison Command (Abn.), Installation Management Agency, Bldg 3-1333 Butner Road, Fort Bragg, NC 28310-5000, USA

Katalin HERBICH, University of Szeged, Department of Geology and Paleontology, H-6722 Szeged Egyetem u.2-6, Hungary

Ferenc HORVÁT, Ferenc Móra Museum, H-6722 Szeged, Roosevelt tér 1-3, Hungary.

Mariann IMRE, University of Szeged, Department of Geology and Paleontology, H-6722 Szeged Egyetem u.2-6. Hungary

Imola JUHÁSZ, Institute of Archeology, Hungarian Academy of Sciences, H-1014 Budapest, Úri u.49, Hungary.
Email: juhasz.imola@vnet.hu

Birgit KEDING, Universität zu Köln, Forschungsstelle Afrika, Jennerstr. 8, 50823 Köln, Germany

Róbert KERTÉSZ, János Damjanich Museum, 5000 Szolnok Kossuth tér 4, Hungary

Lars LARSSON, Institutionen för Arkeologi och Antikens Historia, Lund University, Sandgatan 1, S-223 50 Lund, Sweden.
Email: Lars.Larsson@ark.lu.se

Munyaradzi MANYANGA, Department of Anthropology and Archaeology, 8-3 Humanities Building, University of Pretoria Hatfield Campus, 002 Pretoria, South Africa,
Email: Munya.Manyanga@up.ac.za

Francesco MENOTTI, Institute of Prehistory and Archaeological Science, University of Basel, Spalenring 145, CH-4055 Basel, Switzerland,
Email: francesco.menotti@unibas.ch

Edward MITCHELL, Laboratory of Soil Biology, University of Neuchâtel, Rue Emile Argand 11, CH-2009 Neuchâtel, Switzerland. Email: edward.mitchell@epfl.ch

Sándor MOLNÁR, University of Szeged, Department of Geology and Paleontology, H-6722 Szeged Egyetem u. 2-6, Hungary

José Miguel NIETO, Mineralogy, University of Huelva, Spain

Francisco NOCETE, Department of Prehistory. University of Huelva, Spain

Timothy K. PERTTULA, Archeological & Environmental Consultants, LLC, 10101 Woodhaven Drive, Austin, TX, 78753-4346, USA.
Email: TKP4747@aol.com

Olena V. SMYNTYNA, Chair, Department of Archaeology and Ethnology of Ukraine, Mechnikov National University, Odessa I.I., Ukraine.
Email: smyntyna_olena@onu.edu.ua

Dean R. SNOW, Department of Anthropology, The Pennsylvania State University, 409 Carpenter Building, University Park, PA 16802, USA.
Email: drs17@psu.edu

Dimas Martín-SOCAS, Department of Prehistory. University of La Laguna, Spain

Pál SÜMEGI, Archeological Institute of the Hungarian Academy of Sciences, 1014 Budapest Úri u. 49, Hungary. Mailing Address: University of Szeged, Department of Geology and Paleontology, 6701 Szeged, P.O. Box 658, Hungary,
Email: sumegi@geo.u-szeged.hu

Gabriella SZEGVÁRI, University of Szeged, Department of Geology and Paleontology, H-6722 Szeged Egyetem u.2-6, Hungary

Gábor TIMÁR, ELTE University of Budapest, Department of Geophysics, Space Research Group, Budapest Pázmány Péter sétány 1, Hungary

Anikó TÓTH, Ferenc Móra Museum, H-6722 Szeged, Roosevelt tér 1-3, Hungary

David K. WRIGHT, Department of Archaeology and Art History, College of Humanities, Seoul National University, San 56-1, Sillim 9-dong, Gwanak-gu, Seoul, 151-745, South Korea,
Email: daudi.wright@gmail.com

Paweł VALDE-NOWAK, Institute of History, Department of Archaeology, Gdańsk University, Grunwaldzka 238A, PL 80-952 Gdańsk and Institute of Archaeology and Ethnology, Polish Academy of Sciences, Sławkowska 17, PL-31-016 Kraków, Poland.
Eamil: valde@archeo.pan.krakow.pl

Ralf VOGELSANG, Universität zu Köln, Forschungsstelle Afrika, Jennerstr. 8, 50823 Köln, Germany,
Email: R.Vogelsang@uni-koeln.de

Chapter 1
INTRODUCTION

Maximilian O. BALDIA
The Comparative Archaeology WEB and Southern Methodist University

Timothy K. PERTTULA
Archeological & Environmental Consultants

Douglas S. FRINK
Worcester State College

Comparative Archaeology is a global approach to prehistoric culture change. Its ultimate aim is to promote detailed comparisons between large cultural regions to ascertain the causes for similarities and differences in human development. In this book, the goal is to understand humanity's ability to adjust to climate fluctuations, which result in a constantly changing world. In order to achieve this aim, archaeologists, geologists, climatologists, palynologist, botanists, social/cultural anthropologists, sociologists, historians, and others must bridge the unnatural barriers that exist between the social and the "hard sciences".

All too often, the use of data from the "hard sciences" does not find proper application to human issues. This is especially the case in paleoclimate studies, a rapidly developing field that is dominated by the "hard sciences". However, archaeology is in the unique position to bridge such gaps, especially when long-term issues are concerned. Therefore, the goal of this book is to combine scientific paleoclimate data with archaeological information, in order to determine socio-cultural responses to climate change in the past. The issues addressed here lead to a better understanding of humanities past strategies in dealing with a changing world. Furthermore, the resulting inferences have application in preparing for future climate changes.

The main insight gained from this book is that modern research requires multidisciplinary and nonlinear approaches, to understand the human past. These approaches combine scientific data from various sources with analysis of archaeological information to determine socio-cultural adjustment strategies of human populations to changing climatic conditions. It stresses that culture, environment, and climate are complex and dynamic systems whose organization can be understood through the analysis of their constituent components, and the internal relationships between these components. As separate systems, each can be viewed as interrelated systems, each influencing the other. Climate change influences the physical environment, which in turn places constraints on a human population's strategies and its potential success. Cultural change, however, cannot be predetermined as change or stability will result from the unique structure of and interactions between the constituent systems. Climate change may alter an environment only if such change effects critical balances; if not than there will be no evidence of environmental change. Similarly, different cultures even in the same region may exhibit different degrees of social-cultural adjustment depending on their perceived subsystem vulnerability. A population's adjustment strategy may include modification of various cultural components, including its social organization, technology, economy, and ideology. Such adjustments may result in various degrees of culture change, which are often observable in the archaeological record.

This book presents recent research in the Old and New World, detailing surprisingly strong correlations between climatic oscillations and the character of social and cultural responses exhibited by different human populations in the past. The research results indicate that socio-cultural adjustments to climate change frequently correlates with simultaneous social and technological innovations. Innovations include plant and animal domestication, as well as the punctuated adoption and spread of agriculture, the first use of wheeled vehicles, the construction of large earthen and stone monuments, and perhaps the advent of metallurgy. In other cases, signs of socio-cultural adjustments include religious and social upheavals, warfare, genocide, site abandonment, and population migration. However, the outright collapse of cultural systems, often associated with radical climate change, is not readily demonstrated for the Old and New World cultures under discussion.

The authors examine the range of socio-cultural responses to climate change as evident in the paleoclimate and paleoenvironmental record. The record is derived from tree-rings, lake level and glacial fluctuations, ice core data, pollen analysis, and other climate proxies used in paleoclimate and paleoenvironmental research. In addition, new research approaches are presented. The resulting high quality reconstructions of climate and archaeological data from nomadic hunter-gatherers, proto-horticulturalists, sedentary agriculturists, to early urbanized societies from different parts of the world demonstrate the range of socio-cultural responses to climatic and environmental change. In addition, several authors stress human impact on the environment.

Overall, the book illustrates how the interrelated systems of climate, environment, and culture influence each other. Nonetheless, the interpretation of cultural change and its relationship to climate oscillations depends to some extent on the definition of culture. The traditional concept of the *Kulturkreis Lehre*, promoted by the Vienna School of Ethnology a century ago has been dismissed by most scholars after World War II. Europe. In its place the definition as archaeological culture, divorced of any direct ethnic implications became important in Central Europe. In many areas the culture-historical approach seeks to trace origin and movement of ethnic groups and often tries to define prehistoric boundaries. In Soviet and post-Soviet archaeological literature cultures are defined by a group of sites situated in a peculiar territory during specified periods of time in early prehistory. Cultures are characterized by similarity of artifact complexes, particularly by technology of tool production, composition of tool kits, and tool morphology. In the English speaking countries and Scandinavia the theoretical debates of processual or post-processual archaeology have turned away from the culture-historical approach. In Germany the culture-historical approach has become problematic as archaeological data, including precision dating, have dramatically increased. As a result an effort seems to be underway to preserve the approach without the concept of culture in the traditional culture-historical sense. The North American authors largely deal with social change in cultural regions and replace traditional environmental determinism through a nuanced approach derived from systems theory.

The book is organized geographically and chronologically into three parts and a conclusion. Part I focuses on paleoclimate and socio-cultural change in the New World. The comparative data suggests that different cultures in diverse regions use somewhat different adjustment strategies. Part II provides comparative data from Northeastern and Southwestern Africa, examining the role of climatic and environmental change, in conjunction with social and economic adjustments. Here again, some cultures adjust differently. Nevertheless, there seems to be considerable continuity from the Paleolithic to the earliest state societies, even in harsh environments. Part III contains original archaeological, paleoclimatological, and environmental research from Europe. The chapters in Part III are presented chronologically from East to West and from South to the North.

Part I:
Paleoclimate and Socio-Cultural Change in the New World

Chapter 2

DANGEROUS REGIONS: A SOURCE OF CASCADING CULTURAL CHANGES

Joel D. GUNN

New South Associates, Inc., Mebane, North Carolina, University of North Carolina-Greensboro

William J. FOLAN

Centro de Investigaciones Historicas y Sociales, Autonomous University of Capuche, Mexico

Joseph M. HERBERT

Cultural Resource Program at Fort Bragg

Abstract: *In 1961 Glenn T. Trewartha published The Earth's Problem Climates in which he examined important regional climates around the world. Although the book still provides worthwhile initial assessments of regional climates, a formidable, worldwide observation system has been built in recent decades. It collects global and regional climate information in great detail. Also, increasingly refined cultural chronologies available through archaeologists allow for better understanding of relations between climate and culture change at substantial time depths. This provides understanding of impacts environmental changes have had, and perhaps will have, on human populations. Two examples are provided, one from the Maya Lowlands of the Yucatan Peninsula and the other from the Coastal Plain of North and South Carolina. In both regions, the Bermuda-Azores subtropical high and the Gulf Stream serve as the key links between local climates and global changes. Local geology defines the conditions that make these regions especially sensitive to global changes. The analyses appear to show that although never the sole cause of cultural changes, climate is often the proximate cause. Some regions are more potently linked to global climate than others are. The climatically sensitive regions appear to be locales where changes in global temperatures precipitate shifts across critical cultural thresholds such as from urban-to-town life in the Maya Lowlands and from town-to-village life (horticultural-to-gatherer) in the Carolinas Coastal Plain. Shifts across these critical, subsistence-level boundaries appear in many instances to lead to cascading cultural changes as populations and ideas are flushed out of more sensitive regions into surrounding less sensitive regions.*

INTRODUCTION

In *The Earth's Problem Climates*, Trewartha (1961) discusses climates of the world on a region-by-region basis. His inferences were based on analysis of local precipitation and temperature data previous to that date. His findings are illuminating even today after 50 years of more sophisticated observations of local and global climate. This is the result of his essentially local approach to climate, as well as what can only be described as a gift for seeking out the trends and implications in the data. For the Yucatan Peninsula of Mexico and the southeastern United States, Trewartha provides local perspectives on climate that serve as departures for understanding local climate in the broader scope of the earth system as it is now understood. Trewartha's findings need to be understood as essentially middle 20th century perspectives, but this, as we shall see, is valuable in itself.

In this article, we discuss investigations of two regions, the Yucatan Peninsula or Maya Lowlands, and the Carolinas Coastal Plain. Our intention is to show how regional variations in sensitivity to global climate changes vary significantly as functions of local cultural adaptations. We demonstrate that without an awareness of the local archaeological chronologies extending back in time 3000 years, such sensitivities are unlikely to be detected. As in previous publications, our concern is with the conditions that facilitate the emergence of complex social organizations rather than their collapses.

EARTH SYSTEM DATA

The broader understandings of global climate we now enjoy have been facilitated by worldwide systematic observations of climate instituted during the 1957-1958 International Geophysical Year (IGY) and continued on an ever-expanding basis to the present. This system was designed and organized as Trewartha was publishing his book. Initially these observations were ground-based such as Keelings (1979) measurements of atmospheric carbon dioxide atop the Mauna Loa volcano in Hawaii. At the same time, a worldwide program of daily balloon releases to measure atmospheric pressure, temperature and moisture was begun and continued to the present (Angell 1991:231-47, Angell and Korshover 1975:1007-12). Since the IGY, data collection has been much extended by massive observations from earth orbiting satellites, ocean buoys, and many other means. The *Climate Diagnostics Bulletin* of the NOAA Climate Prediction Center (www.cpc.ncep.noaa.gov) provides a broad perspective on these capabilities. Locally we have found it important that monthly measurements of river discharge and tree rings are important to linking local and global climate (Gunn and Crumley 1991:579-92, Gunn *et al.* 1994:174-97, Gunn *et al.* 1995:3-42, Gunn 1991:393-420).

The title of Trewartha's book is, in the view of archaeologists, a bit of a misnomer. Trewartha focused on the agriculturally most productive "regions" of the world. This reflects the agricultural meteorology that pervaded the discipline of geography at the time (R. Bryson, personal communication, 1986). Trewartha's regions are, in fact, now the most problematic climates of the world because any larger-than-normal changes in climate in any of those highly populated locales would have major, worldwide ramifications. From a cultural perspective, however, these regions would be for the same reasons the cores of "cultural areas" and of civilizations.

The archaeological terms "area" and "region", and the climate term "region", require clarification. Trewartha (1968:241) saw climatic regions as convergences of solar radiation and air masses that were related to "patterns of climate, natural vegetation, and soil" not unlike the modern archaeological concept of landscape. "Cultural area" as defined by Wissler (1906:147-149:148, 1914:447-505) is equivalent to "region" in the climatic literature, which is to say a large, contiguous zone of similar plant and animal life and human culture. Wissler focused on the convergence of cultural and psychological factors erected on physiographic foundations: "cultural areas corresponding to well-defined geographic areas." Wissler's cultural area evolved into Willey and Phillips (1958) definition of archaeological "area." In the Willey and Phillips terminology, archaeological "region" refers to an archaeological sub-area, thus, a smaller unit of analysis than a climatic region. In this article, we will use region in the climate sense.

CULTURE-RELEVANT TIME SCALES

When studying the impact of climate on culture, it is important to keep in mind that cultures possess an internal averaging mechanism in the form of human generations. Each individual human experience a life cycle composed of a youthful learning stage and an adult stage of application of that learning. These generation-scale patterns lend stability to cultures that year-to-year climate changes frequently lack. Minor climate changes tend to be "overwritten" in the archaeological record by intra-generational lifecycles and related cultural inertias. However, as we shall discuss, there are lifecycle-scale patterns in global climate that also formulate approximately half century averaging mechanisms (Tanner 1993:220-31). The earth system itself possesses a certain amount of inertia seated in ocean temperatures that smooth the otherwise irregular progress of year-to-year climate.

These averaging mechanisms allow archaeologists to pose cultural time series as episodes of relative stability separated by short-term, often-swift changes. The swift climate changes are too powerful to be overwritten by the human intergenerational inertia and generate disruptions such as famine, plague, migration, and population collapse (Lutz *et al.* 2000). Such step mechanisms are also observable in the climate record, and sometimes, though not always, the two correspond. The functional, temporal scale of this article is therefore at the lifecycle scale of approximately 50 years. The units of analysis are episodes of culture and climate that can be understood as episodes of relatively coherent patterns of geological, cultural, and atmospheric attributes. A discussion of widely accepted episodes of the Late Holocene (0-4500 BP) is provided in Gunn (1994:1-22).

The essential insights offered by Trewartha and others concerning the climates of the Yucatan Peninsula and southeastern United States are as follows.

YUCATAN PENINSULA OF MEXICO

The Yucatan Peninsula (Fig. 2.1A) is a limestone platform of Cretaceous age that grades from a semi-karst in the south to a full karst in the north (Gunn *et al.* 1994:174-97, Gunn *et al.* 1995:3-42, Murray and Weide Jr. 1967, Pope and Dahlin 1989:87-106, Siemens and Puleston 1972:228-39). The climate grades from tropical rain forest in the southeast to dry subtropical steppe in the northwest (Fig. 2.1A, 2.1B). Trewartha (Trewartha 1961:70-71) outlined the annual cycle of precipitation on this physiographic structure as being divided into a warm season (summer-fall) rainy period and a cool season (winter-spring) dry season (Fig. 2.1B). The onset of the rainy season, usually in early summer, is interrupted by an episode of reduced rainfall in August. This is referred to as the *canicula*, or in the American vernacular as the *dog days of summer*. The rainy season corresponds to the spring-to-fall warming of the Gulf Stream as it flows around the peninsula. Although we have never seen an attempt to explain the canicula, we suggest that it is a result of the sun passing directly over head in August on its return path to the southern hemisphere. This overhead incidence of solar heat warms any exposed sediment surface creating a thermal inversion that forestalls daily generation of rain typical of the wet season. The fall-side canicula is balanced by a spring-side period of intense drought and heat in late April and early May as the sun passes directly overhead in its spring northward progression. This is the driest part of the dry season. Failure to account for the effects of direct insolation and the variable character of surface sediments is, in our view, an important missing link between meteorological weather models and local weather and climate.

Studies subsequent to Trewartha have expanded this basic description elaborating the precise atmospheric and earth-surface mechanisms that control sub-regional variation in the patterns. In a 1976 study of satellite imagery, Williams (Folan *et al.* 1983:453-68, Williams Jr. 1976:15-19) discovered that daily precipitation is brought to the peninsula by a front of moist Trade Winds-born air sweeping across the landscape from east to west. As a consequence, precipitation varies from high in the southeast to low in the northwest of the peninsula (Fig. 2.1B). This defines a diagonal ecotone across the peninsula from northeast to southwest. The diagonal ecotone, we found, shifts with the global average

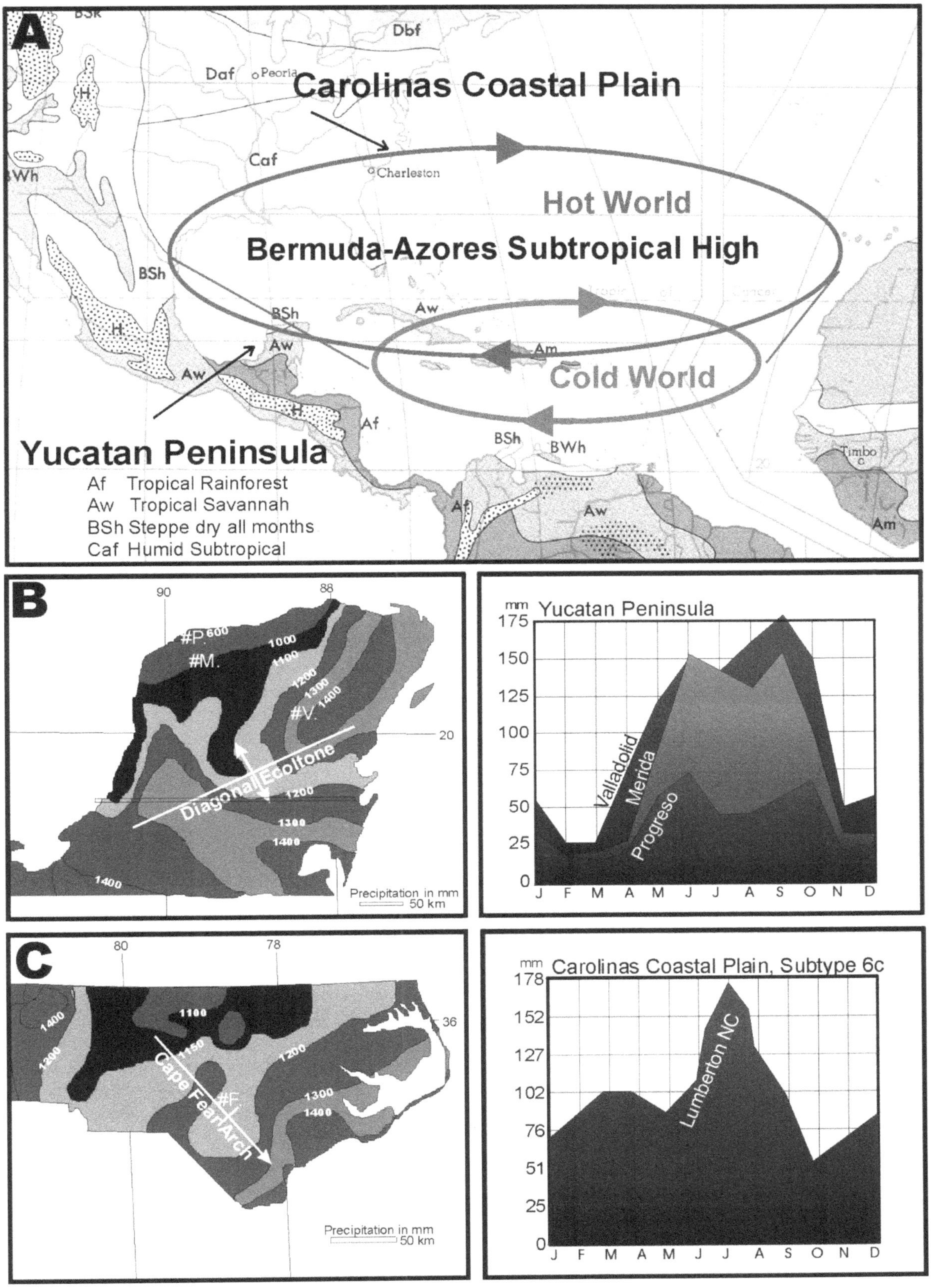

Figure 2.1. A. The Bermuda-Azores Subtropical High provides a common link between the otherwise largely independent regions of the Yucatan Peninsula and the Carolinas Coastal Plain. As the global temperatures warm from cold world to hot world conditions, the high expands in volume and is displaced northward.
B. The Yucatan Peninsula climate regime is one of high precipitation in the south and east and low precipitation in the north and west. The key variable is movement of the annual average position of the diagonal ecotone.
C. The Carolinas Coastal Plain is moist toward the coast and drier inland. The key variable is the lack of moisture retention of the Cape Fear Arch because of its uplift and dissection

temperature under the influence of the subtropical high (Folan *et al.* 1983:453-68, Gill 2000, Gunn *et al.* 1994:174-97, Gunn *et al.* 1995:3-42). Thus, in any given location, it acts as the critical factor in shifting the climate from too wet, i.e., too long of a wet season, to too dry, i.e., too long of a dry season. We will turn below to the question of why it is critical, which is to say essential with respect to the agricultural system.

The details of this intricate climate system are summarized by Folan *et al.* (Folan *et al.* 1983:453-68) with subsequent relevant studies in Gunn *et al.* (1994:174-97, 1995:3-42), Gill (1995, 2000), and Gunn and Folan (2000:263-70). Moist air comes to the peninsula via the Trade Winds that circulate westward along the southern margin of the Bermuda-Azores High. Precipitation is triggered on a daily basis by a so-called double sea breeze effect that sweeps westward across the land mass with daily cloud formations. Interior elevations and atmospheric turbulence propel moist air upward cooling it and causing rains (Folan *et al.* 1983:453-68, Williams Jr. 1976:15-19). The average annual location of the boundary between the dry northwest and the wet southeast shifts depending on the summer and fall location of the Bermuda High and the strength of the easterly Trade Winds. The location of the Bermuda-Azores High and the strength of the Trade Winds varies with global average temperature (Folan *et al.* 1983:453-68, Gill 2000, Neilson 1986:27-34). Higher global temperatures expand the cell northward; thus, the hotter the world the more northerly the wet tropics. Locally, this is reflected on the ground in a longer, wetter rainy season (Gunn *et al.* 1994:174-97, Gunn *et al.* 1995:3-42). The northward extension of the Inter-tropical Convergence Zone (ITCZ), as could be anticipated, is also involved and appears to be correlated with global temperatures (Haug *et al.* 2003:1731-35, Haug *et al.* 2001:1304-08).

The cultural ramifications of this climatic pattern have been worked out using Maya Lowlands archaeological chronology and ethnographic studies of modern populations. The local agricultural methods were developed by the Maya population over the last 3,000 years. They include a highly productive form of tropical gardening (milpa) that thrives on a strong contrast between wet and dry season (just-right dry and wet season lengths) and diminishes in productivity in the face of too little precipitation (shorter and drier rainy season), or too much precipitation (long rainy season and little or no dry season). The system seems to cycle through a 200-300 year scale of variation often resulting in extreme hardships, especially for urban populations (Folan *et al.* 1983:453-68, Gunn *et al.* 1994:174-97, Gunn *et al.* 1995:3-42, Landa 1978[1566]). This cycle is based in part on solar emissions (Gunn *et al.* 1994:174-97, Gunn *et al.* 1995:3-42, Hodell *et al.* 2001:1367-70). A corresponding cycle has been detected in sea level changes (Tanner 1993:220-31:229).

Over the last 3,000 years, the Maya Lowlands have witnessed a cycle of repeated rise and fall of great Maya cities in the interior (Gunn *et al.* 1993). The diagonal ecotone appears to shift not only on an annual basis, but also to follow more long term patterns resulting in the multi-decadal episodes of drought (too long dry seasons) and too wet (too long wet seasons). Embedded in this trajectory are periods of just-right balance between wet and dry seasons. The just-right episodes correspond to the periods of greatest florescence of interior cities. Both too-wet and too-dry episodes correspond to periods of decline.

While the climate parameters associated with the just-right episodes facilitate the progress of urban life, it is not the only factor in its rise or decline. The most famous of the Maya collapses in the 9th century A.D. was during the 9th century extended drought (Brooks 1970[1949], Folan 1981:336-37, Folan *et al.* 1983:453-68, Gunn and Adams 1981:85-100, Haug *et al.* 2003:1731-35, Haug *et al.* 2001:1304-08, Hodell 1995:391-94, Huntington 1917), but the drought was accompanied by a combination of delicately interrelated circumstances including declining soil fertility (Deevey *et al.* 1979:298-305), burgeoning population (Johnston 2003:126-61), and relentless warfare (Demarest 1993:95-111, Landa 1978[1566], Rice 1996:193-206) resulting in a decline in trade across the peninsula (Rovner and Lewenstein 1997). During another episode of collapse in the 3rd century A.D., less complicated demographic factors allowed a rapid recovery from the effects of extended drought and the initiation of the most active period in Maya architectural and urban florescence, whereas recovery from the 9th century drought required centuries and in many sub-regions resulted in permanent abandonment. For the most part, coastal cities continued to be occupied while interior cities were abandoned, probably because cross-peninsular trade routes shifted to sea lanes (Adams 1977, Folan *et al.* 2004).

Needless to say, the periods of extended drought were episodes of severe hardship for interior urban populations. Because of brackish natural water supplies, urbanizing the interior high valleys of the Yucatan Peninsula was an arduous undertaking that required planning and knowledge of engineering and construction of advanced urban water systems (Gunn *et al.* 2002:297-315, Scarborough 1998). The annual cycle of drought resulted in saline natural water supplies through the dry season. Without Maya water systems that involved artificial ponds (aguadas) and underlying perched water tables, year round urban occupation of the interior high valleys would have been impossible. The elements of this water system continued in some villages in the interior such as Pich, Capuche, until wells were dug to supply municipal water systems in the 1960s (Faust 1998, Gunn *et al.* 2003:119-30, Gunn *et al.* 2002:297-315). Social credit in this social system was extended on the basis of participation in the water collecting and conserving activities. Families in the village were organized to collect water in the wet season, and to guard and apportion the water in the dry season. Work groups cleaned the aguada and wells below its floor to tap the underlying perched water table in the dry season. Social relations were maintained with other, less drought sensitive villages through marriage. These villages would become destinations of retreat in the event of extended drought. We view the organization of Calomel, Capuche, arguably the largest urban environment in the Maya Lowlands (Folan *et al.* 1995:310-34), as an aggregate of such village water communities extending over 10 km northeastward from the main acropolis.

Extensive studies of the 55-meter high acropolis at Calomel (Folan *et al.* 1995:310-34) show that Late Classic renovations of Temple Structure II were not completed and the building may have been abandoned. It was then reoccupied and reorganized. In the reactivated city, temples on top of pyramids continued to be religious centers. Both religious and secular activity refocused on the huge Temple Structure II at the center of the acropolis in the form of an additional series of small rooms. Rooms were built on the side zones of the pyramid, apparently to house functionaries and their families. A nearby Classic Period palace like Structure II became a place for votive offerings, usually in the form of anthropomorphic and zoomorphic ceramic ocarinas, whistles, and flutes. They were left on the front platform, stairs, and in the previously collapsed vestibule in the center of the building. The Classic Period distinction between secular and sacred became blurred. While the city has not been studied in sufficient detail to provide demographic structural changes over time, indications to date seem to suggest a substantial influx of population from the north at the outset of extended drought followed by an outflux of populations from the city (Domínguez Carrasco *et al.* 2002:279-304). The movement of such populations can only be understood indirectly, but studies seem to indicate a complicated pattern that resulted in the relocation of northern traits further south (Gill 2000). This pattern was recapitulated in a 1939 drought, which saw the relocation of large population segments from the state of Yucatan in the north to the state of Capuche in the south (Faust 1998). The reason given by the immigrants to Pich was that the drought was of less duration in the more tropical south than in the more arid and subtropical north.

To summarize, Maya Lowlands cities in interior Capuche were established after the development of urban hydraulic systems. There, urban centers flourished during times of optimal climate so long as demographic, soils, warfare and trade parameters were within acceptable bounds. When these non-climatic parameters exceeded acceptable boundaries in the 9th century A.D., a climatic disturbance, a severe drought, culminated urban life in the elevated interior uplands. Urban societies of a different order appeared in better-watered sub-regions such as the coasts and around lakes.

CAROLINAS COASTAL PLAIN OF SOUTHEASTERN UNITED STATES

While the Coastal Plain of the Carolinas lacks the formidable stone architecture of the Maya Lowlands, it too possesses a culture history of significant climate and cultural changes. Here the scale of human endeavor revolves around villages rather than cities. As we shall see, just as in the Maya Lowlands, the coming of villages awaited appropriate climate and technological developments.

The Atlantic Slope of Southeastern United States is a complicated suite of arch and embayment geological structures dating from the late Mesozoic (Horton and Zullo 1991:1-10). The Carolinas section of this geological structure is composed of the Cape Fear Arch near and paralleling the North Carolina boarder with South Carolina (Fig. 2.1C). To the south is the Charleston Embayment and to the north is the Albemarle Embayment. Of special interest to this study is the Cape Fear Arch. Its elevated character has exposed the late Mesozoic and Cenozoic Coastal Plain sediments to dissection making them extremely susceptible to drought.

As can be seen (Fig. 2.1B, 2.1C) the long term climates of the Yucatan Peninsula and coastal Carolinas are surprisingly similar in total annual precipitation. The modern climate of the Carolinas is the most tropical of any in North America outside Mexico (Currie and Paquin 1987:326-27). On the opposite side of the subtropical high the Yucatan Peninsula resides on the subtropical margin of the tropics. In the details of monthly precipitation, however, they are nearly opposite in distribution across the year. As we shall see, opposite patterns but with similarities of impacts is a repeated theme in the comparisons between the two regions.

Trewartha (1961:303-304) outlined the annual cycle of precipitation in the Carolinas sub-region as wet most of the year with a dry episode in October. Peaks of moisture occur in July and March. The July peak is due to an influx of moist tropical air from the Gulf of Mexico combined with sea breeze effects. The winter moisture peak is from Pacific moisture packed by lows that cross though northern Mexico and collide with winter fronts along a line from Texas to the Carolinas. In more recent years, the driest part of the year continues in October and November (Epperson 1971). As in the Yucatan Peninsula, the annual pattern of precipitation is frequently interrupted by a period of reduced rainfall in August, the "dog days of summer." The dog days appear to be a product of the onset of continental air movements as the summer retreats. This pattern continues into fall until winter precipitation of Pacific origin begins.

Geographically, precipitation varies in a complex pattern with more precipitation in the coastal zone and Appalachian Mountains. Precipitation is relatively diminished in the Coastal Plain (Fig. 2.1C). The diminished precipitation in the Coastal Plain during the warm season is due to thermal inversions resulting from heating of air over the flat, light-colored, sandy sediments. This condition could be due to the largely agricultural character of the present day Coastal Plain. Whether it was an enduring condition can be debated. However, ethno historic accounts by early explorers in the 17th century seem to indicate that Native Americans kept the terrain burned off to provide open lands for large game (John Smith in Kupperman 1988, John Lawson in Lefler 1967, Hammett 1992:1-50, 1992:121-35). This would have established a precipitation regime not very different from that of the present with broad upland grass lands providing thermal inversions. The first historical accounts of the Coastal Plain landscape in the 1660s during the Little Ice Ages report precisely such a landscape (Lefler 1967:72-79).

Long term perspectives on climate in the area of the Cape Fear Arch are supplied by a bald cypress tree ring chronology, extending from A.D. 365 to 1985 (Gunn 2002:1-33, Stahle *et al.* 1988:1517-19). They are also covered by ethno historical accounts, spanning the last 500 years, and thus the Little Ice Ages to the present globally warm conditions, by geological observations of episodes of aeolian activity (Daniel 2002:6-11, Gunn and Foss 1997:53-74, Gunn and Foss 1992:1-17, Jester *et al.* 2001:273-302, Markewich and Markewich 1994, Otvos and Price 2001:150-58, Seramur and Cowan 2003:101-18, Soller and Mills 1991:290-308), and by modeling local episodes with global conditions (Gunn 1997:134-51).

Geological studies provide the Late Cenozoic time scale. During globally cold glacial periods, cold dry conditions lead to the aeolianization of the landscape. Remnants of this wind-driven landscape are observable as the hill top blowouts and pocosins, the later being upland ponds and wetlands that fill old Pleistocene depressions in the upland surfaces. The opposite, hot-world conditions are demonstrated by the presence of active sand dunes and hillside blowouts in the Middle Holocene. Reworked aeolianized sediments have been dated to the 6th to 8th millennia B.P. (Markewich and Markewich 1994, Soller and Mills 1991:290-308) and by active hillside blowouts that were subject to human occupation. At Copperhead Hollow (Gunn and Foss 1997:53-74, Gunn and Foss 1992:1-17), a southwest-facing blowout on the Carolinas boarder near Charlotte, a hearth with calcined bone that was buried by aeolian sediments on the lee of the blowout, was dated by AMS to 5,300 B.P. This is similar to the AMS dates for the "Ice Man" in the Tyrolean Alps (Kuchera *et al.* 2000, Rom *et al.* 1999), and thus during a period of Middle Holocene retreat of the Alpine glaciers the extent of which has only reoccurred during the present global warming event.

From the combination of Pleistocene and globally hot Holocene experience, it can be inferred that the too-cold global conditions produces cold droughts on the Cape Fear Arch. The too-hot world produces hot droughts. An in-between global temperature results in a moisture regime that is favorable for sustained human occupation by cultures of a more complex sort. We will now examine the details of these broad patterns through tree rings.

Bald cypress trees in the Black River drainage, a tributary of the Cape Fear River, produced the longest tree ring chronology in eastern North America (Stahle *et al.* 1988:1517-19). The chronology dates to A.D. 365 and thus spans the colder episodes of the Late Holocene, the Vandal and Little Ice Ages minima and the Medieval Maximum episodes (see Gunn 1994:1-22 for summary of Late Holocene episodes). The tree-ring data show a 200-300 year cycle broadly paralleling the Maya Lowland sequence and presumably of similar origin, i.e., solar emissions variations (Fig. 2.2A). This presumption is informed by the thermal inversion forces at work in the Coastal Plain as discussed above, but is a model that bears further examination to understand its precise nature. Among the features of the model that need to be elaborated are at what global temperature is the optimal moisture conditions achieved? We will pursue this question next via the tree ring record.

At briefer time scales, within the 200-300 year time periods, tree-ring data records a persistent cycling averaging 34 years duration (range from 50 to 67 years) between moister and drier conditions (Stahle *et al.* 1988:1517-19). On the elevated Cape Fear Arch, these variations are notably exaggerated resulting in an aeolianized landscape of dunes and blowouts not present in adjacent low lying valleys (Soller and Mills 1991:290-308). When both the long- and short-term cycles are implying drought, sand dunes and hilltop blowouts become active, resulting in a distinctively aeolian landscape. The human occupation signature of this landscape is archaeological sites on blowouts near reliable water sources. The sandy upland deposits provide a dry place to live as they overlie marine and deltaic clay deposits. They also act to conserve wet season water in perched water tables on the clays. This water feeds springs through the dry season for drinking water, and supports nearby wetlands that provide a rich source of aquatic fauna and flora for human consumption (Gunn *et al.* 2003).

The cultural ramifications of this climatic pattern have been worked out using the local archaeological chronology (Gunn 2002:1-33). For most of the last 3,000 years, the Cape Fear Arch, apparently because of its elevated topography and concomitant ecological instability, was inhabited by unimposing horticultural people living in villages and temporary camps. This is exceptional because surrounding sub-regions developed more complex social forms, whose emblematic architecture was accretional mounds of clay and/or stone. These mounds were refreshed and enlarged with new material with each generation of rulers. They are thus an index of sustained human occupation and activity. In Georgia, South Carolina, and marginally in western North Carolina, imposing clay mounds were constructed over hundreds of years by Siouan speaking peoples. To the west in the mountains, Iroquoian speaking populations built clay mounds and in the north stone mounds (Dunham 1994:971, Frierson 2002, Gardner 1993, Gunn 2002:1-33, MacCord 1986).

Evident on the Coastal Plain is a possible exception to the pattern of low-level horticulture and dispersed settlement, typical of the Sandhills during the Woodland period,

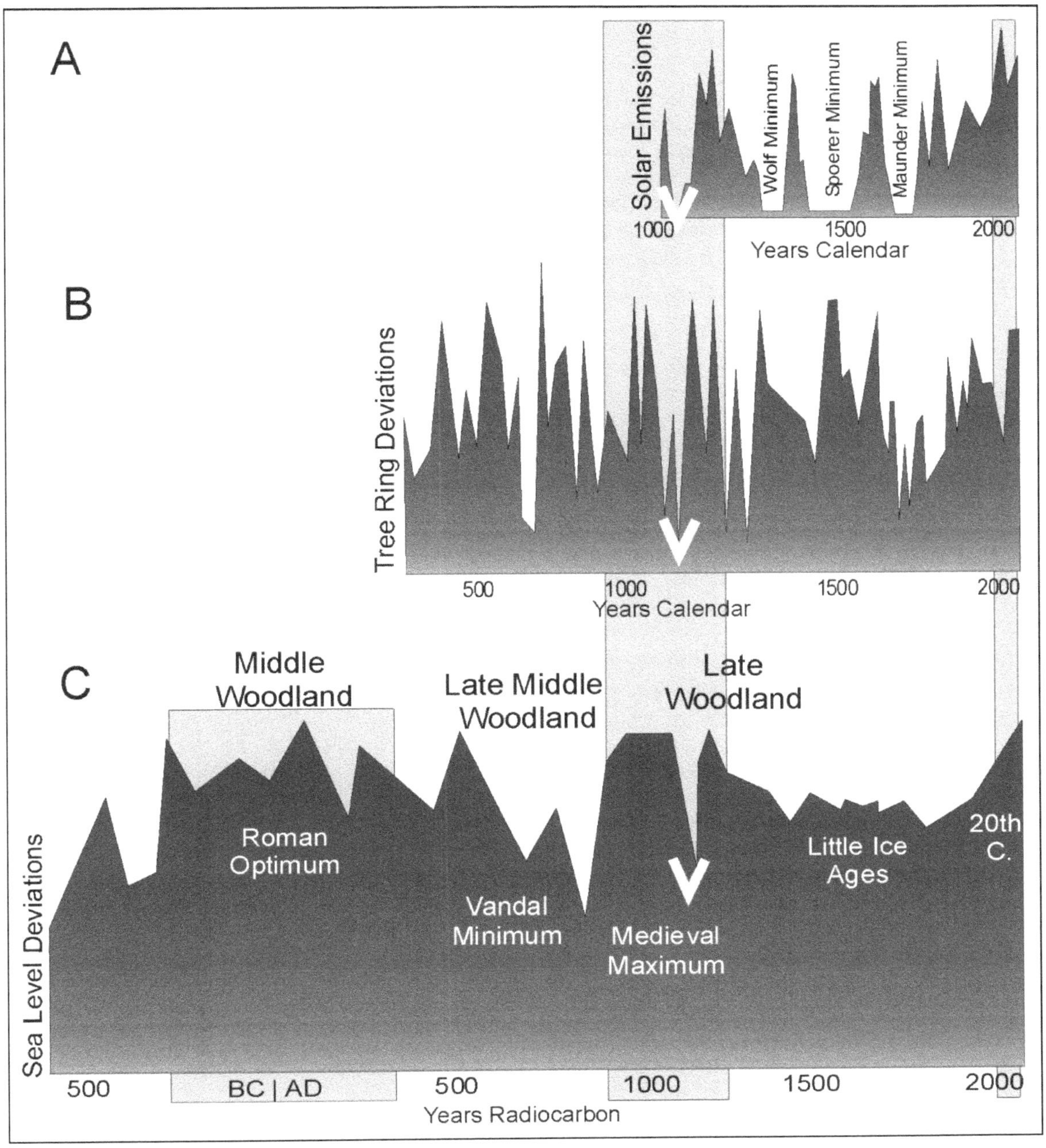

Figure 2.2. A. Solar emissions for the last 1000 years as measured by observations of auroras and sunspots (Eddy 1994:23-36). B. Tree ring deviations from the Black River of North Carolina (Stahle *et al.* 1988:1517-19). C. Sea level deviations for the last 3000 years (Tanner 1993:220-31). The deviations are thought to range over about 3-4 meters. The white V in all three graphs points to a globally cold excursion during the Medieval Maximum warm period at about A.D. 1100

occurred during an episode of elevated global temperatures around A.D. 950 to 1100 (Fig. 2.2). During this episode, the Cape Fear Arch around Fayetteville, North Carolina, witnessed an efflorescence of mortuary customs that included the building of accretional burial mounds on a moderately intensive scale (Irwin *et al.* 1999:59-86, MacCord 1966: 3-45). These accretional mounds, and their apparent association with moderately large Woodland period sites, suggest that an incipient-village or proto-town residential pattern may have emerged at this time (Gunn and Sanborn 2002, MacCord 1966: 3-45: Appendix V). The characteristics of certain ceramics found at the McLean and Buie Mound sites, and the style of stone pipes found in at the McLean and McFayden Mounds, suggests Late Woodland or Mississippian context and possible links with the neighboring fortified village at the Town Creek Mound site, about 90 km to the west. The Town Creek site possesses an accretional clay mound while the mounds near Fayetteville were made of sand, reflecting the complete dominance of sandy sediments in the Sandhills and Coastal Plain. The pattern of accretional mounds and associated artifacts, however, represents a change toward a more complex society than the usual denizens of the

Sandhills and interior Coastal Plain, and suggests the diffusion of culture or immigration into the region during the period in question.

Did climate play a role in this diffusion or colonization effort? Examination of the tree rings and a measure of global temperature, sea level, suggest that it may. The tree rings (Fig. 2.2B) for the period show a pattern of drought of short duration during the period A.D. 950 through 1100 (about 34 years), but extended and more intense droughts during the period A.D. 1100 through 1270 (50-67 years).

Sea level data are drawn from the ca. 50-year interval record constructed by William F. Tanner (Tanner 1993:220-31, 2000:169-70) from an isostatically rebounding beach on the east coast of Denmark (Figure 2C). This record has been found to parallel sea level records on both sides of the Atlantic (Tanner 1993:220-31). The ca. 50-year interval is important because it provides an averaging function for global temperatures similar to that of human generations for cultures as discussed above.

The sea level shows that the period of town/village life on the Cape Fear Arch corresponds to an episode of elevated global temperature. It occurs after globally much colder centuries, the years that produced the 9th century drought in the Maya Lowlands. During this episode between ca. A.D. 600 and 900, the Carolinas Coastal Plain experienced a much-diminished variety of tool types implying a less complex social organization (Gunn 2002:1-33). The period following from ca. A.D. 1100 to 1270 is modestly cooler at the global scale, but has higher-than-present solar emissions. This would have produced the droughts of greater duration recorded by the tree rings through the thermal inversion process. The episode also exhibits reduced settlement activity or abandonment of the Sandhills/Coastal Plain.

The greater complexity of social organization between ca. A.D. 900 and 1100 suggests that the "just right" condition for more complex social development on the Cape Fear Arch falls in the global temperatures range of the early Medieval Maximum. They might have been like those of the middle 20th century described by Trewartha when precipitation was nicely balanced. To the contrary, however, the times before experienced the cold drought phenomenon and those after were subject to hot drought. Archaeological work in the Fayetteville-area Sandhills suggests that population was especially low during the later period. A cooler inflection evident in all three graphs in of Fig. 2.2 (marked by a white V) indicates a sudden cooling event around A.D. 1100. This apparently sharp drought could have set the stage for lesser occupation of the Coastal Plain in the late Medieval Maximum. A key element in this pattern would have been increased spring-summer rainfall producing monsoon conditions as will be discussed below.

The abandonment of the Sandhills during the late Medieval Maximum implies movement of the population to some surrounding sub-region less prone to the droughts of the Cape Fear Arch. The influx of a population, or at least of ideas, from the mound building sub-regions at the beginning of the optimal moisture period implies the opposite process. Such influxes and exfluxes are referred to as "cultural anvil" effects (Fitzhugh 1972, Gunn 1979:257-74, Gunn 2002:1-33). The cultural anvil effect avers that climatically sensitive sub-regions will attract immigration during episodes of favorable conditions, i.e., conditions to which some surrounding sub-regional culture is preadapted. When unfavorable changes occur, the resident population will be crushed as a culture because their new home is degraded and the lands they had formerly occupied were taken over by follower populations when they left. This form of the theory might be thought of as the "strong cultural anvil effect" as its implications are the most radical that can be imagined. However, a weaker version of the idea would be that populations immigrating to a sub-region would maintain social ties to the old sub-region, perhaps as the people of Pich maintain social relations with less drought-prone villages of the valley floor, and are received with sympathy when they are forced to retreat from the more exposed environment.

The most sensitive indicators of human interaction on the Carolinas Coastal Plain are the many independently varying elements that go into the manufacture of ceramics: paste, temper, vessel form and size, surface design and decoration. It might be expected, if the above-discussed dynamics were repeated several times over the last 3000 years of ceramic manufacture, that ceramics on the Coastal Plain would reflect a complex interleaving of these attributes from surrounding sub-regions. These attributes do in fact exhibit a complicated mix of relationships with the boundaries of the Cape Fear Arch sub-region (Anderson 1996, Herbert 1999:37-58). The mode of entry of attributes includes:

- Entering the sub-region and stopping at the opposite border
- Completely crossing the sub-region to the adjoining sub-region
- Stopping at the border and not entering into the sub-region (Gunn 2002:1-33).

Examples of all of these modes exist. However, the invention and dispersal of heat-tolerant clay-grog temper is an interesting phenomenon possibly related to the early Medieval Maximum occupation of the Sandhills. Clay temper is functionally important because it has the same heat expansion coefficient as the vessel in which it is incorporated. Earlier and later vessel designs were based on quartz temper. When quartz is heated past the alpha-beta threshold of 573° C it expands as much as three percent, which in theory would increase the risk of vessel failure during repeated firing. For this reason, quartz would not appear to be the best choice of tempering material for vessels intended for routine use in direct-fire cooking. It is therefore assumed that when clay-tempering technology appears from the south at the beginning of, or somewhat before, optimum climate conditions around A.D. 950 (Herbert 2003, Herbert *et al.* 2002:92-108), its wide adoption across the sub-region (mode 1) may relate to the improved vessel durability that the process conveys. It seems more than coincidental that this shift in technology to methods that produce more durable cooking vessels also occurred at a time when agricultural food production was intensifying, sedentism increasing, and the environment across the Cape Fear Arch assuming "just right" conditions.

Another example of sub-region-related temper change is the introduction of shell-tempered ceramics from the north (Mathis 2000:89-98). This introduction appears to have followed the A.D. 536 climatic event, a period of extreme global cooling in the 15 years following A.D. 536 (Baillie 1994:212-17, Gunn 2000:5-20). Shell tempered ceramics spread along the coast about half way across the sub-region (somewhat mode 1). The shell-tempered ceramics served the same function as clay temper. However, they were introduced by Algonquian speakers moving south along the coast toward warmer climate.

To summarize, occupants of the Cape Fear sub-region were, for the most part, annually mobile hunter-gatherers. Horticultural villages flourished briefly during a time of optimal climate when complex social organizations were already long-present in adjacent sub-regions. Following a climatic disturbance in the middle of the Medieval Maximum, accretional mound building ceased and the sub-region may have been abandoned for a time during a period of hot drought. The construction of accretional mounds continued in surrounding sub-regions.

COMMONALTIES AND CONTRASTS

As the Bermuda High and the Gulf Steam ocean current figure in both the Carolinas Coastal Plain and Yucatan Peninsula climate systems, it is instructive to compare scenarios for the Medieval Maximum for the two regions. As discussed above, both regions appear to have an optimal range of global temperatures within which local climate are best adjusted for human adaptation to the local moisture regime and their food production technology. When global temperatures decline or elevate from local optima, negative impacts are observed on the complexity of the social organization. However, the precise mechanisms of the impacts vary significantly.

In both regions, as global temperatures decline, the precipitation departs from some optimal annual amount and monthly distribution toward dryer conditions. In the Yucatan Peninsula (ca. 18°N), this involves the lengthening of the dry season into summer and the shortening of the wet season toward fall (Fig. 2.3A) On the opposite side of the Bermuda High in the Carolinas, moisture declines in the spring-summer because of a longer duration of cold, dry continental air flow (Fig. 2.3B). Declines in vegetation cover (permitting blowouts) and the reduced growth rates of bald cypress trees support this argument. Since growing season is a more critical issue at this higher latitude (ca. 36°N), the shortening of the growing season and the reduction of spring-summer precipitation is proportionally more important. This is especially the case following the introduction of tropical cultigens including corn (maize) in the 1st millennium A.D.

When global temperatures warm beyond the optimal zone, the effects are quite opposite for the two regions. In the Yucatan Peninsula, precipitation increases because of expansion of the tropical zone outward from the equator, further northward movement of the ITCZ, and northward displacement of the Bermuda-Azores High. This shortens the dry season toward spring and lengthens the wet season. This same movement of the Bermuda-Azores High increases the flow of hot, dry air into the Carolinas creating a hot drought and vitalizing dunes and blowouts. The June-July-August peak in moisture coincides with the maximum annual expression of the subtropical high, which in turn reduces moisture during the critical months of the growing season.

The asynchrony of cultural chronologies of the two regions indicates that the timing of the optima is not at equal global temperatures (Table 2.1). We can deduce the relative global temperatures of optima in the two regions by comparing the timing of cultural florescences compared to global temperature. Global temperature fell irregularly to a nadir below that of the present in the 9th century and then rose to an apogee in the 13th century (Fig. 2.2C). The fluorescence of Maya Lowlands urban settlements in the interior was during episodes of optimal climate in the 6th and 8th centuries A.D., while the manifestation of village/town life in the Cape Fear valley was during the 11th century A.D.

Our modelling of past runoff (Gunn *et al.* 1995:3-42) indicates that the southwestern Maya Lowlands were struck by episodes of first too-cold and then too-hot climates in the 9th and 11th centuries, or too-dry then too-wet. The Carolinas Coastal Plain was in a cold drought during the 9th century, then an optimum in the 11th century A.D., so the Carolinas optimum occurs at a higher global temperature than that of the Yucatan Peninsula. The 11th century Carolinas optimum is at approximately the global temperature of the middle 20th century. The 12th century Carolinas hot drought condition occurred during high solar emissions (Eddy 1994:23-36, see Fig. 2.2A), which would have produced thermal inversions halting precipitation and extending the residence time of the subtropical high over the area in the warm season. This would have been a period broadly equivalent to the 2001 drought, which occurred during a solar maximum and relatively high global temperature compared to the 20th century. A longer duration of the subtropical high residence time saps soil moisture and destroys temperate zone vegetation.

The supporting arguments for the Capuche climatic scenario have been developed in several articles (Gunn *et al.* 1994:174-97, Gunn *et al.* 1995:3-42, Gunn and Folan

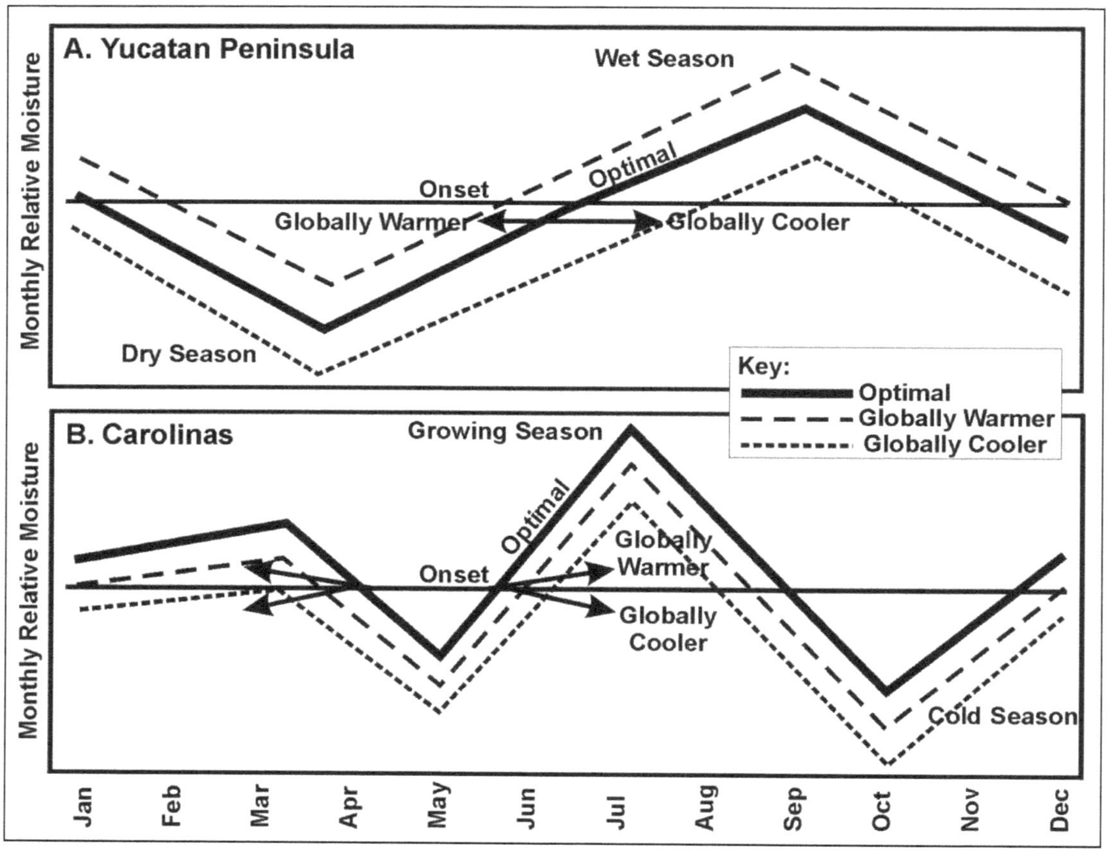

Figure 2.3. Impact of global temperatures departing from local optima. Vertical scales are in relative magnitude of wet season moisture. A. In the Yucatan Peninsula, global cooling shortens the wet season and reduces the magnitude of wet season precipitation. Global warming lengthens the wet season and expands its magnitude of precipitation. B. In the Carolinas, global cooling expands the annual May drought attenuating upland temperate vegetation. Global warming also reduces precipitation during summer because of the earlier onset of the subtropical high

2000:263-70, Gunn *et al.* 2002:297-315). The verification and fine tuning of the Cape Fear Arch scenario will be elaborated here. The modelling discussed above implies that there will be a linear relationship between global temperature and the discharge of rivers in Capuche because the tropics simply expand as the world warms. The information gathered for the Carolinas Coastal Plain, however, implies a curvilinear system with a peak of effective moisture somewhere in the range of the early Medieval Maximum/mid-20th century precipitation range.

The Black River, a tributary of the Cape Fear River, has a tree ring chronology that can be used to gather insights into relationships between global climate variables and the local precipitation/runoff regime. Modern climatological data and the tree ring chronology overlap between 1958 and 1985. This time series has a relatively wide range of global temperature variations. The 1960s and 1970s were globally cool with 1976 being the coldest year in the 20th century approaching the conditions of the Little Ice Ages. On the other hand, the 1950s and 1980s were globally warm.

A matrix plot (Fig. 2.4) of the most important variables in the global and local system shows that the relationships in the global-to-local system are quite complex. The tree ring thickness simply increases with the amount of runoff in May, June, and July, as Stahle *et al.* (1988:1517-19) presumed (r^2=.39, p=.001). However, there is a curvilinear relationship between March runoff and annual tree ring growth and this relationship is also respectably strong (r^2=.26, p=.010). The r^2 (percent of variance accounted for) for the two together as predictors of tree ring growth is .45, or in other words, they define about half of the volume of tree ring growth. Figuring in the global parameters temperature of the Northern Hemisphere troposphere (850-300 mb), solar emissions, and greenhouse gas increases (Year), raises variance accounted for to 66 percent (Table 2.2). Interestingly, in this context the May, June, July discharge (MJJ) drops to insignificance (p=.70) implying that the March discharge sets up the growing season from the point of view of bald cypress trees. Their growth is mostly in May and June (Stahle *et al.* 1988:1517-19).

The sum of this analysis is that there are, as our anecdotal observations suggested, statistically significant, curvilinear relationships between tree ring growth in the Black River and global conditions. The complexity of the system suggests a need for more data and more work on

Table 2.1. Comparative Chronologies from the Southwestern Yucatan Lowlands and the Carolinas Coastal Plain

Year A.D.	Yucatan Peninsula	Carolinas Coastal Plain
775	Late Classic Florescence draws to a close in interior uplands	
800	Calomel 1st Abandonment	
825	Calomel reoccupied by northerners	Late Middle Woodland impoverished assemblages, hunters and gatherers
850		
875	Calomel 2nd Abandonment	
900		Ceramics change from Hanover I to Hanover II
925	Occupation focuses in coasts, southern Lake District, and mountains, interior uplands abandoned	
950		
975		
1000		McLean Mound built, accretional mounds and possible village organization
1025		
1050		
1075		
1100		Middle Medieval Maximum Decline in Global Temps
1125		Decline of occupation on the Cape Fear Arch
1150		

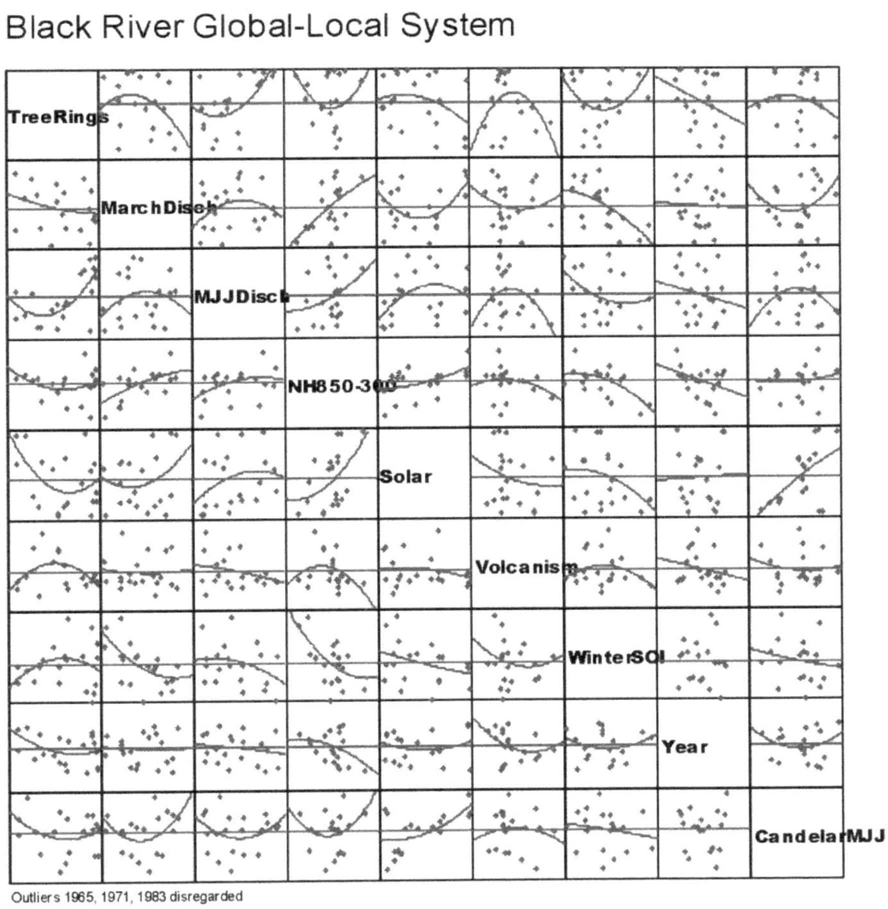

Figure 2.4.

Table 2.2. Multiple Regression for Black River tree rings and local and global parameters (tree rings are transformed by squaring the Z-Score to remove the effects of the curvilinear relationships from the equation)

	r=.81	r^2=.66	1-r^2=.34	Ajd r^2=.57	Sig. = .001
	B	Std Err	Beta	t	Sig.
(Constant)	-114.65	34.30		-3.34	0.00
March Discharge	0.00	0.00	-0.44	-2.85	0.01
May-June-July Discharge	0.00	0.00	-0.06	-0.39	0.70
Temp. N. Hem. 850-300 mb C.	2.71	0.91	0.61	2.98	0.01
Solar Emissions	0.00	0.00	0.24	1.36	0.19
Year	0.06	0.02	0.54	3.40	0.00

the relationships. However, there are known factors at work that may explain the curvilinearity in terms of weather systems. If one thinks of precipitation and discharge in terms of the temperate zone storm track and the influence that global warming has on its position, the curvilinearity is reasonable. Under cold conditions the Bermuda-Azores High retreats south of the Carolinas and the temperate zone storm track follows leaving cold, dry Canadian air and cold drought. As the world warms, the Bermuda High and the storm track moves north in average position until it brings optimum rain. With further warming the storm track moves north leaving the Carolinas under the influence of the dry, hot air of the subtropical Bermuda High. This context provides a framework within which optimal moist conditions and flanking hot and cold droughts can be understood. Recalling that Trewartha pointed out the winter peak in moisture was in March (Fig. 2.1C), it is clear that the manipulation of this rainfall peak is a key feature of the response of vegetation and the landscape to global warming. It would be the late winter-early spring rains that set up the conditions for vibrant spring growth of temperate zone vegetation, as the regression results imply.

CONCLUSIONS

Perhaps the most important element of this analysis for consideration in present day global policy matters is that in each sub-regional human adaptation has a distinctive geology-soils-climate signature. This signature is wrought by the demands of the food production system. Both of the Native American food production systems considered here were based in part on corn (maize) horticulture and a related suite of food plants. However, the differences in corn horticulture are important. As Diamond (1997) points out in his insightful but mercifully brief synthesis of world history, domestication of corn in the western hemisphere was an arduous task that involved long periods of adaptation to highly varied latitudinal habitats, in this case in the tropical Yucatan Peninsula and the temperate Carolinas. We can see in the Yucatan Peninsula case that production in the context of a tropical garden involved strongly contrasting wet and dry seasons with a long period from May to September of optional planting times depending on the weather. In the temperate version of the story, corn has to mature in a dangerously short growing season and is dependent on high summer rainfall. We see precipitous human consequences, repeatedly, when the global conditions dictate changes out of the preferred optimum conditions in either sub-region. In the Yucatan Peninsula, reduction in productivity defined shifts from urban to town/village life or even retreating into the forest. In the Carolinas, shifts were followed by changes from town/village life to hunter-gatherer life. In both case studies, certain changes, notably hot world conditions, resulted in permanent abandonment by socially more complex forms. This occurred even though the impacts were completely opposite in effect with respect to moisture.

Ultimately the question that needs to be addressed with regard to present day populations of the two study areas and global warming is, "Where do we stand at present global temperatures relative to the local ecology and adaptive profile?" In the Carolinas, the year 2001-2002 provided a helpful insight. In the Carolinas, and all along the Atlantic Coast, that year featured the most extended drought in recorded history. This drought resulted from the additive effects of the current warmed-world state and a solar maximum. The conditions were so extreme that water control measures were widespread and some communities were on the verge of needing to truck water for basic needs. This probably implies that those communities were nearing a condition that would require human abandonment or unprecedented measures of readaptation. The area has no subterranean aquifers to rely upon. We might suppose that population movements would have been noted if the condition had continued for another year or so. The condition was reversed as solar emissions lessened toward the end of 2002 and El Niño effects appeared bringing, as it typically does, precipitation to the Southeast. This scenario implies that the solar maximum provided a preview of what will follow as global warming continues its upward surge pushing local conditions permanently over the hot-drought threshold.

In the Yucatan Peninsula, local conditions are more difficult to read. Our model suggests that at some time in the recent past, or near future, conditions of excessively long rainfall will shorten the dry season thus reducing the productivity of tropical gardens. Such a change would be troublesome to rural populations that still depend to some extent on tropical gardens, but the impact on an economy that is not broadly dependent on tropical gardens is less certain (see Faust 1998, Gates 1993 for relevant conditions).

Acknowledgements

The authors would like to thank Maximilian O. Baldia, Timothy Perttula, and Douglas S. Frink for the invitation to participate. Several persons at the Centro de Investigaciones Historicas y Sociales, Universidad Autónoma de Capuche, contributed effort and thought to the manuscripts including Lynda Florey Folan and Maria del Rosario Domínguez Carrasco.

References

ADAMS, R.E.W. 1977 – *Prehistoric Mesoamerica*, Little Brown, Boston, 1977.

ANDERSON, D.G. 1996 – *Indian Pottery of the Carolinas: Observations from the March 1995 Ceramic Dorkshop at Hobcaw Barony*, Columbia, SC, Council of South Carolina Professional Archaeologists, 1996.

ANGELL, J.K. 1991 – Changes in Tropospheric and Stratospheric Global Temperatures 1958-1988, In Schlesinger, M. (ed), *Greenhouse-gas-induced climatic change: A critical appraisal of simulation and observations*, Elsevier Science Publishers B.V., Amsterdam, 1991:231-247.

ANGELL, J.K., and J. KORSHOVER 1975 – Estimate of the Global Change in Tropospheric Temperature Between 1958 and 1973, *Monthly Weather Review* 103, 1975:1007-1012.

BAILLIE, M.G.L. 1994 – Dendrochronology Raises Question About the Nature of the AD 536 Dust-Veil Event, *The Holocene* 3, 1994:212-17.

BROOKS, C.P.E. 1970 – *Climate Through the Ages*, New York: Ernest Benn, Limited, original publisher, Dover Edition, 1949.

CURRIE, D.J. and V. PAQUIN 1987 – Large-scale Biogeographical Patterns of Species Richness of Trees, *Nature* 329, 1987:326-27.

DANIEL, I.R.J. 2002 – Stratified Early-Middle Holocene Remains in the North Carolina Coastal Plain, In J. Eastman, C. Rodning, and E. Boudreaux III, (eds.), *The Archaeology of Native North Carolina*, Southeastern Archaeological Conference, 2002:6-11.

DEEVEY, E.S., D.S. RICE, P.M. RICE, H.H. VAUGHAN, M. BRENNER, and M.S. FLANNERY 1979 – Mayan Urbanism: Impact on a Tropical Karst Environment, *Science* 206, 1979:298-305.

DEMAREST, A.A. 1993 – The Violent Saga of a Maya Kingdom, *National Geographic* 183, 1993:95-111.

DIAMOND, J. 1997 – *Guns, Germs, and Steel: The Fates of Human Societies*, W.W. Norton and Company, New York, 1997.

DOMÍNGUEZ CARRASCO, MD. R., W.J. FOLAN, A.M. LOPEZ, R. GONZOLEZ HEREDIA, H.C. SEU, J.D. GUNN, and L.F. FOLAN 2002 – The State of Calakmul, Campeche: Its Regional Concept. In V.T. Blos, R. Cobos, and M.G. Robertson (eds.), *Memoria de la Tercera Mesa Redonda de Palenque, La Organización Social entre los Mayas Prehispanicos, Coloniales y Modernos*, Coordinadores Conaculta INAH, Universidad Autónoma de Yucatan, Merida, 2002:279-304.

DUNHAM, G.H. 1994 – *Common Ground, Contesting Visions: The Emergence of Burial Mound Ritual in the Late Prehistoric Central Virginia*. Unpublished Ph.D. dissertation, University of Virginia, Charlottesville.

EDDY, J.A. 1994 – Solar History and Human Affairs, *Human Ecology* 22, 1994:23-36.

EPPERSON, D.L. 1971 – *Weather and Climate in North Carolina*, Agricultural Extension Service, North Carolina State University, Raleigh, 1971.

FAUST, B.B. 1998 – *Mexican Rural Development and the Plumed Serpent: Technology and Maya Cosmology in the Tropical Forest of Campeche, Mexico*, Bergin & Garvey, Westport, Connecticut.

FITZHUGH, W.W. 1972 – *Environmental Archeology and Cultural Systems in Hamilton Inlet, Labrador: A Survey of the Central Labrador Coast from 3000 B.C. to the Present*, Smithsonian Institution, Washington, D.C., 1972.

FOLAN, W.J. 1981 – Comments on the Late Postclassic Eastern Frontier of Mesoamerica: Cultural Innovation Along the Periphery. By J. Fox, *Current Anthropology* 22, 1981:336-37.

FOLAN, W.J., J. GUNN, J.D. EATON, and R.W. PATCH 1983 – Paleoclimatological Patterning in Southern Mesoamerica, *Journal of Field Archaeology* 10, 1983:453-68.

FOLAN, W.J., J. MARCUS, S. PINCEMIN, Md. R. DOMÍNGUEZ CARRASCO, L. FLETCHER, and A.M. LÓPEZ 1995 – Calakmul: New Data from an Ancient Maya Capital in Campeche, Mexico, *Latin American Antiquity* 6, 1995:310-34.

FRIERSON, J.L. 2002 – *South Carolina Prehistoric Earthen Indian Mounds*. Unpublished M.A. thesis, University of South Carolina, Columbia, 2002.

GARDNER, W.M. 1993 – *Early/Middle Woodland Mounds in the Upper Shenandoah Valley and Contiguous Regions in West Virginia: Observations on Distribution and Internal Structure*, unpublished paper present at the Middle Atlantic Archaeological Conference, Ocean City, 1993.

GATES, M. 1993 – *In Default: Peasants, the Debt Crisis, and the Agricultural Challenge in Mexico*, Westview Press, Boulder, 1993.

GILL, R.B. 1995 – *The Great Maya Droughts*. Unpublished Ph.D. dissertation, Department of Anthropology, University of Texas at Austin, 1995.

GILL, R.B. 2000 – *The Great Maya Droughts: Water, Life, and Death*, University of New Mexico Press, Albuquerque.

GUNN, J. 1979 – Occupation Frequency Simulation on a Broad Ecotone, In C. Renfrew and K.L. Cooke, (eds.), *Transformations: Mathematical Approaches To Culture Change*, Academic Press, New York, 1979:257-74.

GUNN, J., and C.L. CRUMLEY 1991 – Global Energy Balance and Regional Hydrology: A Burgundian Case Study, *Earth Surface Processes and Landforms* 16, 1991:579-92.

GUNN, J., W.J. FOLAN, and H.R. ROBICHAUX 1993 – *Global Climate and Episodic Urbanism in the Maya Lowlands*, Mexico City. Paper presented at the 1993 Meeting of the International Association of Ethnological and Archaeological Sciences (IUAES).

GUNN, J., W.J. FOLAN, and H.R. ROBICHAUX 1994 – Un Analisis Informativo Sobre la Descarga del Sistema del rio Candelaria en Campeche, Mexico: Reflexciones Acerca de los Paleoclimas que Afectaron a los Antiguos Sistemas Mayas en los Sitios de Calakmul y El Mirador, In W.J. Folan Higgins, (ed), *Campeche Maya Colonial*, Universidad Autónoma de Campeche, Campeche, Mexico, 1994:174-197.

GUNN, J., W.J. FOLAN, and H.R. ROBICHAUX 1995 – A landscape Analysis of the Candelaria Watershed in Mexico: Insights into Paleoclimates Affecting Upland Horticulture in the Southern Yucatan Peninsula Semi-Karst, *Geoarchaeology* 10, 1995:3-42.

GUNN, J., and J.E. FOSS 1997 – Variable Artifact Displacement and Replacement in a Holocene Eolian Feature, In A.C, Goodyear, J.E. Foss, and K. Sassaman, (eds.), *Second international conference on pedoarchaeology*, South Carolina Institute of Anthropology and Archaeology, Columbia, South Carolina, 1997:53-74.

GUNN, J.D. 1991 – Influences of Various Forcing Variables on Global Energy Balance During the Period of Intensive Instrumental Observation (1958-1987) and their Implications for Paleoclimate, *Climatic Change* 19, 1991:393-420.

GUNN, J.D. 1994 – Introduction: A Perspective from the Humanities – Science Boundary, *Human Ecology* 22, Special Issue, 1994:1-22.

GUNN, J.D. 1997 A Framework for the Middle-Late Holocene Transition: Astronomical and Geophysical Conditions, *Southeastern Archaeologists* 16, 1997:134-151.

GUNN, J.D. 2000 – A.D. 536 and its 300 Year Aftermath, In J. Gunn (ed), *The Years Without Summer: Tracing AD 536 and its Aftermath*, BAR, Oxford, 2000:5-20.

GUNN, J.D. 2002 – Steep Shore, Deadly Environment: The Case for a Cultural Anvil Along the Unembayed Atlantic Coast, *North Carolina Archaeology* 51, 2002:1-33.

GUNN, J.D., and R.E.W. ADAMS 1981 – Climatic Change, Culture, and Civilization in North America, *World Archaeology* 13, 1981:85-100.

GUNN, J.D., and W.J. FOLAN 2000 – Three Rivers: Subregional Variations in Earth System Impacts. In The Southwestern Maya Lowlands (Candelaria, Usumacinta, and Champoton Watersheds), In R. Mcintosh, J. Tainter and S. Mcintosh (eds.), *The Way the Wind Blows: Climate, History, and Human Action*, Columbia University Press, New York, 2000:263-270.

GUNN, J.D., W.J. FOLAN, and H.R. ROBICHAUX 2003 – Urbanismo Episódico en las Tierras Bajas de Los Mayas: Un Modelo de Sistema Complejo Para la Agricutural Modern y el Clima Antiguo, In P. Colunga-Garciá Marín and A. Larqué Saavedro (eds.), *Naturaleza y Sociedad en el Área Maya: Pasado, Present y Futuro*, Academia Mexican de Ciencia and Centro de Investigación Científica de Yucatán, Merida, 2003:119-130.

GUNN, J.D. and J.E. FOSS 1992 – Copperhead Hollow (38ct58): Middle Holocene Upland Conditions on the Piedmont-Coastal Plain Margin, *South Carolina Antiquity* 24, 1992:1-17.

GUNN, J.D., J.E. FOSS, W.J. FOLAN, Md.R. DOMÍNGUEZ CARRASCO, and B.B. FAUST 2002 – Bajos Sediments and the Hydraulic System of Calakmul, Campeche, Mexico, *Ancient Mesoamerica* 13, 2002:297-315.

GUNN, J.D., E. SANBORN, K. SERAMUR and L.E. ABBOTT 2003 – *Deposition and Distribution at Horses Grazing: Phase III Data Recovery at 31mr205, Moore County, North Carolina*, Report prepared by New South Associates, Inc., 6150 East Ponce de Leon Avenue, Stone Mountain, Georgia. Submitted to North Carolina Department of Transportation, Archaeology Section, Raleigh, 2003.

GUNN, J.D., and E.E. SANBORN 2002 – *Dimensions of Fall Line Site Function: Survey and Testing the West Fayetteville North Carolina Outer Loop*, New South Associates, Inc., Stone Mountain, Georgia, submitted to North Carolina Department of Transportation, Raleigh, 2002.

HAMMETT, J.E. 1992 – Ethnohistory of Aboriginal Landscapes in the Southeastern United States, *Southern Indian Studies* 41, 1992:1-50.

HAMMETT, J.E. 1992 – The Shapes of Adaptations: Historical Ecology of Anthropogenic Landscapes in the Southeastern United States, *Landscape Ecology* 7, 1992:121-135.

HAUG, G.H., D. GUNTER, L.C. PETERSON, D.M. SIGMAN, K.A. HUGHEN, and B. AESCHLIMANN 2003 – Climate and the Collapse of Maya civilization, *Science* 299, 2003:1731-1735.

HAUG, G.H, K.A. HUGHEN, D.M. SIGMAN, L.C. PETERSON, and U. RÖHL 2001 – Southward

Migration of the Intertropical Convergence Zone Through the Holocene, *Science* 293, 2001:1304-1308.

HERBERT, J.M. 1999 – Prehistoric Pottery Taxonomy and Sequence on the Southern Coast of North Carolina, *North Carolina Archaeology* 48, 1999:37-58.

HERBERT, J.M. 2003 – *Woodland Ceramics and Social Boundaries of Coastal North Carolina.* Unpublished Ph.D. dissertation, University of North Carolina, Chapel Hill, 2003.

HERBERT, J.M., J.K. FEATHERS, and A.S. CORDELL 2002 – Building Ceramic Chronologies with Thermoluminescence Dating: A Case Study from the Carolina Sandhills, *Southeastern Archaeologists* 21, 2002:92-108.

HODELL, D.A., J.H. CURTIS, M. BRENNER, and T. GUILDERSON 2001 – Solar Forcing of Drought Frequency in the Maya Lowlands, *Science* 292, 2001:1367-1370.

HODELL, D.A., J.H. CURTIS, and M. BRENNER 1995 – Possible Role of Climate in the Collapse of Classic Maya Civilization, *Nature* 375, 1995:391-394.

HORTON, W.J., and V.A. ZULLO 1991 – Introduction to the Geology of the Carolinas, In W.J. Horton, and V.A. Zullo (eds.), *The geology of the Carolinas: Carolina geological society fiftieth anniversary*, University of Tennessee Press, Knoxville, 1991:1-10.

HUNTINGTON, E. 1917 – *Maya Civilization and Climatic Change*, Proceedings of the XIXth International Congress of Americanists, December 27-31, 1915, Washington, D.C. 1917:150-164.

IRWIN, J.D., W.C.J. BOYKO, J.M. HERBERT, and C. BRALEY 1999 – Woodland Burial Mounds in the North Carolina Sandhills and Southern Coastal Plain, *North Carolina Archaeology* 48, 1999:59-86.

JESTER, A.H, D.S. LEIGH, and D.I. GODFREY-SMITH 2001 – Chronology of Inland Eolian Dunes on the Coastal Plain of Georgia, USA, *Quaternary Research* 55, 2001:273-302.

JOHNSTON, K.J. 2003 – The Intensification of Pre-Industrial Cereal Agriculture in the Tropics: Boserup, Cultivation Lengthening, and the Classic Maya, *Journal of Anthropological Archaeology* 22, 2003:126-161.

KEELING, C.D. 1979 – *The Influence of the Mauna Loa Observatory on the Development of Atmospheric CO2 Research*, University of California San Diego, Scripps Institution of Oceanography, San Diego, 1979.

KUPPERMAN, K.O. 1988 – *Captain John Smith: A Select Edition of His Writings*, University of North Carolina Press, Chapel Hill.

KUTSCHERA, W., R. GOLSER, A. PRILLER, W. ROM, P. STEIER, E. WILD, M. ARNOLD, N. TISNERAT-LABORDE, G. POSSNERT, S. BORTENSCHLAGER, and K. OEGG1 2000 – Radiocarbon Dating of Equipment from the Iceman. In Bortenschlager, Sigmar and Klaus Oeggl (eds.) *The Iceman and His Natural Environment: Palaeobotanical Results*. The Man in the Ice. 4. 2000:1-9, Springer, Wien and New York. 2000.

LANDA, F.D. 1978 – *Yucatan Before and After the Conquest*, Dover Publications, Inc., New York, 1566.

LEFLER, H.T. 1967 – *A new Voyage to North Carolina (by John Lawson (1709)*, University of North Carolina Press, Chapel Hill, 1967.

LUTZ, W., L. PRIETO, and W. SANDERSON 2000 – *Population, Development, and Environment on the Yucatán Peninsula: From Ancient Maya to 2030*, International Institute for Applied Systems Analysis (IIASA), Luxemburg, Austria, 2000.

MacCORD, H.A., Sr. 1966 – The McLean mound, Cumberland county, North Carolina, *Southern Indian Studies* 18, 1966:3-45.

MacCORD, H.A., Sr. 1986 – *The Lewis Creek Mound Culture in Virginia*, Privately printed, 1994.

MARKEWICH, H.W. and W. MARKEWICH 1994 – *An Overview of Pleistocene and Holocene Inland Dunes in Georgia and the Carolinas: Morphology, Distribution, Age, and Paleoclimate*, U. S. Geological Survey, Atlanta, 1994.

MATHIS, M.A. 2000 – The Middle and Late Woodland Shift on the Central Coast of North Carolina, In Gunn, J (ed), *The years without summer: Tracing ad 536 and its aftermath*, BAR, Oxford, 2000:89-98.

MURRAY, G.E. and A.E. WEIDE, Jr. 1967 – Regional Geologic Summary of Yucatan Peninsula, In A. Weide, Jr., (ed), *Guidebook, field trip to Peninsula of Yucatan*, New Orleans Geological Society, New Orleans, 1967:5-51.

NEILSON, R.P. 1986 – High-Resolution Climatic Analysis and Southwest Geography, *Science* 232, 1986:27-34, 1986.

OTVOS, E.G., and D.M. PRICE 2001 – Late Quaternary Inland Dunes of Southern Louisiana and Arid Climate Phases on the Gulf Coast Region, *Quaternary Research* 55, 2001:150-158.

POPE, K.O. and B.H. DAHLIN 1989 – Ancient Maya Wetland Agriculture: New Insights from Ecological and Remote Sensing Research, *Journal of Field Archaeology* 16, 1989:87-106.

RICE, D.S. 1996 – Paleolimnological Analysis in the Central Petén, Guatemala, In S. Fedick (ed), *The Managed Mosaic: Ancient Maya Agriculture and Resource use*, University of Utah Press, Salt Lake City, 1996:93-206.

ROM, W., R. GOLSER, W. KUTSCHERA, A. PRILLER, P. STEIER, and E.M. Wild 1999 – AMS ^{14}C Dating of Equipment from the Iceman and of Spruce Logs from the Prehistoric Salt Mines of Hallstatt. *Radiocarbon* 41/2:183-197.

ROVNER, I., and S.M. LEWENSTEIN 1997 – *Maya Stone Tools of Dzibilchaltún, Yucatán, and Becán and Chicanná, Campeche*, Tulane University, New Orleans, 1997.

SCARBOROUGH, V.L. 1998 – Ecology and Ritual: Water Management and the Maya, *Latin American Antiquity* 9, 1998:135-159.

SERAMUR, K.C. and E.A. COWAN 2003 – Site Formation Processes of Buried Cultural Horizons in the Sandhills of North Carolina: An Example from the Horses Grazing Site (31mr204), *North Carolina Archaeology* 52, 2003:101-118.

SIEMENS, A.H. and D.E. PULESTON 1972 – Ridged Fields and Associated Features in Southern Campeche: New Perspectives on the Lowland Maya, *American Antiquity* 37, 1972:228-239.

SOLLER, D.R. and H.H. MILLS 1991 – Surficial Geology and Geomorphology, In J. Horton Jr., and V. Zullo, (eds.), *The Geology of the Carolinas*, University of Tennessee Press, Knoxville, 1991:290-308.

STAHLE, D.W., J.G. HEHR and M.K. CLEAVELAND 1988 – North Carolina Climate Changes Reconstructed from Tree Rings: A.D. 372-1985, *Science* 240, 1988:1517-1519.

TANNER, W.F. 1993 – An 8000-Year Record of Sea-Level Change from Grain-Size Parameters: Data from Beach Ridges in Denmark, *The Holocene* 3, 1993:220-231.

TANNER, W.F. 2000 – Beach Ridge History, Sea Level Change, and the A.D. 536 Event, In Gunn, J (ed), *The Years Without Summer: Tracing AD 536 and its Aftermath*, British Archaeological Reports, Oxford, 2000:169-70.

TREWARTHA, G.T. 1961 – *The Earth's Problem Climates*, University of Wisconsin Press, Madison, 1961.

TREWARTHA, G.T. 1968 – *An Introduction to Climate*, McGraw-Hill Book Company, New York, 1968.

WILLEY, G.R. and P. PHILLIPS, 1958 – *Method and Theory in American Archaeology*, University of Chicago Press, Chicago.

WILLIAMS, A.Jr. 1976 – The Interpretation of Rainfall Patterns in Northern Yucatan Utilizing Meteorological Satellite Imagery, *Proceedings of the Association of American Geographers* 8, 1976:15-19.

WISSLER, C. 1906 – Ethnic Types and Isolation, *Science* 23, 1906:147-149.

WISSLER, C. 1914 – Material Cultures of the North American Indians, *American Anthropologist* 16, 1914:447-505.

Chapter 3
RISKY BUSINESS: CADDO FARMERS LIVING AT THE EDGE OF THE EASTERN WOODLANDS

Timothy K. PERTTULA
Archeological & Environmental Consultants, LLC, Austin, Texas, USA

Abstract: *This chapter considers the successes and failures of Caddo farmers during times of rapidly fluctuating climatic conditions between ca. AD 1400-1680. These farming societies lived at the western edge of the eastern Woodlands in Texas, Arkansas, Louisiana, and Oklahoma during repeated periods of droughty conditions, especially a major cool and dry period from ca. AD 1430-1470.*

INTRODUCTION

The prehistoric occupation of the Big Cypress Creek, Sabine River, and Sulphur river basins in northeastern Texas by Titus phase Caddo peoples began during the latter part of the Mississippi period, around AD 1400. These Caddo peoples were contemporaneous with various Plaquemine, Middle Mississippian, and South Appalachian aboriginal groups living across eastern North America. They were a strong and powerful group of peoples (e.g., Early 2000, 2004; Perttula 2002, 2004; Calloway 2003:105-110; Pauketat [2005] for a current summation of the history of the Mississippian peoples). The Titus phase Caddo were farmers, as were other Mississippian groups, living in dispersed communities, and they were active traders, as we know from the wide distribution outside the Caddoan archeological area (Rogers and Sabo 2004: Fig. 1) of decorated Titus phase pottery. The Titus phase Caddo groups in the Big Cypress Creek basin were perhaps the most populous and socially complex of the many Caddo societies living in northeastern Texas at that time. They were the westernmost aboriginal group that was socio-politically akin to middle and late Mississippian polities in the broader southeastern U.S. region (Fig. 3.1).

The Titus phase Caddo communities in the heartland of the Big Cypress Creek basin were experiencing rapid and sustained population growth during times of fluctuating climatic conditions in the 15th, 16th, and 17th centuries. These dynamic farming communities dealt with climatic and subsistence stresses by effecting new means of holding their societies together. They boldly came together into several stronger political communities centered around the establishment of larger mound centers and villages at key nexuses in the Big Cypress Creek basin (Fig. 3.2). In the words of George Sabo (2003:444-445), "Caddo history as enacted … history is neither mute nor static; it is a dynamic component of Caddo culture that people use today – just as their ancestors did in times past – to shape identities and transfer those identities from generation to generation, even in the face of disruption and loss."

THE ENVIRONMENTAL SETTING OF TITUS PHASE COMMUNITIES

The Pilgrim's Pride site on Big Cypress Creek is one of these newly created larger and community-centered Caddo mound and village settlements (Fig. 3.2). These are places where the most important and life-giving ceremonies, rituals, and decisions were made by the social and political elite that guided and organized the changing Titus phase societies living along Big Cypress Creek. The community appears to have been established around ca. AD 1430. Smaller farming households were dispersed for several miles around the Pilgrim's Pride site. Life here was organized around the rhythm of planting and harvesting the cultivated plants, men hunting large game, the rituals and ceremonies of the seasons, and daily life in the household and village settlements.

These Titus phase political communities generally, and the Pilgrim's Pride in particular, are located along and near the modern ecotone between the Pineywoods and the Post Oak Savannah. The latter lies on sandy loam soils on the north side of Big Cypress Creek (Fig. 3.3). The Post Oak Savannah is a narrow strip of woodlands between the Pineywoods to the east and south, and the Blackland Prairie vegetational region to the west, north (Talco Prairie; Fig. 3.3) and northwest, no closer than 20 km away. According to Schmidly (2002:371), the "topography is level to gently rolling and slopes gently from the northwest to the southeast … the post oak region can best be described as an ecotone between the eastern deciduous forest and the tall-grass prairie. The area supports a stunted, open forest dotted with small tall-grass prairies. The dominant plants of the overstory are post oak and blackjack oak and to a lesser extent winged elm and black hickory." The Pineywoods have medium-sized to tall broadleaf deciduous forests in more mesic habitats, and shortleaf and loblolly pines are common on upland fine sandy loam soils with adequate moisture. Smaller areas of tall grass prairie may be present in both communities throughout the region (e.g., Jordan 1981: Fig. 4.1), particularly in drier sandy lands.

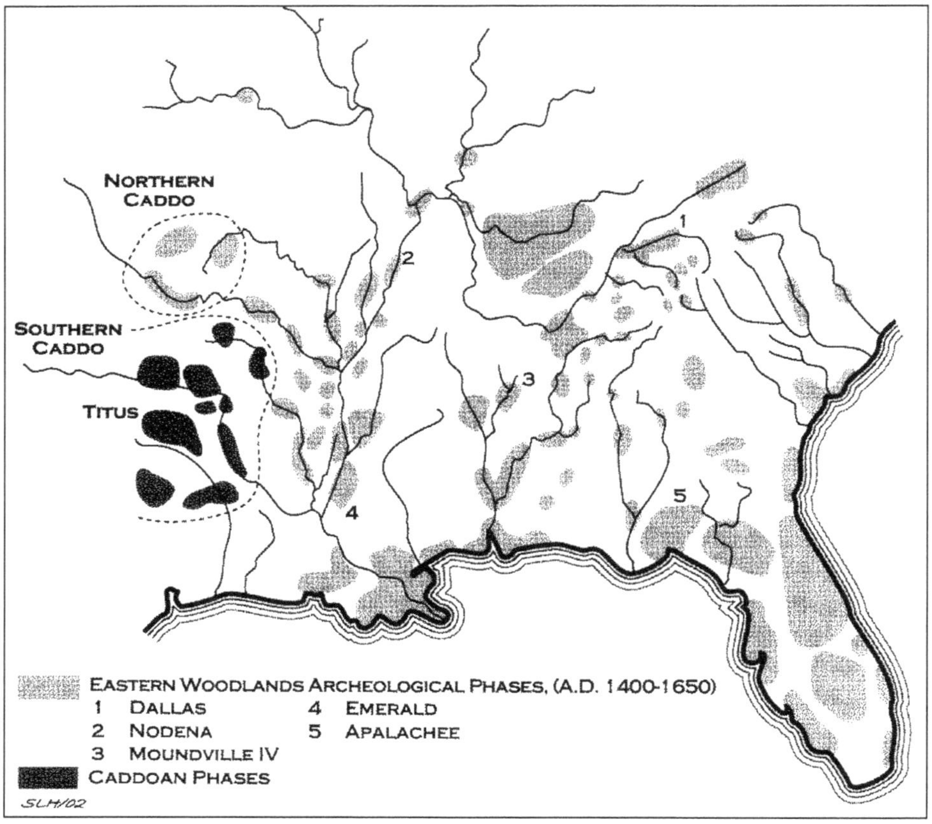

Figure 3.1. Eastern Woodlands archeological phases contemporaneous with the Titus phase, including other phases in the Southern Caddo area, based on Milner *et al.* 2001: Figure 2.2)

MID-19TH CENTURY VEGETATION CONDITIONS

Texas General Land Office (GLO) survey notes from a number of the patented land grant surveys in and around the middle reaches of the Big Cypress Creek valley in Camp and Titus counties, Texas, provide initial environmental data on the vegetation conditions in this part of the Big Cypress Creek basin in the mid-19th century. This is before the area was likely to have been extensively cleared and lumbered (Perttula and Nelson 2002:15-16). Consequently, it may not have been much different then when Caddo groups lived there. The 30+ land survey field notes date from 1837-1854.

The predominant overstory trees in this general locale in the mid-19th century were red oak (*Quercus falcata*), post oak (*Q. stellata*), blackjack oak (*Q. marilandica*), and various species of hickory (*Carya sp.*), along with sweetgum (*Liquidambar styracliflua*). Pine trees must have only occurred in patches, particularly in Camp County, as they only represent only 0.8 percent (Titus County) to 3.2 percent of the marker trees (Table 3.1). The general composition of the forested landscape on both sides of the Big Cypress Creek was an upland woodland of oaks and hickories – with more mesic patches of white oak (*Q. alba*) and red oak – with hardwood forests in the floodplain. The floodplain hardwood forests comprised willow oak (*Q. phellos*), water oak (*Q. nigra*), overcup oak (*Q. lyrata*), maple (*Acer sp.*), sweetgum, ash (*Fraxinus sp.*), elm (*Ulmus americana*), and sassafras (*Sassafras albidum*). There must have been some swampy or marshy, frequently inundated floodplain areas along Big Cypress Creek because of the occurrence of black gum (*Nyssa sylvatica*) or black tupelo (Table 3.1). Pine was not a primary constituent in the forest in the mid-19th century. The pine that did occur (probably shortleaf pine, *Pinus echninata*) grew on the drier soils in the forest, likely in patches mixed with blackjack oak and post oak (Bonnicksen 2000:229). The pine was also likely affected by the frequency and intensity of natural or human-created fires. One 1838 land survey on a large tract of land on the north side of Big Cypress Creek, and west of the Pilgrim's Pride site, had a "little prairie." This was probably an area with poorly drained soils that would have had big (*Andropogon gerardii*) and little bluestem (*Schizachyrium scoparium*), switchgrass (*Panicum virgatum*), and Indiangrass (*Sorghastrum sp.*) (Marietta and Nixon 1984).

The forest composition in the 1830s-1850s appears to have been greatly influenced by the frequency and timing of Indian-set and lightning-ignited fires (Bonnicksen 2000:331, 339). These fires created a mosaic of patches of trees with different tolerances to fire, shade, and moisture. The more-fire-tolerant shortleaf pine was found on drier upland soils, along with the more fire resistant post oak and blackjack oak also dominant on the drier soils in the forest.

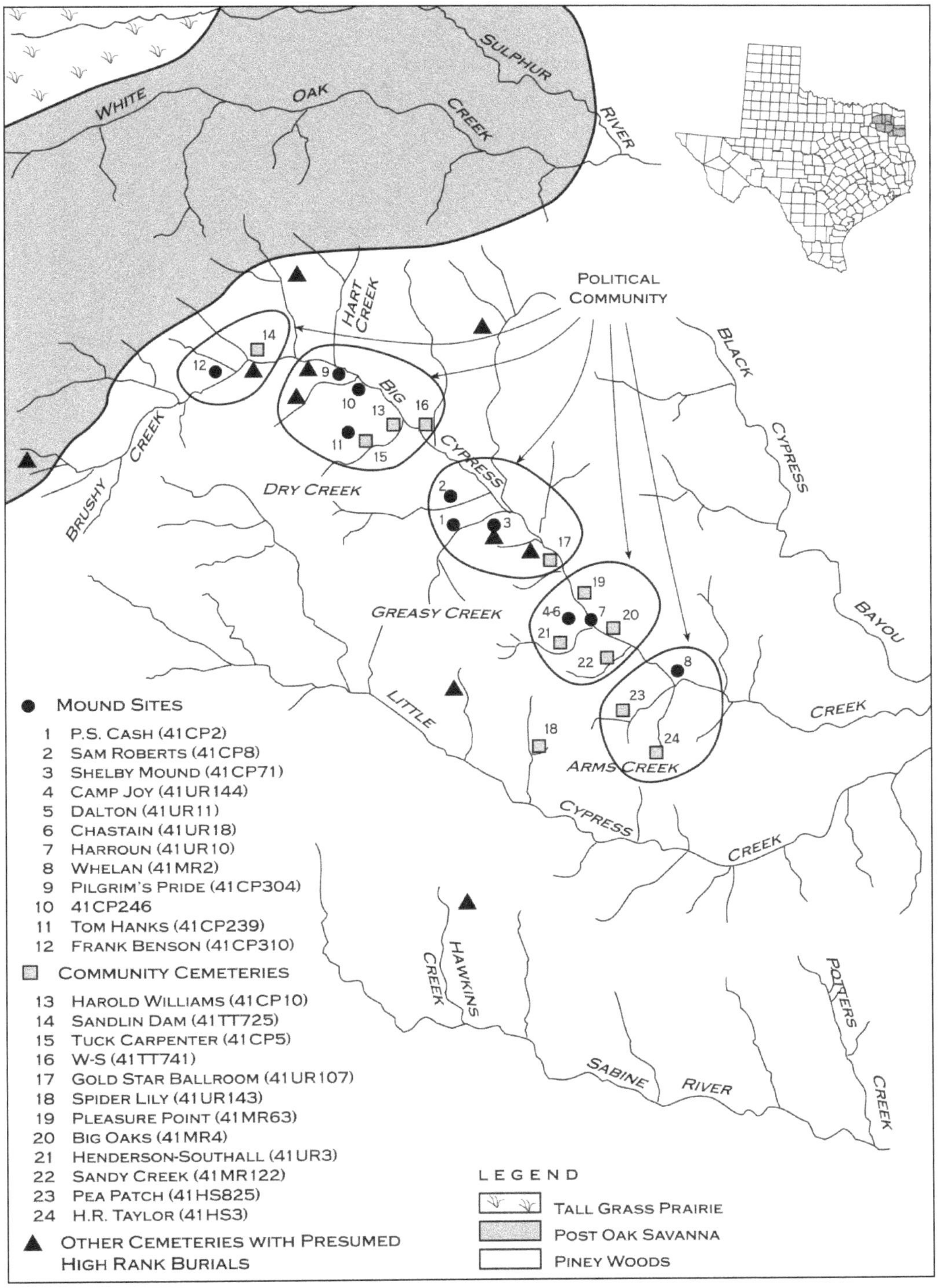

Figure 3.2. Titus phase political communities

Post oak and blackjack oaks comprised between 28 to 35 percent of the tree species mentioned in the area (Table 3.1). These two species were actually more common on the Camp County side of Big Cypress Creek (within the modern boundaries of the Pineywoods, Fig. 3.3) than they were on the Titus County side (within the modern boundaries of the Post Oak Savannah). The post oak and blackjack oaks would have been found on leached soils on poorly drained upland landforms with a low clay content, and there would have had a sparse floor understory cover.

Moister slopes and other upland landforms, along with elevated alluvial landforms, apparently tended to have trees that were moderately tolerant of fire. This included loblolly (*Pinus taeda*), red oak, white oak, and hickory, along with maple, walnut (*Juglans nigra*), and other hardwoods. The white and red oak were nut-bearing trees. This forest mosaic tended to have a greater diversity of species in the canopy than the post oak-blackjack oak or pine forests (Marietta and Nixon 1983). About 21-22 percent of the tree species in Camp and Titus counties tabulated in Table 3.1 included these more

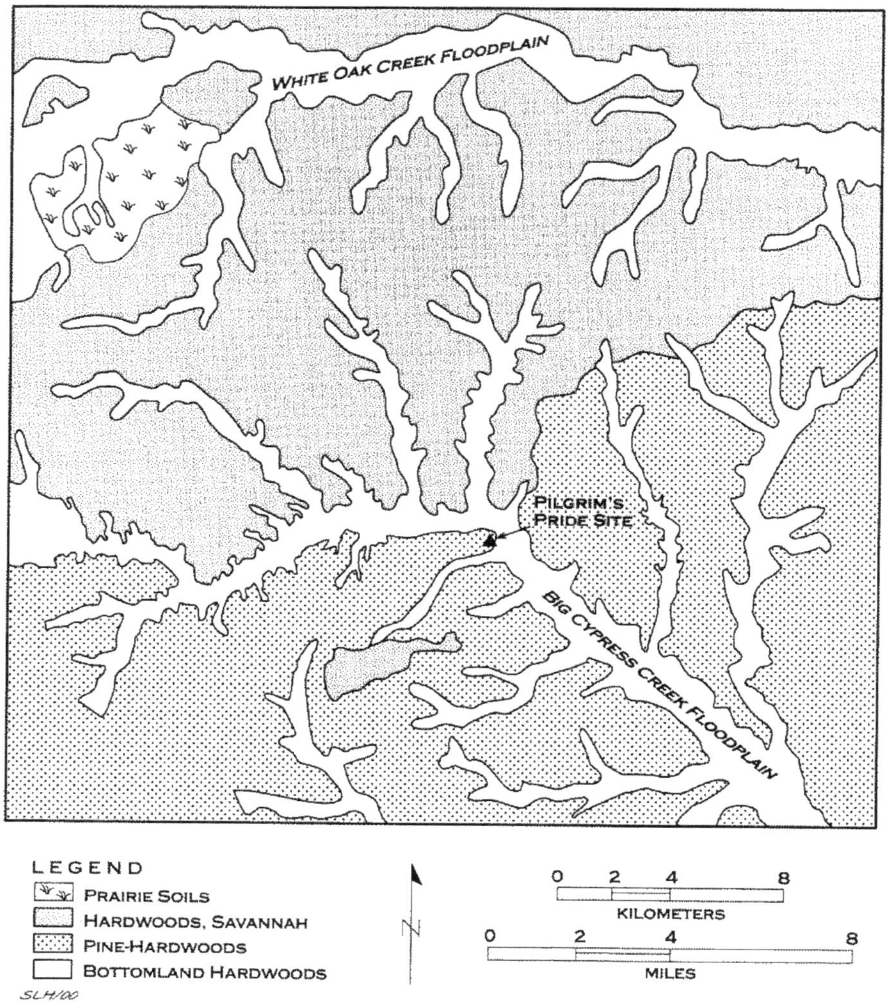

Figure 3.3. Vegetation zones in the vicinity of the Pilgrim's Pride site in northeastern Texas

Table 3.1. Tree species mentioned in General Land Office records for the middle part of the Big Cypress Creek valley

Common Name	**Species name**	**Camp County**	**Titus County**
Post oak	*Q. stellata*	15.9%	19.0%
Blackjack oak	*Q. marilandica*	19.1%	9.1%
Red oak	*Q. falcata*	25.5%	33.1%
White oak	*Q. alba*	1.9%	4.1%
Willow oak	*Q. phellos*	1.9%	1.6%
Water oak	*Q. nigra*	2.5%	2.4%
Overcup oak	*Q. lyrata*	--	0.8%
Hickory	*Carya sp.*	17.9%	17.4%
Black walnut	*Juglans nigra*	--	1.6%
Sassafras	*Sassafras albidum*	0.6%	0.8%
Sweetgum	*Liquidambar styraciflua*	7.6%	4.9%
Ash	*Fraxinus sp.*	0.6%	1.6%
Elm	*Ulmus sp.*	0.6%	1.6%
Maple	*Acer sp.*	1.9%	--
Black gum	*Nyssa sylvatica*	0.6%	0.8%
Pine	*Pinus sp.*	3.2%	0.8%
Number of observations		**157**	**121**

mesic upland forests. The distribution of mesic forests appears to have been comparable on both sides of Big Cypress Creek. Hickory, in particular, preferred moist slopes as well as river bottoms because they are more vulnerable to fires than the oaks and shortleaf pine.

The distribution of sweetgum in mid-19th century Camp and Titus County land records indicate that there were floodplain habitats in the immediate area (Fig. 3.3) that were only occasionally inundated (Nixon *et al.* 1983). Other trees common in such habitats would have included maple, holly (*Ilex sp.*), and American hornbeam (*Carpinus caroliniana*).

It is interesting how few pine trees were noted in the middle reaches of the Big Cypress Creek valley during the 1837-1854 General Land Office surveys, particularly since much of Camp County falls within the modern Pineywoods. This is probably a product of two different, but unrelated factors. First, the land surveys that were specifically examined were relatively close to Big Cypress Creek. This would have excluded much of the higher and drier upland areas of shortleaf pine that oftentimes occurred in parts of Northeast Texas in pure stands with little undergrowth. The second factor is the possibility that the dominance of pine in modern times in what is termed the Pineywoods may well be the product of the cessation of Indian-set fires after the Caddo Indians were removed from the region by the early 1840s. There probably were also more strenuous attempts by farmers after the mid-1850s to fight lightning-ignited fires. As the frequency and intensity of fires diminished in modern times, and fires had not burned for a number of years, the extent of upland sandy loam habitats suitable for pines also increased.

General Land Office field notes indicate that Big Cypress Creek had only a 20-28 foot wide channel in this area, not much different than in modern times. The stream flowed all year-round. The channels of the smaller tributaries ranged from 6-10 feet in width, and many of these (particularly in Titus County) were probably spring-fed, while others only flowed part of the year (Thurmond 1990:16 and Fig. 4).

LATE HOLOCENE ENVIRONMENTAL CHANGE

The Late Holocene period after ca. 5000 years ago appears to have been characterized by fluctuating climates – between moist or dry cycles – that were generally wetter than during the preceding Middle Holocene period. Modeled precipitation histories (Perttula 2005: Fig. 2.3a-c) suggests that the peaks and valleys differed by ca. 100-200 mm through time.

With these climatic and rainfall conditions, Oak-hickory-pine woodlands were probably the principal vegetation in upland habitats in the Big Cypress Creek basin, with a well-developed riverine hardwood forest in the floodplain settings. Supporting the hypothesized drier and warmer cycle in the middle portion of the Late Holocene, the Ferndale Bog pollen record indicates that the peak in pine pollen was between ca. AD 150-1150 (Holloway 1994: Table I.2). Bousman (1998:207) notes one grass spike or peak in the Weakly Bog in Central Texas that dates about AD 350-450, with another between AD 1450-1550. These periods were also slightly colder and drier.

For the last 1000 years or more, dendrochronological or tree-ring records of paleoenvironmental change are the most accurate and temporally sensitive data available on Late Holocene environmental change (e.g., Stahle 1996). Recent tree-ring research in Texas, Arkansas, and Louisiana, as well as the Southeast U.S., by Stahle and Cleaveland (1988, 1992, 1993, 1994, 1995) has compiled significant new information on subtle but changing climatic and rainfall conditions and trends for the general Trans-Mississippi South region (of which the Caddoan area is a part).

Most notably, droughts are common in the region in modern times. There were numerous wet and dry spells and periods of climatic instability between ca. AD 1000-1700 and after, just as there were between 5000-1000 years ago (Stahle and Cleaveland 1988, 1994). Some of the worse droughts may have occurred around AD 1555, 1570, 1595, and 1670, and the period between AD 1549 and 1577 has been suggested to have had the worse droughts in the past 450 years (Stahle *et al.* 1985).

More detailed analyses are available from bald cypress tree-ring chronologies on spring rainfall between AD 997 and 1988 from Big Cypress State Park in northwestern Louisiana (Stahle and Cleaveland 1995; also Tree-Ring Data Bank, IGBP Pages/World Data Center for Paleoclimatology Program, Boulder, Colorado). Year by year changes in prehistoric times indicate that the seven sets of wettest years were between AD 1053 and 1057, 1168 and 1176, 1178 to 1180, 1265 and 1268, 1323 and 1328, 1553 and 1555, and 1584 to 1586. The wettest years in prehistoric times were about a decade from 1168 to 1176 and 1178 to 1180 (Fig. 3.4). These years would likely have been optimal growing years for Caddo horticultural groups, assuming a correlation between crop production and spring precipitation values (cf. Anderson *et al.* 1995:265). The wetter conditions would also likely have led to an increase in the extent of swamp and wetland habitats in much of the Big Cypress Creek basin. There would have also been a concomitant expansion in the carrying capacity of woodland plants and animals in valley and floodplain areas.

Conversely, the driest years in prehistoric and early historic times – between AD 1014 and 1016, 1215 and 1217, 1444 and 1447, 1455 to 1460, 1529 and 1533, 1653 and 1655, and 1697 to 1699 – may well have been periods when food supplies were stressed (see Perttula 2008). The climatic conditions during these times would have put at risk the ability of Caddo groups to produce sufficient food reserves from the cultivation of tropical cultigens. They would also have adversely affected their chances of success in obtained good maize harvests during these extended droughty periods (below). The

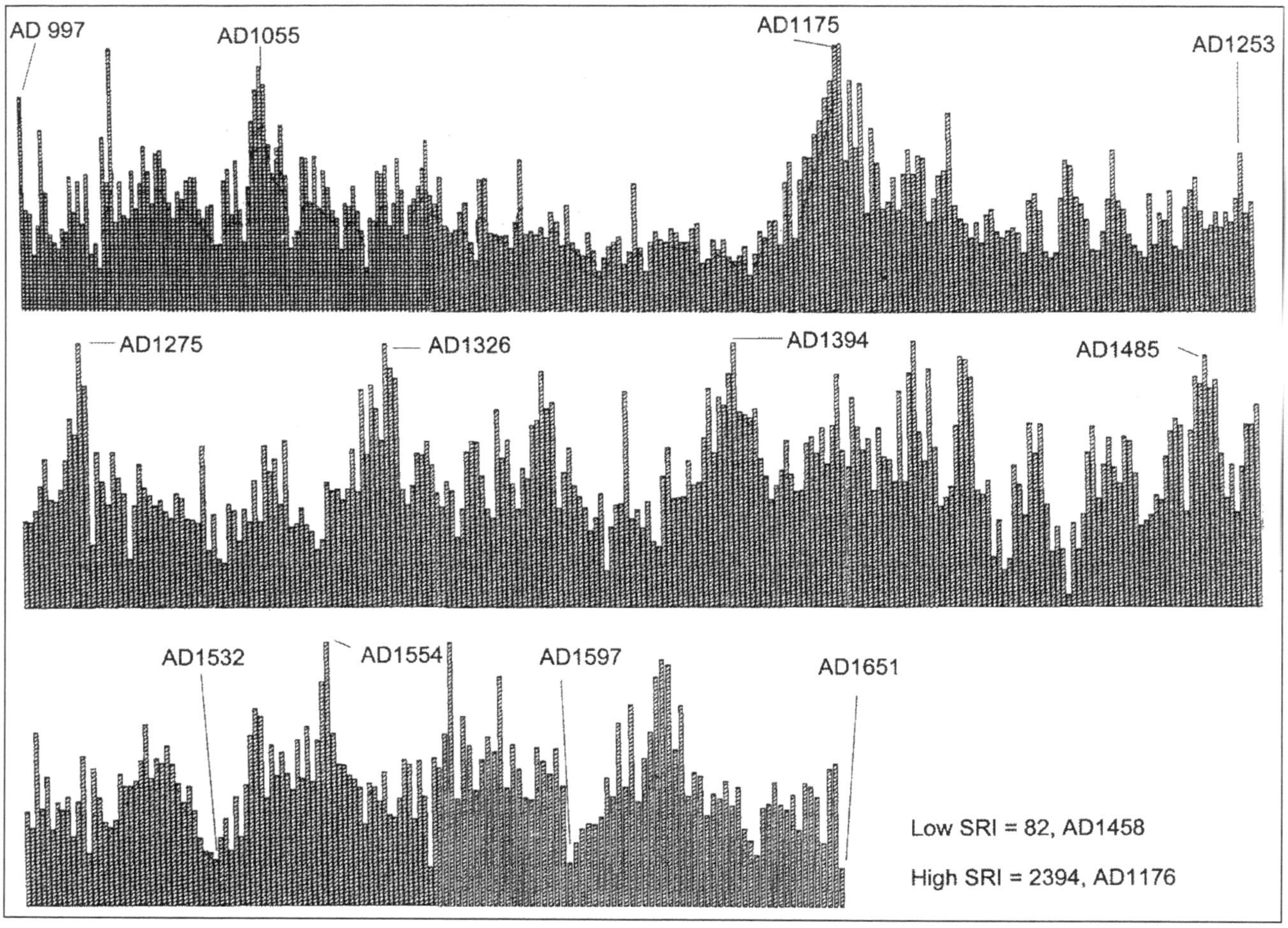

Figure 3.4. Tree-ring sequence, Big Cypress State Park, Louisiana, AD 997-1651

very dry years between AD 1444 and 1460 detected by the tree-ring record (Fig. 3.4) correlate well with the grass spike/drier episode noted by Bousman (1998) from the Weakly Bog pollen record. These droughts probably also affected the constancy of flow in the numerous upland springs in the area, as well as the volume of flow in the Big Cypress Creek basin. This in turn would have influenced the relative quantity of animal and plant foods in floodplain and upland forested habitats.

Looking at the period of wet and dry spells from ca. AD 1000 to 1650, the wetter years (>1400 standard ring width indices [sri]) were more than two times as frequent as the driest and droughty (<560 sri) years (Fig. 3.4). After ca. AD 1430, the wetter years occurred less often, some 55 percent less between AD 1600 and 1700 than in the ca. AD 1200 and 1400 period.

The frequency of very dry years remained rather constant after ca. AD 1430 (and remained relatively constant until the 1790s), but were conversely quite rare between AD 1000 and 1400 (Fig. 3.4). Clearly, then, if the tree-ring data from Big Cypress State Park are relevant to understanding local climatic conditions in the Big Cypress Creek basin, Caddo settlement of the region before AD 1400 took place during an equitable climatic episode when floodplain and upland forests were expanding at the expense of drier habitats. There were comparable spring rainfall amounts during most of a 400-year period. It is only after the early to mid-15th century that drier and cooler conditions probably existed in the Big Cypress Creek basin. There were major periods of drought between AD 1444 and 1447 and between 1455 and 1460. Others occurred in the early 16th century, the mid-17th century, and then with regularity until the latter part of the 18th century. During these times, the Big Cypress Creek region was occupied by Titus phase Caddo groups (Figs 1 and 2).

Changes in solar radiation inferred from atmospheric delta C14 variation (Bradley *et al.* 2003: Fig. 6.13) also provide a useful climatic proxy for paleoenvironmental change. Climatic minima (i.e., cooler and drier times) peaked around AD 1450 and AD 1650, and were probable periods of cultural hardships for the Titus phase Caddo peoples.

CLIMATIC EPISODES, CA. AD 1430-1680

More detailed paleoenvironmental reconstructions of past climatic episodes during the Titus phase Caddo settlement of the Big Cypress Creek basin are based on the previously discussed Big Cypress State Park tree-ring

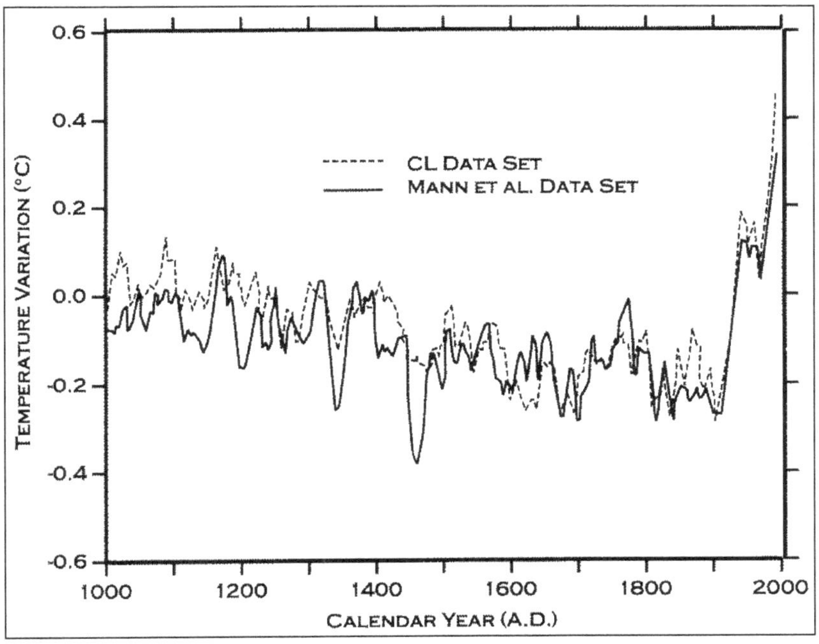

Figure 3.5. Reconstructed temperature variation, AD 1000-2000, after Mann *et al.* (1998) and Crowley (2000)

data base and recent reconstructions of changes in temperature over the last 1000 years (Fig. 3.5) offered by Mann *et al.* (1998) and Crowley (2000). These used a wide range of proxies, such as tree rings, ice cores, and corals (Jones *et al.* 2001:662, Mann 2002).

The mean temperature reconstructions indicate that there has been a general decline in temperature from about AD 1000 (if not earlier) to about 1900, with a rapid warming after that time. Crowley's (2000) studies suggest that prior to 1850, decadal-scale changes in temperature variation are due to low frequency changes in solar irradiance and pulses in volcanism that served as climatic forcing mechanisms. The 11th and 12th centuries were warm, and the 13th century was a time of temperature fluctuations, as were the 14th century and much of the 15th century. Some of the coldest reconstructed temperatures occurred around the mid-14th century, as well as in the mid-15th century, and much of the 17th century was cool. Otherwise, much of the period between ca. AD 1300 and 1580 was relatively warm. After AD 1700, about the time that the Titus phase Caddo peoples abandoned Northeast Texas, temperatures warmed again, until a period of abrupt cooling in the early part of the 19th century. The most recent reconstruction of past temperature variability over the last 1000 years suggests that the 17th century was even colder than previously thought. Much of the 12th, 13th, and 14th centuries were cool (Esper *et al.* 2002, Mann 2002), as "reconstructed temperatures are consistently well below those indicated by all other records" (Briffa and Osborn 2002:2228).

The second part of the climatic data base is the AD 997 to 1988 tree ring width data from Big Cypress State Park. The tree ring data are the proxy for moisture over the last 1000 years in the general East Texas region. The range in tree ring values over the 250 year period of the Titus phase, namely the standard ring indices, are from a low of 82 (in AD 1458) to a high of 2386 (in AD 1578) (Fig. 3.4). Trends from wet to dry are apparent over the course of a millennium.

From these trends in tree ring width and reconstructed mean temperature variation, I have defined six alternating droughty and mesic periods between AD 1430 and 1680 (Table 3.2). The droughty periods date from AD 1430 to 1476, 1525 to 1538, and 1573 to 1602. The generally warmer and wetter periods date from AD 1477 through 1524, from 1539 through 1572, and from 1603 through 1670+. After 1670, the period from 1671 to 1676 was relatively dry and cool.

Based on the mean tree-ring width in the three principal droughty climatic episodes, the most sustained and persistent period of drought was at the beginning of the Titus phase, in the AD 1430 to 1476 climatic episode (Table 3.2). The three peaks of drought conditions in AD 1444 to 1447, 1455 to 1460, and 1472 to 1473 had mean tree ring widths of only 370.5-556.0. This is between 40-60 percent lower than in times of an equitable climate and average growing and moisture conditions. The AD 1455 through 1460 drought was also a notably colder era during the Titus phase.

The other two droughty climatic episodes were also very dry (with mean tree-ring widths ranging between 304 and 487.5), some 50 to 70 percent lower than in average climatic conditions. These drier and colder drought periods, and the four year drought (AD 1651-1655) during the last mesic period, did not generally last as long as the cold and dry pulses during the AD 1430 to 1476

Table 3.2. Reconstructed Climatic Episodes, AD 1430-1680

Climatic episode	Droughty periods	Mean tree-ring width in droughty periods	Mesic periods
AD 1430-1476	X	370.5 (1444-1447)	
		380.5 (1455-1460)	
		556.0 (1472-1473)	
AD 1477-1524			X
AD 1525-1538	X	466.0 (1525-1538)	
AD 1539-1572		487.5 (1597-1598)	X
AD 1573-1602	X	304.0 (1573-1574)	
AD 1603-1660+		425.7 (1651-1655)	X

Note: the lower the mean tree-ring width, the drier the climate. A mean standard ring width of 1000 represents an equitable climate and average growing and moisture conditions

period. While the AD 1525 through 1538 droughty period was quite dry, it was nowhere as severe a drought as the drought in the mid-16th century that Stahle *et al.* (2000:121) consider the "most severe prolonged drought over much of North America for at least the last 500 years." This period of persistent drought occurred between about 1560 to 1590 in parts of Texas and between 1540 and 1580 in northern Mexico. The worst years of the mega-drought were in the mid-1570s in Texas – apparently indicated locally by the two very dry years in 1573 and 1574 – with very low summer precipitation. Reconstruction of the spatial extent of this mid-16th mega-drought by Cook *et al.* (1999) suggests its effects were more severe from southern Texas to the panhandle of Texas, and then north and west into the southwestern U.S. They were less intensely felt in the Caddoan archeological area.

LATE CADDO TITUS PHASE ARCHEOLOGICAL AND ENVIRONMENTAL INTERRELATIONSHIPS

The climatic episodes that can be defined over the 250 year period of the Titus phase (ca. AD 1430 to 1680) provide the opportunity to examine responses made by these Caddo peoples to climatic changes over inter-annual to decadal and multi-decadal intervals. I am interested in knowing if climatic changes disrupted ordinary or traditional cultural responses of Caddo communities. Furthermore, it is important to determine whether Caddo peoples had the means to respond to changes in climatic conditions that may have exceeded normal annual and inter-annual variations.

Since by ca. AD 1300 the Caddo had become a society that was increasingly dependent upon cultivated plant foods. It is reasonable to argue that "as agricultural food production increased in importance, fluctuations in climate and hence potential crop yields had varying impacts on societies at both local and regional scales" (Anderson 2001:143). At the same time, in no way do I mean to suggest that the impacts of climate changes, no matter how deleterious, were the root and deterministic cause of cultural changes among the Titus phase Caddo peoples, or that there is necessarily any connection between the two. Rather, climatic changes at different scales can be reasonably expected to "elicit adaptive responses or, at the very least, destabilize well-established adaptive systems" (Binford 2001:447). In other words, climate changes "influence but do not cause particular responses on the part of human populations" (Miller 2004:137). No adaptive strategy of any peoples is divorced from the range of habitats it exploits to survive and prosper. In essence, therefore, what I want to do here is lay the temporal and climatic framework to consider how changes in habitat variability and climate may have affected the adaptive success and influenced the cultural responses of the Titus phase agricultural Caddo peoples living at the edge of the eastern Woodlands of North America.

There may be some relevant comparisons to be made between the climatic and native history of the Titus phase Caddo peoples and recent archeological, bioarcheological, and paleoenvironmental research in the lower Mississippi valley. During the mid-16th century, societies living there experienced a mega-drought that lasted from the 1540s to the 1580s (Stahle *et al.* 2000, Fisher-Carroll 2001, Burnett and Murray 1993). This lengthy period of drought is thought to have had major effects on these native peoples, leading to a major reduction in reproductive potential, agricultural destabilization, social stresses, and "the abandonment of an entire region within a relatively short period" (Fisher-Carroll 2001:242). How might the droughty periods in Northeast Texas between AD 1430 to 1680 have affected the Titus phase Caddo groups living in a number of Big Cypress Creek political communities?

The droughty climatic episodes occurred about every 35 to 48 years. This was about once during each person's lifetime, given that the average life span of Caddo males and females in Late Caddoan period times was about 40 years of age (Derrick and Wilson 2001: Table 2). Thus, the memories of these droughts – and the cultural ability to learn how to recognize and respond to the signs and changes signaling the onset of droughty conditions (i.e., social memory, McIntosh *et al.* 2000:24-25) – were likely part and parcel of the adaptive strategies and knowledge possessed by the Titus phase Caddo farming peoples in managing subsistence risks. Even so, it is unlikely that they were culturally prepared for the onset of the most intensive droughty period over the last millennium that occurred between AD 1444-1460. Nor did they likely have the crop reserves to successfully withstand such lengthy very cold and very dry conditions.

The more mesic periods between severe droughty conditions also lasted between 35 and 49 years at a time (Table 3.2). These periods were warmer and wetter than times that came before and after. They were periods of more equitable rainfall and increased net productivity and carrying capacity of plants and animals in the oak-hickory Savannah and oak-hickory-pine habitats that were settled by the Titus phase populations. How are these environmental changes linked to subsistence and demographic changes? What are the effects of subsistence changes on reproductive potential during equitable climatic episodes? We will turn to these questions in the remainder of this chapter.

In Late Caddo Titus phase times, the Caddo peoples had a diet that primarily consisted of cultivated plants like maize, beans, and squash (see Perttula 2008). Insuring the success of agricultural pursuits must have been of primary importance in determining the location of individual farmsteads and hamlets, as well as the community centers. This is likely to have been even more the case than it was in the Early or Middle Caddoan periods when Caddo peoples were not apparently quite so dependent upon cultivated plants for their diet, or at least that was the case until ca. AD 1300 to 1350 (Perttula 1996, Dering 2004, 2005). How are these constraints reflected in the spatial distribution of Late Caddo Titus phase sites during times of climatic instability? What we see in this part of the Big Cypress Creek basin (Fig. 3.2) is that the overall settlement pattern was dispersed. This was in conjunction with a heightened emphasis on situating sites along the secondary streams and the spring-fed branches. These areas may have had more dependable water, or more accessible water. It is also likely that fields would have been easier to clear along the more open upland forests than if fields had to be located in the more mesic valleys.

There are more Late Caddo Titus phase sites in this part of the Big Cypress Creek basin than sites found during earlier periods (Perttula 2004: Table 13.2). This first suggests that the regional population was quite a bit higher during the Late Caddoan period (all things being equal, especially the length of time each settlement was occupied and the general shifting of populations across the landscape). There are several clusters of settlements that may represent parts of contemporaneous small communities or villages. Titus phase cemeteries are particularly common in sites located along Big Cypress Creek itself, followed by other cemeteries on a series of tributaries to Big Cypress Creek. The principal tributary creek settings for Titus phase cemeteries are Swauano Creek, Boggy Creek, Dry Creek, Arms Creek, Meddlin Creek, and Greasy Creek (Fig. 3.2). The Late Caddo communities living in the Pineywoods on the tributaries appear to represent recognizable concentrations of settlements, mounds, and community cemeteries that constitute distinct political communities. A political community as used here is a cluster of interrelated settlements and associated cemeteries that are centered on a key site or group of sites distinguished by public architecture (i.e., earthen mounds) and large domestic village areas.

Perttula and Nelson (2003:34) have noted that Late Caddo Titus phase sites are more common south of Big Cypress Creek in the Pineywoods than they are on the north side of the basin in the Post Oak Savanna. Titus phase sites are also less frequent in other stream valleys (such as Little Cypress Creek, White Oak Creek, or streams in the Lake Fork Creek basin) in the Big Cypress, Sulphur River, and Sabine River basins. Much of the upper Sulphur and Sabine River basins were abandoned after ca. AD 1430 (cf. Bruseth 1987, Fields *et al.* 1997:91). Poorly drained and steeper, rockier landforms on the east side of Big Cypress Creek were also apparently not heavily settled by Caddo peoples.

Regional settlement data for the Titus phase does suggest that this pattern in the spatial distribution of sites may be part of a much broader trend in the density of Late Caddo sites between the Titus phase "heartland" and outlying areas also occupied by Titus phase Caddo peoples (Perttula 1998, 2004, Perttula and Nelson 2003). That trend indicates that Titus phase sites – as well as Titus phase sites with mounds and large community cemeteries – are more common across the landscape from the Brushy Creek area downstream along Big Cypress Creek than they are in the Post Oak Savannah immediately north and northeast of Big Cypress Creek. This distribution strongly indicates that the density of Caddo peoples during the Titus phase was more concentrated in the Big Cypress Creek heartland in the Pineywoods, including its many southward-flowing and eastward-flowing tributaries, than it was elsewhere across the landscape then, or in earlier times.

It is not until the onset of the Titus phase about AD 1430 that the Pineywoods were the scene of large numbers of village settlements of Caddo farmers along the western margins of the Post Oak Savanna and Pineywoods ecotone. These villages were established during a major droughty period (ca. AD 1430 through 1476, Table 3.2). At the same time, there is evidence of a dramatic residential movement of Titus phase Caddo peoples (as marked by the absence of sites) from the Post Oak Savanna (e.g., Bruseth 1987:276-277, 280). The movement, redistribution, or emigration of Caddo peoples from more marginal settings may have been the product of a response to periods of stress and subsistence instability during this time of climatic perturbations. Allen (2004:200-201) has noted that populations living in temporally variable environments – environments where the carrying capacity varies widely because of significant climatic perturbations of some duration and relative frequency, as is the case at the western edge of the

Eastern Woodlands – will frequently emigrate during bad years rather than risk extinction or catastrophic population losses.

These villages and political communities apparently flourished for at least 5 to 6 generations, during both dry and more mesic periods. Several of these large village settlements were apparently abandoned in the early to mid-17th century, however. The important villages are situated in similar topographic settings in the Pineywoods, namely along tributary streams near their confluence with Big Cypress Creek, but they are not found in the Post Oak Savanna. They are marked by a higher density of settlements around one premier community center (Fig. 3.2). These components here represent permanent, year-round, settlements of horticultural peoples. The locations that they chose to permanently settle and build structures and other facilities at had to be situated in habitats where suitable sandy soils were nearby that could be worked with simple wood and bone digging tools. The land they built their homesteads and communities on had to be well-drained and elevated above the annual floods along Big Cypress Creek and its tributaries. They also had to be in areas where wood and grass was plentiful for house construction and refurbishing, as well as near fresh drinking water. The fact that the Titus phase settlements are not found in any notable spatial clusters around the larger and more important villages suggests that the many resources that were needed by sedentary Caddo populations to successfully live in the Big Cypress Creek valley could best be exploited by dispersing the groups in a variety of settings. In turn, the religious and political elite at the community centers probably helped develop strategies that controlled the redistribution and exchange of resources in times of need, especially the redistribution of agricultural surpluses.

This dispersed settlement arrangement would have also helped lessen the competition for plant and animal resources, and would not have led to the environmental degradation of suitable habitats by a single large community. This would have been of critical concern as a means of alleviating or minimizing stress during droughty periods for Caddo peoples living along the margins of the Pineywoods, in environmentally risky areas. It would also have permitted the Caddo peoples to take advantage of the diversity in habitats to exploit a number of them. This would have insured that, through resource redistribution and agricultural diversification, the overall community could survive if there were economic difficulties or failures (i.e., local droughts, flooding, fires) in some habitats but not in most of the others within the broader community homelands.

Early (2004:568) has noted that "episodes of drought and climatic instability that have been identified [in the Caddoan archeological area] may have made an agrarian economy untenable in some locations." This seems to have been the case in parts of the upper Sabine and Sulphur River basins after ca. AD 1430. Furthermore, Early (2004:563) has commented that "stresses on arboreal canopy cover and habitat, as well as on plants and animal species sensitive to water availability, are likely to have been significant. Expansion of grass and other open canopy species eastward, and other drought responses can occur rapidly in this ecotone, and human adjustments might have required dietary and subsistence changes of some magnitude."

In the Titus phase, the tropical cultigen maize (*Zea mays L.*) was a dietary staple, and domesticated beans (*Phaseolus vulgaris*) were also an important food source. Nuts and seeds were also gathered, but they appear in some areas to have been of lesser importance in the diet of Caddo peoples than they were between ca. AD 900 and 1350 (Rogers and Perttula 2004:37; Perttula 2008). In fact, the subsistence evidence from Titus phase Caddo sites in the Pineywoods and Post Oak Savanna, as well as elsewhere in the Caddoan archeological area after ca. AD 1400, suggests the rather successful development of an Caddo maize-based economy by about this time. Dering (2005) does make a strong argument, however, that the diet of Titus phase Caddo peoples was a diverse mixture of cultivated foods and gathered wild plants, especially hardwood mast (also Dering 2004). This diversification would have been an appropriate strategy to pursue in times of risk to insure the long-term success of Caddo agricultural economies, even in the most productive locales.

As noted earlier in the discussion of Late Holocene environmental changes, the Late Caddo agricultural lifeway probably was flourishing in parts of northeastern Texas (likely those areas with the highest agricultural potential). This was at least in part due to quite adequate growing season rainfall from the late 14th century through the first quarter of the 15th century, the last quarter of the 15th century and the first quarter of the 16th century, the mid-16th century (AD 1539 through 1572), and then again in the early 17th century. During droughty periods (Table 3.2) that sometimes lasted for several years, the effects of the droughts must have been localized in the Pineywoods. That the effects were local is indicated by the fact that there is a general continuity of Titus phase settlement across many parts of the Sabine and Big Cypress Creek basins that suggests crops were successfully grown and harvested even in these droughty periods.

The long-term storage of plant foods and seed stock – perhaps in above-ground granaries like the one apparently identified at the Whelan site community center (Thurmond 1990:168, Fig. 3.2) – also would have helped to offset losses from poor or failed harvests. Above-ground granaries appear to have been a technological innovation whose purpose was to maintain agricultural stability within local communities. That stability was maintained by minimizing the impact of risk (or at least redistributing the risk) across the community. The first evidence for the use of above-ground granaries is from the late village (ca. AD 1350 to 1375 through 1450) at the Oak Hill Village in the middle Sabine River basin. There, the granaries had no obvious entrance, instead, entrance to the structures was probably by a ladder to an elevated platform (Rogers and Perttula 2004; Perttula and Rogers 2007). One granary there had a second line of post holes on the interior of the structure, probably serving as the supports for an interior compartment with good air circulation and ready access.

Two Titus phase sites in the Little Cypress Creek basin have bison skeletal remains, indicating the exploitation during droughty periods of prairie habitats to the northwest and west of Caddo settlements in the Pineywoods. The occurrence of bison in prehistoric Caddo faunal assemblages is otherwise quite rare in northeastern Texas sites. It is likely that small herds of bison roamed the tall-grass prairie in Late Holocene times. It would have been a considerable trek to the tall-grass prairies in the upper Sulphur River basin (Fig. 3.2) for Titus phase hunters who did not yet have any horses. The high density of arrow points and scraping tools, as well as signs of intensive arrow point manufacture, at the Earspool site (41TT653) in the White Oak Creek basin and Post Oak Savanna (Sherman *et al.* 2002; Perttula and Sherman 2009), suggest that this particular Caddo population was intensively but opportunistically exploiting large game animals.

Stable carbon isotope values have been obtained from Late Caddo and post-1650 Caddo burials along the Red River in northeastern Texas, as well as from the Titus phase, and other Caddo populations living in southwestern Arkansas, and northwestern Louisiana (Perttula 1996:321, Rose *et al.* 1998). Those values on collagen range from –14.8 ± 1.35 o/oo (n=28) from ca. AD 1450 to 1650 Caddo sites and –14.2 ± 1.17 o/oo (n=18) from protohistoric Caddo sites. Generally contemporaneous agricultural populations – such as Cahokia and American Bottoms sites along the Mississippi River (Hedman *et al.* 2002); Iroquois sites in Ontario, Canada (Harrison and Katzenberg 2003); Central Mexico; the Maya region in Guatemala; or from coastal Ecuador (Smalley and Blake 2003:Table 2) in South America – had more enriched isotope values that ranged between –7.0 o/oo to –11.7 o/oo.

These populations had very high maize diets, actually about 15-50% higher than did the Titus phase Caddo and other contemporaneous Caddo groups living along the Red River. While we may be fairly confident that maize was an essential part of the Titus phase Caddo diet, as it was for other Late Caddo groups, it was nowhere near as intensive a reliance as it was for many other aboriginal groups in North America, central America, or even South America. There was a flexibility, diversity, or resilience in the Titus phase diet, that is highlighted by the continued use of a broad mixture of wild and domesticated plants (Dering 2005, Fritz 2000:244).

CONCLUSIONS

For Caddo communities living in certain parts of the Post Oak Savanna and Pineywoods of northeastern Texas, the regional archeological information on the density and distribution of settlements after ca. AD 1430 does not suggest that there was a continual movement of settlements and communities to exploit new habitats during periods of climatic instability. Further, this information does not point to the fact that the Caddo were moving to reoccupy more marginal areas that had been previously settled by Caddo farmers, or might have been physically degraded by forest clearance and hunting impacts on game animals. There simply do not appear to be Caddo sites dating after ca. AD 1430 in some areas (like the upper Sabine River and upper Sulphur River basins) until a period of resettlement in the early 18th century (cf. Scurlock 1962). Instead, the archeological record suggests a considerable movement or emigration of people from west to east across the landscape. This movement occurred in concert with the development of a number of political communities in the Big Cypress Creek basin during Titus phase times.

The pervasive evidence of regional abandonment, along with the eastward movement of Caddo peoples in, and into, the Big Cypress Creek basin, points to a broad and regional (or larger-scale) cause. It is suggested that the root cause of the Caddo abandonment of settlements of many sites occupied around this time is paleoclimatic. Climatic proxies such as tree-ring data, reconstructed northern hemisphere mean annual temperatures, and solar radiation inferred from atmospheric delta ^{14}C variation, all point to increasingly colder and drier conditions beginning around AD 1375, and lasting well into the 15th century. Peak periods of colder temperatures and drier conditions occurred between AD 1440 and 1470, when parts of northeastern Texas (and parts of southwestern Arkansas, see Lockhart 2007) were definitely being abandoned by a number of Caddo groups, while stronger polities were developing in other and less marginal parts of the region.

Repeated crop failures brought on by cooler and drought-like conditions, extending over several years may have had stressful, if not disastrous, results for these Caddo societies that were dependent on agricultural crops. Successful crop production would have been especially problematic for those societies who lived in parts of the region that were more marginal in soil fertility, rainfall, incidence of flooding, or incidence of drought. Living in a region subject to repeated droughts, the Caddo probably developed cropping and storage strategies to successfully sustain themselves through a 1- or 2-year drought. However, the likelihood of continued poor harvests and the lack of storable food reserves over a decadal-scale drought, when linked with evidence for regional abandonment around AD 1450, suggests they were unsuccessful and they abandoned those more marginal areas.

In the face of climatic unpredictability, and risky agricultural practices, Titus phase Caddo societies living along Big Cypress Creek and tributaries in the East Texas Pineywoods developed strategies to manage and minimize environmental and agricultural risks. Their efforts, particularly the creation of distinctive political communities marked by the development of monuments (earthen mounds and large community cemeteries) represent the actions and practices of many different Titus phase Caddo groups to strongly integrate and bond together with each other. This integration was not a form of increasing social complexity as much as it was the constructive actions of people to insure their survival in stressful times.

References

ALLEN, M.S. 2004 – Bet-hedging Strategies, Agricultural Change, and Unpredictable Environments: Historical Development of Dryland Agriculture in Kona, Hawaii, *Journal of Anthropological Archaeology* 23/2, 2004:196-224.

ANDERSON, D.G. 2001 – Climate and Culture Change in Prehistoric and Early Historic Eastern North America, *Archaeology of Eastern North America* 29, 2001:143-186.

ANDERSON, D.G., D.W. STAHLE, and M.K. CLEAVELAND 1995 – Paleoclimate and the Potential Food Resources of Mississippian Societies: A Case Study from the Savannah River Valley, *American Antiquity* 60/2, 1995:258-286.

BINFORD, L.R. 2001 – *Constructing Frames of Reference: An Analytical Method for Archaeological Theory Building Using Ethnographic and Environmental Data Sets*. University of California Press, Berkeley.

BONNICKSEN, T.M. 2000 – *Americas Ancient Forests: From the Ice Age to the Age of Discovery*. John Wiley & Sons, Inc., New York.

BOUSMAN, C.B. 1998 – Paleoenvironmental Change in Central Texas: The Palynological Evidence, *Plains Anthropologist* 43/164, 1998:201-219.

BRADLEY, R.S., K.R. BRIFFA, J. COLE, M.K. HUGHES, and T.J. OSBORN 2003 – The Climate of the Last Millennium. In K.D. Alverson, R.S. Bradley and T.F. Pedersen (eds.), *Paleoclimate, Global Change, and the Future* 2003: 105-141. Springer-Verlag, Berlin.

BRIFFA, K.R., and T.J. OSBORN 2002 – Blowing Hot and Cold, *Science* 295, 2002:2227-2228.

BRUSETH, J.E. 1987 – Late Holocene Environmental Change and Human Adaptive Strategies in Northeast Texas. Unpublished Ph.D. Dissertation, Department of Anthropology, Southern Methodist University, Dallas.

BURNETT, B.A., and K. MURRAY 1993 – Death, Drought, and De Soto. In G.A. Young and M.P. Hoffman (eds.), *The Expedition of Hernando de Soto West of the Mississippi, 1541-1543* 1993: 227-236. University of Arkansas Press, Fayetteville.

CALLOWAY, C.G. 2003 – *One Vast Winter Count: The Native American West before Lewis and Clark*. University of Nebraska Press, Lincoln.

COOK, E.R., D.M. MEKO, D.W. STAHLE, and M.K. CLEAVELAND 1999 – Drought Reconstruction for the Continental United States, *Journal of Climate* 12, 1999:1145-1162.

CROWLEY, T.J. 2000 – Causes of Climate Change Over the Past 1000 Years, *Science* 289, 2000:270-277.

DERRICK, S. and D. WILSON 2001 – The Effects of Epidemic Disease on Caddo Demographic Structure, *Bulletin of the Texas Archeological Society* 72, 2001:91-103.

DERING, J.P. 2005 – Plant Remains from Three Late Caddoan Sites in Camp County, Pilgrim's Pride (41CP304), 41CP316, and Shelby Mound (41CP71). In T.K. Perttula (ed.), *Archeological Investigations at the Pilgrim's Pride Site (41CP304), a Titus Phase Community in the Big Cypress Creek Basin, Camp County, Texas* 2005:325-345. 2 Vols. Report of Investigations No. 30. Archeological & Environmental Consultants, LLC, Austin.

DERING, J.P. 2004 – Archeobotanical Evidence for Agriculture and Wild Plant Use at 41RK214. In Rogers, R. and Perttula, T.K., *The Oak Hill Village (41RK214), Rusk County, Texas* 2004:329-336. Document No. 030083. PBS&J, Austin.

EARLY, A.M. 2004 – Prehistory of the Western Interior after 500 B.C. In R.D. Fogelson (ed.), *Handbook of North American Indians, Volume 14, Southeast* 2004:560-573. Smithsonian Institution, Washington, D.C.

EARLY, A.M. 2000 – The Caddos of the Trans-Mississippi South. In B.G. McEwan (ed.), *Indians of the Greater Southeast: Historical Archaeology and Ethnohistory* 2000:122-141. University Press of Florida, Gainesville.

ESPER, J., E.R. COOK, and F.H. SCHWEINGRUBER 2002 – Low-Frequency Signals in Long Tree-Ring Chronologies for Reconstructing Past Temperature Variability, *Science* 295, 2002:2250-2253.

FIELDS, R.C., M.E. BLAKE, and K.W. KIBLER 1997 – *Synthesis of the Prehistoric and Historic Archeology of Cooper Lake, Delta and Hopkins Counties, Texas*, Report of Investigations No. 104. Prewitt & Associates, Inc., Austin.

FISHER-CARROLL, R.L. 2001 – Environmental Dynamics of Drought and its Impact on Sixteenth-Century Indigenous Populations in the Central Mississippi Valley. Unpublished Ph.D. dissertation, Department of Anthropology, University of Arkansas, Fayetteville.

FRITZ, G.J. 2000 – Native Farming Systems and Ecosystems in the Mississippi River Valley. In D.L. Lentz (ed.), *Imperfect Balance: Landscape Transformations in the Precolumbian Americas* 2000:225-249. Columbia University Press, New York.

HARRISON, R.G. and M.A. KATZENBERG 2003 – Paleodiet Studies using Stable Carbon Isotopes from Bone Apatite and Collagen: Examples from Southern

Ontario and San Nicolas Island, California, *Journal of Anthropological Archaeology* 22/3, 2003:227-244.

HEDMAN, K., E.A. HARGRAVE and S.H. AMBROSE 2002 – Late Mississippian diet in the American Bottom: Stable Isotope Analyses of Bone Collagen and Apatite, *Midcontinental Journal of Archaeology* 27/2, 2002:237-271.

HOLLOWAY, R.G. 1994 – Ferndale Bog: A Record of Late Pleistocene-Holocene Climate and Vegetational Change. In C.R. Ferring (ed.), *Past Environments and Prehistory at McGee Creek Reservoir, Atoka County, Oklahoma* 1994:9-27. McGee Creek Archaeological Project Reports Vol. V/4. Institute of Applied Sciences, University of North Texas, Denton.

JONES, P.D., T.J. OSBORN, and K.R. BRIFFA 2001 – The Evolution of Climate Over the Last Millennium, *Science* 292, 2001:662-667.

JORDAN, T.G. 1981 – *Trails to Texas: Southern Routes of Western Cattle Ranching*, University of Nebraska Press, Lincoln.

LOCKHART, J.J. 2007 – Prehistoric Caddo of Arkansas: A Multiscalar Examination of Past Cultural Landscapes. Ph.D. dissertation, Department of Environmental Dynamics, University of Arkansas, Fayetteville.

MANN, M.E. 2002 – The value of Multiple Proxies, *Science* 297, 2002:1481-1482.

MANN, M.E., R.S. BRADLEY, and M.K. HUGHES 1998 – Northern Hemisphere Temperatures during the past Millennium: Inferences, Uncertainties, and Limitations. *Geophysical Research Letters* 26, 1998:759-762.

MARIETTA, K.L., and E.S. NIXON 1983 – Vegetation Analysis of a Post Oak-Black Hickory Community in Eastern Texas, *Texas Journal of Science* 35, 1983:197-203.

MARIETTA, K.L., and E.S. NIXON 1984 – Vegetation of an Open, Prairie-Like Community in Eastern Texas, *Texas Journal of Science* 36, 1984:25-34.

McINTOSH, R.J., J.A. TAINTER, and S.K. McINTOSH 2000 – Climate, History, and Human Action. In R.J. McIntosh, J.A. Tainter and, S.K. McIntosh (eds.), *The Way the Wind Blows: Climate, History, and Human Action* 2000:1-42. Columbia University Press, New York.

MILLER, N.F. 2004 – Long-Term Vegetation Changes in the Near East. In C.L. Redman, S.R. James, P.R. Fish and, J.D. Rogers (eds.), *The Archaeology of Global Change: The Impact of Humans on Their Environments* 2004:130-140. Smithsonian Institution Press, Washington, D.C.

MILNER, G.R., D.G. ANDERSON, and M.T. SMITH 2001 – The Distribution of Eastern Woodlands Peoples at the Prehistoric and Historic Interface. In D.S. Brose, C.W. Cowan and R.C. Mainfort, Jr. (eds.), *Societies in Eclipse: Archaeology of the Eastern Woodlands Indians, A.D. 1400-1700* 2001:9-18. Smithsonian Institution Press, Washington, D.C.

NIXON, E.S., R.L. EHRHART, S.A. JASPER, J.S. NECK, and J.R. WARD 1983 – Woody, Streamside Vegetation of Prairie Creek in East Texas, *Texas Journal of Science* 35, 1983:205-213.

PAUKETAT, T.R. 2005 – The Forgotten History of the Mississippians. In T.R. Pauketat and D.D. Loren (eds.), *North American Archaeology* 2005:187-211. Blackwell Publishing, Malden, Massachusetts.

PERTTULA, T.K. 2008 – Caddo Agriculture on the Western Frontier of the Eastern Woodlands. *Plains Anthropologist* 53(205):79-105.

PERTTULA, T.K. 2004 – The Prehistoric and Caddoan Archeology of the Northeastern Texas Pineywoods. In T.K. Perttula (ed.), *The Prehistory of Texas* 2004:370-407. Texas A&M University Press, College Station.

PERTTULA, T.K. 2002 – Social Changes among the Caddo Indians in the Sixteenth and Seventeenth Centuries. In E. Ethridge and C. Hudson (eds.), *The Transformation of the Southeastern Indians, 1540-1760* 2002:249-269. University Press of Mississippi, Jackson.

PERTTULA, T.K. 1998 – Late Caddoan Societies in the Northeast Texas Pineywoods. In T.K. Perttula and J.E. Bruseth (eds.), *The Native History of the Caddo: Their Place in Southeastern Archeology and Ethnohistory* 1998:69-90. Studies in Archeology 30. Texas Archeological Research Laboratory, The University of Texas at Austin, Austin.

PERTTULA, T.K. 1996 – Caddoan Area Archaeology since 1990, *Journal of Archaeological Research* 4/4, 1996:295-348.

PERTTULA, T.K. (ed) 2005 – *Archeological Investigations at the Pilgrim's Pride Site (41CP304), a Titus Phase Community in the Big Cypress Creek Basin, Camp County, Texas*. 2 Vols. Report of Investigations No. 30. Archeological & Environmental Consultants, LLC, Austin.

PERTTULA, T.K. and B. NELSON 2003 – Temporal and Spatial Patterns in the Prehistoric Settlement of the Lake Bob Sandlin Area, Big Cypress Creek Basin, Northeastern Texas, *Caddoan Archeology Journal* 13/2, 2003:28-35.

PERTTULA, T.K. and B. NELSON B. 2002 – *Archeological Survey of Lake Bob Sandlin State Park, Titus County, Texas*. Report of Investigations No. 48. Archeological and Environmental Consultants, Austin.

PERTTULA, T.K. and R. ROGERS 2007 – The Evolution of a Caddo Community in Northeastern Texas: The Oak Hill Village Site (41RK214), Rusk County, Texas. *American Antiquity* 72(1):71-94.

PERTTULA, T.K. and D.L. SHERMAN 2009 – *Data Recovery Investigations at the Ear Spool Site (41TT653), Titus County, Texas*. Document No. 070205. PBS&J, Austin.

ROGERS, J.D. and G. SABO III 2004 – The Caddos. In R.D. Fogelson (ed.), *Handbook of North American Indians, Volume 14, Southeast* 2004: 616-631. Smithsonian Institution, Washington, D.C.

ROGERS, R., and T.K. PERTTULA 2004 – *The Oak Hill Village Site (41RK214), Rusk County, Texas*. Document No. 030083. PBS&J, Austin.

ROSE, J.C., M.P. HOFFMAN, B.A. BURNETT, A.M. HARMON, and J.E. BARNES 1998 – Skeletal Biology of the Prehistoric Caddo. In T.K. Perttula and J.E. Bruseth (eds.), *The Native History of the Caddo: Their Place in Southeastern Archeology and Ethnohistory* 1998:113-126. Studies in Archeology 30. Texas Archeological Research Laboratory, The University of Texas at Austin, Austin.

SABO, G. III 2003 – Dancing into the Past: Colonial Legacies in Modern Caddo Indian Ceremony, *The Arkansas Historical Quarterly* 62/4, 2003:423-445.

SCHMIDLY, D.J. 2002 – *Texas Natural History: A Century of Change*. Texas Tech University Press, Lubbock.

SCURLOCK, J.D. 1962 – The Culpepper Site, A Late Fulton Aspect Site in Northeast Texas, *Bulletin of the Texas Archeological Society* 32, 1962:185-316.

SHERMAN, D.L., M.A. NASH, T.K. PERTTULA, and M.T. IRUEGAS 2002 – *Data Recovery at the Ear Spool Site (41TT653), Titus County, Texas*, PBS&J, Austin.

SMALLEY, J. and M. BLAKE 2003 – Sweet Beginnings: Stalk Sugar and the Domestication of Maize, *Current Anthropology* 44/5, 2003:675-703.

STAHLE, D.W. 1996 – The Hydroclimatic Application of Tree-Ring Chronologies. In J.S. Dean, D.M. Meko and T.W. Sweetman (eds.), *Tree Ring, Environment and Humanity: Proceedings of the International Conference, Tucson, Arizona, 17-21 May 1994* 1996:119-126. Radiocarbon, Department of Geosciences, The University of Arizona, Tucson.

STAHLE, D.W. and M.K. CLEAVELAND 1995 – Texas Paleoclimatic Data from Daily to Millenial Time Scales. In J. Norwine, J.R. Giardino, G.R. North and J.B. Valdes (eds.), *The Changing Climate of Texas: Predictability and Implications for the Future* 1995:49-69. GeoBooks, College of Geosciences and Maritime Studies, Texas A & M University, College Station.

STAHLE, D.W. and M.K. CLEAVELAND 1994 – Tree-Ring Reconstructed Rainfall over the Southeastern U.S.A. during the Medieval Warm Period and Little Ice Age, *Climatic Change* 26, 1994:194-212.

STAHLE, D.W. and M.K. CLEAVELAND 1993 – Southern Oscillation Extreme Reconstructed from Tree Rings of the Sierra Madre Occidental and southern Great Plains, *Journal of Climate* 6, 1993:129-140.

STAHLE, D.W. and M.K. CLEAVELAND 1992 – Reconstruction and Analysis of Spring Rainfall over the Southeastern U.S. for the Past 1000 Years, *Bulletin of the American Meteorological Society* 73, 1992:1947-1961.

STAHLE, D.W. and M.K. CLEAVELAND 1988 – Texas Drought History Reconstructed and Analyzed from 1698 to 1980, *Journal of Climate* 1, 1988:59-74.

STAHLE, D.W., M.K. CLEAVELAND and J.G. HEHR 1985 – A 450-Year Drought Reconstruction for Arkansas, *Nature* 316, 1985:530-532.

STAHLE, D.W., E.R. COOK, M.K. CLEAVELAND, M.D. THERRELL, D.M. MEKO, H.D. GRISSINO-MAYER, E. WATSON and B.H. LUCKMAN 2000 – Tree-ring Data Document 16th Century Megadrought Over North America, *Eos* 81/12, 2000:121, 125. Transactions of the American Geophysical Union.

THURMOND, J.P. 1990 – *Archeology of the Cypress Creek drainage basin, Northeastern Texas and Northwestern Louisiana*. Studies in Archeology 5. Texas Archeological Research Laboratory, The University of Texas at Austin, Austin.

Chapter 4
ENVIRONMENTAL CHANGE, POPULATION MOVEMENTS, AND THE ARCHAEOLOGICAL RECORD

Dean R. SNOW
Dept. of Anthropology, The Pennsylvania State University

INTRODUCTION

Environmental change prompts human population expansions and contractions over time and space at scales that are archaeologically detectable. It is also the case that technological innovations can also facilitate population expansions, often at the expense of human populations that do not share the innovations and often without any conditioning environmental change. Various key technological innovations illustrate the process, such as equestrian nomadism, boat transportation, and agriculture. Diamond and Bellwood's recent article on population expansions prompted by the development of agricultural systems provides no fewer than eleven major cases from around the world (Diamond and Bellwood 2003). When environmental change and technological innovation are both in play the consequences quickly become complex. While those consequences are archaeological detectable, the archaeological record is static, and the logic pathways by which environmental, demographic and technological processes led to that static record are subject to much debate and few generally accepted conclusions. Migration continues to be difficult to define and apply in archaeological interpretation because the term covers several often smaller scale processes that together drive large-scale demographic change. Moreover, patterns generated at large scale were not often intentional, even though the many smaller-scale decisions that in aggregate led to the larger patterns often were. Despite these difficulties, cases from North American archaeology show how archaeology can solve large scale demographic problems using modern dating techniques and geographic information systems.

THE IROQUOIAN CASE

The case of Iroquoian expansion in northeastern North America shows that large scale processes involve many small scale migrations (Snow 1995, 1996a, 1996b). The Iroquoian case also illustrates a population expansion that involved both climatic change and technological innovation. The ancestral Iroquoians living mainly in what is now central Pennsylvania had recently acquired maize horticulture by AD 750 (Fig. 4.1). At that time the regional climate was shifting towards the Medieval Maximum, which peaked around AD 1000 (Fig. 4.2). The form of agriculture practiced by the ancestral Iroquoians was particularly well suited to the glaciated soils around Lake Ontario under the warm climatic conditions of that era, so it is not surprising that village relocations that were a necessary part of the shifting swidden agricultural practices of these people tended to be in that direction (Fig. 4.3). They avoided poorer glaciated soils overlying bedrocks that do not typically support rich soils, most notably in New England and northeastern New York. They also avoided more northern areas because of the shorter growing season there. Over the course of a few centuries most Iroquoian came to be scattered mainly around the Ontario drainage basin and unglaciated landscapes to the south were abandoned (Fig. 4.4).

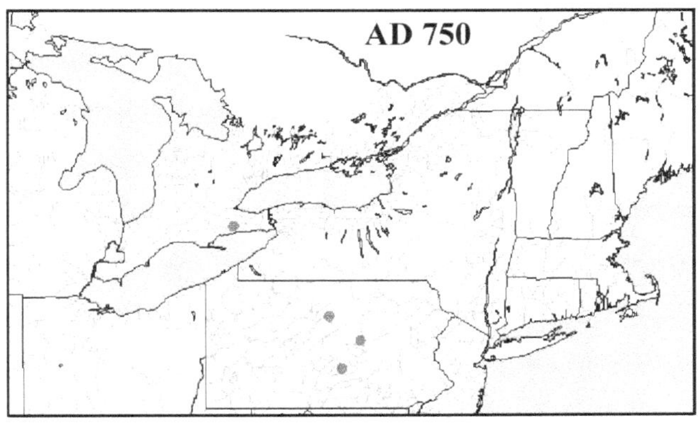

Figure 4.1. Iroquoian village sites (gray dots) in AD 750

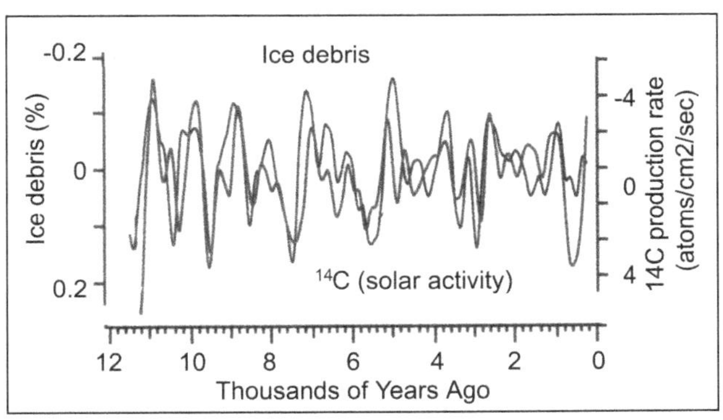

Figure 4.2. Climate curve for North America. Note the peak around 1000 bp (AD 1000)

A SOUTHWESTERN NORTH AMERICAN CASE

As archaeological survey, dating, and mapping techniques get increasingly precise we are able to control

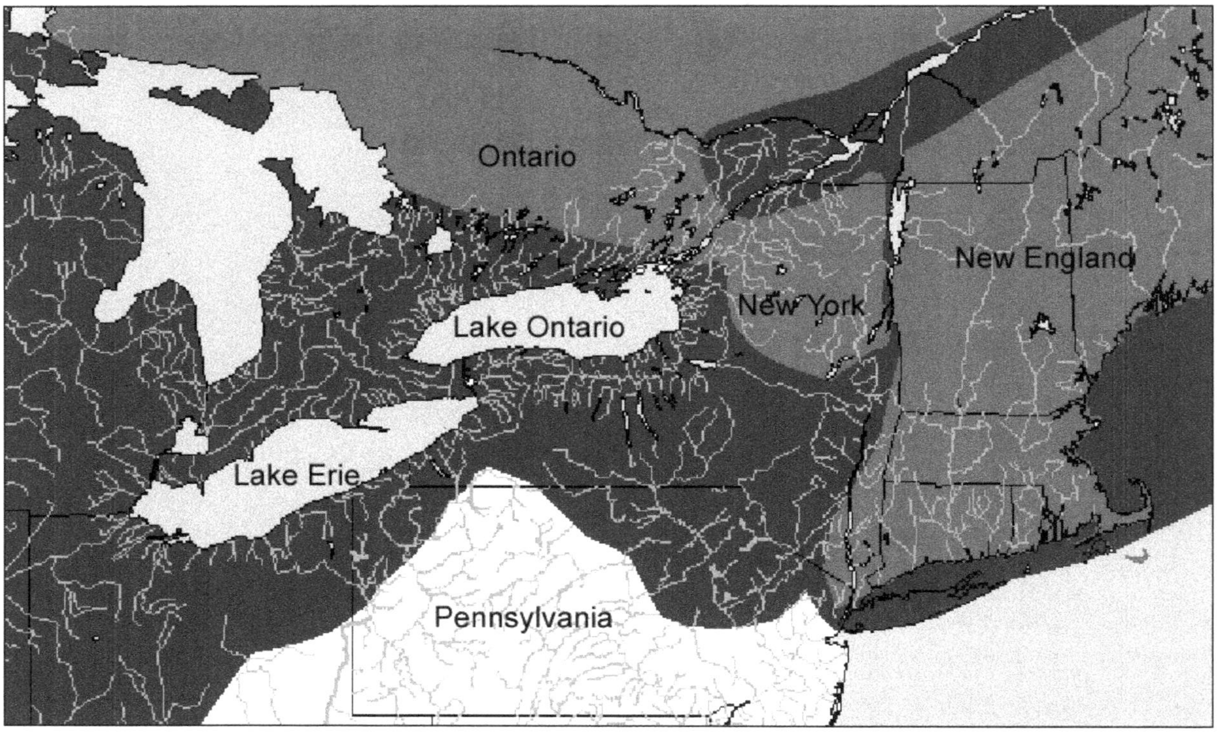

Figure 4.3. Extent of maximum glaciation (dark gray) and locations of unsuitable bedrocks (light gray)

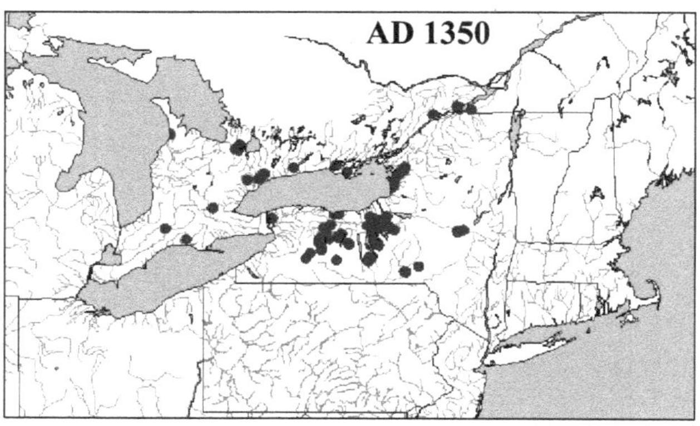

Figure 4.4. Distribution of Iroquoian sites (gray dots) by AD 1350

large data sets well enough to detect population movements and local demographic changes. We no longer automatically assume that the lack of information for some specific period in a particular locale is nothing more than evidence of our ignorance. We can increasingly say with fairly high certainty that some locales were devoid of human occupation, sometimes for long periods. For example, while the Atacama Desert of Chile was occupied by hunter/gatherers between 11000-7500 BC (13,000-9500 bp), it was abandoned for five millennia (7500-2500 BC / (9500-4500 bp) after that. The abandonment was associated with the drying up of the Late Glacial/Early Holocene lakes on the Altiplano (Núñez, *et al.* 2002).

Another example comes from the American Southwest. It is not easy to distinguish routine site relocation from regional abandonment. Nelson and Hegmon (2001) argue that archaeologists have too often mistaken the former for the latter. Abandonment, to the extent that it occurred at all, was cyclical, not a singular event. Van West and Kohler have developed a theoretical model that predicts that Anasazi farmers would have aggregated into large centers and pooled agricultural resources in periods when mean productivity was high (Van West and Kohler 1996). Agricultural productivity was always patchy and unpredictable over space and time in the Southwest, but when it was high overall people were likely to pool their resources as insurance against local shortfalls. Van West and Kohler predicted that when mean production was low Anasazi farmers would be more inclined to take their chances on their own than to share equally in the general misery.

Van West and Kohler identified five periods of relatively high mean productivity. These correlated with episodes of aggregation and resource pooling as their hypothesis predicted. The process is best illustrated by the growth and decline of the Chacoan system of northwestern New Mexico. High mean productivity during the period AD 1100-1129 correlated with an expansion of the Chacoan system. Breakup of the Chacoan system occurred during AD 1130-1179, a period of low mean productivity. Another period of high mean productivity and aggregation followed before final abandonment of the region in the 1270s or 1280s, when mean productivity was as low as it got in centuries.

The cycles of aggregation and dispersal in the Southwest tell us much about human behavior. Enthusiasm for collectivism depends upon the perception and reality of

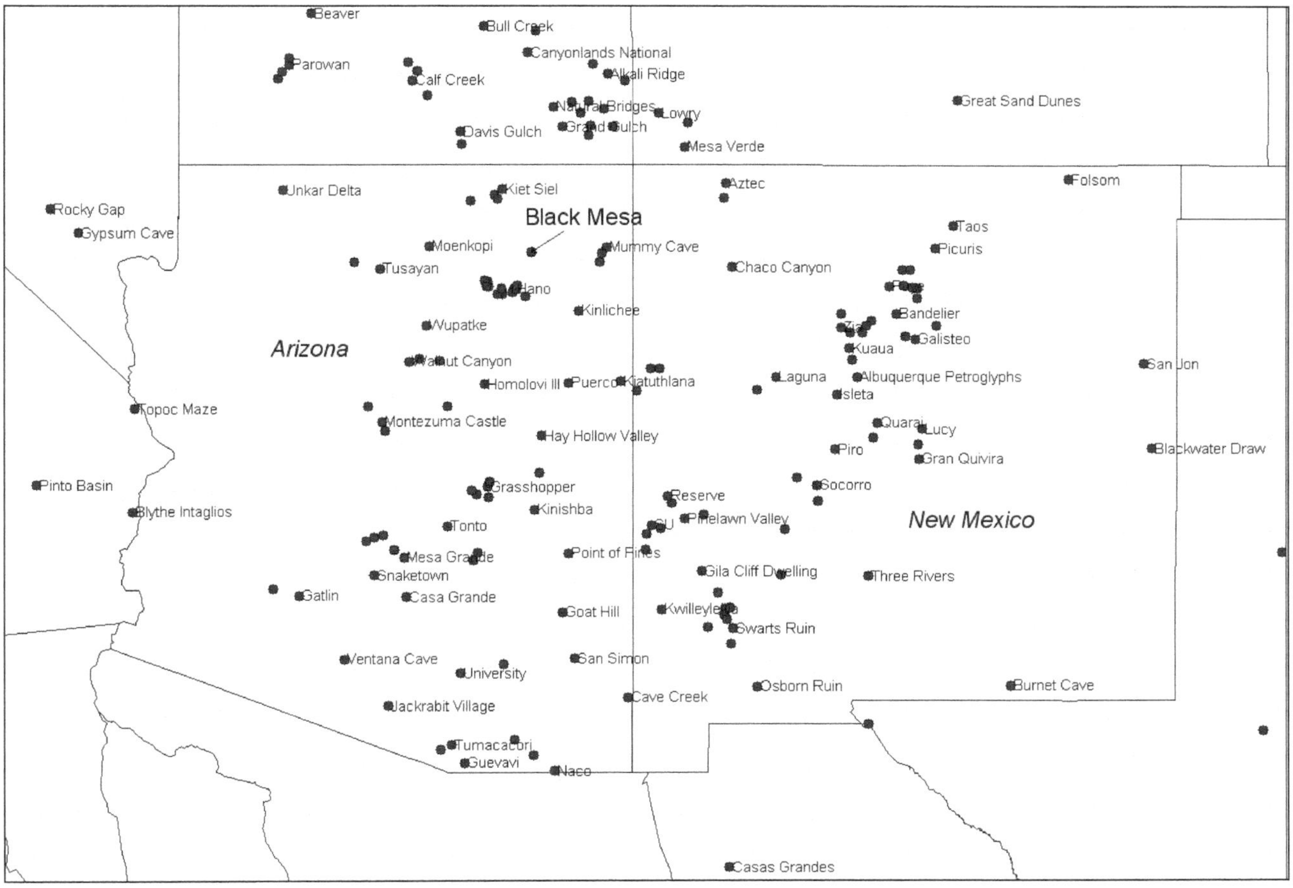

Figure 4.5. The Southwest culture area showing Black Mesa and various major sites

overall well-being and long-term security. When external forces erode overall productivity and raise both immediate and long-term risks, people fall back on smaller settlements and smaller production units. There is indeed strength in numbers, but not when the numbers are unsustainable.

Researchers working in the northeastern region of Arizona got closer still with a detailed dynamic computer model that is credible because it mimics what actually happened very closely (Axtell, *et al.* 2002; Dean, *et al.* 2000). They used tree rings to measure annual climatic conditions from AD 200 to the present, a long sequence with no gaps. Next they used the Palmer Drought Severity Index (PDSI), which is based on careful measurements over the last century, as a bridge between the tree ring data and maize yields. They determined the relationship between maize yields and various soil types under varying conditions recorded by the tree rings. They then applied these measurements to the soil type found in Long House Valley in the Black Mesa area of northeastern Arizona (Fig. 4.5). To make the model work they had to include rules about the mean annual maize consumption per person (160 kg), average family size, the average age that women bore children, average age at death, average longevity of households, and so forth. The first version of the model did not work well, producing a simulated population six times larger than the one the archaeologists had painstakingly reconstructed from exacting field work in Long House Valley. However once means were converted to distributions around means, and values were repeatedly selected at random from those distributions as subroutines of the simulation, the model replicated the measured rise and fall of local Anasazi population very closely. This success convinced the researchers that they did indeed correctly understand that factors underplayed Anasazi population history.

CONCLUSIONS

Five important points emerge from these cases. First, change driven by climate change is often complicated by simultaneous technological adjustments. The Iroquoians seized upon maize agriculture and other innovations as a means to take advantage of new conditions produced by the medieval climatic maximum.

Second, even when technological change does not appear to accompany demographic change brought on by climatic change, as in the Long House Valley case, migration and other demographic consequences often entail some degree of cultural reinvention. In the Southwest cultural innovations such as the rise of the Kachina cult or the spread of Salado ceramics appear to have been consequences of the stresses of local agricultural failures, migration, and the blending of traditions in new multicultural communities. Thus

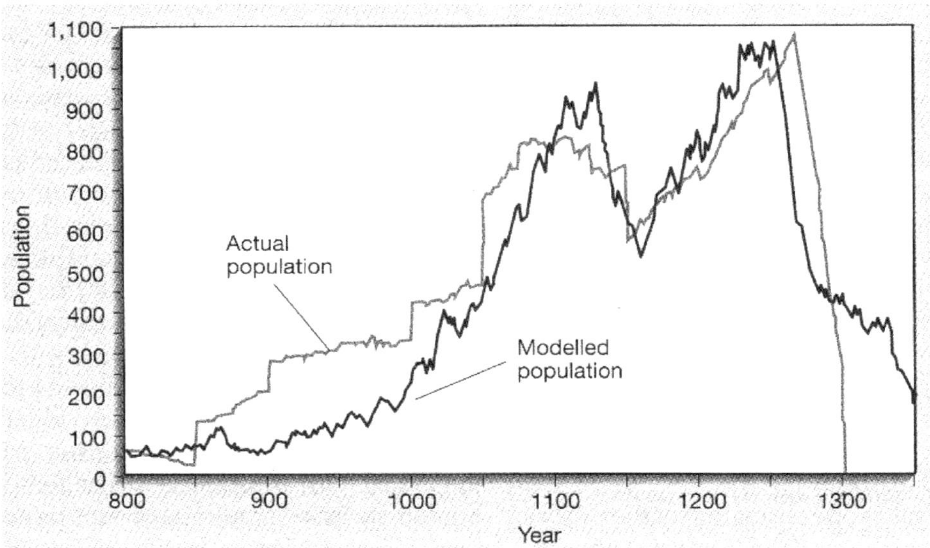

Figure 4.6. Actual and modeled population curves for Longhouse Valley, Arizona (Axtell, *et al.* 2002)

descendant communities sometimes bear little resemblance to their parent communities, making the task of tracking them across time and space a very difficult archaeological problem.

A third point is that our understandings of prehistoric cultural landscapes are now becoming sufficiently complete and detailed for us to describe not just variable population densities over time and space but to also specify regional vacancies. Even rather densely occupied regions were characterized by clusters of settlements separated by sometimes large buffer zones that were not permanently occupied and sometimes infrequently utilized. But it is now clear that there were probably also many zones that were nearly vacant of human presence, at least in certain periods.

A fourth important point has to do with claims of environmental determinism, whether one advocates them or abhors them. I would not claim that Iroquoian expansion into the Lake Ontario basin was directly caused by either climate change or the acquisition of maize agriculture, or the two things together. Instead, I argue that if, 1300 years ago, a group of people living in what is now Pennsylvania had an opportunity to take up maize agriculture and they were looking for suitable soils, the glaciated soils of the Lake Ontario basin were a suitable destination given the emerging climatic conditions of the time. It should be no surprise that taken together the separate decisions of many such communities led to the general drift of the Northern Iroquoian population northward over time. This kind of adaptive shift over space and time, while familiar to animal ecologists, does not equate closely to the more directed, short-term, and intentional meanings usually assigned to the term "migration" when it is used to describe human activity. It also amounts to a more nuanced way to account for and describe cause and effect.

This leads to a closely associated fifth point. People vote with their feet, and while their short-term actions are intentional the larger-scale consequences typically are not. The patterns that emerge at large spatial and temporal scales are not necessarily intentional at all. Indeed, the scale to which patterned intentionality can rise is proportional to the socio-economic complexity of the society being studied. Thus while the Great Wall of China and the large highway systems of Europe and America are all both large scale and the result of intentional decisions, the large scale spatial settlement patterns (Fig. 4.4) was not intentional in the same sense. Again in the same sense, the curve (Fig. 4.6) was not intentionally produced, when Maize production in Long House Valley drove the rise of Anasazi population there until the 12th century. It rose sharply during AD 1000-1130, and then crashed over the next fifty years because of drought and falling groundwater levels. The population rose again after AD 1180, reaching a peak in AD 1275 then crashing again. This time the population fell to such a low level that the Anasazi gave up on the valley and abandoned it entirely by AD 1300. The model indicates that a few hundred Anasazi farmers could have survived in Long House Valley after AD 1300. They probably chose not to for other reasons. These could easily involve nothing more than the size of the critical mass of people needed to maintain traditional rituals.

No matter what people do we have to assume that it seemed like a good idea at the time. In that sense all acts are rational given the circumstances, knowledge, and contingencies of the time and place they were made. We can infer further that all acts are intentional, yet we must also be mindful of unintended consequences, which account for many but by no means all of the patterns we observe in the archaeological record. The larger patterns that emerge as the aggregated consequences of intentional actions are typically not themselves intentional. Failure to appreciate and specify the limits of inferred intentionality in archaeological patterns has caused much of the confusion and disagreement that lingers in our science. In particular, the chapters that make up this volume all deal with the effects of climate and climate change on human adaptation, and we will do well to remember that the people whose behaviors we attempt to study, indirectly and at great temporal remove, were just using the resources they had or could acquire to survive and improve their lot at the margin. We can say that exogenous environmental changes cause changes in human societies, but when we do so we must be prepared to specify our scales and terms much more tightly than we have traditionally done.

References

AXTELL, R.L., J.M. EPSTEIN, J.S. DEAN, G.J. GUMERMAN, A.C. SWEDLUND, J. HARBURGER, S. CHAKRAVARTY, R. HAMMOND, J. PARKER and M. PARKER 2002 – Population Growth and Collapse in a Multiagent Model of the Kayenta Anasazi in Long House Valley. *Proceedings of the National Academy of Sciences* 99, 2002:7275-7279.

DEAN, J.S., G.J. GUMERMAN, R.L. AXTELL, A.C. SWEDLUND, M.T. PARKER and S. McCARROLL 2000 – Understanding Anasazi Culture Change Through Agent-Based Modeling. In T.A. Kohler and G.J. Gumerman (eds.), *Dynamics in Human and Primate Societies: Agent-Based Modeling of Social and Spatial Processes*, 2000:179-205. Oxford University Press, New York.

DIAMOND, J., and P. BELLWOOD 2003 – Farmers and Their Languages: The First Expansions. *Science* 300, 2003:597-603.

NELSON, M.C., and M. HEGMON 2001 – Abandonment is Not as it Seems: An Approach to the Relationship Between Site and Regional Abandonment. *American Antiquity* 66/2, 2001:213-235.

NÚÑEZ, L., M. GROSJEAN and I. CARTAJENA 2002 – Human Occupations and Climate Change in the Puna de Atacama, Chile. *Science* 298, 2002:821-824.

SNOW, D.R. 1995 – Migration in Prehistory: The Northern Iroquoian Case. *American Antiquity* 60/1, 1995:59-79.

SNOW, D.R. 1996a – GIS Applications in North America. In I. Johnson (ed.), *The Coloquia of the XIII Congrès International des Sciences Prèhistoriques et Protohistoriques, Vol. 1 Theoretical and Methodological Problems*, 1966:159-168. A.B.A.C.O. Edizioni, Forli, Italy.

SNOW, D.R. 1996b – More on Migration in Prehistory: Accommodating New Evidence in the Northern Iroquoian Case. *American Antiquity* 61, 1996:791-796.

Van WEST, C., and T.A. KOHLER 1996 – A Time to Rend, a Time to Sew: New Perspectives on Northern Anasazi Sociopolitical Development in Later Prehistory. In M. Aldenderfer and H.D.G. Maschner (ed.), *Anthropology, Space, and Geographic Information Systems*. Oxford University Press, New York.

Part II:

Paleoclimate and Socio-Cultural Change in the Old World: Africa

Chapter 5
CLIMATE, CULTURE, AND CHANGE: FROM HUNTERS TO HERDERS IN NORTHEASTERN AND SOUTHWESTERN AFRICA

Ralf VOGELSANG and Birgit KEDING
Universität zu Köln, Forschungsstelle Afrika, Köln, Germany

Abstract: *We focus on the process of economic change during the Holocene in northeastern and southwestern Africa. Both areas show roughly similar economic sequences with a transition by adoption from foraging to pastoralism. However, differences in the transition are observed in the process of change. In the Sudanese Saharan region of northeastern Africa a complete change of an economic system is suggested. This contrasts with southwestern Africa, where only some elements of the pastoral lifestyle have been adopted. The background for these diverse developments will be discussed based on two case studies, each situated in areas with different climatic conditions. The first study is from the Wadi Howar region in the eastern Sahara. It is an ecologically favored area in a geographically key position, but characterized by dramatic climatic changes during the last 10,000 years. The second case study is from the Opuwo District in northwestern Namibia with a more stable climate. However, the differences of economic transition processes observed in both areas can only partly be explained by different intensities of climatic and environmental changes. In the Wadi Howar region, the transition from foragers to cattle pastoralism seems to be less stimulated by climatic change than by social, demographic, and other factors. However, a second, later change, resulting in increasing economic diversification, seems to be mainly an adaptation to growing aridity. The slower and selected adoption of a new economic system in Namibia can be partly explained by the more static conditions in southwestern Africa. In this area, stimuli like strong climatic and environmental change, as well as intense regional and inter-regional contacts with groups exhibiting different subsistence patterns and social structures, are missing, and new ideas seem to take more time to be fully adopted.*

INTRODUCTION

The discussion focusing on the internal and external factors that stimulated cultural and economic change in prehistoric societies can be traced back at least to the 19th century. Changing theoretical approaches have been formulated and tested. These include the ecological approach, which understands culture as human adaptation to the social and natural environments. Archaeological studies conducted in arid regions concentrate particularly on the influence of climatic and environmental conditions on the adoption or development of a food-producing economy (e.g. Clark and Brandt 1984). Social change is attributed to external pressures rather than to forces inherent in the system itself. In contrast to this deterministic environmental approach, other archaeologists look to historical materialism for their explanations. They argue that the reasons for cultural and economic change must focus on social relations of production and the competitive nature of human society (Bender 1978; Lourandos 1988; Zvelebil 1986:10; Hayden 1990). Finally, the post-processual approach concentrates on the cognition of man and its influence on culture and culture change. The "natural environment" is interpreted as a construct of man (Ingold 1996:118).

Nevertheless, even today neither the "why" nor the "how" – the initiation of cultural change and its subsequent development – are fully understood. Attempts to develop a universal theory seem to have been abandoned. The current debate focuses increasingly on defining combinations and interactions of prime-movers, the specific prevailing conditions (e.g. Gebauer and Price 1992:3), and the importance of the role of decision-making social units, that encourage cultural and economic change (e.g. Dobres and Robb 2000).

In Europe the term Neolithic is closely tied to the debate over economic change. Coined for the Near East, this term is used to define cultures associated with a food-producing economy, pottery, and ground-stone production (Lubbock 1865). However, in a global perspective, especially with respect to Africa, this term is increasingly avoided (Sinclair *et al.* 1993; see Sadr 2003 for a critical point of view). The expanding archaeological database shows a wide range of different economic developments, circumstances, and associated material culture, making this term inappropriate for most regions.

Based on two case studies conducted in arid regions of northeastern and southwestern Africa, some of the differing processes involved in economic change and their possible prime movers will be discussed in this chapter. The regions are the Wadi Howar, Sudan, and the Opuwo District, Namibia. Data for the research stem from archaeological projects taking place in Egypt, Sudan, and Namibia. The projects are run by the multidisciplinary collaborative research centre SFB 389. The research is entitled Arid Climate, Adaptation and Cultural Innovation in Africa (ACACIA). The centre was established by the Deutsche Forschungsgemeinschaft (DFG) at the University of Cologne in 1995. The concentration on economic change in arid environments during the Holocene is common to both projects. The interrelations of economic and social processes are of particular interest. Environmental conditions raise the following question: Are there significant indicators for

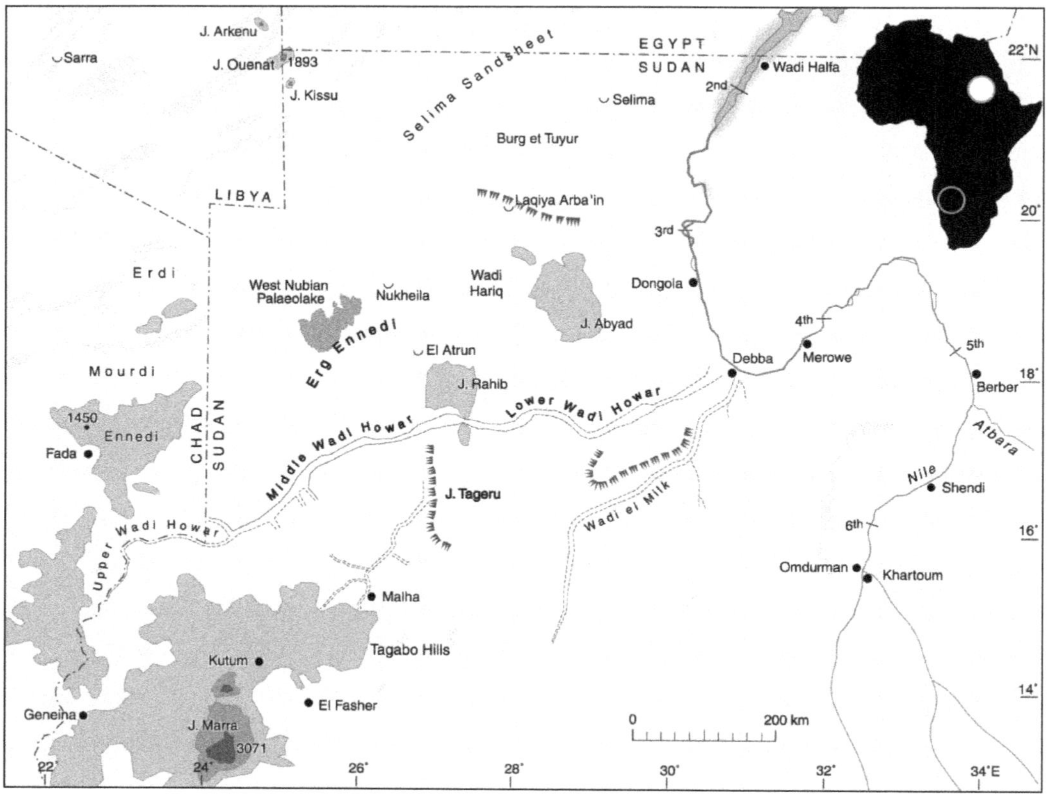

Figure 5.1. Northwestern Sudan and the location of four key areas of investigation of the SFB 389 in the Eastern Sahara: 1: Lower Wadi Howar, 2: Middle Wadi Howar, 3: Ennedi Erg, 4: Djebel Tageru

interaction between climatic change and cultural development, and, if so, are these common features in both northern and southern Africa?

The focus is mainly on the process of pastoral adaptation. Pastoralism is a phenomenon, which began in North Africa at least 5000 years earlier than in the south of the continent. In both case studies, it is assumed that animals were not domesticated in an autochthonous process, but were introduced by migration or diffusion.

RESEARCH ENVIRONMENT

Wadi Howar

The Wadi Howar region is situated in northern Sudan, on the southern fringe of the eastern Sahara (Fig. 5.1). Today, open sand plains, single dune ridges, and the scarcely pronounced Wadi Howar characterize this landscape. It is located within the fully arid desert and thorn savannah with rainfall rates of between 25 mm and 100 mm per year (Simons 1973:505).

Nowadays, the Wadi Howar borders the Eastern Sahara. The exact course of this dry, four to 15 km wide river bed is difficult to reconstruct in its entirety. Running in a west-east direction for over 1,000 km, it links its sources, located in the mountainous regions of Djebel Marra (Sudan), Borkou, and Ennedi (Chad), with the Nubian Nile Valley. At this point, the Wadi Howar joins the river Nile opposite Old Dongola, south of the Third Cataract (Pachur and Kröpelin, 1987; Kröpelin 1993).

Based on geomorphological, hydrological, and geological criteria, the wadi can be subdivided into three sectors (Gabriel *et al.* 1985). Sector 1 consists of the Upper Wadi Howar and comprises the 250 km long source area in the west. Today it consists of drainage channels in the thorny bush savannah. Sector 2 is the middle Wadi Howar, comprising the adjoining 400 km to the east as far as Djebel Rahib. It is a broad sandy plain with dunes and vegetation (Fig. 5.2). The third is the morphologically less pronounced lower Wadi Howar, which stretches over a further 400 km to the Nile Valley. It is almost void of vegetation.

Research data have been obtained from regional investigations along the middle and lower sections of the Wadi Howar and from adjacent areas, i.e., from the Ennedi Erg to the northwest and from the Djebel Tageru to the south (Keding 1997b, 1998-2002). The Ennedi Erg in the western Nubian Basin is located east of the Ennedi Mountains and northwest of the middle Wadi Howar. This former lake region is now a featureless area with occasional dune fields, small depressions, and numerous yardangs (Hoelzmann 1992, 1993). The Djebel Tageru, 40 km south of the Middle Wadi Howar, is a sandstone plateau with an elevation of less than 100 m, stretching over 160 km from north to south. Its eastern slope is a hammada. Its western side is a rugged scarp broken by several narrow valleys (Hoelzmann 1993).

Figure 5.2. View of the middle Wadi Howar from Djebel Rahib seen as a strip of vegetation on the horizon (Photo Kuper 1980)

These areas are characterized today, as in the past, by different ecological communities. The sand-plains of the Ennedi Erg and the Lower Wadi Howar are completely uninhabited, but some camel nomads still exploit the middle and upper Wadi Howar and the Djebel Tageru to the south. Vegetation and wells, both profiting from a high groundwater table, are still a guarantee for pasture and a year-round water supply.

Opuwo District

The region, formerly known as Kaokoland, is situated in northwestern Namibia. It corresponds to the Opuwo magisterial district (Fig. 5.3). To the north the area is demarcated by the course of the Kunene River, which also forms the political border with Angola. The Atlantic Ocean to the west and the stone-less plains of the Ovamboland covered with Kalahari sands to the east are natural borders. In the south, the course of the Hoanib River largely demarcates the magisterial border of the Khorixas district.

The Opuwo district comprises several ecologically different regions: the northern Namib Desert, different types of mopane savannahs, and the banks of the Kunene River (Viljoen 1980, Becker and Jürgens 2000, 2002). As precipitation decreases from 350 mm per year in the east to less than 50 mm per year in the west (Mendelsohn *et al.* 2002:84), the conditions for plant growth, and consequently for pastoralism, also declines. Today, specialized cattle pastoralism is possible, but it requires elaborate pasture management (Sander *et al.* 1998, Bollig 1997a, Bollig and Schulte 1999).

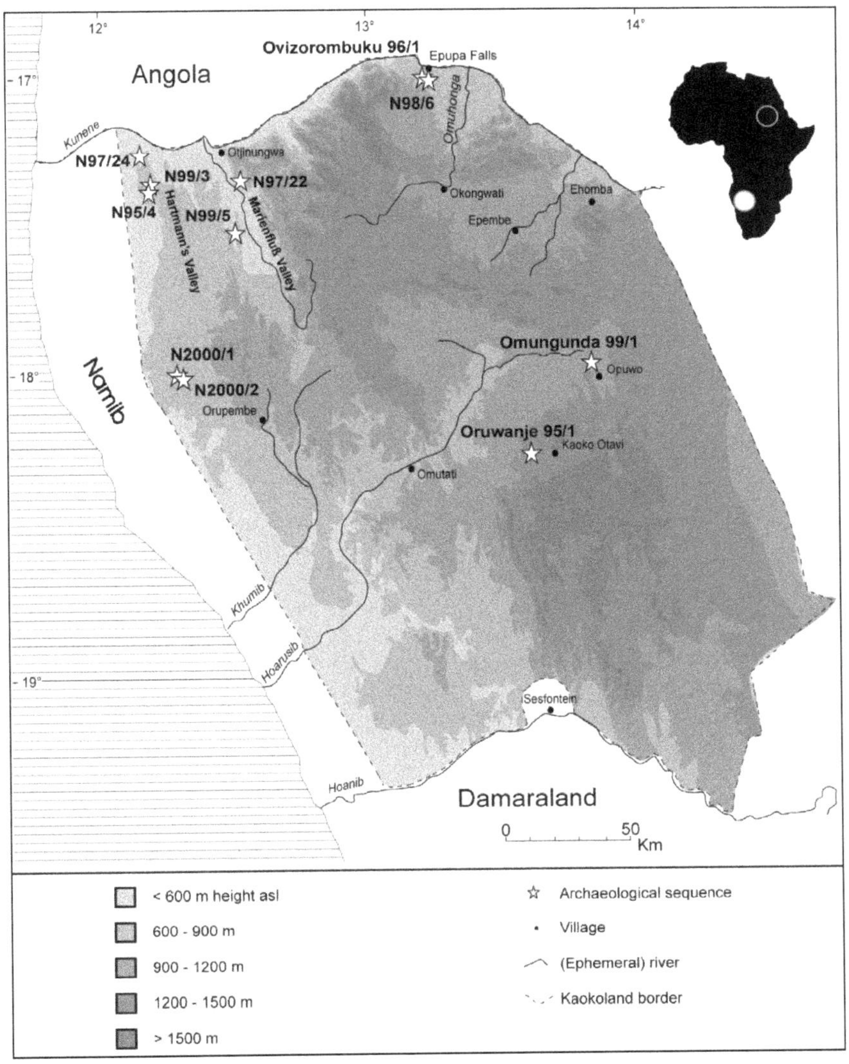

Figure 5.3. Opuwo District research area and location of the excavated sites

REVIEW OF CURRENT RESEARCH

Wadi Howar

Numerous stray finds and sites were reported in the area as early as the 1920s and 1930s (e.g., Newbold 1924; Newbold and Shaw 1928; Frobenius and Rhotert 1934; Shaw 1936:203 ff.). Researchers enthusiastically outlined cultural connections between the Eastern Desert and the Nile Valley (e.g., Hölscher 1955:55; Bietak 1979:126). It was not until the 1980s that systematic archaeological and geomorphological investigations commenced in the Wadi Howar region. Between 1980 and 1993 archaeological fieldwork was carried out by Khartoum University (Mohammed-Ali 1982) and the University of Cologne. Khartoum University covered the upper Wadi Howar, while a team from the University of Cologne worked in its middle and lower sections as part of the project Besiedlungsgeschichte der Ostsahara (B.O.S)[1] (Kuper 1981, 1995, 1999, Richter 1989). Since 1995, the University of Cologne has continued research within the Collaborative Research Centre ACACIA.

Between 1995 and 2000, intense archaeological research by the University of Cologne has yielded more than 1,500 open-air sites in the Wadi Howar region. Unfortunately, these consist mainly of eroded surface sites. Stratigraphic sequences are usually lacking. However, the usually dense layers of pottery, stone-artifacts, and well-preserved bones form a sound basis for reconstructing prehistoric life-ways and their transformations (Fig. 5.4).

Figure 5.4. View of site Djabarona 84/13 with the middle Wadi Howar to the north. The dense artefact scatter, bone- and/or pottery concentrations are typical for some sites of the pastoral phase (Photo Kuper 1995)

The Wadi Howar region was permanently occupied from at least the 6th to the 2nd millennium BC. This is suggested by archaeological data recovered from more than 1500 surveyed sites, small-sized excavations carried out on 20 surface sites, as well as on four settlement dunes producing stratigraphic sequences. Based primarily on pottery classification, the cultural sequence within this time span shows at least two economic transformations (Fig. 5.5) (Keding 1998, Keding and Vogelsang 2001).

The botanical remains are too scarce for any definitive statement about plant utilization. Nonetheless, the associated faunal remains indicate two transitions. There is an economic shift from hunter-gatherers to cattle-herders in the 4th millennium BC. This is followed in the 2nd millennium BC by a change from cattle-pastoralism to diversified pastoralism, with cattle, sheep, and goat (Fig. 5.6) (Van Neer and Uerpmann 1989:331 ff., 335, Berke 2001).

Parallel to these developments there are observable changes in land utilization and material culture (Keding 2004). The transformation processes involved have not yet been traced in detail. However, in both cases a series of relatively rapid and complete changes in the economic systems seem to emerge.

Opuwo District

No archaeological excavations were undertaken in Kaokoland prior to 1995, aside from three small-scale test pits (Kinahan 1981; MacCalman 1972; Smith 1988). The only comprehensive work is an ethnoarchaeological study of historical and recent settlement sites (Jacobsohn 1988, 1995). Since 1995, more than 150 sites have been located within the framework of the ACACIA project of the University of Cologne (Vogelsang 1999, 2001, Vogelsang *et al.* 2002). The project confirms human occupation of the Opuwo District since the Early Stone Age.

Unfortunately, it is extremely difficult to find well-preserved prehistoric open-air sites in the study area. The few known sites are severely damaged by intense erosion. Therefore, useful botanical and faunal macro-remains have not been suitably preserved. Conditions are more favorable for the preservation of these organic materials in rock shelters. This allows the reconstruction of both the subsistence strategy and the environmental conditions. Excavations have been undertaken in eight rock shelters, of which five have comprehensive stratigraphies that cover the entire Holocene. First results point to an initial contact between hunter-gatherer groups and herders about 2000 years ago. In contrast to the situation in the Sahara, at least some of the hunter-gatherers did not change their mode of subsistence, adopting only particular elements.

A MAJOR ASPECTS OF CURRENT RESEARCH

What lies behind these different paths of development as observed in Sudan and Namibia? Why do we have an extensive change of the economic situation in northern Africa and only partial shifts in southern Africa? These questions will be addressed through a comparative analysis of external and internal factors taken from both regions. However, it must be stressed that the basis of our comparisons is performed on archaeological data that

[1] "History of the Settlement of the Eastern Sahara".

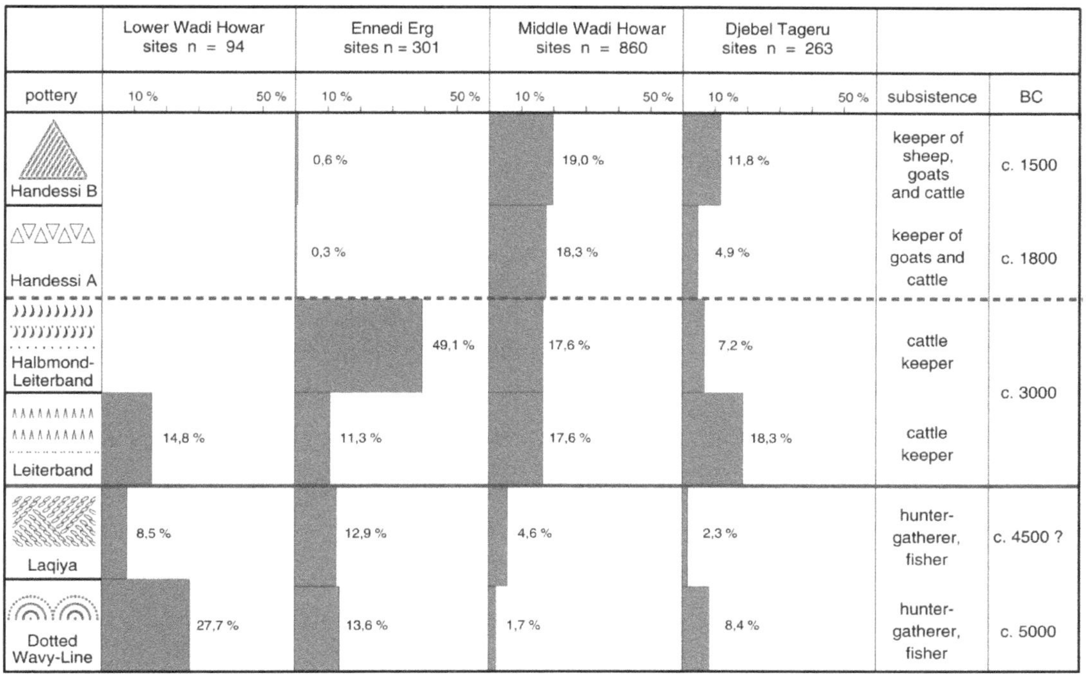

Figure 5.5. Archaeological sequences of the Wadi Howar region and adjacent research areas, based on survey data collected 1995-2000. "n" indicates the total number of sites in each area. Percentages indicate the share of sites with pottery of the different phases. Since sites where no pottery was found, as well as sites which had been reoccupied during different phases, are included, the sum is never 100%

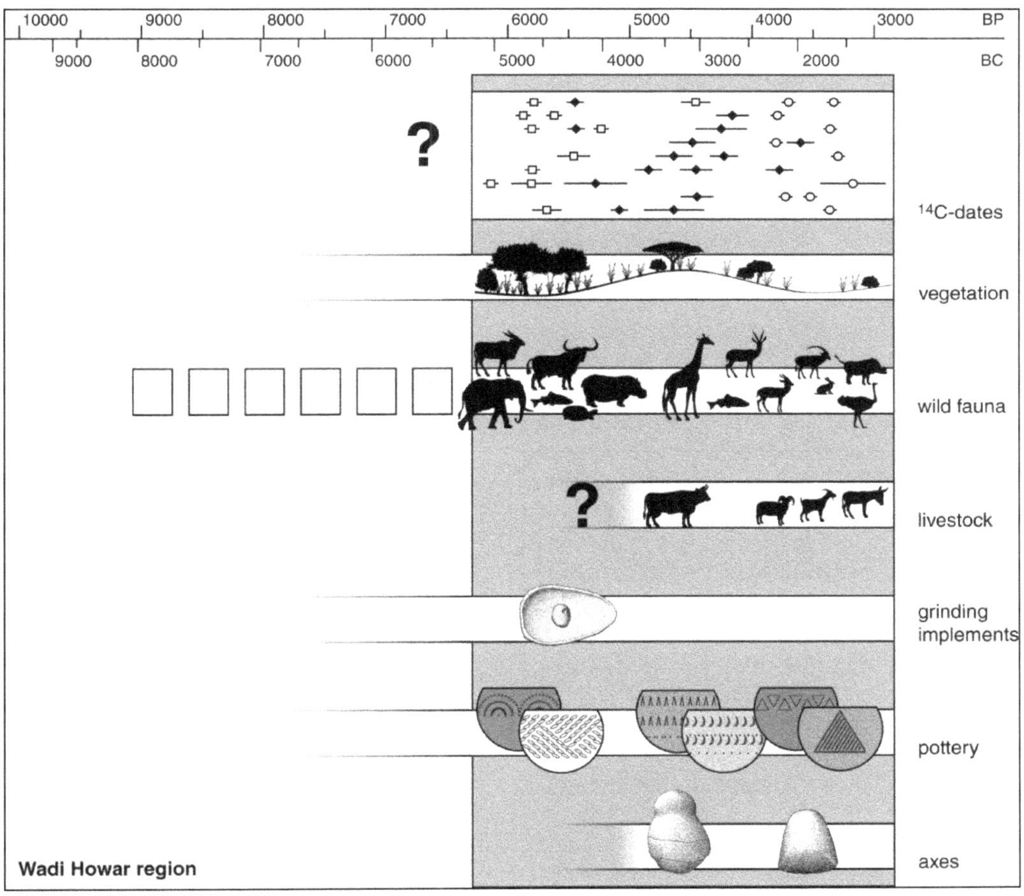

Figure 5.6. The culture sequence in the Wadi Howar region during the Holocene.
□ – sites with pottery of the Dotted Wavy and/or the Laqiya type; hunter-gatherers;
♦ – sites with pottery of the Leiterband and/or the Halbmondleiterband type; cattle-keepers;
○ – sites with pottery of the Handessi A and/or the Handessi B type; sheep, goats and cattle keepers

differs considerably. In the Wadi Howar area, characterized by a dense scatter of open-air sites, stratigraphic deposits are exceptionally rare. The general outline of cultural development is founded on results from numerous surveys as well as excavations undertaken at sites dating to each of the phases from the cultural sequence. In contrast, in the Kaokoland, it is extremely difficult to find well-preserved prehistoric open-air sites – the settlement type most closely related to pastoral nomads. Hence, excavations in rock shelters have yielded the best data for northern Namibia. Therefore, the analysis is based on stratigraphic sequences in Namibia and on spatial analysis of numerous surveys in the Sudan. Consequently, the focus on the process of cultural change is different for each area.

THE PALAEOENVIRONMENTAL CONDITIONS: CLIMATE

Wadi Howar

During the last 10000 years, the Eastern Sahara has experienced considerable climatic and environmental changes. The humid and arid phases vary in intensity and regional expression. Geomorphological (Haynes 1982; Schild 1987; Hoelzmann 1992; Kröpelin 1993; Pachur 1997), botanical (Ritchie and Haynes 1987; Neumann 1989), faunal (Van Neer and Uerpmann 1989; Berke 2001), as well as archaeological (e.g., Wendorf and Schild 1980; Kuper 1981, 1986, 1988, 1995; Close 1987; Schuck 1989; Keding 1997; Jesse 2003) research provide evidence of these changes (see Fig. 5.6).

During the humid period of the early and middle Holocene (ca. 6900-4900 BC, 8000-6000 bp),[2] the Wadi Howar region offered suitable ecological conditions for extensive human occupation. Long sections of Wadi Howar were marked by chains of large, deep, freshwater lakes (Kröpelin 1993) with a wide variety of fish species (Van Neer and Uerpmann 1989:312). The Ennedi Erg to the northwest was a lake region fed by at least three watercourses from the Ennedi Mountains (Hoelzmann 1993:126). The West Nubian Palaeolake with a surface up to 5300 km^2 was its largest stretch of water (Pachur 1997:229; Hoelzmann *et al.* 2001). To the south, the Djebel Tageru was also characterized by stable fluvio-lacustrine environments, with shallow lakes and swamps at its western side (Pachur *et al.* 1990:17; Hoelzmann 1993).

Due to the shifting southern monsoon rains, the vegetation belts moved 500 to 600 km to the north (Neumann 1989). Thus, at the beginning of the middle Holocene, around 5900-5300 BC (ca. 7000-6500 bp), the whole region was characterized by deciduous savannah dominated by Sudanese floral elements – an environment requiring more than 400 mm and up to 1000 mm rainfall per year. *Celtis integrifolia*, *Cratevva adansonii*, and *Terminalia macroptera* lined the borders of lakes. Swamp vegetation was found in deeper locations, and a savannah with a high proportion of trees was typical for the sand dunes (Neumann 1989:140 and Fig. 38).

No charcoal samples suitable for identification exist for the period between ca. 4900-3500 BC (6000-5000 bp). It is assumed that around 4500 BC (ca. 5700 bp) the vegetation belts were still located 400 km further north than they are today. This means that the Wadi Howar was still situated on the northern border of the thorn savannah belt (Neumann 1989:140).

Towards 4350 BC (ca. 5500 bp), increasing aridity spread from the Egyptian Sahara southward. During the following 2000 years, the rain regime and vegetation became similar to today. Rain rates subsequently decreased and the vegetation was transformed from thorn savannah to semi-desert and later to desert. Subsequently, this resulted in a decrease in wood and grass formations (Neumann 1989:140).

Around 3000 BC (ca. 4500 bp) increasing aridity finally reached the Wadi Howar region. Now lakes of the Wadi Howar were lined by *Acacia nilotica, Acacia tortilis, Ziziphus sp.*, and *subsp. Raddiana*. *Grewia tenax* and *Capparidaceen* wood completed the floral spectrum (Neumann 1989:140). The West Nubian Paleolake evolved from a single, large, freshwater lake into a series of smaller segmented ones. In the Ennedi Erg and the middle Wadi Howar lakes remained permanent as demonstrated by faunal relicts from hippopotamus and elephant. However, around 2000/1850 BC (ca. 3500 bp) the lakes dry up (Hoelzmann *et al* 2001 and Fig. 5.11). Charcoal from *Acacia sp.* and *Maerua sp.*, plant impressions on pottery from *Cenchrus bifloras* and *Paniceae sp.* (Zach, personal communication), as well as faunal remains (Berke 2001), demonstrate a sahelian-like environment.

Within a few millennia, the expanding desert conditions from north to south changed the ecological framework of the Eastern Sahara dramatically and challenged the inhabitants to adapt their way of life to the newly prevailing conditions. However, favored by its geographical position and hydrological conditions, the Wadi Howar still remained a potential ecological retreat for some additional hundreds, if not thousands of years.

Opuwo District

In contrast to the marked environmental transformation in northeastern Africa, abrupt climatic changes are not known in northern Namibia during the Holocene. The climate seems to be quite stable, especially during the last 3000 years, our main period of interest (Heine 1995:56). The reconstruction of the Holocene climatic conditions in the Opuwo District is based on anthracological analysis of charcoals from our excavations. They confirm this general picture (Eichhorn and Jürgens 2002, 2003; Vogelsang *et al.* 2002).

[2] The Editors have calculated approximate BC conversions of bp dates in the text and figure using OxCal 3.10 with INCAL04.

The woody vegetation at ca. 11900 BC (12000 bp) is characterized by the lack of typical mopane savannah taxa and the overall dominance of the genus *Acacia*. This indicates that the climate during the terminal Pleistocene in eastern Kaokoland was both drier and cooler than today. This assessment is corroborated by results obtained from marine pollen analysis (Shi and Dupont 1997; Shi *et al.* 1998, 2000). The first evidence of mopane savannah in the research area occurs around 6900 BC (8000 bp). However, the absence of tree species, which are characteristic for the eastern, humid part of the Opuwo district today, still points to relatively dry Early Holocene conditions. This is corroborated by several sets of sedimentological data, which imply aeolian sediment deposition under arid conditions in Namibia (Eitel and Zöller 1995; Eitel and Blümel 1997; Blümel *et al.* 2000; Brunotte and Sander 2000). Relatively dry conditions are also indicated by a low Early Holocene groundwater table (Brook *et al.* 1996, 1999). The rate of aeolian sediment deposition decreases at around ca. 8250-6900 BC (9000-8000 bp). Simultaneously, the semi-arid savannah ecosystems starts to take hold (Blümel *et al.* 2000). Climatic amelioration is indicated by the onset of mopane savannah in the Opuwo District. However, precipitation rates were still lower than today.

The sketchy anthracological data for ca. 6400-1250 BC (7500-3000 bp) point to woody vegetation. This is very similar to the current flora, implying climatic conditions generally comparable to that of today. However, slightly wetter episodes, which did not significantly change the woody taxa composition, may have occurred. In fact, several sets of paleoecological data suggest that the postglacial global climatic optimum was expressed by increased humidity in southwestern Africa east of the coastal desert (Buch 1996; Gingele 1996; Eitel and Blümel 1997; Brook *et al.* 1996, 1999; Brunotte and Sander 2000; Blümel *et al.* 2000; Shi and Dupont 1997; Shi *et al.* 1998).

Between ca. 1240 BC to AD 1430 or 3200 to 530 bp (3000-500 bp), anthracological data from all rock shelters in eastern and central Kaokoland indicate vegetation stability. The composition of woody vegetation at each site is very similar to today. The absence of vegetation shifts can be explained by stable environmental conditions.

To conclude, with the exception of the arid phase between ca. 11900-6900 BC (12000 to 8000 bp), only minor climatic variations are known in Kaokoland during the Holocene. The consequences of these variations are micro-regional shifts of ecological border zones. They do not represent complete alterations of the entire landscape.

THE PALEOENVIRONMENTAL CONDITIONS: ACCESSIBILITY AND LINKAGE

In neither region does domestication represent an autochthonous development. In both cases, it was acquired via external influences linked to diffusion or migration. Therefore, contacts with groups practicing different subsistence patterns must be responsible for the economic transition process. For this reason, the accessibility and linkage of the regions played an important role.

Wadi Howar

From the Early Holocene to the beginning of the Late Holocene, the Wadi Howar region was the east-west lifeline of northeastern Africa. It was the largest drainage system in the Eastern Sahara and connected the Nile Valley with the Chad Basin (Meissner and Schmitz 1983; Pachur and Kröpelin 1987; Kröpelin 1993). Its key location, in combination with its rich ecological resources, favored the (seasonal) use of the same areas by different groups. Climate change caused desert expansion, triggering shifts in settlement location. All this resulted in contact and trade between groups with different subsistence patterns.

In addition, the region was situated near several potential centers of early cattle domestication. At Bir Kiseiba in the Egyptian Eastern Sahara, domesticated cattle are dated to the 6th, and possibly even the 9th millennia BC. In the Central Sahara, domestication is evident by the 5th millennia BC (Gautier 1987). In the Sudanese Nile Valley, cattle pastoralism also occurs at this time (Gautier 1989: Table 1).

The geographical connection with the Nubian Nile valley exposed populations of the Wadi Howar region to numerous cultural influences and stimuli. Here stratified societies, kingdoms, and specialized economies had already developed by the 3rd millennium BC. Direct contacts with the Egyptian Empire led to the introduction of new technologies and ideas into Nubia. As a result, new material items may also have reached the Wadi Howar region. More importantly, there may also have been an increased awareness of different lifeways, favoring a constant appraisal of ideas.

Opuwo District

Due to the inaccessibility of large parts of the region, the Opuwo District is a more remote area than Wadi Howar.

Figure 5.7. The grass-covered Marienfluss valley

Figure 5.8. Descent from the mountainous highland via Van Zyls pass

Prehistoric hunter-gatherers lived outside the sphere of stratified Iron Age societies. The potential settlement areas for herders are, and always were, restricted to the plains, plateaus, and broad valleys (Fig. 5.7). Therefore, retreats for foragers always existed in less hospitable areas: bordering the Namib Desert, and in the mountainous regions with their steep slopes and poor water resources (Fig. 5.8).

To what extent did these different external factors influence the nature of pre-pastoral groups in both research areas?

THE ARCHAEOLOGICAL SEQUENCE: FROM HUNTER-GATHERERS TO HERDERS

Wadi Howar: Pre-pastoral phase (around 4900-4200 BC, 6000-5300 bp)

At present, the oldest Holocene settlement phase discovered in the Wadi Howar region dates to the end of the Holocene humid phase in the 5th millennium BC. This period is characterized by very favorable climatic conditions with a stable fluvio-lacustrine environment throughout the region (Kröpelin 1993; Hoelzmann *et al.* 2001). Precipitation of 450 mm, falling in two annual rainy seasons (Rodrigues *et al.* 2000), permitted savannah vegetation with a large number of trees to develop, along with stable lakes, and accompanying marshland.

This favorable environment, which provided a variety of permanent resources, was incorporated effectively into the lifeways of the local groups. Characteristics are:

- A preference for erecting settlements near lakes and rivers;
- Large amounts of cultural material, including pottery and grinding tools;
- A restricted, probably logistical mobility pattern (Binford 1980; Lourandos 1997:19) with frequent occupation of locations over a long time, perhaps as part of a seasonal cycle;
- A wide range of economic activities, including gathering, and to some degree a specialized, non-opportunistic way of hunting and fishing.

Hunter-gatherer sites are found throughout the whole region (see Fig. 5.5) (Jesse 2003). This is shown by the wide distribution of Dotted Wavy-Line and the slightly later, somewhat more regionalized, Laqiya pottery. By far the majority of these sites are located on large dunes by lakeshores. Their geographic distribution points to a preference to settle near permanent water resources (Keding 2004). One center is situated near the West Nubian Palaeolake in the Ennedi Erg (Hoelzmann *et al.*, 2001) and a second is located in the lower Wadi Howar. However, the Middle Wadi Howar seems to have been too swampy for long-term settlements during this period (Keding 1998, 2006 a) (Fig. 5.9).

The large dimensions of many sites, their dense scatters of occupational refuse, and the quality of the artifacts point to a restricted mobility with long-term stays, or at least repeated stays, at some settlements (e.g., Kent 1993). Although clear habitation structures have not yet been recognized, stone "arrangements," probably used for different purposes, are a common feature.

Both artifacts and features are compatible with the Khartoum Horizon style (Hays 1971). The style in ceramics is distributed east and west of the Wadi Howar region. Affinities are particularly close to the Early Khartoum type culture (Arkell 1949), which was distributed from the Nile valley to Chad (Jesse 2003). The assemblages are mainly composed of pottery, stone artifacts, large amounts of grindstone basins, and animal remains. This assemblage indicates a wide range of economic activities.

The pottery is represented in large quantities. The wide range of vessel forms, including pots, bowls, and plates, point to specialized methods of food preparation and differentiated consumption. There is a remarkably large proportion of fragments of very big, heavy pots with extremely thick-walls (up to 1.6 cm) in some samples, such as Site S98/120 (Keding 2006a). These suggest the storage of seeds or fluids, and underline a more permanent or repeated occupation of some of the sites (cf. the function of pottery as site fixtures by Cribb 1991:75ff.).

The large quantity of various grindstone basins, found at 75% of the sites, attests to an intensive exploitation of plant foodstuffs, especially seeds. Different forms of ground stone implements indicate specialized plant use and processing, such as pestles for pounding and beating,

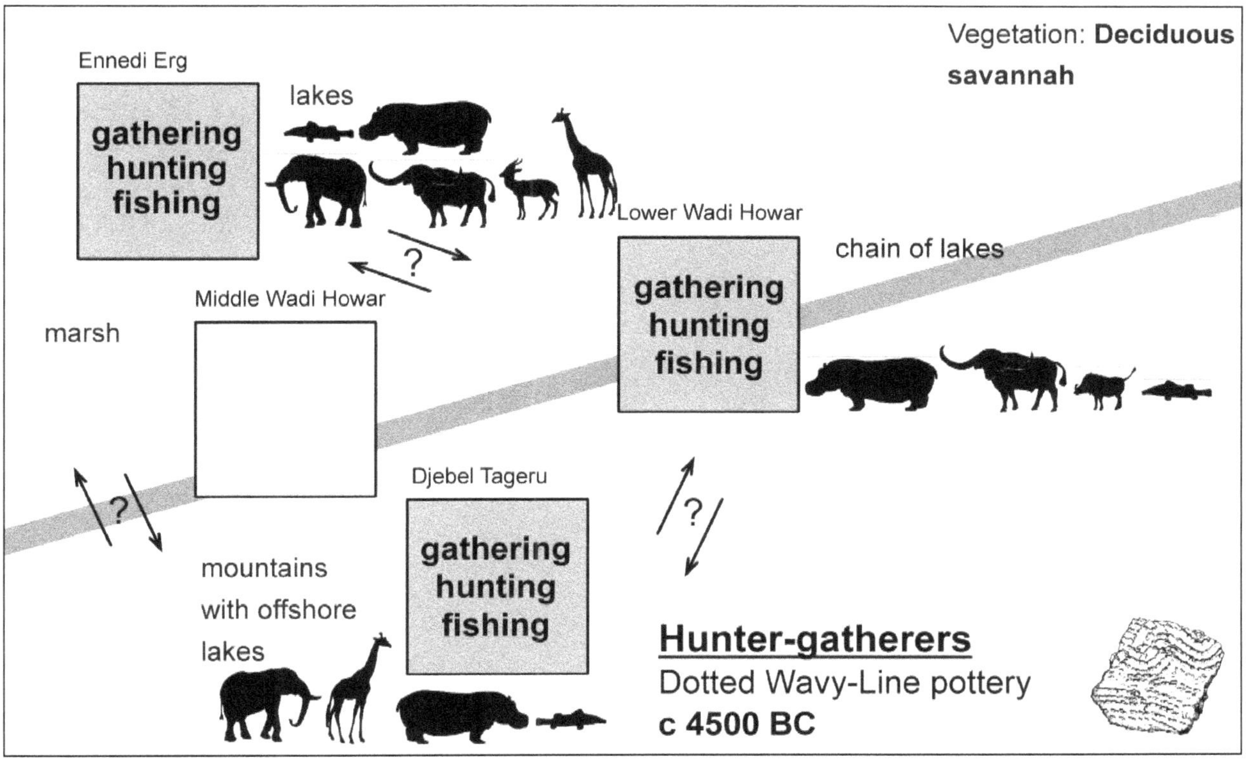

Figure 5.9. Reconstruction of environment and economic activities of hunter-gatherers between ca. 4900-4200 BC (6000-5300 bp) in the Wadi Howar region. Grey line: Wadi Howar. Squares with grey: main focus of regional activities. Arrows: possible mobility patterns. All animals are based on data from archaeological contexts

nut-crackers for opening kernels, and grindstones for crushing and pulverizing plants or grinding seeds.

The stone artifacts are dominated by flakes and segments (Kuper 1981:265; Jesse 2003; Keding 2006a). Together with occasional bone harpoons, they suggest hunting and fishing activities. The analysis of the faunal remains from all four areas of the Wadi Howar region have yielded evidence of big and small game hunting in humid and savannah environments, as well as the gathering of mollusks and amphibians. The spectrum of hunted animals consists mainly of elephant (*Loxodonta africana*), buffalo (*Pelorovis antiquus*), hippopotamus (*Hippopotamus amphibius*), giraffe (*Giraffa camelopardalis*), the giant eland (*Taurotragus derbianus*), as well as warthog (*Phacochoerus aethiopicus*), oryx (*Oryx gazella dammah*) and gazelle (*Gazella dorcas*) (Van Neer and Uerpmann 1989:331; Berke 2001:242). The importance of lake resources is demonstrated by the widespread occurrence of fish remains (Van Neer and Uerpmann 1989). In the West Nubian Palaeolake basin, these even include the large vertebrae of Nile perch (*Lates niloticus*), which can reach a total length of up to 1.5 m (Peters *et al.* 2002:330). Also on the menu were (usually charred) turtle (*Geochelonida sp.*), varan, and probably crocodile and ostrich eggs (Berke 2001:242).

Big and small game hunting took place in the humid and savannah environments of all four study areas of the Wadi Howar region. A closer look at the material does deliver some clues to more specialized subsistence patterns (Keding 2006a). The data described above point to hunting-fishing-gathering populations, which used a wide range of resources in a possibly lightly specialized, non-opportunistic way. An abundance of local resources, available year-round, or in sufficient quantities to be stored for seasonal lean periods, seems to have restricted mobility. This less mobile way of life probably favored the accumulation of material goods (cf. Brown 1989:220). It also favored demographic changes, which ultimately lead to larger group sizes, requiring new forms of social organization.

Opuwo District: Pre-pastoral phase

Hunter-gatherers inhabited the Opuwo District before the introduction of domesticated animals (Vogelsang *et al.* 2002). The classification of their inventories to different Later Stone Age (LSA) phases is mainly based on stone artifact analysis (Figure 5.10). Extensive material from the Pleistocene/Holocene transition is encountered at all of the larger excavations, which yield an indifferent macro-lithic stone artifact industry that is practically void of standardized tools.

In contrast to the abundance of Early Holocene macro-lithic inventories, sites with LSA assemblages of the Middle Holocene 'Wilton' type are rare in the Opuwo District. Not a single LSA open-air surface site could be located in the course of our surveys. These results are attributable to the rare and extremely small stone tools found on these sites, usually known from rock shelter

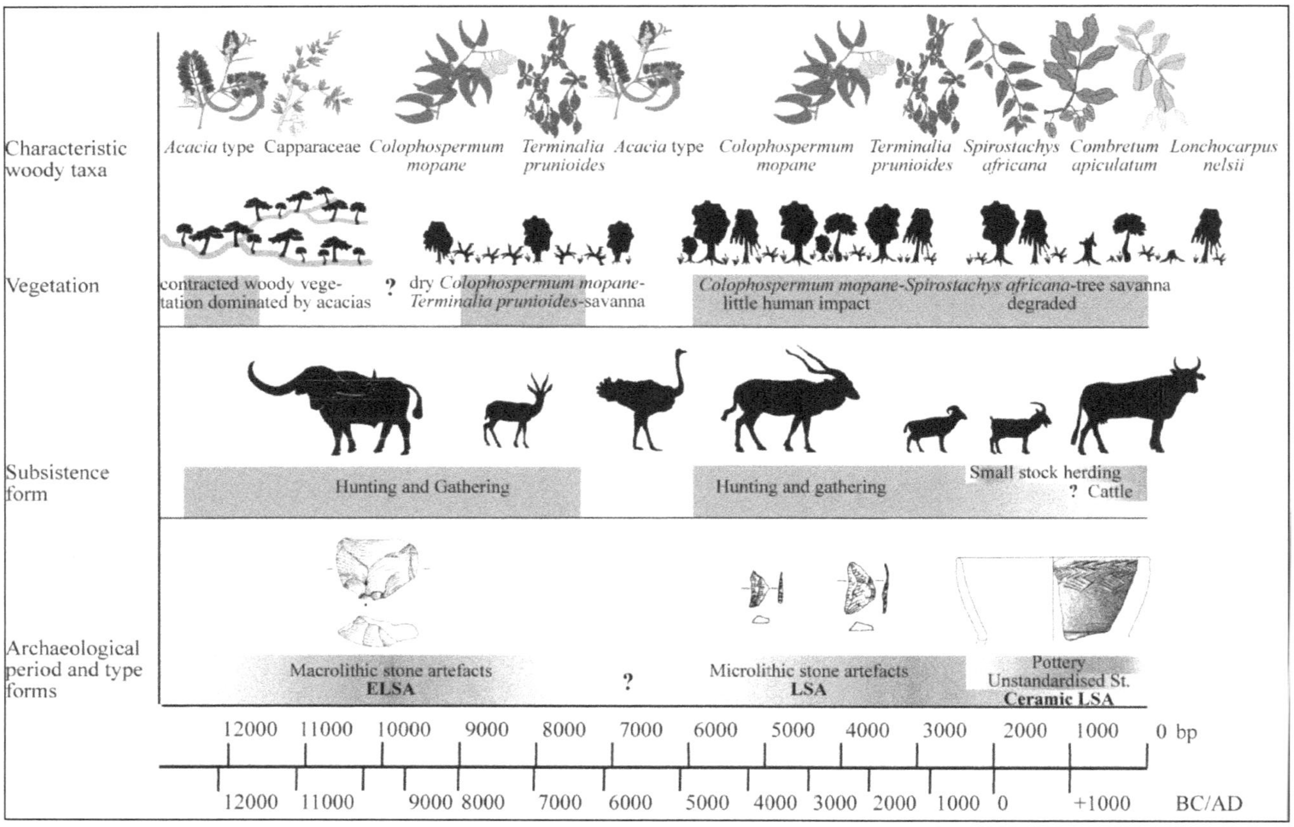

Figure 5.10. Archaeological and archaeobotanical sequence in the Opuwo District during the Holocene

excavations. Characteristic types are micro-lithic tools, especially projectile-insets, such as segments and micro-points. Other finds include small ostrich eggshell beads, bone beads, shale pendants, and bone points. Pottery is not found. Characteristic features of this phase include:

- Small settlement sites, indicating a small numbers of inhabitants;
- The use of non-local raw materials for the production of stone artifacts, indicating large band territories and a high degree of mobility;
- Faunal remains of large numbers of species, indicating non-specialized, opportunistic hunting;
- Macro-botanical remains, illustrating the consumption of collected vegetable foodstuffs.

To sum up, during the pre-pastoral phase in the Opuwo District, the rock shelters were inhabited by highly mobile groups, using a broad spectrum of resources.

Wadi Howar: Pastoral phase (around 4000-2500 BC, 5200-4000 bp)

The transition to pastoralism in the Wadi Howar region is documented by the assemblages of the Leiterband Horizon (Keding 1997a). The Leiterband pottery is decorated with ladder-like bands. In the stratigraphic sequences observed in dune settlements, this horizon follows the Dotted Wavy-Line and Laqiya assemblages. It dates between the 4th and 2nd millennia BC (Fig. 5.5) (Gabriel *et al.* 1985:10 and Fig. 5; Kuper 1988:136, 1995:130, 133; Richter 1989:434; Keding 1997a:35).

Although climate change caused increasing aridity, the study areas still had favorable ecological conditions for human settlement. There were permanent water pools in the middle Wadi Howar and freshwater lakes in the Ennedi Erg, on the western side of the Djebel Tageru. Wadi and lake banks were covered with acacia trees. Green savannahs existed in the hinterland. All this still offered a wide range of resources. However, conspicuous changes in the economic system do become apparent. The Leiterband and the later Halbmondleiterband phases are associated with people practicing a fully developed food-producing economy (Keding 1997a).

Based on the large amounts of recovered cattle bones and the general settlement pattern, the animals of the cattle-keepers probably provided people with ample meat, blood, and possibly milk (Bechthaus-Gerst and Keding in 2007). Other faunal remains indicate fishing and, on a smaller scale, big-game and small game hunting (Van Neer and Uerpmann 1989; Berke 2001). In addition, the numerous grindstones show an intensive exploitation of wild plants. Thus, "traditional" forms of subsistence were still in use, but probably only contributed on a small scale to the daily food requirements (see Fig. 5.6).

The significance of cattle seems to have exceeded the purely economic sphere. A few large sites in the middle Wadi Howar and the Djebel Tageru are characterized by

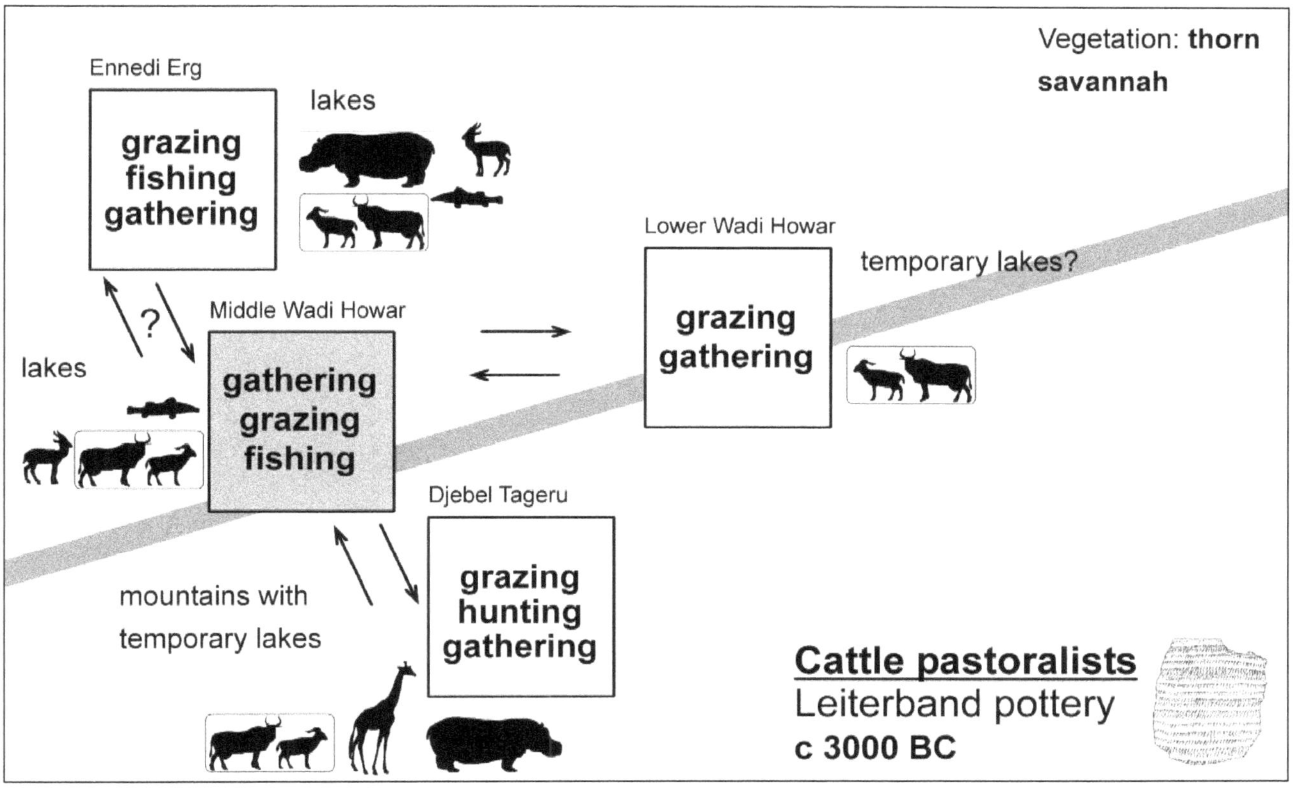

Figure 5.11. Reconstruction of environment and economic activities of cattle keepers between ca. 4000-2500 BC (5200-4000 bp) in the Wadi Howar region. Grey line: Wadi Howar. Squares with grey: primary focus of regional activities. Arrows: possible mobility patterns. All animals are based on data from archaeological contexts

dozens and sometimes even hundreds of pits filled with cattle bones and almost complete pottery vessels (see Fig. 5.4). These imply special, probably ritual activities in which cattle played an important role (Keding 1997a, 1998).

Neither the exact origin of the pastoral economy, nor the process of transition from hunting and gathering to cattle keeping in Wadi Howar, can be traced in detail. Pastoral influences on the region are expected from two main directions (Bate 1953, Krzyzaniak 1974, 1991; Gautier 1989: Table. 1; Peters 1989). One is the northern Sahara, where an incipient pastoralism phase exists from the 5th millennium BC onwards. The other is the Sudanese Nile valley with its fully developed pastoralism attributed to the Khartoum Shaheinab culture of the 5th and 4th millennium BC. Similarities between the Leiterband and the Khartoum Shaheinab cultures are evident. Both complexes share subsistence patterns and characteristic features in material culture, such as pottery forms and decorations (Keding 1997a, 1998).

The appearance of new forms of economy by adoption can be explained by three scenarios. The displacement of the local foraging groups by foreign cattle-herders is one explanation. Another is the assimilation of hunter-gatherers of the Wadi Howar region by foreign herders. The third is diffusion or adoption of cattle-herding by local groups through contacts with cattle-keepers.

The archaeological record of the Wadi Howar region provides no clear answers, but does give some clues about the underlying processes of transformation. Although numerous factors attest to contacts between hunter-gatherers and pastoralists from the Sudanese Nile valley, there are no indications of a displacement of the local foraging groups by newly arriving pastoralists. Therefore, assimilation or diffusion evoked by, perhaps seasonal, contacts with pastoralists from the Nile valley would appear to constitute the most likely explanation for the beginning of cattle-keeping. Pottery analyses of the Wadi Howar region confirm this assumption (Keding 2006b).

The environmental and economic changes resulted in new settlement and mobility patterns, shifting centers of occupation and the obvious variability in site size. There is a clear increase of very small, ephemeral camps, indicative of different and partly seasonal uses in the diverse sub-regions (Fig. 5.11). This suggests increased mobility and a new organization of labor.

The migration cycles in the early cattle-keeping phase include all defined regions of the Wadi Howar. The focal point was the middle Wadi Howar with medium to large dry-season camps, used by sedentary or at least semi-sedentary segments of a pastoral society. In the 2nd millennium BC, when Halbmondleiterban pottery was in use, the trend towards increasing aridity continued. The

Lower Howar is no longer exploited. Occupation is restricted. One center in the middle Wadi Howar features two large base-camps. Another center, the Ennedi Erg, consists of small sized grazing camps.

Additionally, regional and supra-regional transhumance cycles and networks changed orientation. This is reflected in pottery development. While the early stages of Leiterband decorations reveal supra-regional connections with the east, especially with the Khartoum Shaheinab culture, those of the later phases coincide largely with the pottery of the Hohou-type of the Ennedi Mountains to the west of the Wadi Howar (Keding 1998:5 ff.).

Nonetheless, the composition and amount (the archeologically visibility) of the material culture does not alter dramatically. The artifacts comprise primarily pottery, chipped stone artifacts, grinding tools, and, as a new element, Darfur type axes. These reflect supra-regional contacts as well as regional developments.

The very homogenous pottery occurs in large amounts. Any differences between this ware and the pre-pastoralist pottery are more than slight, but relationships in form, technology, and decoration are obvious (Keding 2006b). Stone artifacts are still characterized by a microlithic industry, but they are now dominated by transverse arrowheads and thumb nail scrapers. The use of artifacts from local quartz is supplemented with tools made from chalcedony, which points to newly developed sources (Keding 1997a:199ff.).

Darfur type axes not only signify the importance of woodworking, they also indicate communication patterns. Although the axes are a regional development (see Fig. 5.6), they reflect far-reaching contacts with the Wadi Shaw, situated 500 km to the north, as well as with the Ennedi Mountains 300 km to the west. Their distribution shows a new north-south exchange network, which is roughly comparable to the earlier one of the hunter-gatherer period. In addition, there is a new, more dominant east-west orientated communication network.

Grinding tools are as numerous as in the hunter-gatherer period. These show roughly the known forms, but in a slightly restricted range of varieties, and they reveal no technological change in plant exploitation. If grindstones basins are rated as site fixtures or site furniture (Binford 1980; Cribb 1991:75), their recorded occurrence of 60% in all pastoral sites implies only gradual change in the degree of group mobility and in the operating rights for the repeated access of special locations.

To summarize, the pastoralist evidence in the Wadi Howar region outlines the adoption of a fully developed food producing economy, probably transmitted through contacts with herders from the Sudanese Nile valley. However, only a few of the new cultural traits, which occur contemporaneously, represent further cultural aspects adopted from the Khartoum Shaheinab of the Sudanese Nile valley. These comprise a pottery set, characterized by a selection of vessel forms and decorations that include "counterpoised triangles," as well as plain and diverse dotted zigzag patterns (Keding 1997a). The majority of new traits are regional developments. The most evident features are pits filled with pottery and cattle bones, indicating a special significance of cattle at some sites. The most evident finds are transverse microliths and Darfur type axes, which point to altered activities requiring other tools. Cultural transformations, which seem to be partly attributable to climatic change, are:

- Transformations of some supra-regional networks, e.g., an increasing connection with western regions such as the Ennedi Mountains
- Alterations of occupation and land use patterns, e.g., the shifting of larger sites from the lower to the middle Wadi Howar
- Changes in the spectrum of exploited fish and game species.

Additionally, cultural changes may also be due to economic transformation: the sharp increase of small-sized grazing camps. However, some remaining features were more likely determined by social and ideological components. These include:

The continued use of all four areas of the region at the beginning of the pastoral period. This may point to a group's traditional land use rights, and the knowledge of their attachment to the region.

A reoccupation of 57% of the hunter-gatherer sites.

A high percentage of (probably) semi-permanently or permanently occupied sites. This indicates, on the one hand, a low degree of mobility for a large segment of the pastoral society, and on the other hand, it permits larger group sizes, both features already typical for the former period.

The on-going rights to designate, perhaps even to demarcate, and use a region as well as any special locations.

Again, there is a strong accumulation of material culture, especially at middle to large-sized sites. This may indicate either the repeated (planned) use of the same locations, suggested by site fixtures (cf. Binford 1980, Cribb 1991:75ff.), or it may point to a special significance of material goods as possible mediums of group identity (for the last point compare the role of material culture in some recent pastoral societies in Cribb 1991:82).

A continuation of former, traditional economic activities, such as large and small game hunting, fishing, and intense plant use. While large and small game hunting appears to have played a minor economic role, plant use must still have been important for feeding the group. Also, large game hunting may have held a special social and/or ideological significance.

Opuwo District: Pastoral phase

During the last 3000 years, numerous radiocarbon dates indicate an increase of settlement activities in northwestern Namibia (Vogelsang *et al.* 2002). Around 2000 years ago, hunter-gatherers in the Opuwo District met small livestock herders (see Fig. 5.10). During this period, pottery appears for the first time. However, it is of high quality and does not reflect an initial phase of pottery production. Nonetheless, even after the introduction of pottery and domesticated animals, the assemblage of the stone tools does not change. The sites exhibit low numbers of pottery sherds and the composition of the faunal remains are still dominated by wild species. This means that the sites are not camps of the pastoralists themselves, and indicates that the hunter-gatherer bands merely met the pastoralists. Such contacts are typical in southern Africa.

In the entire region, only one site with early domesticates is regarded as a definite early pastoral site. The extreme scarcity of early pastoral sites can be explained by the difficulties in finding open-air sites with good preservation conditions, especially for bones (Mitchell 2002:237). However, in our opinion, it is also an indication that the pastoralists were small in number and highly mobile (see Sadr 2003 for an alternative interpretation). The high mobility is also reflected by the rapid spread of the new economy, indicated by quite similar radiocarbon dates for early domesticates all over southern Africa (Bousman 1998).

Hunter-gatherers who met pastoralists did not change their way of life, adopting only some elements, although they began to use pottery and may even have learned to make it. However, domesticates were treated more like hunted wild animals and were only rarely used to supplement the diet. In fact, the people did not change their subsistence mode until historical times. Overall, the same features characterize this phase as the pre-pastoral one:

- The same small rock shelters are used, indicating small numbers of inhabitants.
- Stone artifacts are made of non-local raw materials, indicating large band territories and a high degree of mobility.
- A large number of species, but rarely domesticated animals, occur in the faunal remains, indicating non-specialized, opportunistic meat consumption.
- Macro-botanical remains point to the consumption of collected vegetable foodstuffs.

There is no archaeological substantiation for the beginning of the specialized cattle-pastoralism of today. Indeed, there is no evidence of domesticates for the last 1500 years in the Opuwo district. Furthermore, the oral traditions of the cattle keeping Himba tell of their migration from the northeast into Kaokoland around 250 years ago (Bollig 1997b:13). Therefore, it seems that herding was only a short episode in the Late Holocene history in northwestern Namibia.

Wadi Howar: Transition from cattle pastoralism to small-stock pastoralism (ca. 2000/1850 BC, 3500 bp)

During the 2nd and 1st millennium BC sites are characterized by two new geometrically decorated pottery styles, belonging to the Handessi Horizon (see Fig. 5.5) (Keding 1998; Jesse 2006), as floral, faunal and geomorphological data point to an increasingly arid climate. Several botanical impressions from *Cenchrus sp.* found on the surface of Handessi B pottery demonstrate a sahelian-like environment (see Fig. 5.6) (Zach personal communication). The fauna consists only of wild species, including *Gazella dorcas* and *Gazella dama*, giraffe (*Giraffa camelopardalis*), and ostrich, that are adapted to steppe, semi-desert, and desert conditions (Berke 2001). Aquatic species from permanent lakes are completely absent. The growing aridity also reaches the Ennedi Erg at this time, while the vast West Nubian Paleolake completely dries up. These ecological changes coincide with cultural transformations and include:

- A diversification of the subsistence base;
- A decreased mobility;
- A concentration of occupation in a more restricted area;
- A change to north south orientated supra-regional cultural relationships.

Yet, with the exception of the last mentioned point, behavioral patterns do not show any major ruptures. Change occurred gradually within the former cultural framework.

In the Wadi Howar region, the Handessi horizon exhibits pronounced diversification of the subsistence base, which is linked to a confined use of ecological niches. Plant and animal use is intensified. Animal use is broadened by adding goat and sheep to the livestock. However, the traditional tools continue to be employed in plant maintenance, and hunting activities involve maintenance of the still available game. Aquatic animals and reptiles no longer contribute to the food spectrum. The net result is decreased mobility and an occupation increasingly concentrated in the central Wadi Howar. People live in medium to large-sized camps, which were mostly located in the wadi bed or in depressions. The transhumance cycles become smaller, being limited to the middle of the wadi itself and the Djebel Tageru (Fig. 5.12).

The changing ecology also effects the development and the amount of material culture characteristic of the Handessi horizon. During this period new, more complex pottery traditions became established in the Wadi Howar region. Technology, form, and decoration of the vessels give no indications for relationships with the former Leiterband pottery of the cattle keepers (Keding 2006b). The pottery demonstrates connections with Wadi Hariq in the north as well as the Nubian Nile valley (cf. Jesse *et al.*

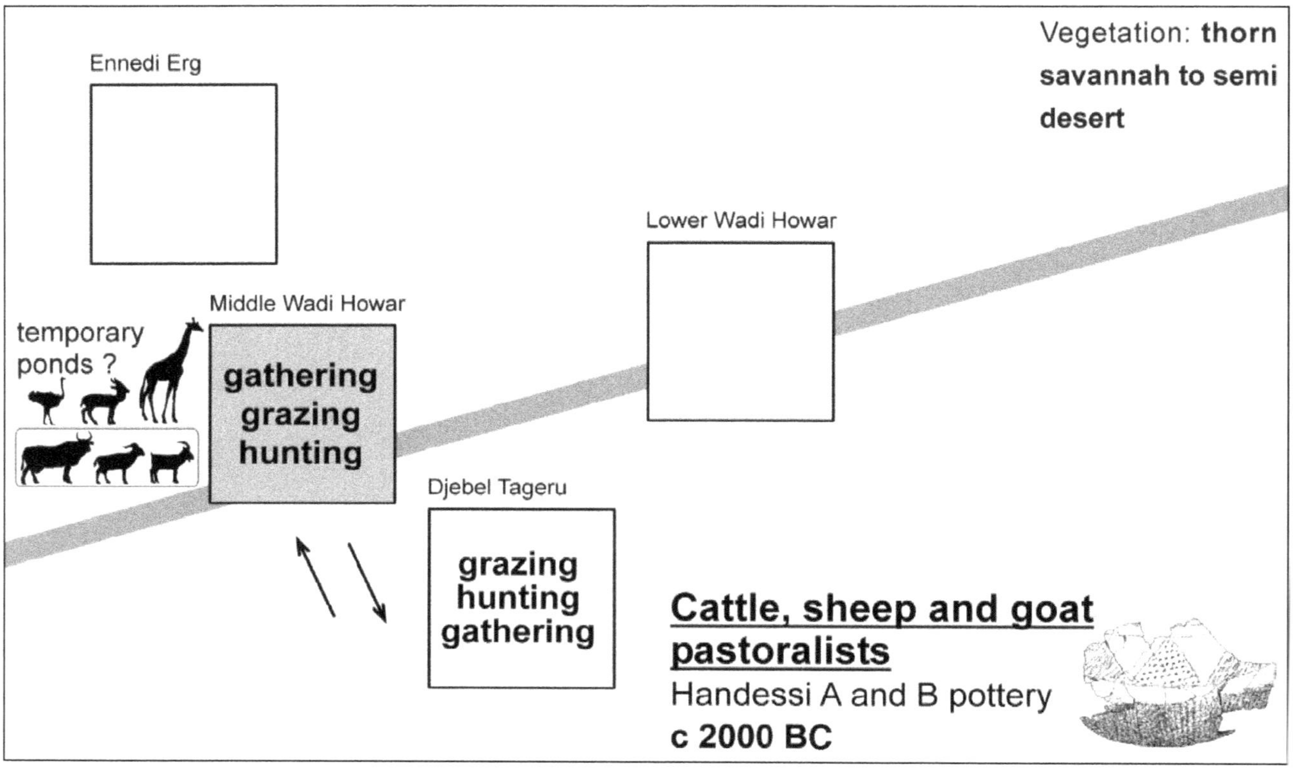

Figure 5.12. Reconstruction of environment and economic activities of goat, sheep, and cattle keepers around ca. 2000/1850 BC (3500 bp) in the Wadi Howar region. The grey line: Wadi Howar. Squares with grey: the primary focus of regional activities. Rounded rectangles: domesticated animals. Arrows: possible mobility patterns. All animals are based on data from archaeological contexts

2004). This indicates north-south oriented networks or population shifts triggered by growing aridity.

There are also noticeable changes in the stone tool technology. Their importance seems to increase, although the variety of grinding tools does not change. Grindstone basins are present at 75% of all recorded Handessi sites, but now they are also found in large quantities in small-sized camps. Contrasting with the complex pottery inventories, the stone artifacts are very simple. Unretouched artefacts made from local quartz dominate. Tools are extremely scarce.

To sum up, the observed cultural transformations seem to have been triggered primarily by environmental changes. Especially the economic change to diverse livestock, integrating sheep and goat with the cattle herds, must be understood as a risk-management strategy. Sheep and goat are better adapted to an arid environment than cattle. In addition, different grazing and browsing patterns, as well as diverse herding cycles of each species, lead to an improved exploitation of an ever decreasing pasture.

From social and ideological viewpoints, changes are less obvious. The middle Wadi Howar was still the center of activities. Judging by site size, people still lived and worked in semi-permanent settlements and grazing camps. Although goat, sheep, and cattle keeping required a new organization of labor, which might be reflected in the increasing number of grazing camps with site fixtures, it does not demand new social or ideological forms. However, increasing population densities might have triggered problems. This is probably reflected in the increasing trend for demarcation with the help of the material culture, such as pottery (Keding 2006b), or features, such as the concentrations of graves (Jesse and Keding 2003).

WHY IS THE ECONOMIC DEVELOPMENT MORE DYNAMIC AND INTENSIVE IN THE WADI HOWAR REGION THAN IN THE OPUWO DISTRICT?

In both regions, there are only gradual changes in the climate during the period of early contacts between foraging and pastoral groups. In fact, in Namibia there are no indications for any major climatic fluctuations, whereas the evidence from the Eastern Sahara points to only slightly drier conditions. Therefore, climatic stress has to be ruled out as a prime mover for economic change. Naturally, the central location of the Wadi Howar region increases the potential for possible change, but in both regions, the extent of the adoption of the new economic system seems to be mainly due to socio-cultural factors, such as social organization and ideology. Cultural anthropological studies indicate that their various forms seem to be differently pre-adapted for the development or adoption of pastoralism and agriculture (e.g. Testart 1982; Wiesner 1982; Woodburn, 1988).

Although use of ethnographic data from contemporary hunter-gatherers must be considered problematic when used as a direct analogy for the reconstruction of the past, anthropological data can give archaeologists new ideas about human behavior and a chance to test theories (David and Cramer 2001:43ff., Wylie 1985).

According to the varying levels of cultural complexity, recent hunter-gatherers can be divided into "simple" (egalitarian) and "complex" societies. Features of egalitarian hunter-gatherer societies are a fluid group structure, sharing, an emphasis on residential mobility, and an immediate-return strategy (Woodburn 1980). This contrasts with the social organization of more complex hunter-gatherers, identifiable by: sedentism or substantially restricted residential mobility, and a "focal exploitation of a particular resource (commonly fish)" (Kelly 1995:302). Other characteristics include ownership of resources, an investment in technology, food storage, and a delayed-return strategy, bounded, discrete, and larger groups with more formal leadership, inherited status, ritual feasting, standardized valuables, and an erosion of the egalitarian ideology (Kelly 1995:302; Zvelebil 1998:8; Lourandos 1997:24). This polarized classification is usually not understood as a dichotomy, but rather as the two ends of a continuum that is neither unilinear nor directional in time (e.g., Woodburn 1988:32). Thus, different societies have diverse degrees of social complexity and structure, which lie between the two extremes (Zvelebil 1998:8).

In Namibia the foraging groups are characterized by high mobility, small group size, and non-specialized resource exploitation. Interpretation of their rock art points to an egalitarian social structure (Lenssen-Erz 1998, 2001:329). As members of an immediate return system, the adoption of a food-producing economy is opposed to their total ideology.[3] Thus, social mechanisms hinder the accumulation of property and, therefore, do not allow herd management.

This contrasts with the Wadi Howar region. There the socio-economic conditions result from a permanent availability of food resources. This leads to restricted mobility, favoring adoption of a food-producing economy.

Research done on contemporary hunter-gatherers shows that one important condition for the transition from a relatively mobile to a more sedentary existence seems to be an abundance of local resources, available year round, or in sufficient quantities to be stored for seasonal lean periods (Kelly 1995:152, 160). Indeed, the Middle Holocene sites in the Wadi Howar region produced evidence of permanent availability and a wide variety of food resources obtainable by hunting, fishing, and gathering. A frequent occupation of the locations over a long time, perhaps as part of a seasonal cycle, is demonstrated by the sometimes enormous size of the sites, the dense accumulation of material items, large pottery vessels, and an abundance of heavy grinding equipment.

Site permanence is judged by some anthropologists to be one of the most important conditions for the development from egalitarian to non-egalitarian social structures (e.g., Keeley 1988, quoted after Kelly, 1995:304). From an economic point of view, this involves reorganization of mobility patterns from residential to logistical (Binford 1980, here used in the sense of Lourandos 1997:20), and a smaller, more limited monitoring area, possibly resulting in resource specialization. In addition, it may include changing attitudes towards property and valued assets. From a social point of view, site permanence leads to larger group sizes and to a need for formal horizontal and/or vertical group structures that do not change constantly as in mobile groups, thus resulting in changing relationships between people with mutual obligations (e.g., Kelly 1995).

Parts of these developments are reflected in the archaeological record of Wadi Howar. There are hints of some degree of resource specialization, in particular calculated fishing in the Ennedi Erg. Site fixtures on many sites may point to resource utilization rights. These would allow repeated visits to a particular place, including the right to leave basic equipment behind. In addition, the frequent occupation of locations over a long period permitted hunter-gatherers to begin to develop accumulations of material culture as well as to produce pottery. This investment in material culture might have prepared a "right" to ownership. Thus, we find a socio-economic pre-adaptation to a food-producing economy, implying change from an immediate to a delayed return system, already in this period.

The above change is followed by a second economic transformation, leading to diversified pastoralism. This is largely triggered by climatic factors. The transformation seems to be a risk management strategy, designed to cope with increasing aridity and its limiting effects on resources and utilizable land. The addition of small livestock to the cattle herds may have required a new herd-management concept, but no transformation of ideology.

The comparison of the two examples shows that external conditions reduce the range of possible developments, while socio-cultural conditions favor particular concepts. Yet further incalculable factors play a role. These might be summarized under the ambiguous term of human agency. They have a determining influence on the final development of cultures (e.g. Dobres and Robb 2000). These factors are, however, hardly detectable in the archaeological material.

References

ARKELL, A.J. 1949 – *Early Khartoum. An Account of the Excavation of an Early Occupation Site Carried*

[3] An immediate-return system as defined by Woodburn (1988) is one where people do not hold valued assets, which represent a yield of return for labour applied over time.

out by the Sudan Government Antiquities Service in 1944-5. Oxford University Press, London, 1949.

BATE, D.M.A. 1953 – The Vertebrate Fauna. In A.J.S. Arkell (ed.), *An Account of the Excavation of a Neolithic Occupation Site Carried out for the Sudan Antiquities Service in 1949-50*. Oxford University Press, London, 1953.

BECHHAUS-GERST, M., and B. KEDING 2007 – Meat or Milk? Archaeological and Linguistic Evidence for the Predynastic use of Cattle in the Eastern Sahara. In W.J.G. Möhlig (ed.), *Cultural Change in the Prehistory of Arid Africa Perspectives of Archaeology and Linguistics*. Proceedings of the SFB Conference, Königswinter 1-3 October 2003. Institut für Afrikanistik, SUGIA, Sprache und Geschichte in Afrika. Rüdiger Köppe Verlag, Köln. 2007:1-25, 2007.

BECKER, T., and N. JÜRGENS 2000 – Vegetation Along Climate Gradients in Kaokoland, North-West Namibia. *Phytocoenologia* 30, 2000:543-565.

BECKER, T., and N. JÜRGENS 2002 – Vegetationsökologische Untersuchungen im Kaokoland, Nord-West Namibia. In M. Bollig, E. Brunotte and T. Becker (eds.), *Interdisziplinäre Perspektiven zu Kultur- und Landschaftswandel im ariden und semiariden Nordwest Namibia*. Kölner Geographische Arbeiten 77. Geographisches Institut der Universität zu Köln, Köln, 2002.

BENDER, B. 1978 – Gatherer-Hunter to Farmer: a Social Perspective. *World Archaeology* 10, 1978:204-222.

BERKE, H. 2001 – Gunsträume und Grenzbereiche: Archäozoologische Beobachtungen in der Libyschen Wüste, Sudan und Ägypten. In B. Gehlen, M. Heinen and A. Tillmann (eds.), *Zeit-Räume. Gedenkschrift für Wolfgang Taute*. Archäologische Berichte 14, Habelt, Bonn, 2001.

BIETAK, M. 1979 – Ceramics of the C-Group Culture: Africa in Antiquity. *Meroitica* 5, 1979:107-127.

BINFORD, L.R. 1978 – Dimensional Analysis of Behavior and Site Structure: Learning from an Eskimo Hunting Stand. *American Antiquity* 43, 1978:330-361.

BINFORD, L.R. 1980 – Willow Smoke and Dogs Tails: Hunter-Gatherer Settlement Systems and Archaeological Site Formation. *American Antiquity* 45, 1980:4-20.

BLÜMEL, W.D., K. HÜSER and B. EITEL 2000 – Landschaftsveränderungen in der Namib: Klimawandel oder Variabilität? *Geographische Rundschau* 5, 2000:17-23.

BOLLIG, M. 1997a – Risk and Risk Minimisation Among Himba Pastoralists in Northwestern Namibia. *Nomadic Peoples* New Series 1/1, 1997:66-89.

BOLLIG, M. 1997b – *"When War Came the Cattle Slept..." Himba Oral Traditions*. History, Cultural Traditions and Innovations in Southern Africa 1, Köppe, Köln, 2001.

BOLLIG, M., and A. SCHULTE 1999 – Environmental Change and Pastoral Perceptions: Degradation and Indigenous Knowledge in two African Pastoral Communities. *Human Ecology* 27, 1999:493-514.

BOUSMAN, C.B. 1998 – The Chronological Evidence for The Introduction of Domestic Stock into Southern Africa. *African Archaeological Review* 15, 1998:133-150, 2001.

BROOK, G.A., J.B. COWART and E. MARAIS 1996 – Wet and Dry Periods in the Southern African Summer Rainfall Zone During the Last 300 Kyr from Speleothem, Tufa and Sand Dune Age Data, *Palaeoecology of Africa* 24, 1996:147-158.

BROOK, G.A., E. MARAIS and J.B. COWART 1999 – Evidence of Wetter and Drier Conditions in Namibia from Tufas and Submerged Speleothems. *Cimbebasia* 15, 1999:29-39.

BROWN, J.A. 1989 – The Beginnings of Pottery as an Economic Process. In S.E. van der Leeuw and R. Torrence (eds.), *What's New? A Closer Look at the Process of Innovation*. Unwin Hyman Ltd., London, 1989.

BRUNOTTE, E., and H. SANDER 2000 – Loess Accumulation and Soil Formation in Kaokoland (Northern Namibia) as Indicators of Quaternary Climatic Change. *Global and Planetary Change* 26, 2000:67-75.

BUCH, M.W. 1996 – Geochrono-Geomorphostratigraphie der Etosha-Region: Nord-Namibia. *Die Erde* 127, 1996:1-22.

CLARK, J.D., and S.A. BRANDT (eds.) 1984 – *From Hunters to Farmers: The Causes and Consequences of Food Production in Africa*, University of California Press, Berkeley.

CLOSE, A. (ed.) 1987 – *Prehistory of Arid North Africa: Essays in Honor of Fred Wendorf*, Southern Methodist University Press, Dallas, 1987.

CRIBB, R. 1991 – *Nomads in Archaeology*. New Studies in Archaeology. Cambridge University Press, Cambridge, 1991.

DAVID, N., and C. KRAMER 2001 – *Ethnoarchaeology in Action*. Cambridge University Press, Cambridge, 2001.

DOBRE, J and J. ROBB (eds.) 2000 – *Agency in Archaeology*. Routledge, London and New York, 2000.

EICHHORN, B., and N. JÜRGENS 2002 – Vegetationsgeschichte und Nutzung pflanzlicher Ressourcen im Kaokoland: Stabilität oder Wandel? In M. Bollig, E. Brunotte and T. Becker (eds.), *Interdisziplinäre Perspektiven zu Kultur- und Landschaftswandel im ariden und semiariden Nordwest Namibia*. Kölner Geographische Arbeiten 77. Geographisches Institut der Universität zu Köln, Köln, 2002.

EICHHORN, B., and N. JÜRGENS 2003 – The contribution of charcoal analysis to the late Pleistocene and Holocene vegetation history of northwestern Namibia. In K. Neumann, A. Butler and S. Kahlheber (eds.), *Progress in African Archaeobotany*. Africa Praehistorica 15, Heinrich-Barth-Institut, Köln, 2003:151-162.

EITEL, B., and W.D. BLÜMEL 1997 – Pans and Dunes in the Southwestern Kalahari (Namibia): Geomorphology and Evidence for Quaternary Paleoclimates. Supplement-Band. *Zeitschrift für Geomorphologie* 11, 1997:73-95.

EITEL, B., and L. ZÖLLER 1995 – Die Beckensedimente von Dieprivier und Uitskot (NW-Namibia): Ein Beitrag zu ihrer paläoklimatischen Interpretation auf der Basis von Thermolumineszenzdatierungen. *Mitteilungen der Österreichischen Geographischen Gesellschaft* 137, 1995:245-254.

FROBENIUS, L., and H. RHOTERT 1934 – Die Ergebnisse der 11. Deutschen (inner-afrikanischen) Forschungsexpedition in die Libysche Wüste und in dem Anglo-ägyptischen Sudan 1933. *Bilderbuchblatt* 4, 1934:39-59.

GABRIEL, B., S. KRÖPELIN, J. RICHTER and E. CZIESLA 1985 – Parabeldünen am Wadi Howar. *Geowissenschaften in Unserer Zeit* 3, 1985:105-122.

GAUTIER, A. 1987 – Prehistoric Men and Cattle in North Africa: A Dearth of Data and a Surfeit of Models, In A. Close (ed.), *Prehistory of Arid North Africa: Essays in Honor of Fred Wendorf*. Southern Methodist University Press Dallas. 1987:183-197.

GAUTIER, A. 1989 – A general Review of the Known Prehistoric Fauna of the Central Sudanese Nile Valley. In L. Kryzaniak and M. Kobusiewicz (eds.), *Late Prehistory of the Nile Basin and the Sahara*, Poznan Archaeological Museum, Poznan, 1889:353-358.

GEBAUR, A.B., and T.D. PRICE 1992 – Foragers to farmers: An Introduction. In A.B. Gebauer, and T.D. Price (eds.), *Transitions to Agriculture in Prehistory*. Monographs in World Archaeology 4, Prehistory Press, Madison, 1992:1-10.

GINGELE, F.X. 1996 – Holocene climatic optimum in Southwest Africa: evidence from the marine clay mineral record. 122 *Palaeogeography, Paleoclimatology, Palaeoecology* 1996:77-87.

HAYDEN, H. 1990 – Nimrods, Piscators, Pluckers and Planters: The Emergence of Food Production. *Journal of Anthropological Research* 9, 1990:31-69.

HAYNES, C.V. 1982 – Great Sand Sea and Selima Sand Sheet, Eastern Sahara: Geochronology of Desertificitation. *Science* 217, 1982:629-633.

HAYS, T.R. 1971 – *The Sudanese Neolithic: A Critical Analysis*. Ph. D. Dissertation, Southern Methodist University, University Microfilms International, Ann Arbor, 1971.

HEINE, K. 1995 – Paläoklimatische Informationen aus südwestafrikanischen Böden und Oberflächenformen: Methodische Überlegungen. *Geomethodica* 20, 1995:27-74.

HÖLSCHER, W. 1955 – *Libyer und Ägypter. Beiträge zur Ethnologie und Geschichte libyscher Völkerschaften nach den altägyptischen Quellen*. Augustin, Glückstadt, Hamburg, 1955.

HOELZMANN, P. 1992 – Palaeoecology of Holocene Lacustrine Sediments Within the West Nubian Basin, SE-Sahara, *Würzburger Geographische Abhandlungen* 84, 1992:59-71.

HOELZMANN, P. 1993 – *Holozäne Limnite im NW-Sudan*. Unpublished Ph.D. Thesis, University of Berlin, Berlin, 1993.

HOELZMANN, P., B. KEDING, H. BERKE, S. KRÖPELIN and H.-J. KRUSE 2001 – Environmental Change and Archaeology: Lake Evolution and Human Occupation in the Eastern Sahara During the Holocene. *Palaeogeography, Palaeoclimatology and Palaeoecology* 169, 2001:193-217.

INGOLD, T. 1996 – Hunting and Gathering as Ways of Perceiving the Environment. In R. Ellen and K. Fukui (eds.), *Redefining Nature. Ecology, Culture and Domestication*. Berg, Oxford, 1996:117-155.

JACOBSOHN, M. 1988 – Preliminary Notes on the Symbolic Role of Space and Material Culture Among Semi-Nomadic Himba And Herero Herders in Western Kaokoland, Namibia. *Cimbebasia* 10, 1988,75-99.

JACOBSOHN, M. 1995 – *Negotiating Meaning and Change in Space and Material Culture: An Ethno-Archaeological Study Among Semi-Nomadic Himba and Herero Pastoralists in North-Western Namibia*. Unpublished Ph.D. Thesis, University of Cape Town, Cape Town, 1995.

JESSE, F. 2003 – *Rahib 80/87: Ein Wavy-Line-Fundplatz im Wadi Howar und die früheste Keramik in Nordafrika*, Africa Praehistorica 16, Heinrich-Barth-Institut, Köln.

JESSE, F. 2006 – Pastoral groups in the Southern Libyan Desert: The Handessi Horizon (ca. 2400 – 1100 BC). In K. Kroeper, M. Chlodnicki and M. Kobusiewicz (eds.), *Archaeology of Early Northeastern Africa. In Memoriam of Lech Krzyzaniak*. Studies in African Archaeology 9. Poznan Archaeological Museum, Poznan, 2006:987-1004.

JESSE, F., and B. KEDING 2002 – Death in the Desert: Burials in the Wadi Howar Region (Eastern Sahara). In Jennerstr. 8 (ed.), *Tides of the Desert: Gezeiten der Wüste. Contributions to the Archaeology and Environmental History of Africa in Honour of Rudolph Kuper*, Africa Praehistorica 14, Heinrich-Barth-Institut, Köln. 2002:277-293.

JESSE, F., S. KRÖPELIN, M. LANGE, N. PÖLLATH and H. BERKE 2004 – On the Periphery of Kerma: The Handessi Horizon in Wadi Hariq, Northwestern Sudan. *Journal of African Archaeology* 2/2, 2004:123-164.

KEDING, B. 1997a – *Djabarona 84/13: Untersuchungen zur Besiedlungsgeschichte des Wadi Howar anhand der Keramik des 3. und 2. Jahrtausends v. Chr.* Africa Praehistorica 9, Heinrich-Barth-Institut, Köln, 1997.

KEDING, B. 1997b – Prehistoric Investigations in the Wadi Howar region: A preliminary Report on the 1995-1996 Season, *Kush* XVII, 1997:33-46.

KEDING, B. 1998 – The Yellow Nile: New Data on Settlement and the Environment in the Sudanese Eastern Sahara. *Sudan and Nubia* 2, 1998:2-12.

KEDING, B. 2002 – Two seasons in the Wadi Howar region (1996-1998): A Preliminary Report. *Kush* XVIII, 2002:89-96.

KEDING, B. 2004 – The Yellow Nile – Settlement Shifts in the Wadi Howar Region (Sudanese Eastern Sahara) and Adjacent Areas Between the Sixth to the First Millennium BC. In T. Kendall (ed.), *Nubian Studies 1998. Proceedings of the Ninth Conference of the International Society of Nubian Studies*, Northeastern University, Boston. 2004:95-108.

KEDING, B. 2006a – Were the Hunter-Gatherers in the Eastern Sahara of the "Delayed Return" Type? An Archaeological Perspective from the Wadi Howar Region, Sudan. In A. Roccati, (ed.), Nubian Studies 2002. *Proceedings of the Tenth Conference of the International Society of Nubian Studies*, Rome 9-14 September 2002. Libreria dello Stato, Rome, 2006:81-94.

KEDING, B. 2006b – Pottery of the Wadi Howar: Traditions, Transformations and their Cultural Implications. In K. Kroeper, M. Chlodnicki and M. Kobusiewicz (eds.), *Archaeology of Early Northeastern Africa. In Memoriam of Lech Krzyzaniak*. Studies in African Archaeology 9. Poznan Archaeological Museum. Poznań, 2006:235-259.

KEDING, B., and R. VOGELSANG 2001 – Vom Jäger-Sammler zum Hirten: Wirtschaftswandel im nordöstlichen und südwestlichen Afrika. In B. Gehlen, M. Heinen and A. Tillmann, (eds.), *Zeit-Räume. Gedenkschrift für Wolfgang Taute*, Archäologische Berichte 14, Habelt, Bonn, 2001:257-282.

KEELEY, L. 1988 – Hunter-Gatherer Economic Complexity and "Population Pressure": A Cross-Cultural Analysis. *Journal of Anthropological Archaeology* 7, 1988:373-411.

KELLY, R.L. 1995 – *The Foraging Spectrum: Diversity in Hunter-Gatherer Lifeways*, Smithsonian Institution Press, Washington.

KENT, S. 1993 – Models of Abandonment and Material Culture. In C.M. Cameron and S. Tomka (eds.), *The Abandonment of Settlements and Regions. Ethno-archaeological and Archaeological Approaches*, Cambridge University Press, Cambridge, 1993:54-73.

KINAHAN, J. 1981 – An Early Pottery Date from Southern Kaokoland. *SWA Scientific Society Journal* 34/35, 1981:43-46.

KRÖPELIN, S. 1993 – Zur Rekonstruktion der spätquatären Umwelt am Unteren Wadi Howar (Südöstliche Sahara/NW-Sudan). *Berliner geographische Abhandlungen* 54. Geomorphologisches Laboratorium der Freien Universität Berlin, Berlin, 1993.

KRYZANIAK, L. 1974 – Kadero (First Season, 1972). *Études et traveaux* 8, 1974:361-366.

KRYZANIAK, L. 1991 – Early Farming in the Middle Nile Basin: Recent Discoveries at Kadero (Central Sudan). *Antiquity* 65:515-532.

KUPER, R. 1981 – Untersuchungen zur Besiedlungsgeschichte der östlichen Sahara. Vorbericht über die Expedition 1980. *Beiträge zur Allgemeinen und Vergleichenden Archäologie* 3, 1981:215-275.

KUPER, R. 1986 – Wadi Howar and Laqiya: Recent field studies into the early settlement of Northern Sudan. In M. Krause (ed.), *Nubische Studien*, Phillip von Zabern, Mainz, 1986: 129-136.

KUPER, R. 1988 – Neuere Forschungen zur Besiedlungsgeschichte der Ost-Sahara. *Archäologisches Korrespondenzblatt* 18, 1988:127-142.

KUPER, R. 1995 – Prehistoric Research in the Southern Libyan Desert: A Brief Account and Some Conclusions of the B.O.S. Project. *Cahier de Recherches de l'Institut de Papyrologie et d'Égyptologie de Lille* 17, 1995:123-140.

LENSSEN-ERZ, T. 1998 – The Third Gender: Human Gender-Related Patterns of Activity in the Rock Paintings of the Brandberg, Namibia. In A. Banks, H. Heese and C. Loff, (eds.), *The Proceedings of the Khoisan Identities and Cultural Heritage Conference*, Infosource, Capetown, 1998:146-152.

LENSSEN-ERZ, T. 2001 – Gemeinschaft, Gleichheit, Mobilität: Felsbilder im Brandberg, Namibia, und ihre Bedeutung: Grundlagen einer Textuellen Felsbildarchäologie. *Africa Praehistorica* 13, Heinrich-Barth-Institut, Köln, 2001.

LOURANDOS, H. 1988 – Palaeopolitics: Resource Intensification in Aboriginal Australia and Papua New Guinea. In T. Ingold, D. Riches and J. Woodburn (eds.), *Hunters and Gatherers 1. History, Evolution and Social Change.* Berg, Oxford, 1988:148-60.

LOURANDOS, H. 1997 – *Continent of Hunter-Gatherers: New Perspectives in Australian Prehistory*. Cambridge University Press, Cambridge.

LUBBOCK, J. 1865 – *Pre-historic Times, as Illustrated by Ancient Remains, and the Manners and Customs of Modern Savages*. Williams and Norgate, London, 1865.

MacCALMAN, H.R. 1972 – The Otjinungwa Valley site: A Middle Stone Age occurrence on the South West Africa/Angola border, *Cimbebasia* 2/2, 1972:66-80.

MEISSNER, B., and H.-J. SCHMITZ 1983 – Zur Kartierung alter Entwässerungssysteme in der Sahara mit Hilfe von Fernerkundungs-Daten: Am Beispiel des Nordwest-Sudan. In F.K. List, and B. Meissner (eds.), Beiträge zur Fernerkundung der Erde an der Freien Universität Berlin. *Berliner Geowissenschaftliche Abhandlungen* (A) 47, Dietrich Reimer, Berlin, 1983:87-93.

MITCHELL, P. 2002 – *The Archaeology of Southern Africa*. University Press Cambridge, Cambridge, 2002.

MOHAMMED-ALI, A.S.A. 1982 – *The Neolithic Period in the Sudan, ca. 6000-2500 B.C.* BAR International 139. British Archaeological Reports, Oxford, 1982.

NEUMANN, K. 1989 – Zur Vegetationsgeschichte der Ostsahara im Holozän. Holzkohlen aus prähistorischen Fundstellen (mit einem Exkurs über die Holzkohlen von Fachiogonboulo, Niger). In R. Kuper (ed.), *Forschungen zur Umweltgeschichte der Ostsahara*. Africa Praehistorica 2, Heinrich-Barth-Institut, Köln, 1989:13-183.

NEWBOLD, D. 1924 – A Desert Odyssey of a Thousand Miles. *Sudan Notes and Records* 7, 1924:43-92.

NEWBOLD, D., and W.B.K. SHAW 1928 – An Exploration in the South Libyan Desert. *Sudan Notes and Records* 11, 1928:103-194.

PACHUR, H.-J. 1997 – Der Ptolemäus-See in Westnubien als Paläoklimaindikator. *Petermanns Geographische Mitteilungen* 141, 1997:227-250.

PACHUR, H.-J., S. KRÖPELIN, P. HOELZMANN, M. GOSCHIN and, N. ALTMANN 1990 – Late Quaternary Fluvio-Lacustrine Environments of Western Nubia, 120. *Berliner Geowissenschaftliche Abhandlungen* 1, 1990:203-260.

PACHUR, H.-J., and S. KRÖPELIN 1987 – Wadi Howar: Paleoclimatic Evidence from an Extinct River System in the Southeastern Sahara. *Science* 237, 1987:298-300.

PETERS, J. 1989 – The faunal Remains from Several Sites at Jebel Shaqadud (Central Sudan): A Preliminary Report. In L. Kryzaniak and M. Kobusiewicz, (eds.), *Late Prehistory of the Nile Basin and the Sahara*. Poznan Archaeological Museum, Poznan, 1989:469-472.

PETERS, J., N. PÖLLATH and A. von den DRIESCH 2002 – Ichthyological Diversity in the Holocene Palaeodrainage Systems of Western Nubia. In Jennerstr. 8 (eds.), Tides of the Desert: Gezeiten der Wüste. Contributions to the Archaeology and Environmental History of Africa in Honour of Rudolph Kuper, *Africa Praehistorica* 14, Heinrich-Barth-Institut, Köln, 2002:325-335.

RICHTER, J. 1989 – Neolithic sites in the Wadi Howar (Western Sudan). In L. Kryzaniak and M. Kobusiewicz, (eds.), *Late Prehistory of the Nile Basin and the Sahara*. Poznan Archaeological Museum, Poznan, 1989:431-442.

RITCHIE, J.C., and C.V. HAYNES 1987 – Holocene vegetation zonation in the Eastern Sahara. *Nature* 330, 1987:645-647.

RODRIGUES, D.P., I. ABELL and S. KRÖPELIN 2000 – Seasonality in the Early Holocene Climate of Northwest Sudan: Interpretation of *Etheria elliptica* Shell Isotopic Data. *Global and Planetary Change* Special Issue 26/1-3, 2000:181-187.

SADR, K. 2003 – The Neolithic of Southern Africa, *Journal of African History* 44, 2003:195-209.

SANDER, H., M. BOLLIG and A. SCHULTE 1998 – Himba Paradise Lost: Stability, Degradation and Pastoralist Management of the Omuhonga Basin (Namibia). *Die Erde* 129, 1998:301-315.

SCHILD, R. 1987 – Unchanging Contrast? The Late Pleistocene Nile and Eastern Sahara. In A. Close, (ed.), *Prehistory of Arid North Africa: Essays in Honor of Fred Wendorf.* Southern Methodist University Press, Dallas, 1987:13-27.

SCHUCK, W. 1989 – From Lake to Well: 5000 Years of Settlement in Wadi Shaw (Northern Sudan). In L. Kryzaniak and M. Kobusiewicz (eds.), *Late Prehistory of the Nile Basin and the Sahara, Poznan Archaeological Museum* Poznan. 1989:431-442.

SHAW, W.B.K. 1936 – An Expedition in the Southern Libyan Desert. *Geographical Journal* 87, 1936:193-221.

SHI, N., and L.M. DUPONT 1997 – Vegetation and Climatic History of Southwest Africa: A Marine Palynological Record of the Last 300,000 Years. *Vegetation History and Archaeobotany* 6, 1997:117-131.

SHI, N., L.M. DUPONT, H.-J. BEUG and R. SCHNEIDER 1998 – Vegetation and Climate Changes During the Last 21,000 Years in S.W. Africa Based on a Marine Pollen Record. *Vegetation History and Archaeobotany* 7, 1998:127-140.

SHI, N., L.M. DUPONT, H.-J. BEUG and R. SCHNEIDER 2000 – Correlation Between Vegetation in Southwestern Africa and Oceanic Upwelling in the Past 21,000 years. *Quaternary Research* 54, 2000:72-80.

SIMONS, P. 1973 – Der Osten der Sahara: Die Nilwüste. In H. Schiffers, (ed.), *Die Sahara und ihre Randgebiete. Darstellung eines Naturgroßraumes, III: Regionalgeographie (Die Landschaften)*. Afrika Studie 62. Weltforum, München, 1973.

SINCLAIRE, P., T. SHAW and B. ANDAH 1993 – Introduction. In T. Shaw, P. Sinclair, B. Andah and A. Okpoko (eds.), *The Archaeology of Africa*. Routledge, New York and London, 1993.

SMITH, A.B. 1988 – Excavation of rockshelter at Otjitati in Kaokoland. *Cimbebasia* 10, 1988:101-105.

TESTART, A. 1982 – *Les Casseurs-Cueilleurs ou L'Origine des Inégalités*. Klincksieck, Paris, 1982.

Van NEER, W., and H.-P. UERPMANN 1989 – Palaeocological Significance of the Holocene Faunal Remains of the B.O.S. Missions. In R. Kuper (ed.), *Forschungen zur Umweltgeschichte der Ostsahara.* Africa Praehistorica 2, Heinrich-Barth-Institut, Köln. 1989:307-341.

VILJOEN, P.J. 1980 – *Veldtipes, Verspreiding van die Grooter Soogdiere, en Enkele Aspekte van die Ekologie van die Kaokoveld.* Unpublished Master of Science Thesis, University of Pretoria, Pretoria.

VOGELSANG, R. 1999 – Archaeological Survey in the Kaokoveld, Namibia. *Nyame Akuma* 50, 1999:22-25, 1980.

VOGELSANG, R. 2001 – Archaeological Investigations in the Kaokoland: Survey and Excavations in 1999 and 2000. *Nyame Akuma* 54, 2001:25-28.

VOGELSANG, R., B. EICHHORN, and J. RICHTER 2002 – Holocene Human Occupation and Vegetation History in Northern Namibia. *Die Erde* 133, 2002:113-132.

WIESNER, P. 1982 – Beyond Willow Smoke and Dogs Tails: A Comment on Binford's Analysis of Hunter-Gatherer Settlement Systems. *American Antiquity* 47/1, 1982:171-178.

WOODBURN, J. 1980 – Hunters and Gatherers Today and Reconstruction of the Past. In E. Gellner, (ed.), *Soviet and Western Anthropology*. Duckworth, London, 1980:95-117.

WOODBURN, J. 1988 – African Hunter-Gatherer Social Organization: Is it best understood as a Product of Encapsulation? In T. Ingold, D. Riches and J. Woodburn (eds.), *Hunters and Gatherers*. Berg, Oxford, 1988:31-64.

WYLIE, A. 1985 – The Reaction Against Analogy. *Advances in Archaeological Method and Theory* 8, 1985:63-111.

ZVELEBIL, M. 1998 – What's in a Name: The Mesolithic, the Neolithic, and Social Change at the Mesolithic-Neolithic Transition. In M. Edmonds and C. Richards (eds.), *Understanding the Neolithic of North-Western Europe*, Cruithne Press, Glasgow, 1998:1-35.

Chapter 6
FITS AND STARTS: WHY DID DOMESTICATED ANIMALS 'TRICKLE' BEFORE THEY 'SPLASHED' INTO SUB-SAHARAN AFRICA?

David K. WRIGHT
Dept. of Archaeology and Art History, Seoul National University, Seoul, South Korea

Abstract: *Two theories account for the relatively slow spread of pastoralism south of the Lake Turkana Basin between 2000 and 1000 BC The first theory posits that arid climates ca. 2000 BC impeded the spread of domesticated animals south of the equator until pluvial conditions set in after 1000 BC. The second theory argues that epizootic diseases were the inhibiting factors. Movement of domesticated cattle into eastern Africa between 2000 and 1000 BC appears restricted to "green corridors" where access to water and fodder are available. Data from Tsavo, Kenya indicates it was likely a combination of both factors.*

INTRODUCTION

The origins and spread of domesticated animals remains one of the most debated topics in archaeology today. With advances in dating techniques and faunal analysis, site-level interpretation of animal procurement and consumption has provided useful insights into when and where domesticated animals occur. Special attention has focused on occurrences of domesticates in virgin habitats because it is believed that understanding movements of early pastoralists will address the remaining questions of how and why domestication occurs in the first place.

Animal herding remains an important component of human subsistence in many dryland environments of Africa. Much of the continent possesses soils that are incapable of supporting intensive agriculture, but pastoralists have successfully occupied these regions and maintain large herds of goats, sheep, camels and cattle in spite of erratic rainfall and xeric plant cover. Therefore, isolating the chronology and mechanisms whereby domesticated animals were introduced to the region is of special significance. In the face of a rapidly changing cultural and climatic milieu, identifying the backdrop of early animal husbandry provides insight on the adaptive challenges prehistoric pastoralists faced.

This chapter presents results from archaeological work undertaken in Tsavo East National Park, Kenya along the Galana River and interpretations of the mechanisms that catalyzed the spread of domesticated animals across the southern latitudes of Africa. Survey and excavations in 2001 and 2004 identified settlements adjacent to the modern-day channel of the Galana River that date from 4000 BC to AD 600. One of these sites, Kahinju (HeJt9), provided evidence of early domesticated cattle (*Bos taurus*) in eastern Africa. This find is significant to the extent that it broadens the presumed range of early domesticated animals in eastern Africa prior to 1000 BC. Archaeological discoveries of livestock in East Africa dating to before 1000 BC have been restricted to the Great Rift Valley and adjacent Central East African Highlands (Ambrose 1998, Barthelme 1985, Bower 1991, Marshall 2000). However, the unearthing of domesticated cattle molars at Kahinju extends the known limits of early cattle in East Africa some 200 km east of the Central East African Highlands. The presence of early livestock in the coastal lowlands of the Indian Ocean give credence to the notion that regional aridity at ca. 2000 BC prevented domesticates from moving too far from a predictable resource base such as a riparian or highland environment. The evidence for limited adoption of domesticated animals in Tsavo after 1000 BC relative to other areas in East Africa may support the hypothesis that epizootic diseases were barriers for keeping large herds of animals, or is indicative of a broader, cultural phenomenon.

EARLY OCCURRENCES OF DOMESTICATED ANIMALS IN NORTHEASTERN AFRICA

Figure 6.1 shows the locations of published archaeological sites that demonstrate the earliest confirmed occurrences of domesticated animals as they spread southward through Northeast Africa. Domesticated cattle in Africa (*Bos primagenius* cf. *taurus*) are first identified in the Fayum region of western Egyptian Sahara prior to 6000 BC (Wendorf and Schild 2001). Ovicaprids (goats [*Capra* sp.] and sheep [*Ovis* sp.]) were introduced to the African continent via the Levant >5000 BC (Close 1992). Early animal husbandry in Africa revolved around maintaining herds of cattle supplemented by foraging, which became the dominant mode of food production in Africa for thousands of years (Marshall and Hildebrand 2002, McDonald 1998). Typically, pastoralism is first identified in the North African archaeological record at sites that are located adjacent to wetlands in aridifying environments (Smith 2005). By 5500 BC, a mortuary complex involving buried megaliths and cattle bones is widely distributed across the Sahara (di Lernia 2006) indicating the entrenchment of livestock within the regional culture. However, as the Sahara became drier and less hospitable for human habitation, pastoralists began to disperse westward and southward in search of green pastures.

Ceramic-using cattle herders are found >4000 BC near the West Nubian Paleolake in the modern day Wadi

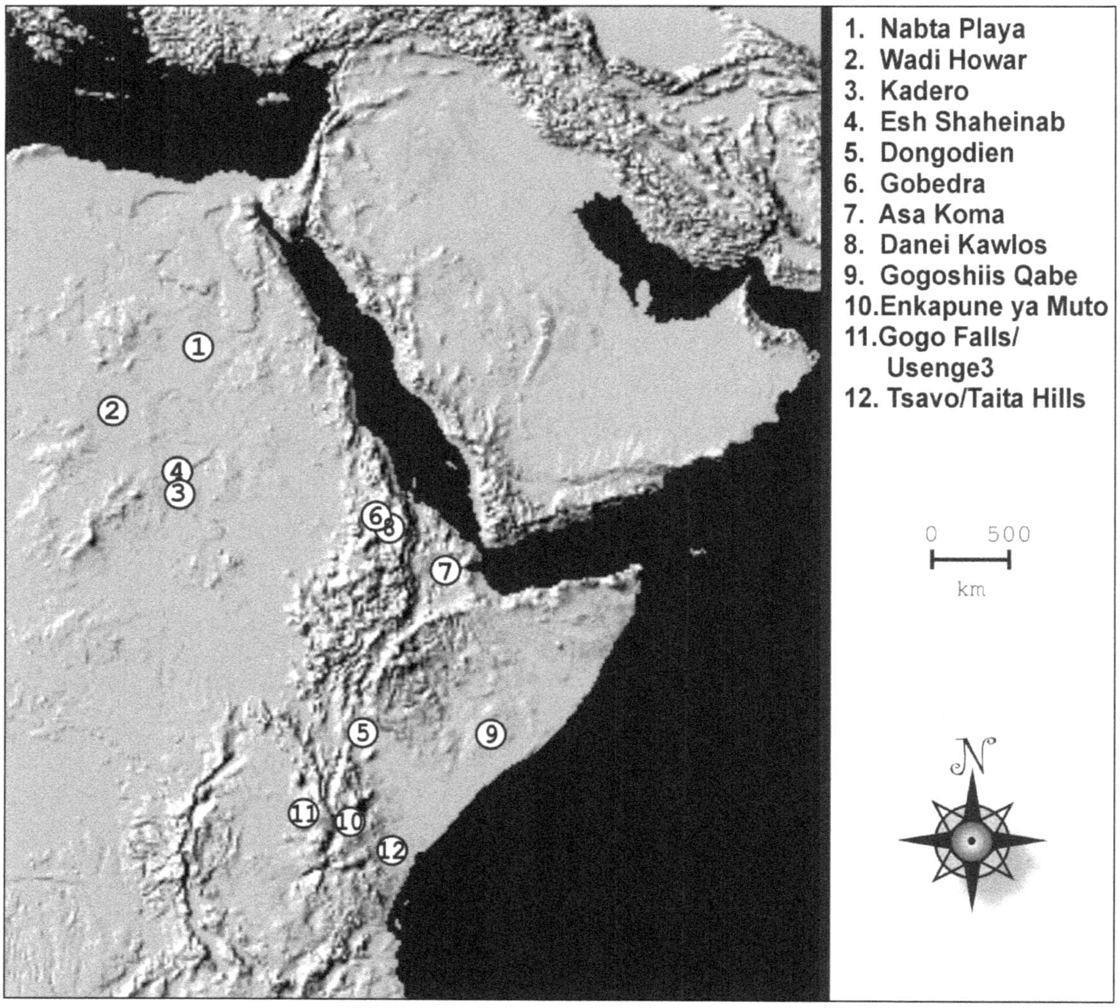

Figure 6.1. Locations of Early Domesticated Animals in Northeastern Africa

Howar region (Hoelzmann *et al.* 2001). Near the confluence of the Blue and White Niles, cattle appear at the sites of Kadero and Esh Shaheinab 3000 BC (Krzyzaniak 1978, 1980, Marks *et al.* 1985). Between 4000 and 2000 BC, domesticated animals appear in increasingly higher quantities in faunal assemblages from archaeological sites in the region (Hoelzmann *et al.* 2001, Krzyzaniak 1978, 1980), indicating a general shift in regional subsistence strategies concurrent to increasing cultivation of sorghum (Haaland 1995). In Wadi Howar, archaeological sites tend to be situated on the edges of small lakes in grassland ecosystems and demonstrate increasingly ephemeral occupations concurrent to falling precipitation levels (Hoelzmann *et al.* 2001).

Domesticated cattle are documented at the site of Dongodien adjacent to Lake Turkana dates ca. 2000 BC (Barthelme 1985, Owen *et al.* 1982). Ceramic evidence has been used to argue that domesticated animals were introduced allochthonously into eastern Africa from the Upper Nile River Valley (Barthelme 1985, Marshall *et al.* 1984). Stylistically, decorations and body styles on ceramic sherds recovered from the Lake Turkana region bear increasing resemblance to Nile Valley and eastern Saharan pottery after evidence for the presence of domesticated animals is confirmed >2000 BC. However, it remains unclear whether pastoral people migrated from the north while fleeing a desiccating Saharan environment (Barthelme 1985, Gifford-Gonzalez 1998:189, Marshall *et al.* 1984) or if livestock was raided from established pastoral communities (Marshall 1994, 2000). In either event, an economy that relied substantially on domesticated cattle and other livestock takes root in the Lake Turkana Basin after 2000 BC and pastoralism remains the central focus of the subsistence economy of that region to this day (McCabe 1994, 2006, Robbins 2006).

Early pastoral communities in the Horn of Africa are generally part of a broad-based agro-pastoral-forager complex and are situated in highland rockshelter environments (Phillipson 1993, 2005). Phillipson (1977) records a "domesticated ox" from an archaeological stratum at Gobedra Rockshelter, northern Ethiopia with an uncalibrated radiocarbon date of "856 BC ± 53 years" [*sic*]. Domesticated cattle are recorded from the site of Asa Koma in Djibouti dated to ca. 1800 BC along the Dagadle River ca. 45 km east of Lake Abbe (Lesur 2004). Domesticated cattle are also recorded at 1500 BC from the highland site of Danei Kawlos in northeastern

Ethiopia (Marshall and Negash 2002). Finally, at Gogoshiis Qabe Rockshelter at the Buur Heybe Inselberg in southern Somalia, domesticated cattle and ovicaprids are recovered in a pre-1500 BC archaeological stratum (Brandt 1986). Complex hunter-gatherers located within diverse topographic and resource rich ecosystems provided the backdrop for the "*acculturation*" of indigenous people within food production systems rather than direct migratory diffusion of pastoralism into the area (Lesur 2007). In southern Somalia and the Ethiopian Highlands, large portions of the population presently engage in agropastoralism with nomadic pastoralists inhabiting arid and semiarid lowlands (Desta and Coppock 2004, Mukhtar 1996).

The site of Enkapune ya Muto (EYM), located in the Central East African Highlands (2400 m.a.s.l.), possesses evidence for domesticated animals prior to 1000 BC. Ambrose (1998) reports faunal remains from a domesticated caprine at EYM that correlate to a stratigraphic layer with a ^{14}C age calibrated to 2620 – 2450 BC and a stratum dated to 1770 – 1610 BC has both cattle and an ovicaprid (see also Marean 1992). At Gogo Falls near Lake Victoria, Karega-Mũnene (2002:101) believes that domesticated caprines may be present in 1400 BC contexts, although the stratigraphic context relative to the radiocarbon age generated indicate that this is a maximum constraining age. Lane *et al.* (2007) document the occurrence of at least one ovicaprid in the faunal assemblage from Usenge 3 near Lake Victoria dating to ca. 1700 – 1400 BC. Prior to 1000 BC, domesticated animals are found only in limited numbers with wild taxa comprising the majority of the faunal assemblages at the sites tested south of Lake Turkana to this point (Gifford-Gonzalez 1998, Marshall 2000, Marshall and Hildebrand 2002).

After 1000 BC, numerous archaeological sites appear across equatorial Africa in which domesticated animals are the key component of the sites inhabitants' diet. These sites are widely distributed across central and western Kenya and Tanzania and remained the primary form of subsistence in the region until the introduction of horticulture to the region after AD 500 (Marshall and Hildebrand 2002). This archaeological tradition is typically referred to as the Pastoral Neolithic (PN), which is characterized by Later Stone Age microliths, ceramics and a subsistence economy with a heavy reliance on domesticated animals, most particularly cattle (Bower and Nelson 1978). PN cultures inhabit a wide range of environments across East Africa and are not a unified cultural manifestation, but a broad category in which similar technologies, subsistence activities and exchange networks are lumped (Ambrose 1984b, Bower 1988, 1991, Gifford-Gonzalez 1998, Marshall 2000).

BARRIERS TO DOMESTICATION

Bower (1991) asserts that there was a "trickle" of domesticated animals into eastern Africa south of Lake Turkana prior to 1000 BC. After 1000 BC, domesticated animals "splashed" into the region with faunal assemblages showing a high proportion of cattle and caprines relative to wild animals (Bower 1991). Two primary hypotheses have been advanced to explain the slow spread of domesticated animals south of Lake Turkana between 2000 and 1000 BC. One theory suggests that climatic conditions in East Africa >2000 BC grew increasingly arid, discouraging pastoralists from grazing their animals far afield from a permanent water source (Ambrose 1982, 1984a, 1984b). Another theory argues that livestock diseases such as *trypanosomiasis* ("African sleeping sickness"), Foot and Mouth disease, East Coast Fever, Rift Valley Fever and Bovine Malignant Catarrhal Fever were pervasive in the region thus restricting the range available for pastoralists to exploit (Gifford-Gonzalez 1998:190–195, 2000, Smith 1984). Finally, the importance of cultural factors (e.g. taboos, mores, etc.) must be considered as a potential barrier to a community considering a reorganization of their subsistence economy.

Regional paleoclimatic proxy data indicates that from 4000 to 1000 BC, the climate in East Africa transitioned from pluvial to more arid conditions. Regressions in numerous East African lakes are recorded after 4000 BC (Beuning *et al.* 1997, Butzer *et al.* 1972, Gasse 2000, 2002, Gasse and Van Campo 1994, Owen *et al.* 1982, Richardson 1972, Ricketts and Johnson 1996, Russell and Johnson 2005) indicating that a regional reduction in precipitation was occurring during the time of the earliest detectable cattle herding occupations across East Africa. Evidence from Lake Abbe and Lake Abyata, Ethiopia show precipitous declines in lake levels after 2000 BC with incremental recharges after 1000 BC (Gasse and Van Campo 1994). Multivariate sediment cores from Lake Edward on the Congo-Uganda border detects a pluvial period centered around 1200 BC following a drought at 1600 BC and preceding a drought at 800 BC (Lærdal *et al.* 2002, Russell and Johnson 2005). Episodic drought and pluviation is likewise noted in δ^{18}O records from calcite extracted from sediment cores in Lake Turkana (Johnson 1996, Johnson *et al.* 1991, Ricketts and Johnson 1996). A sediment core from Kiluli Swamp, Mount Kenya reflects drier than present conditions from about 2000 to 200 BC with formation of wetlands thereafter until AD 1500 (Olago *et al.* 2003). Pollen sequences from Lake Victoria show a progressive decline of semi-deciduous forest with advances in *Capparidaceae* and *Gramineae* arid-adapted grassland species from 2100 until 1000 BC with temporary readvances of semi-deciduous forests in the basin from 1000 until 200 BC (Ssemmanda and Vincens 2002). Finally, a sediment core from Lake Naivasha indicates that the water body was almost completely desiccated at 1000 BC following a high stand that ended around 3600 BC (Butzer *et al.* 1972).

Other regional climate proxies reflect a shift to more arid conditions focused around 2000 BC Ambrose and Sikes (1991) detect a 300 m vertical advance in savanna (C_4) grasses at the expense of C_3 canopy during the Middle to Late Holocene as measured by a series of elevation-dependent transects in the Mau Escarpment in the Kenyan Rift Valley. Additionally, a strong spike in eolian dust trapped in ice cores on Mount Kilimanjaro is dated to 2000 BC, and indicates a short, but severe regional drought (Thompson *et al.* 2002). The aggregate evidence indicates that the period between 2000 and 1000 BC was regionally drier than was witnessed during the middle Holocene. Unfortunately, the resolution of climate proxies presently available in the region is not sufficient to isolate the effects of precipitation on specific vegetation regimes, which means we must also look at archaeological data to inform aspects of the environmental record. For example, ovicaprids consume 82 ml H_2O/kg/24hr, while humpless cattle breeds (e.g. *B. taurus*) consume an average 161 ml H_2O/kg/24hr compared to 75 ml H_2O/kg/24hr for the humped *B. indicus* cattle breeds (MacFarlane *et al.* 1971). With this equation, we can assume that archaeological sites with evidence for domesticated *B. taurus* cattle would necessarily be close to a water source, whereas sites with *B. indicus*[1] and ovicaprids could potentially be more distant from water (Bundt 1979).

The view that epizootic diseases were barriers to the entry of domesticated animals into equatorial Africa is primarily advanced in Gifford-Gonzalez (1998, 2000) and Smith (1984). This theory should not be viewed as necessarily antithetical to the notion that a desiccating climate was an obstacle for migrating pastoralists or animals to overcome. Climate change can catalyze the evolution of epizootic pathogens and cause the ranges of diseases expand and contract with unpredictable alacrity (Bollig 2006:123, Harvell *et al.* 2002). For example, trypanosomiasis presently represents a major epizootic challenge for pastoralists across Africa. Tsetse flies (*Glossina* sp.) are the vectors for the transmission of trypanosomiasis infections and thrive in warm and wet climates. Habitats for tsetse in East Africa generally occur in human disturbance zones and humid areas with ample tree cover for puparia development and shady areas for adults to rest (Pollock 1982a, 1982b, Rogers and Robinson 2004:140). The limit of distribution of tsetse flies is strongly correlated to areas >500 mm annual rainfall (Leak 1999:81). Tsetse flies do not flourish in temperature regimes <16° C and >32° C (Hargrove 2004). As such, many highland environments in East Africa provide poor habitats for tsetse species due to the low overnight temperatures and limited canopy in high altitude areas, whereas lowland riparian zones are well-suited for infestation. No single epizootic disease can be viewed as the catch-all barrier explaining the 1000-year sputter of domesticated animals from their Lake Turkana harbor because they evolve and disappear so quickly (e.g. Bird *et al.* 2007, Domingo *et al.* 2003). We must instead look at where domesticated animals first appear and eventually take root as a permanent fixture of the subsistence economy and evaluate whether a conscientious effort was made on the part of early herders to avoid contact with epizootic diseases.

[1] The introduction of *B. indicus* to eastern Africa is believed to have occurred >AD 700 (Epstein 1971, MacHugh *et al.* 1997) when transoceanic commerce between Africa and Asia was underway.

Alternatively, the 1000-year trickle of domesticated animals from the Turkana Basin into equatorial and the Horn of Africa may have been due to cultural factors independent of external environmental variables. Grassland foraging communities such as the Waata of the Tsavo, Kenya region (Ville 1995), the !Kung San of southern Africa (Jones *et al.* 1994, Lee 1979) and Hadza of Tanzania (Jones *et al.* 1992, Mabulla 2003, Marlowe 2002) were or remain reticent to adopt domesticates in spite of close contact with pastoral communities. Subsistence behavior is a product of a complex process of learning through cultural transmission and is imbued with mores and traditions that strongly connect individuals and communities together (Henrich and McElreath 2003). Cognitive behavioral studies suggest that the natural human penchant for maintaining equilibrium or 'stasis' in outlook and activity is a product of cultural conditioning ('behavioral reinforcement'), and frequently overrules any desire to change (Dowd 1999, Dowd and Courchaine 2002). Among the !Kung San and Hadza, cultural traditions continue to fiat the convenience of culturally assimilating into their agro-pastoral neighborhood. With this in mind, we must be aware that there may simply have not been a social or environmental impetus for the pure foraging communities to adopt pastoralism or for Turkana Basin pastoralists to migrate away from their known resource base between 2000 and 1000 BC. Additionally, linear one-size-fits-all models explaining transitions to domestication are not as fruitful as holistic approaches incorporating a broad array of data and theoretical perspectives (Verhoeven 2004).

PREVIOUS WORK AND GEOGRAPHICAL SETTING IN TSAVO NATIONAL PARK

Archaeological investigations undertaken in Tsavo East National Park in 2001 and 2004 augmented earlier studies conducted by Drs. Chapurukha M. and Sibel B. Kusimba (2000), Peter Thorbahn (1979), David Collett (1985) and Robert Soper (1966). In the present study, PN sites were found eroding from alluvial terraces exposed from fluvial downcutting occurring >AD 1000 (Wright *et al.* 2007). Dating of archaeological horizons was done by AMS ^{14}C ages of carbon-based materials and Infrared Stimulated Luminescence (IRSL) age control on sediments of the alluvial terraces that straddle the present-day channel of the Galana River.

The Galana River is comprised of laterally accreting point bars consisting primarily of silts and fine sands. The system is fed by seasonal runoff and springs located along the Yatta Plateau, from the Central East African Highlands in the vicinity of Nairobi and from Mount Kilimanjaro (Fig. 6.2). The Galana River is the only permanent water source in the Tsavo region and is estimated to run dry, or nearly so, once every three years (Wijngaarden and Engelen 1985:44). Periods of no water flow lasting >90 days occur once every 50 years (Wijngaarden and Engelen 1985:44). However, groundwater sources along the end-Tertiary metamorphic and sedimentary basement systems permanently feed

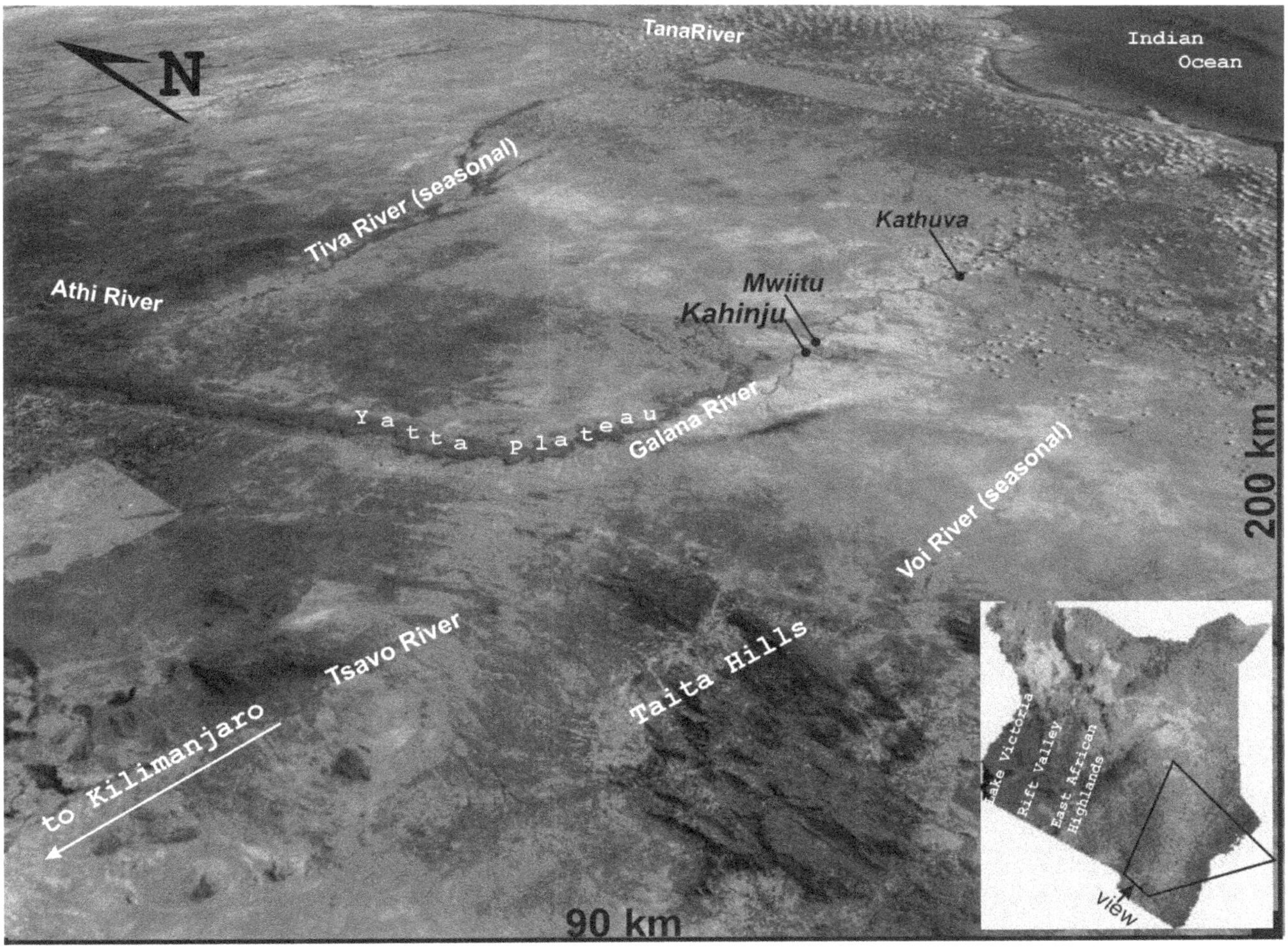

Figure 6.2. Satellite View of Taita Hills, Tsavo and Coastal Lowlands
(Image Rendered from Google Earth 2008)

riparian forest growth along the margins of the Galana River and among intermittent drainages within the catchment even during the dry years.

Distribution of tsetse flies within Tsavo is primarily restricted to riparian forests and dense *Acacia* sp. and *Commiphora* sp. tree cover (e.g. Majiwa *et al.* 1993, Mihok *et al.* 1992). Through intensive land management and modern veterinary intervention, the trypanosome parasite along the Galana River can be managed, but even with these techniques complete success in suppressing incidences of disease has not yet been achieved (Baylis and Nambiro 1993, Dolan *et al.* 1992, Irungu *et al.* 2002, Makumi *et al.* 2000). We can therefore assume that prehistoric pastoralists of the Galana River basin would have reckoned with the potentially devastating impact of tsetse flies on their herds, and weighed the potential for loss of livestock against the benefits incurred from settling along this fluvial system.

ARCHAEOLOGICAL METHODS AND ANALYSIS

In 2001 and 2004, 15-m interval pedestrian reconnaissance was undertaken within Tsavo East National Park, Kenya covering 5234 ha. The survey documented 81 archaeological sites, most of which were cairns (n=68), but six were PN-era sites located adjacent to the banks of the Galana River (Fig. 6.2, Wright 2005a). Subsurface testing focused on three of the sites identified during survey, and trenches were placed non-randomly in locations deemed highly probable to yield archaeological deposits and avoid incidental contact with wildlife.

Three sites, Kahinju (HeJt9), Kathuva (HeJu5) and Mwiitu (HeJt35) were test excavated to culturally sterile levels. Kahinju was by far the largest site in terms of exposure and was tracked 400 linear meters along the scarp face of the Galana River. The sites of Kathuva and Mwiitu were less exposed by fluvial downcutting and therefore estimating their relative sizes was a difficult task. However, surface finds for both sites ranged between 200 and 300 linear meters. Artifact distributions on the surface of the sites appeared to have similar material characteristics, in terms of decorative ceramic attributes, stone tool raw material selection and tool production techniques in addition to the relatively similar stratigraphic positioning of artifact erosion locales. The excavation strategy was oriented to obtain a geochronology on the basis of being able to reconstruct subsistence patterns independent of dietary preferences, technological capacity, timeframe and intensity of

Figure 6.3. Excavating Step Trenches into Terraces at Kathuva

occupations that could be compared with secondary environmental proxy data. To this end, field research employed the following techniques:

1. Exposing profile views of eroded terraces and digging trenches into terrace sediments to recover in situ archaeological artifacts. This task was accomplished by excavating step trenches into the terrace stratigraphy and allowed for controlled excavations using native soil units (Fig. 6.3).
2. One 1-x-2 m test unit dug behind the river bluff at each site to collect artifacts in situ, assist in reconstructing the prehistoric landscape and verify the vertical provenience of artifacts excavated in geological steps are contemporaneous with artifacts now buried deep below the ground surface.
3. Collection of Optically Stimulated Luminescence (OSL) and AMS radiocarbon (^{14}C) samples for absolute dating of levels.
4. Full collection and provenience of cultural material in situ when possible.
5. Sieving through 6.25-mm screen and full collection of artifacts in all levels where human occupation was suspected or confirmed.
6. Collection of flotation samples in confirmed cultural levels and control samples in non-cultural levels.
7. Detailed site maps using a David White LT8-300 Level-Transit and stadia rod at Kathuva and a total station provided by the British Institute in East Africa produced the maps for Kahinju and Mwiitu.

Table 6.1. Infrared stimulated luminescence ages on fine-grained (4-11 μm) polymineral extracts from alluvial sediments recovered at the sites of Kahinju (Kh), Kathuva (Kt) and Mwiitu (Mw), Kenya

Laboratory number	Site & unit	Th (ppm)[a]	U (ppm)[a]	K_2O (%)[b]	A value[c]	H_20 (%)	Dose rate (mGy/yr)[d]	Equivalent dose (Gy)[e]	IRSL age
UIC921	Kt 2	5.3 ± 0.9	3.0 ± 0.4	2.46 ± 0.02	0.07 ± 0.01	15±5	3.86 ± 0.19	13.30 ± 0.12	1450 BC ± 290
UIC943	Kh 5	11.2 ± 1.4	2.0 ± 0.5	2.55 ± 0.02	0.08 ± 0.01	10±5	4.56 ± 0.23	13.51 ± 0.19	960 BC ± 280
UIC944	Kh 4	9.0 ± 1.3	3.1 ± 0.5	2.43 ± 0.02	0.07 ± 0.01	15±5	4.26 ± 0.21	12.78 ± 0.21	1000 BC ± 260
UIC1006	Kh-sterile	6.5 ± 0.8	2.1 ± 0.3	2.62 ± 0.02	0.10 ± 0.01	5±2	4.49 ± 0.22	4.22 ± 0.06	AD 1060 ± 70
UIC1007	Kh 4	8.6 ± 1.2	1.9 ± 0.4	2.68 ± 0.02	0.04 ± 0.00	10±5	3.97 ± 0.20	12.56 ± 0.27	1160 BC ± 280
UIC1035	Kh 4	6.3 ± 0.9	2.4 ± 0.4	2.74 ± 0.02	0.05 ± 0.01	30±5	3.55 ± 0.18	11.37 ± 0.14	1200 BC ± 260
UIC1036	Kh 2	7.1± 0.9	1.9 ± 0.3	2.58 ± 0.02	0.05 ± 0.01	30±5	3.14 ± 0.16	16.21 ± 0.20	3160 BC ± 410
UIC1037	Kh 2	6.4 ± 0.8	2.1 ± 0.3	2.44 ± 0.02	0.06 ± 0.01	30±5	3.12 ± 0.15	15.52 ± 0.23	2970 BC ± 400
UIC1038	Kh 1	7.2 ± 0.8	1.9 ± 0.3	2.47 ± 0.02	0.06 ± 0.01	30±5	3.15 ± 0.16	18.76 ± 0.24	3960 BC ± 480
UIC1068	Kh 3	10.1 ± 1.3	2.2 ± 0.5	2.53 ± 0.02	0.08 ± 0.01	30±5	3.22 ± 0.16	12.47 ± 0.11	1870 BC ± 330
UIC1069	Kt 3	7.4 ± 1.1	2.4 ± 0.4	2.73 ± 0.02	0.06 ± 0.01	15±5	4.15 ± 0.21	12.94 ± 0.16	1120 BC ± 270
UIC1070	Kt 2	9.0 ± 1.3	3.2 ± 0.5	2.42 ± 0.02	0.09 ± 0.01	15±5	4.72 ± 0.24	17.00 ± 0.13	1600 BC ± 320
UIC1071	Kt 1	8.1 ± 1.2	2.5 ± 0.4	2.70 ± 0.02	0.05 ± 0.01	35±5	3.36 ± 0.17	16.21 ± 0.26	2830 BC ± 390
UIC1153	Kt 4 or 5	5.8 ± 0.8	2.5 ± 0.3	2.58 ± 0.02	0.07 ± 0.01	10±5	4.25 ± 0.21	7.31 ± 0.07	AD 280 ± 150
UIC1392	Mw 3	10.5 ± 1.5	2.3 ± 0.5	2.45 ± 0.02	0.11 ± 0.01	5±2	5.08 ± 0.25	5.18 ± 0.08	AD 980 ± 80
UIC1393	Mw 2	8.1 ± 1.0	2.1 ± 0.4	2.31 ± 0.02	0.15 ± 0.01	15±5	4.36 ± 0.22	13.17 ± 0.14	1020 BC ± 270

[a] Thick-source alpha-count rate; U and Th concentrations calculated from alpha count rate, assuming secular equilibrium.
[b] Percent K determined by ICP-MS at Activation Laboratories, Ontario, Canada.
[c] Measured alpha efficiency factor (a) as defined by Aitken and Bowman (1975).
[d] Dose-rate includes a contribution from cosmic radiation from Prescott and Hutton (1994).
[e] Optical stimulation was by infrared emissions (880 ± 80 nm) from a ring of 30 diodes with an estimated energy delivery of 17 mW/cm^2. The resultant blue emissions (with three, 1-mm-thick Schott BG-39 and three, 1-mm-thick Corning 7-59 glass filters that transmit; <10% transmission below 390 nm and <10% above 490 nm) from the sediments were measured. Samples were excited for 90 s, and the resulting IRSL signal was recorded in 1 s increments. The IRSL signal was measured at least 1 day after preheating at 160 °C for 5 hr. Errors are at ±1 one sigma and calculated by averaging the errors across the time range.

Table 6.2. Accelerator mass spectrometry (AMS) radiocarbon ages from archaeological sites, Galana River, Kenya

Sample number	Site & unit	Material	$\Delta^{13}C$ ‰	Laboratory ^{14}C age, yr ± 1σ	2σ age calendar yr.[a]
AA51442	Kh 5	wood	-19.4	2527 ± 84	810–410 BC
AA51443	Kh 4	charcoal	-25.7	2987 ± 57	1390–1050 BC
AA51444	Kh 3	charcoal	-23.7	3423 ± 55	1890–1610 BC
AA51445	Kh 4	charcoal	-24	2973 ± 52	1320–1020 BC
AA51447	Kt 4	charcoal	-28.1	1486 ± 39	AD 530–650
AA51448	Kt 4	charcoal	-27	1526 ± 61	AD 420–640
AA51449	Kt 5	charcoal	-25.4	1671 ± 49	AD 250–440

[a] Calendar correction of laboratory ^{14}C ages following Reimer *et al.* (2004)

The sediments at all sites are ferralic inceptisols using the USDA nomenclature (Soil Survey Staff 1999, Wright 2005a) or calcic Cambisols using FAO (Wijngaarden and Engelen 1985). Tables 6.1 and 6.2 present the composite age estimates generated from three of the six PN archaeological sites identified in 2001 and 2004. Detailed analyses of the geochronology of these sites are provided elsewhere (Wright *et al.* 2007). For the purposes of the present discussion, age control from these locations demonstrate that human occupations of the floodplain of the Galana River occur as early as about 4000 BC and continued successively until ca. AD 700. The evidence for this settlement pattern comes from multiple overlaying strata of alluvium and anthropogenic colluvium separating archaeological horizons at the sites tested (Fig. 6.4). After ca. AD 1000, a period of fluvial incision ensues and settlements are not found proximal to the active fluvial channel, but are instead more dispersed and ephemeral throughout the surrounding open plains.

Artifact analysis was conducted in the National Museums of Kenya, Nairobi and included sorting, washing, weighing and cataloguing of >10,000 individual finds. The diagnostic archaeological assemblages from the sites tested show little overall variability in the types of artifacts manufactured and used through time (Wright 2005a, 2007). Early occupation levels (before 1700 BC) have few cultural remains but include non-diagnostic ceramic sherds, microliths manufactured on quartz and faunal remains from (presumably) wild bovids. Later occupations (after 1700 BC) include ceramics typical of pastoral settlements elsewhere in East Africa, limited evidence for exchange with the Swahili Coast by AD 700 (Wright 2005b) and microliths manufactured on quartz and chert. Fauna from the post-1000 BC archaeological strata in Tsavo show that wild animals remain an important component of the subsistence regime into the late prehistoric period. This phenomenon stands in contrast to faunal assemblages from sites excavated in the central and western portions of East Africa where there is evidence for an almost wholesale adoption of domesticates after 1000 BC with the near absence of evidence for predation on wild animals (Bower 1991, 1997, Gifford-Gonzalez 1998, Marshall 1990, 2000, Robertshaw 1990).

Faunal analysis of *in situ* archaeological deposits focused on identifying minimum numbers of animal taxa present at the Tsavo sites. Faunal elements were recovered and sorted in the field by site, unit and depth. Paul Kyenze Watene, a senior staff archaeologist at the National Museums of Kenya in Nairobi, identified and classified diagnostic faunal elements which were later grouped into Minimum Numbers of Individuals (MNI) present (Klein and Cruz-Uribe 1984, Reitz and Wing 1999) at each site tested. MNI of taxa identified at three of the Tsavo sites are presented in Table 6.3 by time period due to the relatively small diagnostic faunal assemblage at each site. Broad agreement of luminescence and AMS ^{14}C ages within 1 σ allow for a time-staggered view of the Middle to Late Holocene settlement sequence.

Faunal elements have been recovered from archaeological deposits indicating that at least two *B. taurus* individuals were present at the site of Kahinju dating to ca. 1700 BC (Fig. 6.5, Table 6.3). These data indicate that the spread of herding techniques in eastern Africa extended into the riparian areas of the Kenyan Coastal Lowlands region earlier than was previously thought. Although these are significant finds for regional understandings of the timeframe of the introduction of domesticated animals into eastern Africa, they elucidate broad patterns in the settlement ecology of early pastoralists coeval to the presence of a generally more-arid-than-present landscape between 2000 and 1000 BC. The fauna assemblage reflects a 4700+ year sequence in which settlement along the banks of the Galana River included a diverse subsistence economy dedicated to the predation of resources within the riparian zone. The presence of numerous faunal elements not easily transported long distances such as vertebrae, ribs and skull fragments suggests that animals were procured close to habitation areas (Wright 2005a). The faunal assemblage from Tsavo reflects the "schlep effect," whereby elements that are the easiest to carry while yielding the most amount of meat are increasingly likely to be returned to a base camp as the distance one has to travel with meat increases (Binford 1981, Bunn *et al.* 1988, O'Connell and Hawkes 1988, Redding 1994). The >100 kg terrestrial and freshwater mollusks in the faunal assemblage, many of which were burned with cut marks,

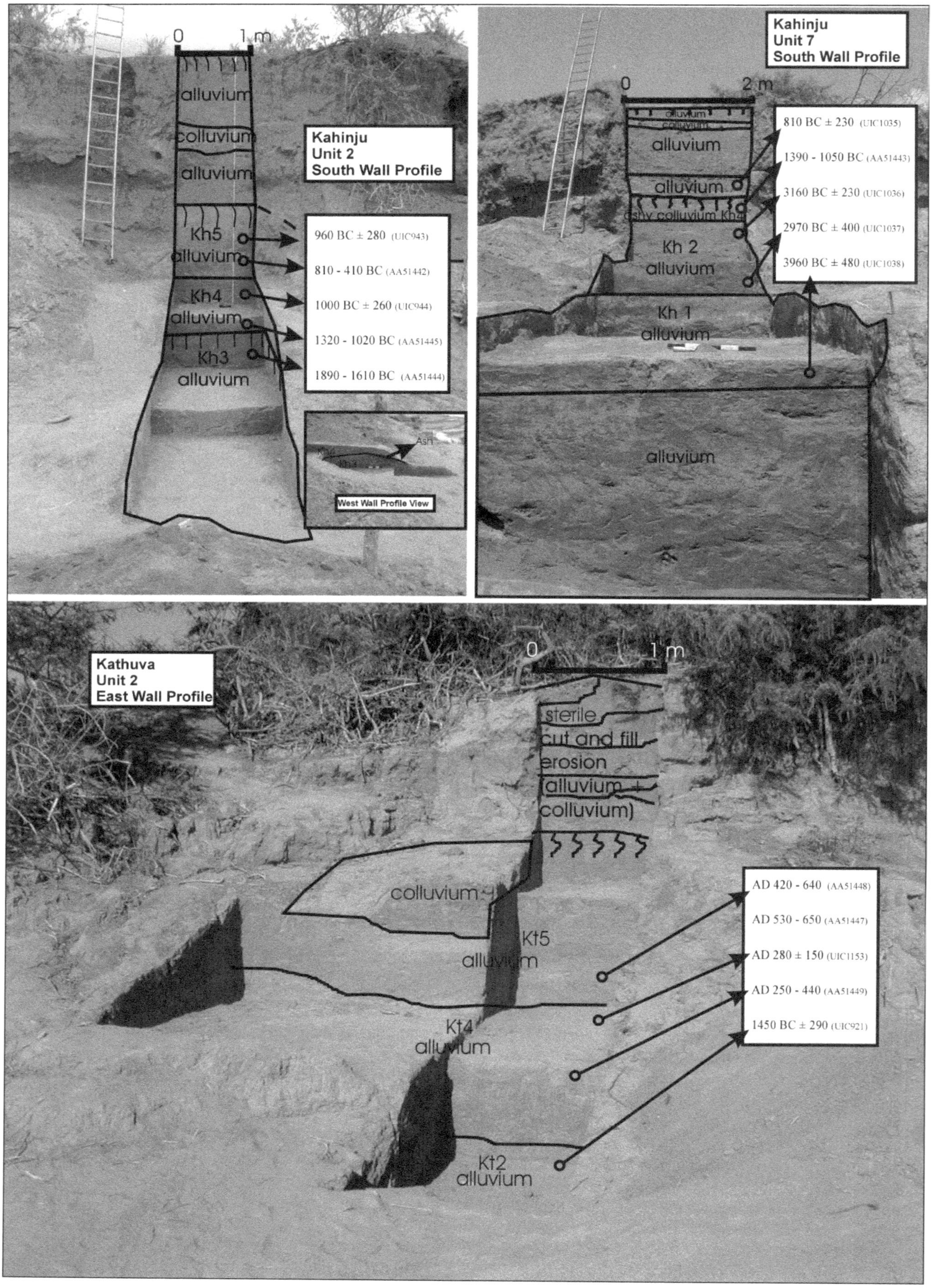

Figure 6.4. Selected Stratigraphic Sections from Kahinju and Kathuva

Table 6.3. Complete identifiable faunal counts (MNI) from Kahinju and Kathuva organized by timeframes of occupation. Faunal elements organized into bovid weight classes and categorized following Brain (1981: Table 2) by Paul Kyenze Watene at the National Museums of Kenya, Nairobi

Occupation	taxa	MNI
Kahinju – ca 4000 BC	Bovid 2	1
Kahinju – ca 4000 BC	Bovid 3	1
Kahinju – ca 4000 BC	*Madoqua* sp.	1
Kahinju – ca 3000 BC	Bovid 2	1
Kahinju – ca 3000 BC	Bovid 4	1
Kahinju – ca 1700 BC	*Aepyceros melampus*	1
Kahinju – ca 1700 BC	*Bos taurus*	1
Kahinju – ca 1700 BC	Bovid 2	1
Kahinju – ca 1700 BC	Bovid 3	1
Kahinju – ca 1700 BC	Bovid 4	1
Kahinju – ca 1700 BC	*Hippopotamus amphibius*	1
Kahinju – ca 1700 BC	*Leo leo*	1
Kahinju – ca 1000 BC	Bovid 2	1
Kahinju – ca 1000 BC	Bovid 4	1
Kahinju – ca 1000 BC	*Equus* sp.	1
Kahinju – ca 1000 BC	*Gazella granti*	1
Kahinju – ca 1000 BC	lizard	1
Kahinju – ca 1000 BC	suidae	1
Kahinju – ca 1000 BC	*Syncerus caffer*	1
Kahinju – ca 600 BC	*Bos taurus*	1
Kahinju – ca 600 BC	Bovid 2	1
Kahinju – ca 600 BC	Bovid 3	1
Kahinju – ca 600 BC	Bovid 4	1
Kahinju – ca 600 BC	*Madoqua* sp.	1
Kahinju – ca 600 BC	*Equus* sp.	1
Kahinju – ca 600 BC	fish	1
Kahinju – ca 600 BC	*Hippopotamus amphibius*	1
Kathuva – ca 1500 BC	*Aepyceros melampus*	1
Kathuva – ca 1500 BC	*Alcelaphinae* sp.	1
Kathuva – ca 1500 BC	Avian	1
Kathuva – ca 1500 BC	*Bos taurus*	1
Kathuva – ca 1500 BC	Bovid 1	1
Kathuva – ca 1500 BC	Bovid 2	2
Kathuva – ca 1500 BC	Bovid 3	1
Kathuva – ca 1500 BC	Bovid 4	1
Kathuva – ca 1500 BC	*Crocodylus nilotocus*	1
Kathuva – ca 1500 BC	*Madoqua* sp.	1
Kathuva – ca 1500 BC	fish	1
Kathuva – ca 1500 BC	monkey	1
Kathuva – ca 1500 BC	*Syncerus caffer*	1
Kathuva – ca 1500 BC	tortoise	1
Kathuva – ca 1100 BC	Bovid 2	1
Kathuva – ca 1100 BC	*Madoqua* sp.	1
Kathuva – ca AD 450	Bovid 2	1
Kathuva – ca AD 450	Bovid 3	1
Kathuva – ca AD 450	*Madoqua* sp.	2
Kathuva – ca AD 550	*Bos taurus*	1
Kathuva – ca AD 550	Bovid 1	1
Kathuva – ca AD 550	Bovid 2	1
Kathuva – ca AD 550	Bovid 3	1
Kathuva – ca AD 550	*Madoqua* sp.	2
Mwiitu – <1000 BC	Bovid 1	1
Mwiitu – 1000 BC–AD 1000	*Neotragus moschatus*	1
Mwiitu – 1000 BC–AD 1000	rodent	1

Figure 6.5. Mandible of *Bos taurus* from Kh3 Occupation Horizon

is further indication of a localized resource procurement strategy.

The ceramic assemblage can be generally characterized as concordant with Narosura tradition ceramics found across East Africa during the PN (Bower *et al.* 1977, Wandibba 1980). Sherds were generally from large, straight-sided or globular jars and open bowls, although the majority of the finds were too fragmentary to be diagnostic of a specific shape (Fig. 6.6, Fig. 6.7). Overall, the assemblage reflects general cultural connections with contemporaneous pastoral communities located in the Central Rift Valley, East African Highlands and Lake Victoria Basin. The sizes of the reconstructable vessels also show investment in the production or acquisition of large, bulky ceramics that would have been difficult to transport long distances.

DISCUSSION

Why did the southward spread of domesticated animals largely stall in the Lake Turkana Basin for ca. 1000

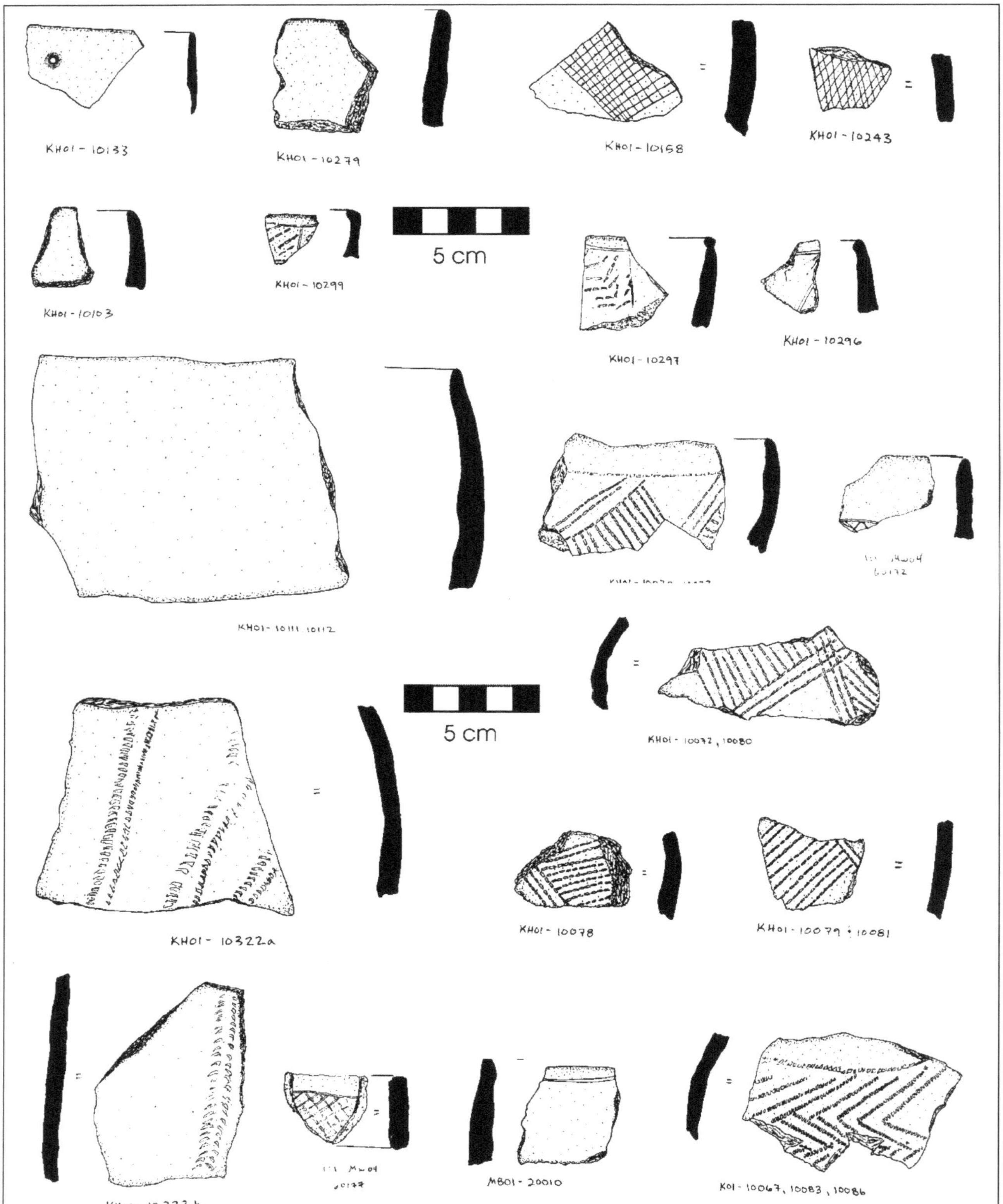

Figure 6.6. Selected Ceramic Sherds from Tsavo

years? Archaeological evidence from equatorial Africa and the Horn indicates that limited numbers of domesticated animals did, in fact, trickle afield as Bower (1991) indicates. Prior to 1000 BC, predation on wild plants and animals was the primary component of the subsistence regime, however a limited number of archaeological sites in Ethiopia, Somalia and Kenya had small numbers of ovicaprids and cattle. The hallmark of early pastoralism in eastern Africa was the heterogeneity of resources that were exploited. This stands in stark contrast to the post-1000 BC facies where domesticated animals, and most particularly cattle, dominate faunal

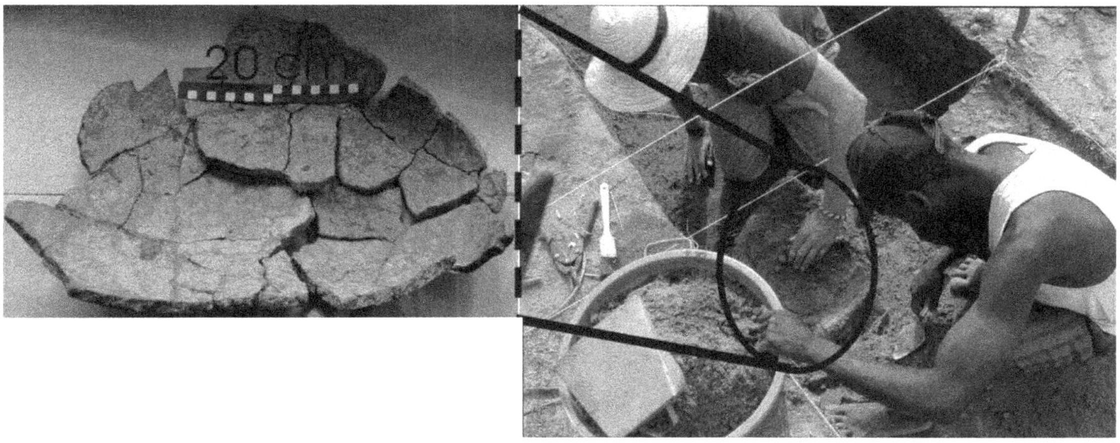

Figure 6.7. Large pot excavated from Kahinju

assemblages and presumably the diets of prehistoric people were more centrally organized around pastoralism (Gifford-Gonzalez 1998, Marshall 1990, 2000). What changed after 1000 BC to encourage domesticated animals to spread from Lake Turkana either by raiding (Marshall 1994), trade or migration/diffusion?

We must first review the context of the sites in which domesticated animals are found prior to 1000 BC. Sites in the Horn and at EYM in Central Kenya are in rockshelters in which resource variability along vertical transects would have been possible even in the most arid of conditions. The historically documented case of the Basseri of southwestern Iran serve as a prime example of how vertical transhumance buffers against resource scarcity. The Basseri are documented as annually endeavoring on long distance migrations between lowland winter and highland summer residences moving their animals >500 km annually over extreme terrain (Barth 1956, 1961, 1964, Digard 1981, Irons 1972, Salzman 2004). This trek was most famously captured by filmmaker Merian C. Cooper in the 1925 documentary *Grass: A Nation's Battle for Life*. Transhumance of this nature provides diversity of forage for grazing ruminants and ensures that overexploitation of a specific area does not occur. Additionally, desiccation events tend to be localized—focused on regions or climes. During such events, resource zones in highlands tend to constrict, but vegetation regimes tend to remain intact and exploitable when lowland resources are scarce. Modern day Rendille and Samburu pastoralists similarly utilize the slopes of the Mt. Marsabit and Mt. Kulal to feed their animals during periods of drought (Fratkin 1986, McCarthy and di Gregorio 2007). Vertical transhumance can provide a needed resource buffer for pastoralists who inhabit marginal ecosystems, but as fertile land is increasingly utilized for agriculture, access to these traditional 'bleeder valve' regions are now a source of conflict (Fratkin 1994).

Due to their generally cooler temperatures, highland environments are not especially fertile grounds for the evolution or spread of epizootic diseases. Lowland, wet, tropical environments that play host to a wide variety of insects are more commonly so-called 'hot zones' for the emergence of new diseases (Peterson 2008). However, the most common denominator for propagation of epizootic diseases is high numbers of animals in close quarters with each other and in contact (directly or indirectly) with wild animals (Cleaveland *et al.* 2001). In the case studies of early domesticated animals in highland environments, rockshelters typically house animals and humans in close quarters. In rockshelters excavated in the Taita Hills in eastern Kenya, animal pens are situated in adjacent or the same rock formations as domestic living areas (Kusimba and Kusimba 2007). No direct evidence of close-quarter animal pens has been recovered for early domesticated animals in the Horn or East Africa, but rockshelters are generally not conducive to dispersed sleeping accommodations.

The sites located near Lake Victoria and in Tsavo have the common denominator that a permanent source of water was available at the times in which the sites were occupied. The fluvial aggradational sequence along the Galana River in Tsavo is recorded to at least 4000 BC indicating that the river was present at the time of the sites' occupation. Even in periods of profound aridity, riparian zones persist in Tsavo (Wijngaarden and Engelen 1985) and desiccation of Lake Victoria is not recorded during the Middle or Late Holocene (Johnson *et al.* 2000, Stager and Johnson 2000). Given the evidence for regional aridity in eastern Africa ca. 2000 BC presented above, it should not be surprising that early cattle herders were situated adjacent to water bodies or in highland environments where vertical transhumance could be used to buffer against climate perturbations.

The idea that early pastoral sites are located within so-called 'green corridors' is not new (Ambrose 1984a, 1984b, Onyango-Abuje and Wandibba 1979). The 2000 BC aridification of the East African landscape has been documented in the paleoclimatic record since the 1960s (Butzer *et al.* 1972, Hamilton 1982, Owen *et al.* 1982, Richardson 1966, 1972). The impact of this event should not be oversimplified to the extent that arid and pluvial periods are cyclic, localized and annually variable. Although the overall precipitation budget for the region was lower than during the middle Holocene (and perhaps even lower than today), there is no evidence for a regional transformation to a xeric or desert landscape on the scale of what occurred in the eastern Sahara from 6000 to 2000 BC.

That said, we must evaluate the nuances of the data of early cattle occurrences in eastern Africa in order to assess the disease and culture hypotheses as potential contributing factors for the "1000-year trickle." First, when cattle appear in faunal assemblages from 2000 to 1000 BC, they represent a small percentage of the MNIs analyzed (Bower 1997, Gifford-Gonzalez 1998, Marshall 2000). Wild game remains the mainstay of the meat consumed during this period. After 1000 BC, faunal assemblages analyzed from many archaeological sites in central and western Kenya East Africa reflect the near absence of wild animals as a meat source (Bower 1991, 1997, Gifford-Gonzalez 1998, Marshall 1990, 2000, Robertshaw 1990), although this pattern is not borne out everywhere (Ambrose 1984b, Dickson *et al.* 2004, Gifford-Gonzalez 2003, Kusimba 2003, Mehlman 1989, Mutundu 1998, Robbins *et al.* 1977). In the instances of western Nubia (Hoelzmann *et al.* 2001), the middle Nile Valley (Hassan and Gross 1986) and Lake Turkana (Barthelme 1985), pastoralism entered the region as a complete cultural complex. Archaeological sites in equatorial Africa dated from 2000 to 1000 BC do not exhibit wholesale supplanting of indigenous tool technologies, ceramics or subsistence strategies indicating that a very different phenomenon was occurring. Marshall (1994) suggests that raiding may have been the mechanism that first brought cattle into equatorial Africa.

Following 1000 BC, the Pastoral Neolithic (PN) cultures, characterized by heavy dependency on domesticated animals, ceramic and stone tool use, are a sea change as they "splashed" into the East African Highlands, Rift Valley and Lake Victoria Basin (Bower 1991). Linguistic evidence indicates that Khoi-speaking hunter-gatherers were pushed to the periphery by the expanding Cushitic-speaking pastoralists (Ehret 1998). In many instances, though, it seems likely that the indigenous populations that the Cushites were encountering as they pushed south of the Lake Turkana Basin already had limited experience with handling domesticated animals and adopting a pastoral lifestyle may not have been as drastic a paradigm shift as previously thought. The so-called Elmenteitan phase settlements recognized primarily in the western Rift Valley inhabited rockshelter environments, hunted small- and medium-sized game and kept mostly small ruminants as livestock (Ambrose 1984b). A similar pattern has been noted in the Taita Hills (Kusimba *et al.* 2005) and in Laikipia, central Kenya (Dickson *et al.* 2004) also dated to >1000 BC. Along the Galana River in Tsavo, the diagnostic faunal assemblage is relatively low, but evidence for a radical transformation of the subsistence economy exclusively geared toward herding domesticated animals does not appear to have occurred >1000 BC. Instead, the focus of the communities along the Galana River remains on procuring endoaquatic resources based on the large ceramic assemblage, as well as types of animals and faunal elements found in the archaeological record. Ceramics and elements such as vertebrae and skull fragments are not easily transportable long distances and suggest a localized resource procurement strategy.

Though murky, due to the low data set of sites available for analysis dating from 2000 to 1000 BC, the pattern appears to demonstrate a period of 'experimentation' with domesticates in equatorial and the Horn of Africa that preceded a wider-scale adoption of a transhumant pastoral lifestyle in large swaths of the region >1000 BC. Migrations of Cushitic pastoralists from the Lake Turkana Basin into equatorial Africa likely occurred based on the corpus of material culture and subsistence characteristics (Ambrose 1984a), but wholesale conquest and supplanting of indigenous people and subsistence traditions did not occur (Bower 1991). The same phenomenon is noted contemporaneously in the Horn of Africa (Lesur 2007). A threshold was crossed at 1000 BC preceded by 1000 years of slow permutations of domesticates south of the Lake Turkana Basin and eastward from the Nile Valley.

What role did disease potentially play in inhibiting the movement of domesticates southward and eastward from Lake Turkana? Increased incidences of epizootic diseases have been succinctly linked to climate change and anthropogenic landscape disturbance (Bollig 2006, Dukes and Mooney 1999, Patz *et al.* 2000, Purse *et al.* 2005), however the salience of linking modern impacts to prehistoric events is questionable given the rate and scale of modern disturbance. Climate change is, however, more generally associated with extreme weather events that can accelerate the spread of insect- and water-borne diseases (Epstein 2001). If the paleoclimatic data is correct and pastoralists ca. 2000 to 1000 BC were forced into highland and riparian zones in response, such areas can become suitable breeding grounds for epizootic vectors. Highland areas can become rife with disease due to the close-quartering of humans, livestock and frequent encounters with other ruminants. Riparian zones, generally speaking, are more prone to epizootic diseases due to festering fecal bacteria in slow water pools, changes in habitat and soil loss after cattle intensively inhabit a stream bank and close interaction of livestock with wild bovines (Ambrosia *et al.* 1989, Belsky *et al.* 1999). The fact that domesticated animals did not become the mainstay of human subsistence in Tsavo during prehistoric times may reflect the fact that this habitat is ill-suited for maintaining large herds of domesticated animals.

The archaeological evidence of very limited instances of livestock trickling away from the Nile River Valley and Lake Turkana into the Horn and equatorial Africa between 2000 and 1000 BC may be interpreted as either a series of failed attempts to introduce this new variant to a biologically unsuitable habitat (epizootic barriers) or as a hand-cuffed resource (culturally or environmentally) that took 1000 years to take root. What variable may have possibly changed after 1000 BC to allow for transhumant pastoralism to become the dominant mode of subsistence for PN people for thousands of years thereafter is not known. Although minor pluviation events are recorded following the 2000 BC desiccation event, the overall climate of eastern and particularly the Horn of Africa remains semi-arid in most of the places inhabited by pastoralists. The bimodal rainfall pattern generated from Indian Ocean monsoons has been present since the Younger Dryas (YD), but weakening since ca. 3500 BC (Hong *et al.* 2003, Overpeck *et al.* 1996). However, this weakening may have also corresponded with more linear responses to insolation forcing once the glacial boundary conditions subsided (Overpeck *et al.* 1996). Marshall (1990) posits that changes in bimodal rainfall were key to the origins of specialized livestock husbandry in East Africa.

Onyango-Abuje (1977, in Bower 1991) argues that epizootic barriers, such as trypanosomiasis, exist in present day East Africa, therefore their role in inhibiting prehistoric pastoralism is dubious. However, livestock diseases are moving targets and tracking the locations and impacts of modern pathogens is difficult even with modern technology. We cannot presume to know the potential impacts of prehistoric epizootic diseases with present understandings of paleopathology in archaeological assemblages. Perhaps the use of plant-based medicines to treat epizootics (Ole-Miaron 2003, Tabuti *et al.* 2003, van der Merwe *et al.* 2001) was introduced >1000 BC to enable livestock to travel from their Lake Turkana refuge. Perhaps the shift in settlement among PN people with large flocks of domesticated animals to open air sites >1000 BC (Marshall 1990) limited the exposure of cattle to epizootics. Gifford-Gonazalez (2000) and Smith (2005) suggest that trypano-resistant livestock selected for along the margins of the tsetse fly zone were crucial for the ultimate penetration of domesticated cattle into the southern latitudes. This process involves many generations of breeding and numerous fits and starts before the genetic formula allowed cattle to move safely into the tsetse corridors. Ultimately, these are matters of speculation at this point, but epizootic diseases as they occurred hand-in-hand with climate change must be viewed as potentially important inhibitors to the introduction of domesticated livestock into the Horn and equatorial Africa.

CONCLUSION

In the final analysis, it is presently impossible to choose between the alternative models explaining the slow trickle of domesticated animals into the Horn and equatorial Africa from 2000 to 1000 BC. Paleoenvironmental proxies do indicate lake level drops and an increased geographical range for xeric vegetation during this period. The locations of archaeological sites with evidence for small numbers of domesticated animals within green corridors lend credence to the notion that prehistoric settlement was focused on exploiting resource diversity. It is difficult to judge whether this was a drastic change from earlier hunting and gathering land use strategies, though. Regionally more pluvial conditions in the Middle Holocene allowed for a broader range of settlement, but residential site selection has always involved positioning the community adjacent to the most amounts of natural resources avoiding exposure to enemies or harmful biota.

The paucity of domesticated animals within archaeological fauna assemblages in East Africa outside Lake Turkana for the 1000-year period may also be a response to cultural or epizootic barriers for stock acquisition as well. The decision to keep animals in close quarters to where you live may not have come easily to people accustomed to protecting themselves from wild animals and hunting them for their survival. Domesticated animals can attract predators, which may have been initially unnerving to those experimenting with animal husbandry. Domesticated animals also bring diseases, which can erase benefits of time investment and threatens the health and well-being of those who are keeping the animals.

On balance, a combination of factors was likely in operation slowing the advance of domesticated animals into the Horn and equatorial Africa between 2000 and 1000 BC. As more data becomes available, we can continue to test the efficacy of the different models and weigh the factors at play during this crucial period. The implications of this type of research from a cultural-historical perspective inform our understanding of the evolution of present-day ecosystems. Pastoralism remains an important socio-economic force in much of Africa and elucidating its roots and pathways provides an evolutionary framework for which deeper insight into ways in which the modern system functions.

References

AMBROSE, S.H. 1982 – Archaeological and Linguistic Reconstructions of History in East Africa. In C. Ehret and M. Posnansky (eds.), *The Archaeological and Linguistic Reconstruction of African History* 1982: 104-157. University of California Press, Berkeley.

AMBROSE, S.H. 1984a – *Holocene Environments and Human Adaptations in the Central Rift Valley, Kenya.* Ph.D. dissertation, University of California, Berkeley. University Microfilms, Ann Arbor.

AMBROSE, S.H. 1984b – The Introduction of Pastoral Adaptations to the Highlands of East Africa. In J.D. Clark and S.A. Brandt (eds.), *From Hunters to Farmers: The Consequences of Food Production in Africa* 1984:212-239. University of California Press, Berkeley.

AMBROSE, S.H. 1998 – Chronology of the Later Stone Age and Food Production in East Africa. *Journal of Archaeological Science* 25:377-392.

AMBROSE, S.H., and N.E. SIKES 1991 – Soil Carbon Isotope Evidence for Holocene Habitat Change in the Kenya Rift Valley. *Science* 253/5026:1402-1405.

AMBROSIA, V.G., K.G. LINTHICUM, C.L. BAILEY and P. SEBESTA 1989 – Modelling Rift Valley Fever (Rvf) Disease Vector Habitats Using Active and Passive Remote Sensing Systems. *Geoscience and Remote Sensing Symposium, 1989. IGARSS'89. 12th Canadian Symposium on Remote Sensing., 1989 International* 5:2758-2760.

BARTH, F. 1956 – Ecological Relationships of Ethnic Groups in Swat, North Pakistan. *American Anthropologist* 58:1079-1089.

BARTH, F. 1961 – Nomads of South Persia–the Basseri Tribe of the Khamseh Confederacy. Little Brown & Co., Boston.

BARTH, F. 1964 – *Nomads of South Persia.* Univeritetsforlaget, Oslo.

BARTHELME, J.W. 1985 – Fisher-Hunters and Neolithic Pastoralists in East Turkana, Kenya. BAR International Series 254, Oxford.

BAYLIS, M., and C.O. NAMBIRO 1993 – The Responses to *Glossina pallidipes* and *G. longipennis* (Diptera: Glossinidiae) to Odour-Baited Traps and Targets at Galana Ranch, Kenya. *Bulletin of Entomological Research* 83:145-151.

BELSKY, A.J., A. MATZKE and S. USELMAN 1999 – Survey of Livestock Influences on Stream and Riparian Ecosystems in the Western United States. *Journal of Soil and Water Conservation* 54:419-431.

BEUNING, K.R.M., M.R. TALBOT and K. KELTS 1997 – A Revised 30,000-Year Paleoclimatic and Paleohydrologic History of Lake Albert, East Africa. *Palaeogeography, Palaeoclimatology, Palaeoecology* 136:259-279.

BINFORD, L.R. 1981 – Bones: Ancient Men and Modern Myths. Academic Press, New York.

BIRD, B.H., M.L. KHRISTOVA, P.E. ROLLIN, T.G. KSIAZEK and S.T. NICHOL 2007 – Complete Genome Analysis of 33 Ecologically and Biologically Diverse Rift Valley Fever Virus Strains Reveals Widespread Virus Movement and Low Genetic Diversity Due to Recent Common Ancestry. *Journal of Virology* 81/6:2805-2816.

BOLLIG, M. 2006 – Risk Management in a Hazardous Environment: A Comparative Study of two Pastoral Societies. Springer, New York.

BOWER, J.R.F. 1988 – Evolution of Stone-Age Food Producing Cultures in East Africa. In J. Bower and D. Lubell (eds.), *Prehistoric Cultures and Environments in the Late Quaternary of Africa* 1988:91-113. BAR International Series 405, Nairobi.

BOWER, J.R.F. 1991 – The Pastoral Neolithic of East Africa. *Journal of World Prehistory* 5:49-82.

BOWER, J.R.F. 1997 – East African Pastoralists. In J.C. Vogel (ed.), *The Encyclopedia of Precolonial Africa* 1997:205-210. Alta Mira, Walnut Creek, California.

BOWER, J.R.F., and C.M. NELSON 1978 – Early Pottery and Pastoral Cultures of the Central Rift Valley, Kenya. *Man* 13:554-566.

BOWER, J.R.F., C.M. NELSON, A.F. WAIBEL and S. WANDIBBA 1977 – The University of Massachusetts' Later Stone Age/Pastoral 'Neolithic' Comparative Study in Central Kenya: An Overview. *Azania* 12:119-146.

BRAIN, C.K. 1981 – The Hunters or the Hunted? An Introduction to African Cave Taphonomy. University of Chicago Press, Chicago.

BRANDT, S.A. 1986 – The Upper Pleistocene and Early Holocene Prehistory of the Horn of Africa. *African Archaeological Review* 4/1:41-82.

BUNDT, C. 1979 – Wo ist Vorn? Sinn und Unsinn Entwicklungspolitischen Eingreifens bei Ostafrikanischen Hirtvölkern. *Sociologus* 29:21-60.

BUNN, H.T., L.E. BARTRAM and E.M. KROLL 1988 – Variability in Bone Assemblage Formation from Hadza Hunting, Scavenging, and Carcass Processing. *Journal of Anthropological Archaeology* 7:412-457.

BUTZER, K.W., G.L. ISAAC, J. RICHARDSON and C. WASHBURN-KAMAU 1972 – Radiocarbon Dating of East African Lake Levels. *Science* 175:1069-1076.

CLEAVELAND, S., M.K. LAURENSON and L.H. TAYLOR 2001 – Diseases of Humans and Their Domestic Mammals: Pathogen Characteristics, Host Range and the Risk of Emergence. *Philosophical Transactions: Biological Sciences* 356/1411:991-999.

CLOSE, A. 1992 – Holocene Occupation of the Eastern Sahara. In F. Klees and R. Kuper (eds.), *New Light on the Northeast African Pastoralists* 1992:155-183. Heinrich-Barth Institut, Cologne.

COLLETT, D.P. 1985 – *The Spread of Early Iron Producing Communities in Eastern and Southern Africa.* Ph.D. dissertation, University of Cambridge, Cambridge. University Microfilms, Ann Arbor.

DESTA, S., and D.L. COPPOCK 2004 – Pastoralism under Pressure: Tracking System Change in Southern Ethiopia. *Human Ecology* 32/4:465-486.

di LERNIA, S. 2006 – Building Monuments, Creating Identity: Cattle Cult as a Social Response to Rapid Environmental Changes in the Holocene Sahara. *Quaternary International* 151/1:50-62.

DICKSON, D.B., F.B. PEARL, G.-Y. GANG, S. KAHINJU and S. WANDIBBA 2004 – Site Reconnaissance in the Kipsing and Tol River Watersheds of Central Kenya: Implications for Middle and Later Stone Age Land-Use Patterns. *African Archaeological Review* 21/3:153-191.

DIGARD, J.-P. 1981 – *Techniques des Nomades Baxtyâri D'iran.* Editions de la Maison des Sciences de l'Homme, Paris.

DOLAN, R.B., P. STEVENSON, H. ALUSHULA and G. OKECH 1992 – Failure of Chemoprophylaxis against Bovine Trypanosomiasis on Galana Ranch, Kenya. *Acta Tropica* 51:113-122.

DOMINGO, E., C. ESCARMÍS, E. BARANOWSKI, C.M. RUIZ-JARABO, E. CARRILLO, J.I. NÚÑEZ and F. SOBRINO 2003 – Evolution of Foot-and-Mouth Disease Virus. *Virus Research* 91/1:47-63.

DOWD, E.T. 1999 – Why Don't People Change? What Stops Them from Changing? An Integrative Commentary on the Special Issue on Resistance. *Journal of Psychotherapy Integration* 9/1:119-131.

DOWD, E.T., and K.E. COURCHAINE 2002 – Implicit Learning, Tacit Knowledge, and Implications for Stasis and Change in Cognitive Psychotherapy. In R.L. Leahy and E.T. Dowd (eds.), *Clinical Advances in Cognitive Psychotherapy* 2002:325-344. Springer Publishing, New York.

DUKES, J.S., and H.A. MOONEY 1999 – Does Global Change Increase the Success of Biological Invaders? *Trends in Ecology & Evolution* 14/4:135-139.

EHRET, C. 1998 – An African Classical Age: Eastern and Southern Africa in World Prehistory History, 1000 BC to AD 400. University Press of Virginia, Charlottesville.

EPSTEIN, H. 1971 – The Origin of the Domestic Animals of Africa. Africana, New York.

EPSTEIN, P.R. 2001 – Climate Change and Emerging Infectious Diseases. *Microbes and Infection* 3/9:747-754.

FRATKIN, E. 1986 – Stability and Resilience in East African Pastoralism: The Rendille and the Ariaal of Northern Kenya. *Human Ecology* 14/3:269-286.

FRATKIN, E. 1994 – Problems of Pastoral Land Tenure in Kenya: Demographic, Economic and Political Processes among Maasai, Samburu, Boran, and Rendille 1950-1990. African Studies Center, Boston.

GASSE, F. 2000 – Hydrological Changes in the African Tropics since the Last Glacial Maximum. *Quaternary Science Reviews* 19:189-211.

GASSE, F. 2002 – Diatom-Inferred Salinity and Carbonate Oxygen Isotopes in Holocene Waterbodies of the Western Sahara and Sahel (Africa). *Quaternary Science Reviews* 21/7:737-767.

GASSE, F., and E. Van CAMPO 1994 – Abrupt Post-Glacial Climatic Events in West Asia and North Africa Monsoon Domains. *Earth and Planetary Science Letters* 126:435-456.

GIFFORD-GONZALEZ, D. 1998 – Early Pastoralists in East Africa: Ecological and Social Dimensions. *Journal of Anthropological Archaeology* 17:166-200.

GIFFORD-GONZALEZ, D. 2000 – Animal Disease Challenges to the Emergence of Pastoralism in Sub-Saharan Africa. *African Archaeological Review* 17/3:95-139.

GIFFORD-GONZALEZ, D. 2003 – The Fauna from Ele Bor: Evidence for the Persistence of Foragers into the Later Holocene of Arid North Kenya. *African Archaeological Review* 20/2:81-119.

HAALAND, R. 1995 – Sedentism, Cultivation, and Plant Domestication in the Holocene Middle Nile Region. *Journal of Field Archaeology* 22/2:157-174.

HAMILTON, A.C. 1982 – Environmental History of East Africa. Academic Press, London.

HARGROVE, J.W. 2004 – Tsetse Population Dynamics. In I. Maudlin, P.H. Holmes and M.A. Miles (eds.), *The Trypanosomiases* 2004:139-180. CABI Publishing, Wallingford, UK.

HARVELL, C.D., C.E. MITCHELL, J.R. WARD, S. ALTIZER, A.P. DOBSON, R.S. OSTFELD and M.D. SAMUEL 2002 – Climate Warming and Disease Risks for Terrestrial and Marine Biota. *Science* 296/5576:2158-2162.

HASSAN, F.A., and G.T. GROSS 1986 – Chronology of the Khartoum 'Mesolithic' and 'Neolithic' and Related Sites in the Sudan: Statistical Analysis and Comparisons with Egypt. *African Archaeological Review* 4:83-102.

HENRICH, J., and R. McELREATH 2003 – The Evolution of Cultural Evolution. *Evolutionary Anthropology* 12/3:123-135.

HOELZMANN, P., B. KEDING, H. BERKE, S. KROPELIN and H.J. KRUSE 2001 – Environmental Change and Archaeology: Lake Evolution and Human Occupation in the Eastern Sahara During the Holocene. *Palaeogeography, Palaeoclimatology, Palaeoecology* 169/3-4:193-217.

HONG, Y.T., B. HONG, Q.H. LIN, Y.X. ZHU, Y. SHIBATA, M. HIROTA, M. UCHIDA, X.T. LENG, H.B. JIANG, H. XU, H. WANG and L. YI 2003 – Correlation between Indian Ocean Summer Monsoon and North Atlantic Climate During the Holocene. *Earth and Planetary Science Letters* 211/3-4:371-380.

IRONS, W. 1972 – Variation in Economic Organization: A Comparison of the Pastoral Yomut and the Basseri. In W. Irons and N. Dyson-Hudson (eds.), *Perspectives on Nomadism* 1972:88-104. E.J. Brill, Leiden, the Netherlands.

IRUNGU, P., S.O. NYAMWARO and D.K. MASIGA 2002 – Financial Implications of Rearing of Sheep and Goats under Natural Trypanosomosis Challenge at Galana Ranch, Kenya. *Tropical Animal Health and Production* 34/6:503-513.

JOHNSON, T.C. 1996 – Sedimentary Processes and Signals of Past Climate Change in the Large Lakes of the East African Rift Valley. In T.C. Johnson and E.O. Odada (eds.), *The Limnology, Climatology and Paleoclimatology of the East African Lakes* 1996:367-412. Gordon and Breach, Amsterdam.

JOHNSON, T.C., J.D. HALFMAN and W.J. SHOWERS 1991 – Paleoclimate of the Past 4000 Years at Lake Turkana, Kenya, Based on the Isotopic Composition of Authigenic Calcite. *Palaeogeography Palaeoclimatology Palaeoecology* 85/3-4:189-198.

JOHNSON, T.C., K. KELTS and E.O. ODADA 2000 – The Holocene History of Lake Victoria. *Ambio* 29/1:2-11.

JONES, N.B., K. HAWKES and P. DRAPER 1994 – Foraging Returns Of !Kung Adults and Children: Why Didn't !Kung Children Forage? *Journal of Anthropological Research* 50/3:217-249.

JONES, N.G.B., L.C. SMITH, J.F. O'CONNELL, K. HAWKES and C.L. KAMUZORA 1992 – Demography of the Hadza, an Increasing and High Density Population of Savanna Foragers. *American Journal of Physical Anthropology* 89/2:159-181.

KAREGA-MŨNENE 2002 – *Holocene Foragers, Fishers and Herders of Western Kenya.* BAR International Series 1037. Cambridge Monographs in African Archaeology. BAR Publishing, Oxford.

KLEIN, R.G., and K. CRUZ-URIBE 1984 – The Analysis of Animal Bones from Archaeological Sites. University of Chicago Press, Chicago.

KRZYZANIAK, L. 1978 – New Light on Early Food Production in the Central Sudan. *Journal of African History* 2:159-172.

KRZYZANIAK, L. 1980 – The Origin of Pastoral Adaptation in the Nilotic Savanna. In R.E. Leakey and B.A. Ogot (eds.), *Proceedings of the 8th Panafrican Congress of Prehistory and Quaternary Studies* 1980:265-267. The International Louis Leakey Memorial Institute for African Prehistory, Nairobi.

KUSIMBA, C.M., and S.B. KUSIMBA 2000 – Hinterlands and Cities: Archaeological Investigations of Economy and Trade in Tsavo, Southeastern Kenya. *Nyame Akuma* 54:13-24.

KUSIMBA, C.M., S.B. KUSIMBA and D.K. WRIGHT 2005 – The Development and Collapse of Precolonial Ethnic Mosaics in Tsavo, Kenya. *Journal of African Archaeology* 3/2:345-365.

KUSIMBA, S.B. 2003 – African Foragers: Environment, Technology, Interactions. Alta Mira, Walnut Creek, CA.

KUSIMBA, S.B., and C.M. KUSIMBA 2007 – Intensification and Protohistoric Agropastoral Systems in East Africa In T.L. Thurston and C.T. Fisher (eds.), *Seeking a Richer Harvest: The Archaeology of Subsistence Intensification, Innovation, and Change* 2007:217-233. Springer, New York.

LÆRDAL, T., M.R. TALBOT and J.M. RUSSELL 2002 – Late Quaternary Sedimentation and Climate in the Lakes Edward and George Area, Uganda – Congo. In E.O. Odada and D.O. Olago (eds.), *The East African Great Lakes: Limnology, Palaeolimnology and Biodiversity* 2002:429-470. Kluwer, New York.

LANE, P., C. ASHLEY, O. SEITSONEN, P. HARVEY, S. MIRE and F. ODEDE 2007 – The Transition to Farming in Eastern Africa: New Faunal and Dating Evidence from Wadh Lang'o and Usenge, Kenya. *Antiquity* 81/1:62-81.

LEAK, S.G.A. 1999 – Tsetse Biology and Ecology: Their Role in the Epidemiology and Control of Trypanosomosis. CABI Publishing, Wallingford, UK.

LEE, R.B. 1979 – *The !Kung San.* Cambridge University Press, Cambridge.

LESUR, J. 2004 – Exploitation De La Diversité Faunique Et Début Du Pastoralisme Dans La Corne De L'afrique À L'holocène, Unpublished Ph.D. dissertation. Université de Paris I, Paris.

LESUR, J. 2007 – *Chasse Et Élevage Dans La Corne De L'afrique Entre Le Néolithique Et Les Temps Historiques.* Cambridge Monographs in African Archaeology. BAR International Series 1602, Oxford.

MABULLA, A. 2003 – Archaeological Implications of Hadzabe Forager Land Use in the Eyasi Basin, Tanzania. In C.M. Kusimba and S.B. Kusimba (eds.), *East African Archaeology: Foragers, Potters, Smiths, and Traders* 2003:33-58. University of Pennsylvania Museum of Archaeology and Anthropology, Philadelphia.

MacFARLANE, W.V., B. HOWARD, H. HAINES, P.J. KENNEDY and C.M. SHARPE 1971 – Hierarchy of Water and Energy Turnover of Desert Mammals. *Nature* 234/5530:483-484.

MacHUGH, D.E., M.D. SHRIVER, R.T. LOFTUS, P. CUNNINGHAM and D.G. BRADLEY 1997 – Microsatellite DNA Variation and the Evolution, Domestication and Phylogeography of Taurine and Zebu Cattle (*Bos taurus* and *Bos indicus*). *Genetics* 146:1071-1086.

MAJIWA, P.A., M. MAINA, J.N. WAITUMBI, S. MIHOK and E. ZWEYGARTH 1993 – Trypanosoma (Nannomonas) Congolense: Molecular Characterization of a New Genotype from Tsavo, Kenya. *Parasitology* 106/2:151-62.

MAKUMI, J.N., P. STEVENSON and C.H. GREEN 2000 – Control of *Glossina Longipennis* (Diptera: Glossinidae) by Insecticide-Treated Targets at Galana Ranch, Kenya, and Confirmation of the Role of *G. Longipennis* as a Vector of Cattle Trypanosomiasis. *Bulletin of Entomological Research* 90/5:397-406.

MAREAN, C.W. 1992 – Hunter to Herder: Large Mammal Remains from Hunter-Gatherer Occupation at Enkapune ya Muto Rockshelter, Central Rift, Kenya. *African Archaeological Review* 10:65-127.

MARKS, A.E., A. MOHAMMED-ALI, J. PETERS and R. ROBERTSON 1985 – The Prehistory of the Central Nile Valley as Seen from Its Eastern Hinterlands. *Journal of Field Archaeology* 12:261-278.

MARLOWE, F. 2002 – Why the Hadza Are Still Hunter-Gatherers. In S. Kent (ed.), *Ethnicity, Hunter-Gatherers, and The "Other"* 2002:247-275. Smithsonian Institution Press, Washington, D.C.

MARSHALL, F. 1990 – Origins of Specialized Pastoral Production in East Africa. *American Anthropologist* 92/4:873-894.

MARSHALL, F. 1994 – Archaeological Perspectives on East African Pastoralism. In E. Fratkin, K.A. Galvin and E.A. Koth (eds.), *African Pastoralist Systems: An Integrated Approach* 1994:17-43. Lynne Rienner Publishers, Boulder.

MARSHALL, F. 2000 – The Origins and Spread of Domestic Animals in East Africa. In K.C. MacDonald and R.M. Blench (eds.), *The Origins and*

Development of African Livestock: Archaeology, Genetics, Linguistics and Ethnography 2000:191-221. UCL Press, London.

MARSHALL, F., and E. HILDEBRAND 2002 – Cattle before Crops: The Beginnings of Food Production in Africa. *Journal of World Prehistory* 16/2:99-143.

MARSHALL, F., and A. NEGASH 2002 – Early Hunters and Herders of Northern Ethiopia: The Fauna from Danei Kawlos and Baati Ataro Rockshelters. Paper presented at the Society of Africanists Archaeologists 16th Biennial Conference, University of Arizona, Tucson.

MARSHALL, F., K. STEWART and J.W. BARTHELME 1984 – Early Domestic Stock at Dongodien in Northern Kenya. *Azania* 19:120-129.

McCABE, J.T. 1994 – The Failure to Encapsulate: Resistance to the Penetration of Capitalism by the Turkana of Kenya. In C.P. Chang and H.A. Koster (eds.), *Pastoralists at the Periphery: Herders in a Capitalist World* 1994:197-211. University of Arizona Press, Tucson.

McCABE, J.T. 2006 – Cattle Bring Us to Our Enemies: Turkana Ecology, Politics, and Raiding in a Disequilibrium System. University of Michigan Press, Ann Arbor.

McCARTHY, N., and M. di GREGORIO 2007 – Climate Variability and Flexibility in Resource Access: The Case of Pastoral Mobility in Northern Kenya. *Environment and Development Economics* 12/3:403-421.

McDONALD, M.M.A. 1998 – Early African Pastoralism: View from the Dakhleh Oasis (South Central Egypt). *Journal of Anthropological Archaeology* 17:124-142.

MEHLMAN, M.J. 1989 – *Later Quaternary Archaeological Sequences in Northern Tanzania.* Ph.D. dissertation. University of Illinois at Champaign-Urbana, Urbana. University Microfilms, Ann Arbor.

MIHOK, S., E. MUNYOKI, R.A. BRETT, J.F. JONYO, D. ROTTCHER, P.A.O. MAJIWA, E.K. KANGETHE, H.F.A. KABURIA and E. ZWEYGARTH 1992 – Trypanosomiasis and the Conservation of Black Rhinoceros (Diceros-Bicornis) at the Ngulia Rhino Sanctuary, Tsavo-West National-Park, Kenya. *African Journal of Ecology* 30/2:103-115.

MUKHTAR, M.H. 1996 – The Plight of the Agro-Pastoral Society of Somalia. *Review of African Political Economy* 23/70:543-553.

MUTUNDU, K.K. 1998 – Ethnohistoric Archaeology of the Mukogodo in North-Central Kenya: Contemporary Hunter-Gatherer Subsistence and the Transition to Pastoralism in Secondary Settings. PhD dissertation, Washington University, St. Louis. University Microfilms, Ann Arbor.

O'CONNELL, J.F., and K. HAWKES 1988 – Hadza Hunting, Butchering, and Bone Transport and Their Archaeological Implications. *Journal of Anthropological Research* 44:113-161.

OLAGO, D.O., F.A. STREET-PERROTT, R.A. PERROTT and E.O. ODADA 2003 – Late Holocene Sedimentology and Palaeoenvironment of Kiluli Swamp, Mount Kenya. *African Journal of Science and Technology* 4/2:12-23.

OLE-MIARON, J.O. 2003 – The Maasai Ethnodiagnostic Skill of Livestock Diseases: A Lead to Traditional Bioprospecting. *Journal of Ethnopharmacology* 84:79-83.

ONYANGO-ABUJE, J.C., and S. WANDIBBA 1979 – The Palaeoenvironment and Its Influence on Man's Activities in East Africa During the Latter Part of Upper Pleistocene and Holocene. In B.A. Ogot (ed.), *Ecology and History in East Africa: Proceedings of the 1975 Conference of the Historical Association of Kenya* 1979:24-40. Kenya Literature Bureau, Nairobi.

OVERPECK, J., D. ANDERSON, S. TRUMBORE and W. PRELL 1996 – The Southwest Indian Monsoon over the Last 18,000 Years. *Climate Dynamics* 12/3:213-225.

OWEN, R.B., J.W. BARTHELME, R.W. RENAUT and A. VINCENS 1982 – Palaeolimnology and Archaeology of Holocene Deposits North-East of Lake Turkana, Kenya. *Nature* 298:523-529.

PATZ, J.A., T.K. GRACZYK, N. GELLER and A.Y. VITTOR 2000 – Effects of Environmental Change on Emerging Parasitic Diseases. *International Journal for Parasitology* 30/12-13:1395-1405.

PETERSON, A.T. 2008 – Biogeography of Diseases: A Framework for Analysis *Naturwissenschaften* 95/6:483-491.

PHILLIPSON, D.W. 1977 – The Excavation of Gobedra Rock-Shelter, Axum: An Early Occurrence of Cultivated Finger Millet in Northern Ethiopia. *Azania* 12:53-82.

PHILLIPSON, D.W. 1993 – The Antiquity of Cultivation and Herding in Ethiopia. In T. Shaw, P.J.J. Sinclair, B. Andah and A. Okpoko (eds.), *The Archaeology of Africa: Food, Metals and Towns* 1993:344-357. Routledge, London.

PHILLIPSON, D.W. 2005 – *African Archaeology.* Cambridge University Press, Cambridge, UK.

POLLOCK, J.N. (ed.) 1982a – *Ecology and Behaviour of Tsetse.* Volume 2. Food and Agriculture Organization, Rome.

POLLOCK, J.N. (ed.) 1982b – *Tsetse Biology, Systematics and Distribution Techniques.* Volume 1. Food and Agriculture Organization, Rome.

PURSE, B.V., P.S. MELLOR, D.J. ROGERS, A.R. SAMUEL, P.P.C. MERTENS and M. BAYLIS 2005 – Climate Change and the Recent Emergence of Bluetongue in Europe. *Nature Reviews Microbiology* 3:171-181.

REDDING, R.W. 1994 – The Vertebrate Fauna. In S.C. Herbert (ed.), Tel Anafa I, I: Final Report on Ten Years of Excavations at a Hellenistic and Roman

Settlement in Northern Israel 1994:279-322. University of Missouri, Colombia.

REIMER, P.J., M.G.L. BAILLIE, E. BARD, A. BAYLISS, J.W. BECK, C. BERTRAND, P.G. BLACKWELL, C. BUCK, G. BURR, K.B. CUTLER, P.E. DAMON, R.L. EDWARDS, R.G. FAIRBANKS, M. FRIEDRICH, T.P. GUILDERSON, K.A. HUGHEN, B. KROMER, F.G. McCORMAC, S. MANNING, C. BRONK RAMSEY, R.W. REIMER, S. REMMELE, J.R. SOUTHON, M. STUIVER, S. TALAMO, F.W. TAYLOR, J. van der PLICHT and C.E. WEYHENMEYER 2004 – IntCal04 Terrestrial Radiocarbon Age Calibration, 0-26 Cal Kyr Bp. *Radiocarbon* 46:1029-1058.

REITZ, E.J., and E.S. WING 1999 – *Zooarchaeology*. Cambridge Munuels in Archaeology. Cambridge University Press, Cambridge.

RICHARDSON, J.L. 1966 – Changes in the Level of Lake Naivasha, Kenya During Post-Glacial Times. *Nature* 209:290-291.

RICHARDSON, J.L. 1972 – Paleolimnological Records from Rift Lakes in Central Kenya. *Palaeoecology of Africa* 6:131-136.

RICKETTS, R.D., and T.C. JOHNSON 1996 – Climate Change in the Turkana Basin as Deduced from a 4000 Year Long Delta O-18 Record. *Earth and Planetary Science Letters* 142/1-2:7-17.

ROBBINS, L.H. 2006 – Lake Turkana Archaeology: The Holocene. *Ethnohistory* 53/1:71-93.

ROBBINS, L.H., S.A. McFARLIN, J.L. BROWER and A.E. HOFFMAN 1977 – Rangi: A Late Stone Age Site in Karamoja District, Uganda. *Azania* 12:209-233.

ROBERTSHAW, P.T. 1990 – *Early Pastoralists of South-Western Kenya*. British Institute of Eastern Africa, Nairobi.

ROGERS, D.J., and T.P. ROBINSON 2004 – Tsetse Distribution. In P.H. Holmes, M.A. Miles and I. Maudlin (eds.), *The Trypanosomiases* 2004:139-180. CABI Publishing, Wallingford, UK.

RUSSELL, J.M., and T.C. JOHNSON 2005 – A High-Resolution Geochemical Record from Lake Edward, Uganda Congo and the Timing and Causes of Tropical African Drought During the Late Holocene. *Quaternary Science Reviews* 24/12-13:1375-1389.

SALZMAN, P.C. 2004 – Pastoralists: Equality, Hierarchy, and the State. Westview Press, Boulder, CO.

SMITH, A.B. 1984 – Environmental Limitations on Prehistoric Pastoralism in Africa. *African Archaeological Review* 2:99-111.

SMITH, A.B. 2005 – African Herders: Emergence of Pastoral Traditions. Alta Mira Press, Walnut Creek, CA.

SOIL SURVEY STAFF 1999 – *Soil Taxonomy: A Basic System of Soil Classification for Making and Interpreting Soil Surveys*. 2nd ed. United States Department of Agriculture, Natural Resources Conservation Service, Washington, D.C.

SOPER, R.C. 1966 – *Archaeological Sites in Tsavo East National Park*. Manuscript on file, British Institute in Eastern Africa, Nairobi.

SSEMMANDA, I., and A. VINCENS 2002 – Vegetation Changes and Their Climatic Implications for the Lake Victoria Region During the Late Holocene. In E.O. Odada and D.O. Olago (eds.), *The East African Great Lakes: Limnology, Palaeolimnology and Biodiversity* 2002:509-523. Boston, New York.

STAGER, J.C., and T.C. JOHNSON 2000 – A 12,400 ^{14}C Yr Offshore Diatom Record from East Central Lake Victoria, East Africa. *Journal of Paleolimnology* 23:373-383.

TABUTI, J.R.S., S.S. DHILLIONA and K.A. LYE 2003 – Ethnoveterinary Medicines for Cattle (*Bos indicus*) in Bulamogi County, Uganda: Plant Species and Mode of Use. *Journal of Ethnopharmacology* 88/2-3:279-286.

THOMPSON, L.G., E. MOSELY-THOMPSON, M.E. DAVIS, K.A. HENDERSON, H.H. BRECHER, V.S. ZAGORODNOV, T.A. MASHIOTTA, P.-N. LIN, V.N. MIKHALENKO, D.R. HARDY and J. BEER 2002 – Kilimanjaro Ice Core Records: Evidence of Holocene Climate Change in Tropical Africa. *Science* 298:589-593.

THORBAHN, P.F. 1979 – *The Precolonial Ivory Trade of East Africa: Reconstruction of a Human-Elephant Ecosystem*. Ph.D. dissertation, University of Massachusetts, Amherst. University Microfilms, Ann Arbor.

van der MERWE, D., G.E. SWAN and C.J. BOTHA 2001 – Use of Ethnoveterinary Medicinal Plants in Cattle by Setswana-Speaking People in the Madikwe Area of the North West Province of South Africa. *Journal of the South African Veterinary Association* 72/4:189-196.

VERHOEVEN, M. 2004 – Beyond Boundaries: Nature, Culture and a Holistic Approach to Domestication in the Levant. *Journal of World Prehistory* 18/3:179-282.

VILLE, J.-L. 1995 – The Waata of Tsavo-Galana. *Kenya Past and Present* 27:21-27.

WANDIBBA, S. 1980 – The Application of Attribute Analysis to the Study of Later Stone Age/Neolithic Pottery Ceramics in Kenya. In R.E. Leakey and B.A. Ogot (eds.), *Proceedings of the 8th Panafrican Congress of Prehistory and Quaternary Studies Nairobi September 1977* 1980:283-285. The International Louis Leakey Memorial Institute for African Prehistory, Nairobi.

WENDORF, F., and R. SCHILD 2001 – Geoarchaeology of the Holocene Climatic Optimum at Nabta Playa, Southwestern Desert, Egypt. *Geoarchaeology* 16/1:7-28.

WIJNGAARDEN, W.v., and V.W.P. v. ENGELEN 1985 – *Soils and Vegetation of the Tsavo Area*. Geological Survey of Kenya, Nairobi.

WRIGHT, D.K. 2005a – *Environment, Chronology and Resource Exploitation of the Pastoral Neolithic in Tsavo, Kenya.* PhD dissertation, University of Illinois at Chicago, Chicago. University Microfilms, Ann Arbor.

WRIGHT, D.K. 2005b – New Perspectives on Early Regional Interaction Networks of East African Trade: A View from Tsavo National Park, Kenya. *African Archaeological Review* 22/3:111-140.

WRIGHT, D.K. 2007 – Tethered Mobility and Riparian Resource Exploitation among Neolithic Hunters and Herders in the Galana River Basin, Kenyan Coastal Lowlands. *Environmental Archaeology* 12/1:25-47.

WRIGHT, D.K., S.L. FORMAN, C.M. KUSIMBA, J. PIERSON, J. GOMEZ and P. TATTERSFIELD 2007 – Stratigraphic and Geochronological Context of Human Habitation Along the Galana River, Kenya. *Geoarchaeology* 22/7:25-47.

Chapter 7
SOCIO-CULTURAL RESPONSES TO A CHANGING ENVIRONMENT: THE SHASHE-LIMPOPO VALLEY SINCE CA. AD 900

Munyaradzi MANYANGA
Dept. of Anthropology and Archaeology, University of Pretoria Hatfield Campus, Pretoria, South Africa

Abstract: *The paper considers archaeological research and environmental reconstruction in Zimbabwe's southern Lowveld, including the Shashe-Limpopo drainage. It incorporates southern Africa's first known state system, Mapungubwe (11th and 12th century AD). Viewed against a background of recent droughts, floods, and survival strategies, the Shashe-Limpopo valley seems inhospitable to humanity since prehistoric times. The general perception is that it is hot and dry. Therefore, it appears not to be conducive to human habitation. However, research confirms that the valley exhibits human activity since 30000 years ago. In addition, the ecological conditions and practices of the present inhabitants offer an alternative explanation on how prehistoric communities could have interacted with this fragile landscape. It is concluded that the valley supported large populations through floodplain utilization even when the environment had deteriorated.*

INTRODUCTION

Zimbabwe's southern lowveld has yielded a wide range of archaeological sites, some of them extensive in size. A number of the sites show close affinities with those that have been recorded on the South African and Botswana side of the valley. This suggests that a related interacting community once occupied both sides of the Shashe-Limpopo rivers. Viewed against the background of recent droughts, floods and survival strategies, the Shashe-Limpopo drainage basin is viewed as a treacherous landscape that has not been preferred by human populations since prehistoric times. This has had a bearing on both archaeological and historical research in the area. Contrary to this general belief, research has confirmed that the valley has always been abuzz with human activity. While the valley was at times affected by unfavorable environmental condition, it continued to attract human settlement from the earliest hunter-gatherers to the current inhabitants of the area. The cumulative data, which has accrued over the years, confirms earlier suggestions that the Shashe-Limpopo drainage basin offers alternative ways to cope with this hash environment.

RESEARCH BACKGROUND

Archaeological research in southern Zimbabwe is as old as the subject itself in southern Africa. However, before 1990, research was mostly done as an afterthought to research and interest at Mapungubwe, a Zimbabwe culture and World Heritage site in the Limpopo valley. It is acknowledged to be the first state system in southern Africa. Much of the archaeological knowledge on the Zimbabwean side of the valley comes from the work of the Rhodesian schools' expeditions that sought not only to study the archaeology of the Limpopo valley, but also to investigate other issues relating to the environment and biodiversity. Early reports on the archaeology of southern Zimbabwe where done by the Rhodesian Schools' Exploration Society (RSES) which carried out a number of expeditions beginning in the 1950s. Expeditions that covered Zimbabwe's southern lowveld include the Mateke Hills (1958), Buffalo Bend (1961), Sentinel Ranch (1961), Shashi (1966), and Maramani (1967). They recorded a wide range of sites dating from the Stone Age to more recent historical periods. The recorded Stone Age sites belong mostly to the Middle and Late Stone Age. A number of rock art sites were also recorded, of which some are summarized by Eastwood *et al.* (1995). The only Stone Age site excavated in the research area is the Mpato Rock shelter, which was excavated and reported by Cooke and Simmons (1969). These school explorations also recorded farming community sites, which have been related to other sites in the Shashi-Limpopo valley, such as Scroda, K2 and Mapungubwe.

Some individuals also had interest in the valley. Following his participation in the Shashi Expedition (1966), Garlake (1968) further examined Mapela Hill and concluded that it had close affinities with Mapungubwe. The 1970s and 1980s saw a general decline in archaeological research in Zimbabwe's southern Lowveld. It is only in the 1990s that the area attracted research on archaeological consultancies, because of tourism development (Eastwood *et al.* 1995). The work presented here is part of ongoing institutional research in southern Zimbabwe conducted by the University of Zimbabwe in collaboration with the National Museums and Monuments Office of Zimbabwe. It is now part of the African Archaeology Network, a regional project in eastern and southern Africa, funded by the Department for Research Co-operation (SAREC) of the Swedish International Development Agency (SIDA). The scope of the project is to understand settlement dynamics and subsistence in Zimbabwe's southern lowveld.

Research Strategies

The research strategy comprised archaeological surveys and excavations that were concerned with current environmental conditions and observations of subsistence strategies of the area. Five sites have been excavated in the research area. These have provided dates and cultural material within undisturbed contexts, valuable to the archaeological analysis. Environmental data comes from geological survey and time series data for rainfall. The climate information comes from major meteorological centers in southern Zimbabwe, collected by the Meteorological Office, in Harare. The climate data is combined with the archaeological survey and excavation data, and analyzed using Geographical Information Systems (GIS) packages (Katsamudanga 2001). Further observations are also made by the author on conditions during the past seven years in which the research area experienced both severe drought and excessive flooding, which called for different responses by current inhabitants of the area. These recent observations, which can also be traced to historical times, offered a different perception of the research area in terms of subsistence strategies associated with flooding in this dry and hot region of southern Zimbabwe.

Surveys

Extensive surveys were conducted in the research area. The area covered is so large that it called for the application of a sampling strategy. Where sampling had to be done, stratified random sampling was employed. The targeted areas included the river valleys, the sandstone or basalt hills, and the open flat areas overlooking the river valleys. Taking into consideration terrain, current land use, and vegetation patterns, chosen survey units consisted of 1 km by 1 km grid squares.

The surveys were conducted by systematically walking over a targeted grid square and inspecting for archaeological sites. The procedure in the surveys began with field walking and site location using 1:50 000 survey maps and GPS. This was followed by site documentation using site survey recording forms and photography. Depending on the site, in situ recording of important artifacts and features was also carried out. Since the research was also interested in establishing inter-site comparisons, surface collections (especially of pottery) were taken when feasible. The data collected formed the basis for dating and characterization of the sites.

Site Types

Sites were classified on the bases of surface finds and features. Pottery and/or architecture were the key determinants in site classification for Iron Age sites, while stone tool types or art were useful in determining categories for Stone Age sites. The Iron Age sites were further classified into specific ceramic traditions that have been established in the southern African sub-region. The range of sites shows that the southern lowveld had a long history of human occupation beginning as early as the Early Stone Age and lasting up to the present. Whether this presents a case of population continuity or the arrival of new people is an issue that will probably become better known after further detailed archaeological and anthropological investigations.

The Stone Age

Very few Early Stone Age sites were recorded and these were characterized by isolated stone implements such as cleavers, hand axes, and flaked cobbles. Other sites had stone implements with Middle Stone Age characteristic, which portray flaking industries. These occurred on rocky areas, especially those associated with dolerite.

The Late Stone Age is well represented both in the form of stone implements and art. These sites were mostly noted in rock shelters. The rock art consists of engravings and paintings on sandstone. This period appears to have attracted earlier research, especially at Mpato cave, which was excavated by Cooke and Simmons (1969). Ed Eastwood *et al.* (1995) also conducted a preliminary survey and characterization of the paintings at Nottingham and Sentinel ranches. The majority of the art was executed by hunter-gatherer societies. Geometric designs, animal, and human figures are depicted in the art. These include stylized male and female human figures, elephant, rhino, giraffe, warthog, buffalo, kudu, eland, sable, zebra, baboon, fish, and birds. The art also exhibits a number of symbols. Most common are the V-shapes and slash marks. Some of the engravings seem to portray footprints of medium- and large-sized bovids. Identifiable prints include antelope and zebra hoofs. There are also human palm prints. Some of the cut marks appear to be consistent with the shape and form of common facial markings associated with the Venda; a dominant ethnic group in southern Zimbabwe and northern South Africa. Not all the art in the area was executed by Late Stone Age hunter-gatherers. Some of it was created by Bantu speakers. A number of the paintings and engravings are associated with grain bins, an observation that made Eastwood *et al.* (1995) conclude that these might be associated with farming communities.

Farming Community Sites

Farming community sites comprised the majority of sites recorded during the surveys and hence form the main discussion of this paper. They were classified into two broad categories, namely the Early and the Later Farming community periods. However, very few Early Farming community sites were observed and the few that were recorded belonged to the terminal phase of the early period. They are characterized by Zhizo pottery. Zhizo sites are dated to the 8th-10th century AD. These were open sites with pottery that is decorated with both comb stamping and incising. The majority of sites recorded in the research area are Later Farming period sites, especially those that relate to K2, Mapungubwe and Zimbabwe phases. They date to between the 10th-19th centuries AD. The biggest concentration was noted in the area around the Shashi-Limpopo confluence and adjacent areas (Fig. 7.1).

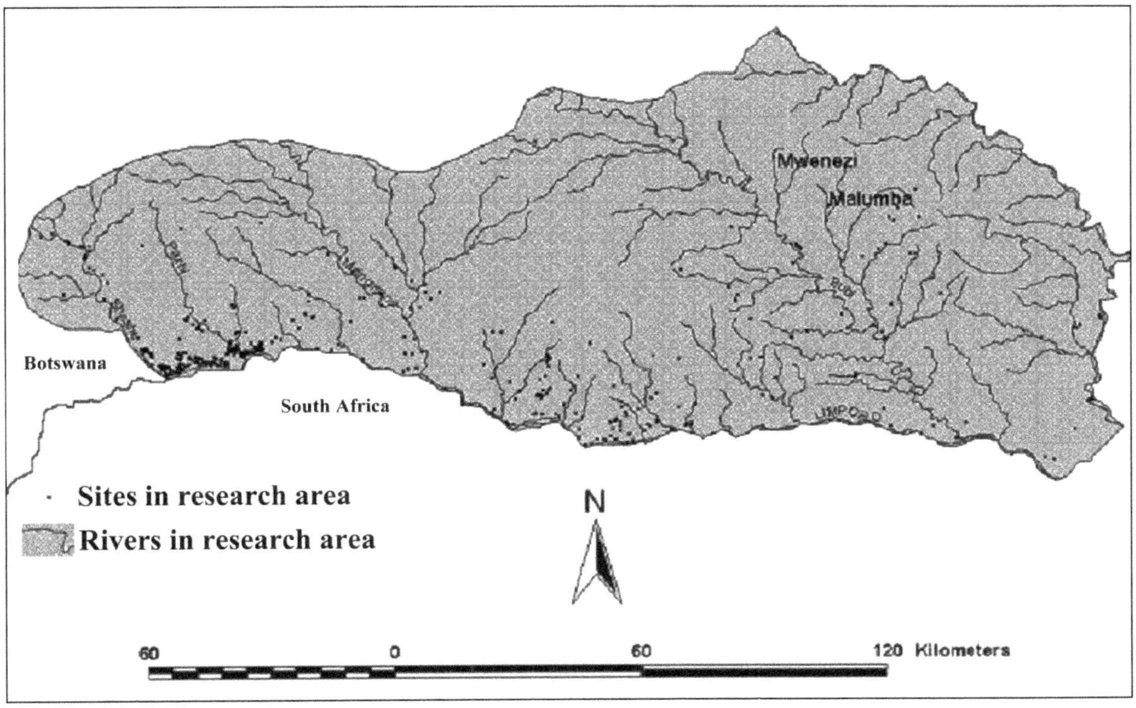

Figure 7.1. Map of the research area showing archaeological site concentration along major river valleys

Table 7.1. Age calibrated for southern hemisphere with the Pretoria program (Talma and Vogel 1993, updated 2000)

Lab Reference	Sample Designation	bp	Calibrated Date AD (and Intercept)
8959	Mutshilashokwe Tr1 Le8	910 ± 60	1047-1097; 1136 (1184) 1236
8996	Mutshilashokwe Tr5 Le5	670 ± 60	1291 (1308, 1373) 1403
8969	Tshobwani Tr 1 Le 6	790 ± 60	1236 (1274) 1291
9000	Tshobwani Tr 2 Le 4	890 ± 60	1157 (1202) 1256
Ua 13192	Malumba Tp2 Le 3	990 ± 75	1010 (1040)1180
Ua 13191	Malumba Tp1 le 4	690 ± 95	1275 (1300)1410
Ua 11802	Mwenezi Farm St 4 le3	1250 ± 75	700 (790) 900

K2 sites were mostly mound sites enclosed by walls, having a central kraal. Some of the Mapungubwe phase sites were similarly constructed, but others were found on hilltops with walling reflecting the establishment of a stratified society. However, not all sites with Mapungubwe pottery show evidence of social stratification. The sites are located on a ridge overlooking the Shashe-Limpopo floodplain. Other sites are located on sandstone ridges that run parallel to the Shashe-Limpopo floodplain – a dominant feature that characterizes much of the Limpopo valley. Most of the sites are commoner residences, cattle outposts, or agricultural villages with central cattle byres. Three sites have been excavated. Radiocarbon analyses obtained from these sites yield dates of between AD 900-1400 (Table 7.1).

The Zimbabwe Culture

The essence of Zimbabwe culture is social stratification and associated ideology (Huffman 2000:14). The culture sequence in the Shashe-Limpopo began with the Mapungubwe period (AD 1220-1290), followed by Great Zimbabwe (AD 1290-1450), and Khami (AD 1450-1820). The latter is thought to have broken up into other known historical groupings, of which the Shashe-Limpopo representations are the Venda. Surveys in the Mateke Hills and the Bubi, Mwenezi, and Limpopo valleys confirm that a Zimbabwe culture is well represented in the Shashe-Limpopo drainage basin (Manyanga 2001). A cluster of stone-walled sites were noted just east of Beitbridge and along the Mzingwani River, while isolated sites occur along the sandstone ridges overlooking the Limpopo valley at Nottingham and Sentinel farms. Most of these Zimbabwe sites are located on hilltops with relatively flat tops. The walls are freestanding, circular with radiating walls that appear to have demarcated the use of space at the site. The stones are dressed and coursing is evident. Important in the construction plan is the incorporation of natural features, such as boulders and hill cliffs, in the overall building plan.

Table 7.2. Age calibrated for southern the hemisphere with the Pretoria program (Talma and Vogel 1993), updated 2000

Lab Reference	Sample Designation	bp	Calibrated Date AD (and Intercept)
Pta 9000	Village 16 Tr 1	420 ± 45	1449 (1478) 1518; 1584-1624
Ua 11800	Mwenezi Farm Tr1 le 2	800 ± 70	1220 (1270) 1290
Ua 11801	Mwenezi Farm Tp19 le 1	215 ± 65	After 1655
Pta 8999	Sentinel Ridge	130 ± 20	1698-1722, 1818-1837 (1889, 1918) 1928

Khami period sites are well represented in the research area. They extend to both the Botswana and South African parts of the valley. They have been noted in Botswana and in areas just to the south of the Limpopo. The Soutpansberg has the highest known cluster of these sites. A number of sites have also been noted at Sentinel and Nottingham farms, and at River Ranch. These comprise both commoner and elite sites with characteristic panel and band pottery. The elite residences have elaborate retaining walls on sandstone ridges with livestock kraals just below the hill. The Khami period is seen as the forerunner to the historical Venda; a dominant group and culture in southern Zimbabwe and northern South Africa. The current stronghold of the Venda is the Nzhelele valley and the Soutpansberg in northern South Africa. Most Venda traditions among these communities trace their origins to the Shona people north of the Limpopo. Venda society today is an amalgam of different groups, such as the Ngona and Shona, who are thought to have been in the Shashe-Limpopo drainage basin since the 13th and 14th centuries AD. (Loubser 1990:24-25). These groups traded and interacted with the Sotho-Tswana groups to the south of the Soutpansberg and traded with the coast leading to the development of the Venda language. A distinct Venda chiefdom is thought to have originated from Zimbabwe following the dispersion of the Tsingo ruling polity from central and southern Zimbabwe. They established themselves south of the Limpopo in the Nzhelele valley around AD 1700 (von Sicard 1952:10, Loubser 1990:28). Some of the sites recorded in the survey area have been associated with the Venda. The modern inhabitants of the area relate a number of stone walled sites along the Mzingwani and Mushilachokwe rivers to Venda history. Pottery from these sites can be classified as Khami, particularly because of the panel and band ware. Two Zimbabwe Culture sites have been excavated and radiocarbon analyses have yielded calibrated dates that range between AD 1200-1900 (Table 7.2).

Among the most interesting archaeological finds from the research area are the Grain storage facilities, generally referred to as grain bins (Fig. 7.2). Surveys in the Shashe-Limpopo valley have revealed the existence of grain bins in caves situated at the bottom of the tree line on the sandstone ridges.

Near the Shashe-Limpopo confluence six sites were recorded and one of these sites has a cluster of 26 grain bins in a single cave. In total, 49 grain bins were noted from the six sites. The grain bins varied in size with the

Figure 7.2. 19th Century Grain Storage Facilities, Shashe-Limpopo Valley. These grain storage facilities are still in a good state of preservation and residues of sorghum and millet have been recovered from them

largest measuring 1.5 m in diameter and 1.8 m in height. The smallest measured 0.3 m in diameter and 0.4 m in height. The bins were built using mopane wood poles and reeds, which were tied together, using tree cambium rope. A wet dhaka plaster was then added to the frame. The caves are fenced off using mopane poles. This fence has a well-defined entrance. Residues of sorghum and millet occur in some of the grain bins.

To date, the survey's database for the Shashe-Limpopo basin has more than 300 sites. The survey covers a distance of about 20 km from the Shashe-Limpopo area. Most of the sites are associated with rivers and alluvial corridors (Fig. 7.1). Despite the climatic deterrents today, it appears that the valleys have always attracted or accommodated human settlement. The past climatic history of southern Africa has been reconstructed and this has been related to the archaeology of the region (Huffman 1996: van Waarden 1998, Jonsson 1998, Manyanga 2001). The time between AD 900-1300, referred to as the Medieval Warm period, is considered the most favourable, because the climate was warmer and wetter. Climatic conditions became variable and unpredictable after AD 1300, and generally conditions deteriorated to those that exist today. If human settlement in the valley continued after AD 1300, what was the basis of this resilience in the face of deteriorating climatic conditions?

THE INTERPRETIVE FRAMEWORK

The concentration of archaeological sites between AD 900-1800 in the Shashe-Limpopo drainage basin has been demonstrated. The general perception of the climate in this drainage basin has been that it is hot and dry and, therefore, not conducive to human habitation. Most rivers in this region drain into the valley, which in some areas falls below 400 m above sea level. The rainfall is less than 400 mm per annum and varies both within and between seasons (Anderson *et al* 1993). Peak temperatures during the summer are often in the high 30s on the Celsius scale (Kelly and Walker 1976: 554). The low rainfall and high temperatures make it impossible to grow even drought-resistant crops through rain-fed agriculture.

In spite of the harsh conditions, archaeological data, especially on the Zimbabwean side, show that the drainage basin was continuously occupied by people from the Stone Age to recent times. The general interpretive framework in the valley has always explained the concentration of people and their prosperity during the K2 and Mapungubwe phases within the context of climatic change (Huffman 1996: van Waarden 1998, Jonsson 1998, Manyanga 2001). This period is viewed as being wetter and thus affording the inhabitants the opportunity to pursue subsistence strategies that would be less successful under the present drier climatic conditions. Therefore, the valley is thought to have been abandoned for higher ground with the onset of drier conditions. This supposedly resulted in the collapse of the valley state (Mapungubwe) and the subsequent rise of Great Zimbabwe on the Zimbabwean Plateau (Huffman 1978, Maggs 1984). This idea, popularized in the 1970s and 1980s, is not supported by the archaeological data outlined above.

While the end of the Medieval Warm epoch (AD 900-1300) signaled the onset of drier and unpredictable weather patterns, archaeological data point to the fact that the valley was not totally abandoned and continued to support human populations. How communities managed to survive within the context of deteriorating environmental conditions can best be understood by observing historical and current subsistence strategies in the valley: human responses that have allowed communities to survive in an arid environment.

Recent observations in the valley associated with flooding enhanced by cyclone ELINE in 2000 show that events within the valley are affected more by developments outside it rather than what happens in the valley itself. Persistent rain in February and March 2000 resulted in flooding of the Shashe and Limpopo rivers and their tributaries. Tropical cyclone ELINE started out as a tropical disturbance in the Indian Ocean, drifting westward to mainland southern Africa. In the process, it developed into an intense tropical cyclone. Despite the fact that it weakened as it drifted westwards over the southern African mainland, a large amount of rain was still recorded over the northern parts of South Africa (i.e., Zimbabwe and southern Botswana). This resulted in excessive flooding, particularly in the Shashe-Limpopo Valley. While it is acknowledged that this was an exceptional episode, it brought about interesting observations on the nature of flooding in the Shashe-Limpopo drainage basin. The floodplains were flooded with excessive water in the Shashe and Limpopo rivers, and the excessive force in the main rivers blocked the tributaries, resulting in flooding of the tributaries upstream as well. This swelling was further enhanced by the meandering nature of most rivers in the Shashe-Limpopo drainage basin.

In such dry areas like the Shashe-Limpopo drainage basin, flooded areas became useful to agro-pastoralists and even hunting and gathering communities. Flooding brings valuable water and fertile alluvial soils. Whether permanent or ephemeral, the supply of water to these floodplains means that these areas are biologically highly productive and less susceptible to drought than the surrounding dry lands (Brown 1997:38). These conditions offer opportunities for local communities to conduct farming activities, fishing, hunting, and gathering, as well as seasonal grazing exploitation of other floodplain resources for subsistence purposes. Today local populations utilize the Shashe and the Limpopo river valley floodplains as crop fields and as areas for seasonal grazing (Fig. 7.3). Any crop fields established without this consideration cannot sustain growth without irrigation. The importance of the Shashe-Limpopo floodplains for purposes of agriculture has also been used by modern commercial farmers in the valley today. Most irrigated fields in the Shashe-Limpopo are located on flat areas where the river meanders. These are areas susceptible to flooding. The value of the floodplains is that they are important receiving areas for moisture and nutrients. Since the area under discussion always has a deficit in the amount of rainfall needed to sustain crop cultivation, these floodplains remain islands of hope for the local communities. Figure 7.3 demonstrates a successful maize crop near the Shashe-Limpopo confluence that was recorded in January 2002. The crop mostly depends on floodplain moisture, since the annual precipitation for the research area of less than 400 mm is well short of the minimum needed to sustain the crop.

Flooding episodes associated with the Limpopo valley have also been noted in the recent past when the catchments of the Shashe and Limpopo rivers received fairly heavy rains. Hanisch (1980) points out that in a strong rain season the Shashe and the Limpopo burst their banks, resulting in floodplain inundation. This brings with it a strong possibility of floodplain agriculture. The Shashe-Limpopo Valley, with a favorable catchment area, would have served as an important focal point for human groups in the past. These conditions brought populations together in this restricted areas, leading to the development of socio-political complexity in the Shashe-Limpopo drainage basin. Control and access to productive floodplains would have been critical under these circumstances.

Historical period sites are also located on the edge of the floodplain and large quantities of grain were stored there as can be noted from the grain storage facilities. The finding of these 19th century grain storage facilities has revived interest in the subject of pre-colonial food

Figure 7.3. A maize field on the Shashe-Limpopo floodplain at the Shashe-Limpopo confluence January 2002. These fields are carefully located such that they benefit from the moisture in the floodplain (photo by the author)

security, as well as the socio-political conditions that might have resulted in their construction. While 19th century conflicts pose as strong candidates for the circumstances leading to their construction, one has also to examine the vulnerability of the regions between the Zambezi and the Limpopo, both in terms of drought and occasional flooding that impact negatively on agricultural practices.

Colonial administrators, ethnographers, and archaeologists have reported the existence of grain storage facilities located mostly in cave or rock shelters on the Zimbabwe Plateau since the early 20th century. Given widespread accounts related to the Nguni, particularly Ndebele and Ngoni raids in the 19th century, these storage facilities are inextricably connected with them. They are thus considered part of a refuge culture, which is thought to have resulted in widespread movement across the plateau, with settlements being located on hilltops and mountains. There is a strong possibility that the grain bins played an important role in securing the food supply. Therefore, discussions about grain bins have focused on the need for concealed places against cereal raiders. However, the greatest threat to grain is moisture and pests. The caves provide concealment, but it has been overlooked that they also provide pest-free conditions and protection against moisture. Therefore, due to invariable flooding, this 19th century storage practice was probably a response to unpredictable crop yields from one year to the other. After all, good storage of cereals guaranteed the community food supply.

The Zimbabwe period sites Mapungubwe and K2 are oriented towards the rivers and the floodplains. The sites are mostly on natural mounds that resemble natural terraces. In other instances, they are located on hill edges or hilltops. Here sites formed a dense cluster of settlements, making it difficult to define household units. It appears as if settlements were located on those areas with the least chance of flooding, while at the same time affording the communities the chance to exploit the floodplains as fields or grazing areas. A number of sites show close affinities with those that have been recorded on the South African and Botswana side of the valley, where opportunities for flooding and floodplain agriculture has been noted (Hanisch 1980, Huffman 2000).

DISCUSSION

It is apparent from the surveys and the excavations that Zimbabwe's southern Lowveld has a long history of occupation from the beginning of the Stone Age to recent times. Some sites seem to reflect continuity in settlements over a long period, a case that was demonstrated by K2, Mapungubwe and Zimbabwe period settlements. The Leopard's Kopje culture is represented by the Zhizo, K2, and Mapungubwe phases. These are the most successful periods of occupation in the valley. These sites are similar in setting, settlement layout, and stratification. Settlements are developed around a central cattle byre. The settlement pattern is repeated over a wide area, giving the impression that it reflects the major ideological framework of the time. The depth of the archaeological deposits on them shows that these sites were occupied for a long time by people who lived in large household units. The fact that three or more kraals can be noted at a single site shows that people lived in large communities mostly based on kinship ties.

Two patterns appear to have emerged in the settlement patterning in the Shashe-Limpopo drainage basin. There are settlements that developed around circumscribed environments, as is the case of the Mateke Hills (see Manyanga *et al.* 2000, Manyanga 2001). The Mateke Hills have unique local environmental conditions. They receive higher precipitation that attracted prehistoric settlements, thereby explaining the concentration of sites in the area. Further down in the Shashe-Limpopo drainage basin, a different scenario is observed. Sites are located on the sandstone ridges overlooking the floodplain. Flooding increased due to a wetter catchment area, making it a regular feature of the Shashe-Limpopo landscape. Increased regularity of floods may have attracted prehistoric communities to the drainage basin, as is reflected by the number of sites that occur around the river valley floodplains. This settlement practice is still in existence today in the Maramani communal areas, where floodplains are utilized as crop fields. Past communities also could have made use of floodplain agriculture, because sites are oriented above river terraces overlooking the floodplains.

The archaeological sites reflect a common permanent feature in all settlements dating from AD 900-1800. This feature is the presence of grain storage facilities that appear in various forms. At K2, Mapungubwe and Zimbabwe period sites, these appear as raised grain bins sitting on stone foundations. These facilities become more elaborate in the historical period, when details relating to construction are well preserved (Fig. 7.2). Throughout occupation of the valleys, evidence for these storage facilities demonstrates a conscious human response to unpredictable food resources from one year to another. Floodplain agriculture may have been an option but it is likely that flooding was not equally good every year. Therefore, conditions may not have met the communities' food requirements on a yearly basis, resulting in the need for grain storage. This demonstrates that food security has always been a challenge in the Shashe-Limpopo drainage basin. Survival in the harsh climate made planning ahead a necessity and required a clear mastery by communities of the opportunities offered by the valley.

CONCLUSION

The discussion raised in this paper argues for a cautious approach in assessing the Shashe-Limpopo landscape, especially with reference to its ability to provide subsistence. The general perception of the valley, that it is hot and dry, is true, but that does not exclude human habitation because the socio-cultural response of the people that lived there was to adapt to changing climate conditions when necessary. The archaeological record shows that despite the heat and dryness, the valley accommodated human settlement in both prehistoric and historic times. Central to the survival options appears to have been the possibility for floodplain agriculture along the Limpopo, the Shashe, and their tributaries – a practice that meets the subsistence needs of some communities in the Limpopo valley even today. Use of floodplains is the most plausible explanation for continued settlement of the valley in the face of changing environmental conditions. While the importance of floodplain agriculture still needs further elaboration, the practice is certainly not unique in southern Africa. It has always been common practice along the Zambezi River in northern Zimbabwe, the Limpopo River in Mozambique, and along the Chobe River in northern Botswana.

Acknowledgements

The field research was supported by funding under the HRAC project and now AAN and the University of Zimbabwe's Faculty of Arts Research grants, which were funded by SIDA/SAREC. Also I thank the numerous University of Zimbabwe students, who sacrificed their valuable vacation time to help with the fieldwork. The production of site plans and maps was done with the assistance of G. Chikwanda, M. Chifamba, and S. Katsamudanga. My thanks also go to Gilbert Pwiti, Innocent Pikirayi, Webber Ndoro, Alfred Tsheboeng, McEdward Murimbika and Paul Sinclair, colleagues who endured the heat in the valley and have shared ideas relating to subsistence strategies in the Shashe-Limpopo valley with me during the large number of field visits.

References

ANDERSON, I.P., and P.J. BRINN 1993 – The Physical Resource Inventory of Communal Lands of Zimbabwe: An Overview. Overseas Development Administration, Chatham, *Natural Resources Institute Bulletin 60.*

COOKE, C.K. 1958 – *A Preliminary Survey of the Archaeology of the Mateke Hills Area.* Unpublished Report of Rhodesian Schools' Exploration Society, Matabeleland.

COOKE, C.K. 1961 – *Archaeology Report.* Unpublished Report of the Rhodesian Schools Exploration Society-Buffalo-Bend Expedition, Matabeleland.

COOKE, C.K. 1966 – *Archaeology Report.* Unpublished Report of the Rhodesian Schools Exploration Society-Shashi Expedition, Matabeleland.

COOKE, C.K., and H.A.B. SIMONS 1969 – Mpato Shelter: Sentinel Ranch, Limpopo River, Beitbridge, Rhodesia: Excavation Results. *Arnoldia Rhodesia* 4/18, 1969:1-9.

EASTWOOD, E. *et al.* 1995 – *The Rock Art of Nottingham and Sentinel.* Unpublished Report Compiled for Sentinel Limpopo Safaris, Nottingham Estates and Border Ridge Farm.

FOUCHE, L. (ed.) 1937 – *Mapungubwe: Ancient Bantu Civilisations on the Limpopo,* Cambridge University Press, Cambridge.

GARDNER, G.A. 1963 – *Mapungubwe. Volume I,* Van Schalk, (ed.), University of Pretoria, Pretoria.

GARLAKE, P. 1966 – *Iron Age Archaeology.* Unpublished Report of the Rhodesian Schools Exploration Society, Shashi Expedition.

GARLAKE, P. 1967 – *Iron Age Archaeology.* Unpublished Report of the Rhodesian Schools Exploration Society, Maramani Expedition.

GARLAKE, P. 1968 – Test excavations at Mapela Hill, near the Shashi River, Rhodesia. *Arnoldia* 3, 1968:1-29.

HALL, M. 1987 – *The Changing Past: Farmers, Kings and Traders in Southern Africa, 200-1860,* David Philip, Cape Town.

HANISCH, E. 1980 – *An Archaeological Interpretation of Certain Iron Age Sites in The Shashi/Limpopo Valley.* Unpublished MA Thesis, University of Pretoria, Pretoria.

HUFFMAN, T.N. 1977 – Zimbabwe: Southern Africa's first town. *Rhodesian Prehistory* 7/15, 1977: 9-14.

HUFFMAN, T.N. 1978 – The Origins of the Leopards Kopje: An 11th Century *Difaquane. Anoldia Rhodesia* 8/23, 1978:1-23.

HUFFMAN, T.N. 1993 – Broederstroom and the central Cattle Pattern. *South African Journal of Science* 89, 1993:220-7.

HUFFMAN, T.N. 1996 – Archaeological Evidence for Climatic Change During the Last 2000 Years in Southern Africa. *Quaternary International* 33, 1996:55-60.

HUFFMAN, T.N. 2000 – Mapungubwe and the Origins of the Zimbabwe culture. *The South African Archaeological Society Goodwin Series* 8, 2000:14-29.

HUFFMAN, T.N., and J.C. VOGEL 1991 – The Chronology of Great Zimbabwe. *South African Archaeological Bulletin* 46, 1991:61-70.

JONSSON, J. 1998 – *Early Plant Economy in Zimbabwe*, Department of Archaeology and Ancient History, Uppsala.

KATSAMUDANGA, S. 2001 – Geographical Information Systems (GIS) Applications in Archaeological Research: An Investigation into Settlement Patterning in the Shashe-Limpopo Valley. Unpublished Honors Dissertation, University of Zimbabwe.

KELLY, R., and B. WALKER 1976 – The Effects of Different forms of Land Use on the Ecology of a Semi-arid Region in South-eastern Rhodesia. *Journal of Ecology* 64, 1976:553-76.

KUPER, A. 1980 – Symbolic Dimensions of the Southern Bantu Homestead. *Africa* 50/1, 1980:8-23.

LOUBSER, J.H.N. 1990 – Oral Traditions, Archaeology and the History of Venda *Mitupo*. *African Studies* 49/2, 1990:13-42.

MAGGS, T. 1984 – The Iron Age South of the Zambezi. In Klein, R.G (ed.), *Southern African Prehistory and Palaeoenvironments*, Balkema, Rotterdam.

MAGGS, T. 2000 – African Naissance: An Introduction. *The South African Archaeological Society Goodwin Series* 8, 2000:1-3.

MANYANGA, M. 2001 – *Choices and Constraints. Animal Resource Exploitation in South Eastern Zimbabwe C. AD 900-1500*, Department of Archaeology and Ancient History, Uppsala.

MANYANGA, M. 2003 – Settlement Patterning in the Shashe-Limpopo Valley: Reflections from Surveys in Maramani and Lower Mzingwani Area. In Chami, F., G. Pwiti, and C. Radimilahy (eds.), *People, Climatic Change, Trade and Modes of Production in Sub Saharan Africa,* Dar es Salaam University Press, Dar es Salaam.

MANYANGA, M., I. PIKIRAYI, and W. NDORO 2000 – Coping with Dryland Environments: Preliminary Results from Mapungubwe and Zimbabwe Phase Sites in the Mateke Hills, south eastern Zimbabwe. *The South African Archaeological Society Goodwin Series* 8, 2000:66-77.

PWITI, G. 1997 – The Origins and Development of Stone Building Cultures of Zimbabwe. In Dewey, W.J (ed.), *Legacies of Stone: Past and Present, Vol. 1*, Royal Museum for Central Africa, Brussels.

ROBINSON, K. 1961 – The Archaeology Of The Sentinel Expedition 1961. Unpublished Report of the Rhodesian Schools Exploration Society, Sentinel Expedition.

ROBINSON, K. 1967 – Archaeology Report 2. Unpublished Report of the Rhodesian Schools Exploration Society, Maramani Expedition.

SICARD, H. von 1961 – Ruins and their Traditions on the Lower Mzingwani and the Beitbridge Area. *The Southern Rhodesian Native Affairs Department Annual* 34, 1961:8-30.

SINCLAIR, P. 1997 – Human Responses and Contribution to Environmental Change. In B. Sitter-Liver and C. Uehlinger (eds.), *Partnership in Archaeology: Perspectives of a Cross-Cultural Dialogue,* Fribourg University Press, Fribourg.

SUMMERS, R. 1960 – *Environment and Culture in Southern Rhodesia,* Cambridge University Press, Cambridge.

TSHEBOENG, A. 2001 – Late Iron Age Human Responses and Contribution to Environmental Change in the Shashi-Limpopo River Basin: Eastern Botswana. In Chami, F., G. Pwiti, and C. Radimilahy (eds.), *People, Contacts and the Environment in the African Pas*t, Dar es Salaam University Press, Dar es Salaam.

TYSON, P.D., and R. LINDSEY 1992 – The Climate of the Last 2000 Years in Southern Africa. *Holocene* 2, 1992:271-8.

WAARDEN, C. van 1998 – The Later Iron Age. In P. Lane, A. Reid and A. Segobye (eds.), *Ditswa Mmung: The Archaeology of Botswana*, Pula Press, Gaborone.

VOIGT, E.A. 1983 – *Mapungubwe: An Archaeozoological Interpretation of an Iron Age Community*, Transvaal Museum, Pretoria.

Part III:
Paleoclimate and Socio-Cultural Change in the Old World: Europe

Chapter 8
MESOLITHIC SETTLEMENTS OF THE UKRAINIAN STEPPES: MIGRATION AS SOCIOCULTURAL RESPONSE TO A CHANGING WORLD

Olena V. SMYNTYNA
Mechnikov National University, Odessa, Ukraine

Abstract: *Change in archaeological cultures of the Late Palaeolithic and Mesolithic in the Ukrainian is usually attributed migration. Migration can be regarded as a population movement to distant territories, colonization is a long-lasting migration, and relocation is a relatively rapid movement. Two different variants of relocation can be distinguish: a) migrations that give rise to significant enlargement or total change of the culture carriers' habitat, b) movements inside one foraging territory, often expressed as seasonal migrations. In all cases, the main historical function of migrations is to ensure the survival of a specific population. Analysis of migrations in the steppe zone of Ukraine during the Mesolithic does not support the thesis that the migrant culture immediately and totally transforms in the new habitat. As a rule, transformation happens later and, thus, is a result of a complicated multistage interaction between the autochthonous and recently arrived population. In this case, migrations can be regarded as a specific form of ethnic contact. Recent debates related to climatic change and sea level fluctuations are beginning to add yet another dimension to our understanding of migration.*

INTRODUCTION

Contemporary prehistory, archaeology, ethnology, and cultural anthropology tend to interpret migration as one of the four basic genres of human activity along with habitation, storage, and creation (Klein 1978:110). From the very beginning of the systematic studies of early prehistoric hunter-gatherer cultures, there was no doubt that Late Palaeolithic and Mesolithic groups changed their territory in time and space. In most cases, these regular movements are explained by the necessity for a secure food base. Contemporary archaeology should try to reconsider this oversimplified interpretation by taking into account new theoretical and empirical information.

The term "migration" is an integral part of professional terminology of contemporary archaeologists, ethnologists, cultural anthropologists, and related disciplines. The concept seems to be so clear that some encyclopedias and dictionaries consider it unnecessary to define and interpret this concept. Nevertheless, more careful analysis of the term "migration" indicates that the application of this concept differs depending on the spatial and chronological scale. The ecological, economic, ethnic, and social consequences also differ.

In particular, prehistoric studies of migration are explored when it is a question of group movement versus movement of the entire culture. Cultural transmission without human substrate displacement can hardly be traced archaeologically by prehistorians and cultural anthropologists. Since every artefact is the result of targeted activity by a particular person or group, changes in artefact distribution area traditionally are connected with changes of the artefact creators' habitat. At the same time, only data for physical anthropology can exclude (mitigate) doubts about population displacement. However, such data are extremely scarce for the Paleaolithic and Mesolithic.

The process of identifying migration is still under discussion and differing perspectives exist. Most important among them is the opposition between individual versus group character of migration. Associated with this problem is the detection of social, ethnic, or economic backgrounds of the migrant groups' formation.

Taking into account specific features of the contemporary source base of early prehistoric migration studies, two types of migration (colonization and relocation) seem to be the most logical distinction for sociocultural analysis.

COLONIZATION, OR DURABLE AND OFTEN LONG-DISTANCE MIGRATION

Colonization is regarded as a long-term, and in most cases, long-distance and repeated movement of human collectives (never individuals) to empty unsettled areas. The colonization of America and Australia are traditional examples of Upper Palaeolithic migrations. For the Mesolithic, comparable colonization occurred in Northern Europe, i.e. the Baltic region, Scandinavia, Greenland, etc.

Colonization exhibits a typical undulating multi-stage and multilevel character, allowing the possibility of interpreting the phenomena with its historical, cultural, ethnic, economic, geographic, and ecological consequences (Ghumilev 1989:168). Colonization, being one of the basic features of general human evolution, causes a long-lasting effect (human impact) on the landscape of the natural human habitat (Kottak 1991).

First, colonization signifies the origin of the landscape itself as a phenomenon resulting from human activity (Muir 1999:60-61). Nowadays, one can observe a wide variety of forms of such landscape interpretations. They have been elaborated in the framework of traditional and post-modern geography. Most common of them are notions of the cultural landscape as elaborated by C. Sauer and developed by O. Schluter, and the concept of landscape resulting from a series of sequent occupancies as proposed by D. Whittlesey. Historically in Soviet science, such interpretations of landscape origin were discussed within the framework of anthropogenic landscape theory. During the last decade the notion of landscape as an artefact was conceptualized by Darvill, Criado, Boado, and others, signifying similar ideas in post-modern modernism (Smyntyna 2001a:147, Smyntyna 2002:181-184).

To some degree colonization in turn influenced the economic system, the social and ethnic structure, mental activity, psychology, and ideology of its inhabitants. Contemporary prehistory proposes at least three main approaches to the conceptualization of the inner relationships of hunter-gatherer societies with their new natural habitat (originating from their activity in the landscape): "man as nature creature", "man as nature creator" and "man in nature" (Smyntyna 2004:49-52).

The gradual growth in popularity of ideas on mutual and interdependent evolution of nature, and society can be observed by the turn of the century. The fundamental theoretical background for this idea was already created by the beginning of twentieth century in F. Ratzel's anthropogeography and in V. Vernadsky biosphere theories. Increased knowledge about climate, relief, flora, and fauna of prehistoric times became available during the middle of twentieth century thanks to the active development of environmental archaeology, geoarchaeology, and palaeogeography. They provided the fundamental empirical background for the theoretical conceptualization of the "man in nature" idea in Western European and American archaeology, prehistory, and palaeogeography. In Soviet science, such ideas were reflected in the research activity of proponents of the so-called palaeoenvironmental approach to prehistoric studies developed by S. Bibikov. Identification of deep causes of change that took place in nature and society was regarded as integral to systems theory, and was introduced to the study of prehistory, cultural anthropology, and geography. The most widespread concepts in this theory are adjustment (J. Sonnerfeld, A. Goodman), sustainability (G. Bruntland, D. Wobster), and co-evolution (G. Shvebs, M. Feldman). These concepts, as well as the mechanisms of their application to the case studies of hunter-gather communities, have been discussed elsewhere (Smyntyna 2004).

Relocation, or one-time targeted displacement

As a form of migration, relocation is relatively strictly defined in time. It implies isolated changes of habitat by a group, or by individuals, for a particular purpose. This sort of migration can be subdivided into at least two kinds of relocation. One is transmigration, or displacement, that leads to the complete change of living space and foraging territory. The other is seasonal migration, i.e. a movement inside the group's living space.

Transmigration, or living space change

This sort of human relocation is usually traced relatively easily in archaeology. There are two archaeological phenomena. One consists of change to a different zone in the location of the site distribution in a particular archaeological culture.[1] The other is the detection of sites attributed to other cultures on the territory of the traditionally inhabited area. A glance at the maps of the archaeological cultures' site distribution in contemporary Ukraine during the Dryas III/Boreal shows that transmigration was not rare among prehistoric hunter-gatherers (Smyntyna 2001b Fig. 39, 76). Since the origin of the study of prehistory at the end of 6th century, specialists in different branches have tried to detect causes of transmigration and its role in the survival of society. During the last decade, ecologico-demographic explanations have become popular.

Accordingly, balance disturbance, hunting intensity and scale, reproductive potential of the environment, and population density were the main agency that caused transmigration, or at least expansion of living space among Late Paleaolithic and Mesolithic hunter-gatherers. The Soviet science conceptualization of this idea is connected with S. Bibikov's concept of a crisis in the hunter-gatherer economy. Similar thoughts have been discussed in Western archaeology by the processual school of Ethnoarchaeology headed by L. Binford, using the concepts of adaptation and foraging territory (Bibikov 1969, MacArthur and Pianka 1966). This fits well with the processes reconstructed for the steppe zone of Ukraine.

One example of transmigration is the Bilolissya culture that originated at what is now the Ukrainian part of the Lower Danube region. Bilolissya culture, being typical of the Middle Danubian culture sphere at the end of Dryas III, came here from the neighboring territory of Romania. The people followed the aurochs – their main hunting species (Smyntyna 2001b:155-157, Fig. 59). They preserved their traditional tool kits, subsistence, and procurement systems in the new territory until the beginning of the Preboreal phase of the Holocene.

Unfortunately, it is rather difficult to define the character of the Bilolissya culture's population displacement (Smyntyna 2007). It appears to have left no serious impact on the later cultural history of the region. There is

[1] Editors' note: The notion of "archaeological culture" presented here is used mainly in Soviet and post-Soviet archaeological literature. In it cultures are defined as a group of sites situated in peculiar territory during specified period of time in early prehistory. Such cultures are characterized by similarity of artifact complexes, particularly by technology of tool production, composition of tool kits, and tool morphology.

no archaeologically traceable contact with neighboring cultures of that time. Based on the number of migrants and the duration of the occupancy, this phenomenon could, therefore, be interpreted as infiltration. For the basic unit of the Middle-Danubian culture sphere it might be regarded as ethnomigrational separation (Bromley 1983:236). At the same time, the absence of Palaeolithic sites and artefacts in the territory allows the assumption that this process can be interpreted as colonization with respect to the particular ecological region (niche) of occupation. However, this is an atypical rapid form of migration. The necessary setting for this process was formed by two circumstances. One is the growing subsistence crisis in the Middle Danube region. The other is the aurochs distribution in the Lower Danube region that gave the migrants a chance to maintain the same subsistence strategy in the new territory. Recent research in the Black Sea basin at the Pleistocene-Holocene boundary provide additional detail regarding the connection of this population displacement as it relates to sea-level changes (Smyntyna 2008:168). In fact, the "Great Black Sea Flood" hypothesis put forward by W. Ryan and B. Pitman has intensified research, resulting into useful inquiries into the degree of human adjustment necessary to the probable climatic changes that caused the flood (cf. Yanko-Hombach *et al.* 2007).

Most probably, analogous changes of living space are traceable in the north-western part of the Ukrainian Polissya region. Similarity of the fauna complex dominated by reindeer, and a landscape consisting of a pine and birch forest interspersed with swampy and steppe-like spaces, made this territory attractive for the culture bearers of the Sviderian and Ahrensburg cultures. These cultures transmigrate here from the neighboring territory of Poland several times beginning with Dryas II. Unlike the Bilolissya culture bearers, the newcomers stayed here for a longer period (until the middle of the Preboreal). Throughout this time, they preserved their mode of life based on collective hunting of reindeer. The flint industry of most sites displays traits of both traditions of tool processing.

It is equally logical to assume that it was the traditional hunting and gathering economy that caused transmigration of part of the Molodova culture from Middle Dniester region to the Crimean foothills at the end of Dryas II. Later, at the end of Proboreal – beginning of Boreal, the same situation can be supposed for the appearance of the Komornica and Janislavice tradition in the Ukrainian Polissya region (Smyntyna 1999b:34).

Many researchers underline the special role of demographic agency in the process of transmigration. Primary attention is given to population pressure on the foraging territory (Dolukhanov 1979:16, Freeman 1971). At the same time, it should be stressed that increasing population density caused the hunter-gatherer economic crisis to come to a head all by itself.

Analysis of Boreal (Late Mesolithic) site distribution lets us distinguish a special form of transmigration that led to habitat changes, i.e. the oncoming relocation. This kind of migration is exemplified by the dispersion of the Grebeniki culture, found between the Carpathians and the Dniestr Valley in the Ukraine around 6000 BC. It migrated into the Lower Dnieper region and the Crimean foothills, accompanied by the Kukrek population movement to Lower Dniester area, and the Girs`ko-crimean culture's dispersion into the Lower Dnieper region (Smyntyna 1999b:35, Fig. 4). Analogous processes may have occurred in the boreal and forest-steppe zones during the second half of the Boreal.

A vital need for broadening foraging territory seems to be necessary but, taken alone, does not diminish actualization of such transmigration model. Steady, durable, and peaceful character of "oncoming" relocation could be guaranteed only by mutual interest of all migration participants in these cross-ethnic intercultural contacts. Ethnographic evidences testify that such forms of ethnic interaction are aimed to increase survival rates and improve basic secure and sustainable development of all sides of this process. It could be achieved with the help of knowledge, skills, and source-base borrowed from the other cultures taking part in the cross-ethnic contacts. If such contacts are accompanied with oncoming movements of different ethnic cultural transmitters, their logical and inevitable result is the origin of new culture or new forms of social integration. For hunter-gatherers societies it is often supposed that oncoming movements could bring about formation of specific forms of tribe (the so-called "blinking tribe"). It displays itself at the moments when representatives of different cultures meet and it disappears when they separate for a while (Ghirenko 1991).

In most cases, the "oncoming" movement was durable and peaceful. Today, the only known evidence of war-like conflicts and armed resistance to this process is found in the collective burials at the Dnieper rapids, but the sociocultural interpretation remains an acute issue (Lillie 1997).

Although such implications of "counter-relocation" seem quite logical, it is also possible that we face another sort of targeted relocation, i.e. seasonal migration that will be examined later. At the same time, one should not ignore the inherent desire for "a hunters journeys". This may arise from simple curiosity and the search for a "better life". This Most likely occurred together with the tendency to broaden the living space, strengthening of the subsistence base, eliminating personal conflict and tension, and other factors connected with behavioral and livelihood peculiarities of the early prehistoric population (Krupnik 1989:220, Jolly and Plog 1987:5).

Worth mentioning are additional conditions that promoted not only "oncoming" but also one-way transmigration. These include the absence of insuperable natural frontiers, availability of a variety of ecological niches in the new territory, and pre-adaptation or the potential capability of individuals or groups to survive and adapt in new habitats (Stein and Rowe 1989:132,

134). It should be stressed that these conditions also contribute in some measure to colonization process.

Analysis of the consequences of migration can be traced through the data of Mesolithic settlements of steppe zone of Ukraine. It is thought that the data indicate that migrant cultures transform rather quickly in a new habitat. To the contrary, in most cases migration appears to be the only possibility to preserve the traditional ethnic culture and subsistence systems as indicated by tool kits and technology of flint processing in cultures such as Bilolissya, Swiderian, Ahrensburg, Molodova, etc. This is so, because for hunter-gatherers the main criteria of new habitat choice was only the similarity in landscape with the usual well-known environment and climate.

As a result, transmigration regarded as living space change for purposes of increased territory could be regarded as a necessary measure permitting further evolutionary development through human impact, thus permanently changing the environment over time. The culture of the newcomers began to change gradually. It was influenced by multi-stage contacts with the local autochthonous population and adaptation to the local landscape. Such processes usually led to modification of both the recently arrived and the previous population, and resulted in the formation of a new archaeological culture. In such context, transmigration for living space change can be interpreted as a specific form of intercultural contact, as well as a peculiar agency of cultural genesis.

Seasonal migrations

Seasonal migration can be simplistically defined as periodically repeated population movements within the same living space. Originally, this was conceptualized as the "residential mobility" theory proposed by representatives of processual "New Archaeologists" headed by L. Binford (Binford 1980:16). Its proponents believe that residential mobility is linked to temporal and spatial variability of resources, such as meat, fish, and vegetables (Bahn 1997:245-257, Kelly 1983:277-306). Residential mobility theory is connected with the social agency of seasonal migration (Anthony 1990:895-914, Freeman 1971:512-521). Important information about the mechanisms and conditions of this sort of migration can also be obtained with the help of middle range theory with the site catchment area and optimal foraging theory based on the patch choice model (Vita-Finzi and Higgs 1970:1-37, MacArthur and Pianka 1966:603-609). Most seasonal migration studies in Western European and American archaeology are connected with elaboration of this context zone system that begins with the place of occupation (permanent or temporary). Binford distinguishes between foraging and strategic radius, broadened space, and visiting zones (Binford 1980:4-20). Another useful concept is that of the "landscape of habit" (Gamble 1998:426-449).

In contemporary "Eastern" (i.e., post-Soviet) prehistory this kind of population movement has a comprehensive empirical definition related to seasonal site occupation, although a theoretical framework for seasonality still awaits conceptualization. A first steps in this direction is S. Bibikov`s assumption of winter occupation of the Crimean Mountains versus a summer habitation on the Kerch Peninsula (Bibikov 1971:18-19). Another important idea discussed in this context is connected with the "cluster" occupation system, originating occasionally during favorable phases (Smyntyna 1999a).

In spite of these efforts, one must admit that the absence of organic remains in the majority of Mesolithic sites of steppe zone in Ukraine, as well as the impossibility to demonstrate synchronicity of most sites provides little chance to detect this type of population movement empirically. The general observations made by representatives of the different approaches to seasonal migration, including ethnographic case studies, suggest that they are not connected with any peculiar phase of a yearly activity cycle, but rather with spatial distribution of the food base and the character of the particular habitat. In such context, seasonal migrations could be interpreted as an important element of food procurement. In theory, this kind of migration is an integral part of the subsistence system and of the human activity in general.

CONCLUSION

Analysis of the archaeological evidences for movements of Mesolithic hunter-gatherers provides the possibility to distinguish long lasting migrations (colonization) and the comparatively rapid movements (relocation). We can trace two variants in the latter: a) migration resulting in a significant enlargement or total change of the habitat (transmigration), and b) movements inside the foraging territory that are often called seasonal migrations.

Such variability in concrete forms of population displacement, and differences in their scale and historical as well as sociocultural consequences, implies differences in agencies. These differences caused their necessity and regularity. In some cases, such as the colonization process, migrations are regarded as a condition of landscape formation, and the elaboration of a special sort of culture (totally depended on nature, including climate and environmrent, deeply influencing its natural habitat through human impact or trying to provide common creative acts together with nature). Transmigration is a more common form of cultural preservation in a new territory. However, "oncoming" movement within the same ecological environment is one of the forms of ethnic contact. Seasonal migrations are one of the many ways of food procurement, and thus an important element of the subsistence system.

The imprecise understanding of sociocultural consequences for early prehistoric migrations requires a deeper conceptualization of many problems that inevitably arise in any migration case study. Most important among them are the lack of archaeological criteria for migration, identification of the migration subject, and detection of differences between cultural transmission and human displacement.

Living space exploitation with its emphasis on territory and temporal scheduling requires a purposeful hierarchy of human activities. This concept opens an opportunity to delineate new aspects of migration theory. Clarification of this theory should consider the following points:

- Migrations operate a various scales that must be analyzed in accordance with the temporal, territorial, and purposeful frameworks of the migrant's living space and exploitation structure;
- Migrations need to be considered in relation to the economic, social, and ritual activity of early prehistoric communities and;
- Definitions of migration must take into account the route and destination of the migrants, food and raw material procurement, ethnic traditions, marriage and kinship systems, specific features of ritual activity, world outlook, and religious beliefs etc.

Finally, in the future the discussion of migration will utilize more scientific environmental and climatic data, to enhance our understanding of migration as a form of human adjustments in the face of climatic change. The current debate about the impact of Black Sea sea-level changes on human populations is indicative of this trend.

References

ANTHONY, D. 1990 – Migration in Archaeology: The Baby and the Bathwater. *American Anthropologist* 92, 1990:895-914.

BAHN, P. 1977 – Seasonal Migration in The South-West France During the Last Glacial Period. *Journal of Archaeological Science* 4, 1977, 245-257.

BIBIKOV, S. 1969 – Nekotorye Aspekty Paleo-economicheskogo Modelirovaniya Paleolita, [Some Aspects of Paleoeconomic Simulation of Palaeolithic]. *Sovetskaya Archeologia* 4, 1969:5-22.

BIBIKOV, S. 1971 – Plotnost` Naseleniya i Velichina Okhotnichiih Ugodiy v Paleolite Kryma, [Population Density and Width of Foraging Territory at the Crimean Palaeolithic]. *Sovetskaya Archeologia* 4, 1971:11-22.

BINFORD, L. 1980 – Willow Smoke and Dog's Tails: Hunter-Gatherers Settlement System and Archaeological Site Formation. *American Antiquity* 45/1, 1980:4-20.

BROMLEY, Y. 1983 – *Ocherki Teorii Etnosa*, [Outlines of the Ethnos Theory]. Nauka, Moskwa.

DOLUKHANOV, P. 1979 – *Geografia Kamennogo Veka,* [Stone Age Geography]. Nauka, Moskwa.

FREEMAN, M. 1971 – A Social and Ecological Analysis of Systematic Female Infanticide Among the Netsilik Eskimo. *American Anthropologist* 73/5, 1971:512-521.

GAMBLE, C. 1998 – Palaeolithic Society and Release from Proximity: A Network Approach to Intimate Relations. *World Archaeology* 29, 1998:426-449.

GHIRENKO, N. 1991 – *Sociologia Plemeni*, [Sociology of Tribe]. Nauka, Moskwa.

GHUMILEV, L. 1989 – *Etnogenez i Biosphera Zemli*, [Ethnogenesis and Biosphere of the Earth]. Leningradskogo Gosudarstvennogo Universiteta, Izdatelstvo, Leningrad.

JOLLY, C., and F. PLOG 1987 – *Physical Anthropology and Archaeology*, McGraw-Hill, New York.

KELLY, R. 1983 – Hunter-Gatherer Mobility Strategies. *Journal of Archaeological Research* 39/2, 1983:277-306.

KLEIN, L. 1978 – *Archeologicheskie Istochniki*, [Archaeological Sources]. Izdatelstvo Leningradskogo Gosudarstvennogo Universiteta, Leningrad.

KOTTAK, C. 1991 – *Anthropology: An Exploration of Human Diversity*, McGraw-Hill, New York.

KRUPNIK, I. 1989 – *Arkticheskaya Etnoekologiya*, [Arctic Ethnoecology]. Nauka, Moskwa.

LILLIE, M. 1997 – Women and Children in Prehistory: Resource Sharing and Social Stratification at the Mesolithic-Neolithic Transition in Ukraine. In Moore, J., and E. Scott (eds.) *Invisible People and Processes: Writing Gender and Childhood into European Archaeology*, Leicester University Press, London and New York.

MacARTHUR, R., and E. PIANKA 1966 – On Optimal Use of a Patchy Environment. *American Naturalist* 100, 1966:603-609.

MUIR, R. 1999 – *Approaches to Landscape*, Macmillan, London.

SAUER, C. 1963 – The Morphology of Landscape. In J Leighly (ed.), *Land and Life: A Selection from the Writings of Carl Otwin Sauer,* University of California Press, Berkley.

SMYNTYNA, O.V. 1999a – K Probleme Tipologii Mesoliticheskih Pamyatnikov, [To the Problem of Mesolithic Sites Typology]. *Stratum Plus* 1, 1999:239-256.

SMYNTYNA, O.V. 1999b – Migratsii Naselenia i Sposob Kulturno-Istoricheskoi Adaptatsii, [Migrations of Population and Mode of Cultural-Historic Adaptation]. *Vita Antiqua* 2, 1999:31-37.

SMYNTYNA, O.V. 2001a – Landshaft v Konteksti Paleoekologichnogo Pidhody do Vyvchennya Kultury, [Landscape in the Context of Paleoecological Approach to Culture Studies]. *Zapyski Istorychnogo Fakultety / Odes'kiy Natsionalnyi Universitet I.I.*, *[Notes of Faculty of History, Odessa National University I.I.*,], Mechnikov, 1999:143-152.

SMYNTYNA, O.V. 2001b – *Zonalnist Rannyopervis-nuch Cultur: Doslidjennya, Facty, Hipotezy,* [Zonal Features of Early Prehistoric Cultures: Investigations, Facts, Hypothesis]. Astroprint, Odessa.

SMYNTYNA, O.V. 2002 – Landshaft v Zakhidno-Evropeis`Kiy Archeologii: Suchasni Pidkhody do Kontseptualizatsiyi, [Landscape in West-European Archaeology: Contemporary Approaches to Conceptualizations]. In D. Kozak, (ed.). *Novi Tekhnologiyi v Archeologiyi*, Lviv, 2002:177-188.

SMYNTYNA, O.V. 2004 – The Environmental Approach to Prehistoric Studies: Approaches and Theories. Theme Issue: Environment and History: History and Theory. *Studies in the Philosophy of History* 42/4, 2004:44-59.

SMYNTYNA, O.V. 2007 – Late Mesolithic of the Ukrainian Part of the Lower Danube Region: New Perspectives of Human Adaptation and Interpretation of Natural Environments. IGCP521. Black Sea-Mediterranean Corridor during the last 30ka: Sea level change and human adaptation. Istanbul 2005, IGCP 521, *Quaternary International* 167-168, 2007:114-120.

SMYNTYNA, O.V. 2008 – Transmigrations as a Mechanism of Living Space Exploration in the Northwestern Black Sea Region at the Pleistocene – Holocene boundary. In V. Yanko-Hombach ed., *Extended Abstracts of 4th Plenary Meeting and Field Trip of Project IGCP 521 "Black Sea – Mediterranean Corridor During the Last 30 ky: Sea Level Change and Human Adaptation.* Springer. Dordrecht, 2008:167-169.

STEIN, Ph., and B. ROWE 1989 – *Physical Anthropology*, McGraw-Hill, New York.

VITA-FINZI, C., and E. HIGGS 1970 – Prehistoric Economy in the Mount Carmel Area of Palestine: Site Catchment Analysis. *Proceedings of Prehistoric Society* 36, 1970:1-37.

WHITTLESEY, D. 1929 – Sequent Occupancy. *American Association of Geographers* 19, 1929:162-166.

YANKO-HOMBACH, V., A. GILBERT, N. PANIN, and P. DOLUKHANOV (eds.) 2007 – *The Black Sea Flood Question. Changes in Coastline, Climate and Human Settlement*. Springer, Dordrecht.

Chapter 9
THE EARLY MEGALITHS OF SW ATLANTIC EUROPE AND THE INFERENCE OF THE SOCIO-ECONOMIC ORGANIZATION OF THEIR BUILDERS (8TH – 6TH MILLENNIUM BC)

David CALADO
The Portuguese State Institute of Architectonic Heritage (IPPAR)

Francisco NOCETE and José M. NIETO
University of Huelva

Dimas MARTÍN-SOCAS and Maria D. CÀMALICH
University of la Laguna

Abstract: *An intensive surface survey was undertaken to identify sites with standing stones in a 50 km² area near the town of Lagos, southern Portugal. Seventeen large settlements with menhirs were detected during the survey. Statistical analysis of lithic artifacts suggest that these settlements date before the transition from the 6th to the 5th millenniums BC, corroborating findings for the central Portuguese standing stones. This chronology is further supported by archaeological excavations at Quinta Queimada, a settlement site with standing stones, where a typical early Mesolithic EMM (Mediterranean Epipaleolithic Microblade Industry) assemblage was found in association with the menhirs, as well as by the OSL dates from a menhir at the Quinta Queimada site. The upper half of a thick soil layer overlaying and sealing the implantation pit of the menhir was OSL-dated to the transition of the 5th to the 4th millennium BC (Shfd 2013: 3975±175 BC), providing a solid date ante quem for the erection of the standing stone. The erection of the menhir is substantially older and was dated by OSL to 7145±445 BC (Shfd 2014), indicating that the standing stone was, with 95.4% probability, erected between the early 8th and late 7th millenniums BC. The use of geographical models to understand the dense pattern and spatial distribution of settlement sites with standing stones suggest that SW Atlantic Europe probably had widespread complex hunter-gatherers or early food producer communities without domesticates with a high level of residential stability during the 8th millennium BC, and certainly by the transition from the 7th to the 6th millennium BC. These communities seem to have been responsible for a heavy economic-related impact on the environment.*

INTRODUCTION AND CHRONOLOGICAL SETTING

Following our investigations into the origins of social complexity in SW Europe, a systematic survey was carried out over a 50 km² sector in SW Portugal. During the survey, 17 sites with standing stones were identified (Fig. 9.1). Some of them were previously known, but the vast majority are newly recorded sites. Surprisingly, all the groups of standing stones seem to be linked to settlements with lithic assemblages predating the middle of the 5th millennium BC (Calado 2000a, 2000b, 2003, Nocete 2001). Other researchers have identified similar chronological patterns for the standing stones of Algarve (Gomes and Cabrita 1997) and central Portugal (Calado, M. 2002, 2003). Related ^{14}C dates from short-lived samples of Iberia is given by López de Pablo and Gómez Puche (2009).

The inferred chronology was substantiated during excavations at the settlement site of Quinta Queimada, where a typical early Mesolithic, Mediterranean Epipalaeolithic Microblade Industry (Fortea 1973) in association with the menhirs was identified. Optically Stimulated Luminescence (OSL) dating was conducted on soils associated with an *in situ* standing stone at Quinta Queimada. The OSL results from a soil sample collected inside the undisturbed and sealed menhir`s implantation pit yielded an age of 7145±445 years BC (Shfd 02014), suggesting that, with 95.4 % probability, the soil filling the implantation pit was last exposed to light between 7983 and 6203 BC. Thus, the erection of at least one of the standing stones at Quinta Queimada must be no younger than 6200 BC (Bateman 2002, Calado *et al*. 2003a).

The chronometric results from Quinta Queimada have parallels in several, still unpublished, 6th millennium BC calibrated radiocarbon ages obtained by Tavares da Silva and Joaquina Soares. They come from a hearth overlying a layer covering and sealing a fallen menhir at the settlement site of Vale Pincel, located approximately 125 km north of Quinta Queimada (Calado *et al*. 2003a). Similar radiocarbon dates have been obtained from two other sites associated with standing stones in Portugal. A charcoal sample recovered from inside the implantation pit of a, presumed younger type, Alentejo, standing stone at Meada, in central Portugal, yielded a calibrated radiocarbon age (UtC-4452: 6022±40 bp)[1] from the transition of the 6th to the 5th millenniums BC (Oliveira 2000). Also, Varela Gomes obtained two calibrated radiocarbon dates from the second half of the 6th

[1] Editors' note: UtC-4452: 6022±40 bp, 68.2% probability = 4960 (68.2%) 4840 BC; at 95.4% probability = 5020 (95.4%) 4790 BC.

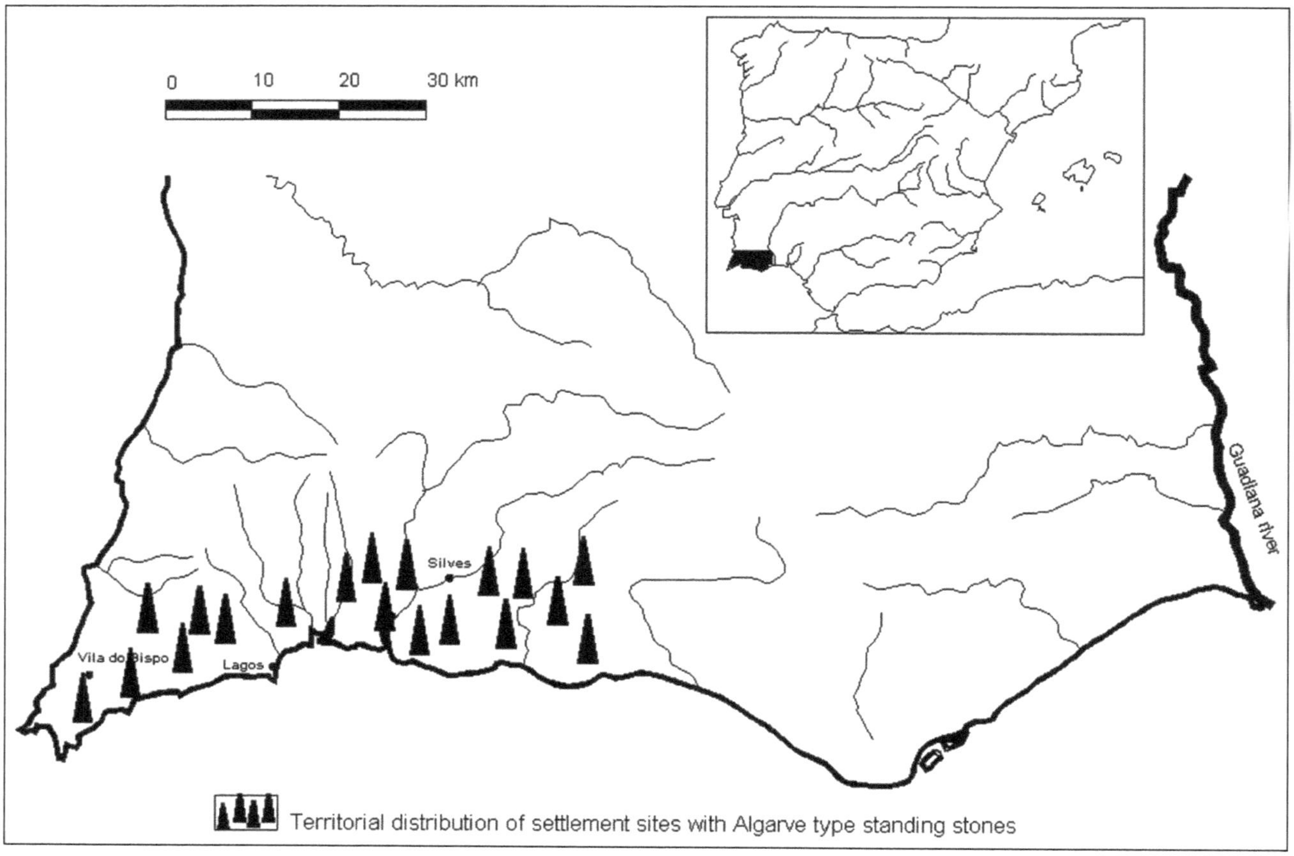

Figure 9.1. Lagos, southern Portugal, survey area and standing stones

millennium BC (Icen-645: 6440±60 bp;[2] Icen-873: 6560±70 bp).[3] They are from a hearth associated with a menhir, as well as associated with typical early Neolithic ceramics, including Cardial Ware, at Padrão, a large settlement site with the apparently oldest Algarve type standing stones.

Increasing and reliable evidence is building for the erection of standing stone in Portugal from at least the late 7th millennium BC. Furthermore, these very old standing stones in the Algarve are associated with settlement sites (Gomes and Cabrita 1997, Calado, D. 2000a, 2000b, 2003, Ramos 2000, Nocete 2001, Martín-Socas *et al.* 2003, Calado *et al.* 2003d) that are characterized by a pattern of high degree of residential stability (Calado, D. 2000a, 2000b, Nocete 2001, Calado *et al.* 2003b, 2003d).

STRUCTURE OF THE SETTLEMENT

Analysis of the spatial distribution of settlement sites with standing stones was conducted in a 50 km² sector. The Delaunay triangulation method was used and resulted in a convex hull with an area of 24 km² and a perimeter of 21.7 km for a total of 17 sites. The nearest neighbors are less than 900±440 meters apart (Calado, D. 2000a, 2000b). A similar pattern of distribution was identified for the 7th and 6th millenniums BC (late Mesolithic/ early Neolithic) settlement sites without menhirs in the Spanish provinces of Huelva (Nocete 2001) and Cádiz (Ramos 2000). It revealed a regular pattern of territorial occupation common to the whole of the Extreme SW Atlantic Europe, between Gibraltar and the western coast of the Algarve, extending throughout the Mesozoic, Cainozoic, and Quaternary geologic periods (Ramos 2000).

Southern Portuguese settlement sites with menhirs vary between 6 and 40 hectares in area (Calado 2000a, 2000b). At Milrei – Padrão the site extends for roughly 1.5 km. The large settlement sites with standing stones in Extreme SW Atlantic Europe appear to have been true permanent villages characterized by a dispersed scatter of house compounds with some similarities to those of various still existing farmer-herdsman societies, such as those of the Sidama people of Ethiopia. Although this pattern of *intra-site* occupation is particularly difficult to understand and interpret via archaeological intervention (Binford 1983), it is identifiable in the specific case of the Algarve by incorporating the standing stones with the settlement sites.

The earliest phases of the settlement sites with standing stones of Algarve appear to predate the earliest

[2] Editors' note: Icen-645: 6440±60 bp, 68.2% probability = 5480 (68.2%) 5360 BC; at 95.4% probability = 5510 (95.4%) 5300 BC.

[3] Editors' note: Icen-873: 6560±70 bp, 68.2% probability = 5610 (9.6%) 5590 BC, 5570 (58.6%) 5470 BC; at 95.4% probability = 5630 (83.6%) 5460 BC, 5450 (11.8%) 5370 BC.

indications of domesticated cereals in SW Iberia. The earliest evidence comes from the Murciélagos and Toro caves (Calado *et al.* 2004), in Andalusia, Spain. These sites have been radiocarbon dated from the middle of the 6th millennium BC onwards. Similarly, the pollen diagrams for Lagoa Travessa, in the Alentejo coast, Portugal, indicate the earliest cereal pollen from the second half of the 6th millennium BC (Mateus 1992, Calado *et al.* 2004). This would suggest that the rise of a high degree of residential stability and social complexity precedes cereal agriculture in the SW Atlantic Europe. Such a generalist pattern, proposed by Godelier (1981) and Aikens (1981), today seems to be confirmed as a common and widespread phenomenon (Chang 1981, Price and Brown 1985, Calado *et al.* 2003b, 2004, Okada 2003).

TERRITORY AND SOCIAL ORGANIZATION

In addition to the *intra-site* analysis, the standing stones of the Extreme SW Atlantic Europe have been shown to be a powerful and singular instrument for *inter-site* territorial analysis. Their uniformity of manufacture and recurring symbolism enables the inference of a 8th or 7th millennium BC Atlantic Europe society reproduced in symbolic artifacts. The standing stones themselves define a divisional cultural trait in the common material complex, typifying the entire SW Atlantic coast of Iberia. We understand the distribution of the western Algarve`s menhirs as reflecting specific and uniform elements of a common superstructure shared by its builders, an *emic* manifestation of a them *vs.* us dichotomy that fully correspond to the demands of self-perpetuation, unity, interaction and identification, as postulated by Barth (1969) for determination of ethnic distinction. As such, it seems to replicate a common ideology and identity throughout a region of almost 1,000 km², apparently reflecting a concrete territory akin to a supra-local polity (Johnson and Earle 1987, Nocete 2001) organized along lineages (Dupre and Rey 1973, Meillassoux 1977, Nocete 1984, Plog 1990, Gailey and Patterson 1998).

Driscol utilized standing stones symbols to identify patterns of social organization in Scotland (Driscol 1988). However, the society described by Driscol is very different in social organization and time than the ones from western Algarve. The Pictish symbols of Scotland's standing stones are interpreted to be a coercive expression emanated by the local socio-political elite to articulate the relation in space and time between the forces of production, the serfs, with the means of production, the land, and as a way to control the social and natural resources reflecting the ideology of a hierarchical society. To the contrary, the six thousand years older standing stones and symbols from SW Algarve are not linked to a class-related appropriation inequality of the means of production and productive forces. Due to the supposed non-existence of widespread socio-centric exchange circuits, cereal agriculture and appropriation of the means of production, the menhirs and symbols must be understood as a classificatory code within the villages. We interpret them as reflecting the different totemic lineages of the inhabitants of the various house-compounds as well as the prestige and the consensual domain over the social labour and territory through the unequal appropriation of the magical means of reproduction of the universe and life by the leading social agents (Asombang 1999, Godelier 1981, Southall 1999).

The standing stones of the Extreme SW Atlantic Europe, specifically in Western Algarve, exhibit some unique features. All of them are explicit sculptures of phalluses, with clear delineation of the gland and urethra. The size and weight of the menhirs in a particular settlement may vary considerably, with standing stones of less than one thousand kilograms found alongside giants weighing more than ten times that weight. The symbols decorating the stones are consistently of only four different types, and express primarily feminine sexual attributes: female breasts (*Mama*), vulvas with one or more of the labia open (*Vulva B*), vulvas with the outer labia closed (*Vulva A*), and undulated patterns (*Onda*). There appears to be no special relation between an explicit type of symbol and particular settlement site. Each site may include a great number of standing stones with randomly distributed symbols. However, each menhir bears only one type of symbol, suggesting the internal division of the *intra-site* space by a (Wallace 1966) communal cult expressed in a totemic ritual of solidarity, identifying each basic cluster of people through symbols engraved on objects. The shape of these objects denotes an intrinsic value of an intensification rite (Calado *et al.* 2003b), a structuring element of the ideology of producer societies (Brögger 1977). The recurrence of the same four types of symbols in numerous settlement sites indicates what seem to be four huge, probably leaderless, lineages that, like the Naga (Friedman 1977), spread throughout the territory of the menhirs builders (Calado *et al.* 2003b).

One of the symbols, the *Mama*, is very common in westernmost areas, becoming very rare, or even non-existent, in the very eastern part of the menhirs builders' territory. Also, settlement sites/villages with standing stones become scarcer to the east, with the eastern boundary located in the middle of the Algarve. The combination of the progressive disappearance of the settlement sites with standing stones along with the progressive scarcity of the *Mama* symbol as one moves east, may reflect a phenomenon of multiple fission of the primordial social units that began to erect the standing stones in their large villages. Population fission is interpreted as a way to resolve conflicts and demographic pressure among societies having a low level of social stratification and the capacity for territorial expansion (Kent 1989). Because of the specific characteristic evidence for territoriality, boundaries and ethnicity projected onto artifacts that ought to be integrated in the sphere of the superstructure of a society, the Algarve's standing stones seem to represent the earliest records evidencing a social phenomenon of population fission in Atlantic Europe.

RAW MATERIALS AND NATURAL OUTCROPS

The analysis of the lithic items from the settlement site of Quinta Queimada show that, apart from the bracelets made of black shale (the nearest natural outcrops are roughly 25 km away); all of the artifacts found are made from rocks that were collected close to the settlement site. Even the standing stones at the different settlements are made from material chosen for its hardness and splitting quality, usually an oolithic, crinoidal, and micritic limestone. In the few cases where outcrops of limestone do not exist near the settlement sites, other types of rocks local to the area were used for standing stone construction. It may thus be concluded that, with the exception of the black shale bracelets, there is no evidence for finished items or raw materials being acquired from distant locations.

The variety of imported items introduced into socio-centric-intricate exchange circuits is a distinctive feature of hierarchical and political centralized societies (Brunfield and Earle 1987, Bayman 2002), and is amply demonstrated in the archaeological records of Eastern North America in the early part of our era (Binford 1983), and in SW Atlantic Europe from the 3rd millennium BC onwards (Nocete 2001). This phenomenon must not be confused with the egocentric model of dispersal through exchange of restricted sets of objects, as observed by Sharp, among the Yir Yoront of Australia (Sharp 1964), and well documented in Europe and elsewhere since late Palaeolithic times (Lyons *et al.* 2003). Thus, the evidence does not sustain the hypothesis that the menhirs builder's society was hierarchical or politically centralized.

In the specific case of the Algarve, the evidence from Quinta Queimada points towards a model of mainly local self-exploitation of the raw materials for artifact production. The only exogenous objects are the bracelets in black shale, which outcrops one-day's journey away. Alternatively, the origin of the black shale could be from farther to the north or northeast, from where black shale outcrops in the Palaeozoic strata of Alentejo or Andalusia have been identified. The finding of black shale bracelets in the middle 6th millennium BC early Neolithic levels of the Nerja and Carigüela caves as well as in the superficial level of Toro cave (Martín-Socas *et al.* 2004), in the Spanish Andalusia, supports this possibility. If the source of the black shale bracelets proves to be from the Palaeozoic strata of Alentejo or Andalusia, than the data can be interpreted as evidence of long egocentric exchange lines of restricted sets of status objects.

SETTLEMENT STRUCTURE AND IMPACT UPON THE ENVIRONMENT

It has been understood that the first large-scale human-induced impact on the environment in extreme SW Europe was the result of the introduction of agriculture. However, recent data calls into question this traditional view.

The earliest evidences for cereals (*T. durum, T. aestivum, T. monococcum, T. dicoccum, H. sativum)*, obtained from the caves of Murcielágos and Toro, in inland Andalusia, Western Iberia, date to the middle of the 6th millennium BC (Calado *et al.* 2004, Martín-Socas *et al.* 2004, Martín-Socas *et al.* 2004). Use of domesticated, or able to be domesticated, specimens of possible native origin, pig (*Sus domesticus*), ox (*Bos taurus*) and horse (*Equus sp.*), together with non-endemic species, goat (*Capra hircus*) and sheep (*Ovis aries*) was found at the 6th millennium BC coastal settlement site of El Retamar (Lazarích *et al.* 2002), near Cádiz. However, the wild species, red deer (*Cervus elaphus*), rabbit (*Oryctolagus cuniculus*) and hare (*Lepus capensis*), seem to have constituted the major part of the diet (Cáceres 2002). The data from El Retamar, where secondary exploitation of domesticated specimens in comparison to wild animals is indicated, can be related to the pollen diagrams of Lagoa Travessa, where the production of cereals in extreme SW Europe was marginal until approximately the middle of the 5th millennium BC. This suggests that the production economy based on cereals and domesticated animals was marginal in extreme SW Atlantic Europe before approximately 4500 BC and probably nonexistent before about 5500 BC (Calado, Nieto, and Nocete 2003b, Calado *et al.* 2004). However, evidence from river estuarine cores and pollen diagrams establish a much earlier human impact upon the environment.

Isotopic analysis from *tapes decussata* shells from the Odiel River shell middens (Álex *et al.* 2004), and sediment cores from the Guadiana River (González-Vila *et al.* 2003), attest to major incidents of soil erosion starting in approximately the 8th millennium BC with a sharp increase during the middle 7th millennium BC (Fig. 9.2). The nearly complete destruction of the ancient soils related in time to the use of the settlement sites is identified in the standing stones villages (Calado, D. 2000a, Calado *et al.* 2004). Additionally, thousands of heavy Mirense stone axes, characterized as early Mesolithic (*Epipalaeolithic*) hoes by Cardoso and Gomes (1997), are spread throughout the surrounding fields. Similarly, pollen cores from Lagoa Travessa indicate heavy deforestation since the middle of the 7th millennium BC. These events are temporally related to the rise of a dense network of settlement sites identified in extreme SW Atlantic Europe, and appear to reflect profound human impact on the environment (Nocete 2001, Calado *et al.* 2004). In other words, the pristine pinewood forest evident in the SU81-18 maritime pollen core since approximately 21,000 BC (Turón *et al.* 2003), was devastated by man during the 8th and 7th millenniums BC, perhaps with help from natural causes (Calado *et al.* 2004).

Pollen charts also suggest that this man-induced deforestation was not chaotic, but selective. The disappearance of pinewood forest evident at Lagoa Travessa apparently did not affect fruit trees, oak (*Quercus*), and olive (*Olea*), suggesting some form of primitive forest management that could be expanded to other floral species (Calado, Nieto, and Nocete 2003b,

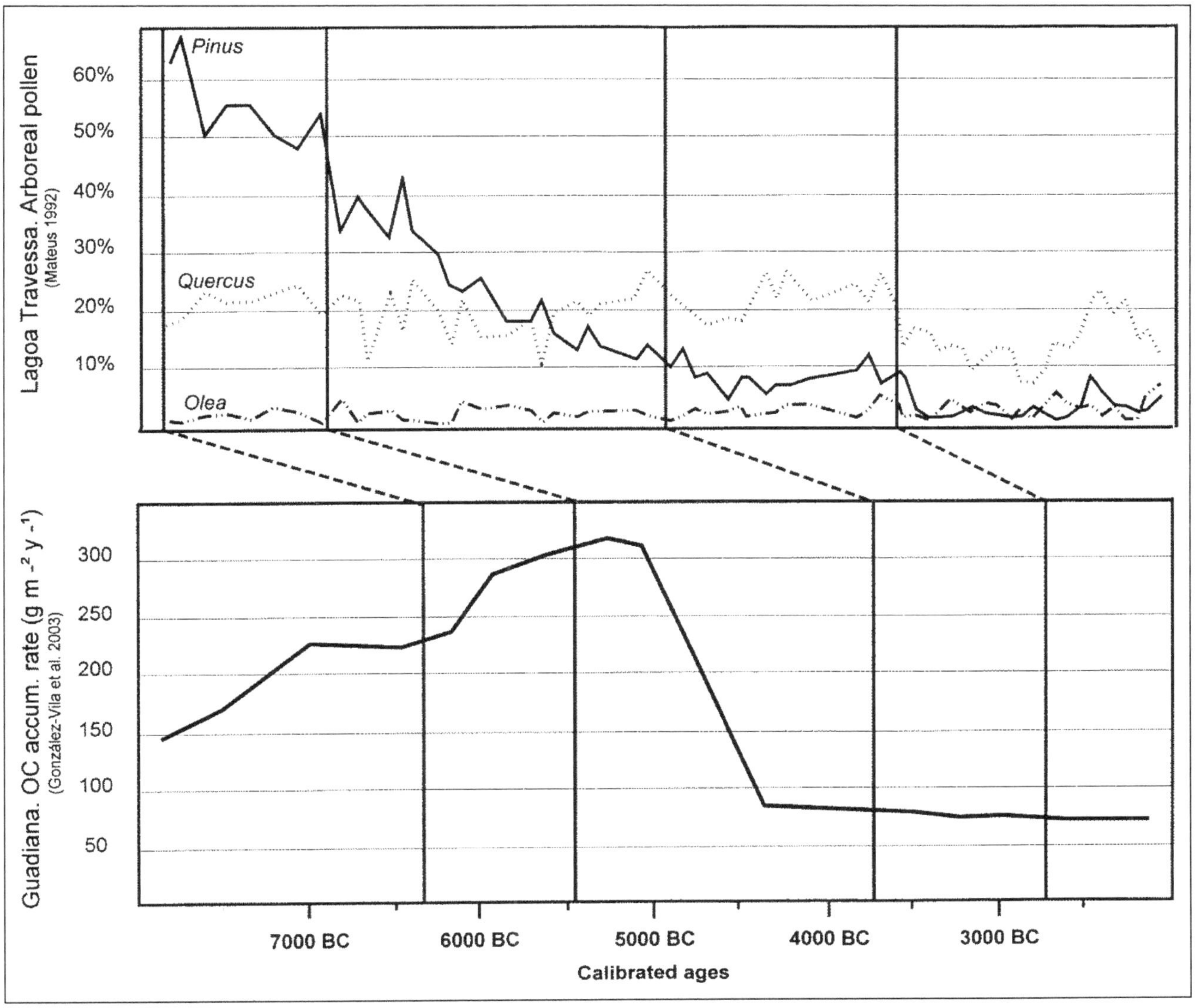

Figure 9.2. Pollen diagram from Lagoa Travessa and sediment cores from the Guadiana River

Calado *et al.* 2004). The recent identification of acorn phytoliths in grindstones at the Early Neolithic settlement site of Los Barruecos (Cerrillo 2005) in central Spain, seems to corroborate this hypothesis. Acorns were widely used for food during the prehistory of SW Atlantic Europe, and olives have always been an important part of the Mediterranean diet (Calado, Nieto, and Nocete 2003b). A significant quantity of dwarf palm (*Chamaerops humilis*) silica phytoliths in the soil samples collected from inside the menhir's implantation pit at Quinta Queimada, suggests little climatic change since erection of the standing stone. The dwarf palm has been, and continues to be, widely used today throughout the Mediterranean basin for production of ropes, baskets, mats, and brooms. Also, the stem of the plant is edible and consumed by North African peasants. Many other local plants, like the wild apple (*Malus sylvestris*) and the hazel (*Corylus avellana*), could also have been used (Calado, Nieto, and Nocete 2003b). Thus, native plants could have played a major role in the economy of prehistoric human communities before and after the introduction of cereals, which are only marginally represented in the pollen diagrams from Lagoa Travessa until much later periods.

A similar picture arises from the faunal analysis. Exploitation of domesticated, or able to be domesticated, specimens of possible native origin, pig (*Sus domesticus*), ox (*Bos taurus*), and horse (*Equus sp.*), together with non-endemic species, such as goat (*Capra hircus*) and sheep (*Ovis aries*) are evident at the 6th millennium BC coastal site of El Retamar. However, wild specimens, such as red deer (*Cervus elaphus*), rabbit (*Oryctolagus cuniculus*), and hare (*Lepus capensis*) appear to have constituted the major portion of the diet (Cáceres 2002).

When analysing the archaeological evidence from the Neolithic period in SW Europe, the existence of incongruence in existing "neolithisation" models is clear. First, at the time when the "first farmers" were still arriving by land (Ammerman and Cavalli-Sforza 1984, Cavalli-Sforza 1996) or sea (Zilhão 2000) to the southwest shores of Atlantic Europe, the local populations were already living in a tight network of large settlement sites and erecting standing stones decorated with totemic lineage symbols. This is at variance with the traditional characterization of nomadic bands. Secondly, although exploitation of domesticated animals of possible native origin is well documented, we lack evidence of semi-wild/semi-domesticated specimens transitional between pig and wild boar or the ox and aurochs. Third, evidence from pollen analysis does not back up the traditional idea that cereals were the catalyst of the "neolithization" process in SW Europe. Cereals do seem to have played a minor role in the local economy and do not become a focus until later times.

Recent excavations at Quinta Queimada suggest that this site was abandoned well before the middle of the 6th millennium BC. Heavy soil erosion characterizes the site during its use, and similar erosive processes are evident at most of other settlement sites with standing stones (Calado, D. 2000a). Regionally, evidence for a period of heavy erosion, starting before 7000 BC and ending quite abruptly at the beginnings of the 6th millennium BC, correlates with the development of numerous settlement sites in the Tinto River estuary, in Huelva, Andalusia (Nocete 2001). This apparent relation between a tight network of settlement sites and heavy erosion is taken as evidence for anthropogenic pressure on the environment (Nocete 2001, Calado, Nieto, and Nocete 2003b, Calado *et al.* 2004). Pollen diagrams from SW Atlantic Europe support the sediment data from the Guadiana River. Ocean core SU 81-18 (Turón *et al.* 2003), recovered off the SW Portuguese coast, shows a regional pollen record for the last 23,000 years. The percentages of *Pinus* pollen in the diagram are relatively constant between 20,900 ±330 BC and 7410±130 BC, crossing various warmer, colder, more humid and dry periods. After 7410±130 BC, the percentage of *Pinus* pollen suddenly registers a heavy drop. Due to the resistant characteristics of this old *Pinus* forest to climate change, it is difficult to explain the heavy recession of the species solely to climatic changes between the Pre Boreal to Boreal climatic periods. Pollen diagram from Lagoa Travessa (Mateus 1992), SW Portugal, corroborates these results (Fig. 9.2). The progressive decrease of *Pinus* pollen is recorded from the base of the column, with calibrated dates of the middle of the 7th millennium BC (GrN 12692: 7580±70 bp).[4] A second drop in *Pinus* percentage is observed during the middle of the 6th millennium BC. Both declines in *Pinus* pollen fall well within the Atlantic climatic period, suggesting the independence between the decimation of the pristine pinewood forest and eventual minor climatic alterations. This argument upholds the anthro-ecological studies previously presented by Jean Louis Vernet for the Algarve region, for a very ancient profoundly humanized environment already existing during the middle Neolithic period is suggested (Straus *et al.* 1992).

Traditional research concerning the origins of the Neolithic, in Spain and Portugal, has been too focused on non-endemic plant and animal species, and has consistently downplayed the possibility of the important economic role of native flora and fauna (Calado *et al.* 2004). Nevertheless, existing data suggests the relative high level of social complexity reflected in the settlement sites with menhirs do not have its origins in cereal farming and stock-rearing, activities. These activities appear to have been marginal prior to 4500 BC and whose earliest undisputed evidence in SW Europe dates from the middle of the 6th millennium BC, at least 700 years after the erection of one of the Quinta Queimada settlement site standing stones.

Hunter-gatherers societies, developing high levels of social complexity and a sedentary life way, appear to be a common occurrence in mankind's history. However, further research is necessary to understand the Holocene pre-cereal societies of SW Atlantic Europe. Ought this society to be characterized just as complex hunter-gatherers? We suggest instead that the enormous impact that these polities seem to have had upon the environment as societies with a broad-spectrum economy and high level of management of the natural resources must be understood as a model of social behavior for production intensification. The ultimate result is like that of cereal agriculture. Such societies are better understood as what Bruce Smith (2001) characterizes as low-level food production economies without domesticates.

The powerful social classification that the menhirs and their symbols may reflect is a pattern of critical resource shortage (Saxe, Gall 1977, Wobst 1977) that initiated a dialectic model of symbolic recognition as a result of frictions between dissimilar ethnical entities, but also, would intensify inner inconsistencies (Wolf 1990, Scott 1985). In other words, the menhirs would identify the society of the standing stone builders against foreign societies, delimiting a factual territory where access to natural resources would be severely restricted. On the other hand, the implementation of lineage identification through symbols inscribed on the menhirs would break internal social cohesion along lines of factual or mythical descent. Evidence from the archaeological record indicates that the vast majority of standing stones villages were abandoned well before the middle of the 5th millennium BC. Thus, the massive erection of standing stones seems to reflect the final stages of a society before its utter disarticulation (Calado *et al.* 2004).

CONCLUSION

A dense network of very large permanent settlement sites with standing stones decorated with recurrent symbols has been identified in extreme SW Atlantic Europe. The OSL chronologies of Quinta Queimada indicates that at least one of these settlements dates from a period between 7983 BC and 6203 BC, and is therefore substantially older than the first evidences for domesticated cereals and animals in SW Iberia.

The distribution of settlement sites, menhirs, and symbols, allow the earliest indication for concrete

[4] Editors' note: GrN-12692: 7580±70 bp, 68.2% probability = 6510 BC (68.2%) 6370 BC; 95.4% probability = 6600 BC (87.8%) 6340 BC, 6320 BC (7.6%) 6250 BC.

territories, boundaries, ethnicity, lineages, fission, social complexity, and sedentary communities during the Late Mesolithic for Atlantic Europe. The rise of high residential stability and social complexity precedes the introduction of cereals and stock rearing in SW Iberia. Sediment and pollen diagrams from the extreme SW Atlantic Europe exhibit large scale forest recession and erosion of soils for the same time period, suggesting a strong anthropogenic impact on the environment.

The development of territoriality and population concentration in SW Iberia seems to have been an independent historical process not biased by contacts with Neolithic societies. The evidence from settlement sites with standing stones calls into question traditional assumptions of primary diffusion, agricultural intensification, trade, and land irrigation, as explanations for territoriality and population concentration.

References

ALEX, E., F. NOCETE, J. NIETO, R. SÁEZ and M. BAYONA 2004 – Estudio del Impacto Ambiental de la Metalurgia Prehistórica en el Andévalo Onubense: Contaminación de Aguas, Deforestación y Erosión. In F. Nocete (ed.), *Odiel: Proyecto de Investigación Arqueológica para el Análisis de la Desigualdad Social en el Suroeste de la Península Ibérica*. Junta de Anadalucía Consejería de Cultura 1, Sevilla, 2004:325-341.

AMMERMAN, A., and L. CAVALLI-SFORZA 1984 – *The Neolithic Transition and the Genetics of Population in Europe. Unpublished*, Princeton University Press.

ASOMBANG, R. 1999 – Sacred Centers and Urbanization in West Central Africa. In S. Keech McIntosh (ed.), *Beyond Chiefdoms: Pathways to Complexity in Africa*, Cambridge University Press, Cambridge, 1999:80-87.

BARTH, F. 1969 – Introduction. In Barth, F. (ed.), *Ethnic Groups and Boundaries. The Social Organization of Culture Difference*, Scandinavian University Press, Oslo, 1969:9-38.

BATEMAN, M. 2002 – *Quartz Optical Dating Report. Quinta da Queimada Project*. Unpublished Report,. SCIDR, University of Sheffield.

BAYMAN, J.M. 2002 – Hohokam craft economies and the materialization of power. *Journal of Archaeological Method and Theory* 9, 2002:69-95.

BENDER, B. 1978 – Gatherer-hunter to farmer: a social perspective. *World Archaeology* 10/2, 1978:204-22.

BINFORD, L. 1983 – *Em busca do Passado*. Publicações Europa-América, Lisboa.

BRUNFIELD, E.M., and T. EARLE 1987 – *Specialization, Exchange and Complex Societies*, Cambridge University Press, Cambridge.

BRÖGGER, J. 1977 – Socio-economic Structures and the Form of Religion. *Thenemos* 13, 1977:7-30.

CÁCERES, I. 2002 – Estudio de los Restos Óseos de la Fauna Terrestre en el Asentamiento de El Retamar. In J. Ramos, and M. Lazarích (ed.), *El Asentamiento de "El Retamar" (Puerto Real, Cádiz). Contribuición al Estudio de la Formación Social Tribal y a los Inicios de la Economía de Producción en la Bahía de Cádiz*, Universidad de Cádiz, 2002:175-192.

CALADO, D. 2000a – Menhires y Poblados: Interfluvial Bensafrim – Odiáxere, Lagos, Portugal. Unpublished Thesis, Universidad de Huelva, Huelva.

CALADO, D. 2000b – Poblados con Menhires del Extremo SW Penínsular. Notas para sú Cronología y Economía. Una Aproximación Quantitative. *Revista Atlántica-Mediterránea de Prehistoria y Arqueología Social (RAMPAS)* 3, 2000:47-99.

CALADO, D. 2003 – Poblado con Menhires del Extremo SW Peninsular: Notas Para sú Cronologia y Economia, Una Aproximación Quantitaiva. *XELB: Revista de Arqueologia, Arte, Etnologia e História* 4, 2003:25-52.

CALADO, D., J.M. NIETO and F. NOCETE 2003a – Quinta da Queimada, Lagos, Portugal. Datação do Momento de Erecção de um Monumento Megalítico Através de Luminescência Óptica de Cristais de Quartzo (OSL). Puerto de Santa María, Cádiz, *V Congreso Ibérico de Arqueometría. Libro de Resúmenes de Actas*, Cádiz, 2003:167-68.

CALADO, D., J.M. NIETO and F. NOCETE 2003b/2004 – Menires, Símbolos e Organização Social: O Extremo SW Peninsular. In M. Calado (ed.), *Sinais de Pedra. Actas do 1º Colóquio Internacional Sobre Megalitismo e Arte Rupestre na Europa Atlântica*, Fundação Eugénio de Almeida, Évora 2003/2004.

CALADO, D., F. NOCETE, M.D. CÀMALICH, D. MARTÍN-SOCAS, J.M. NIETO, E. ALEX, and M. BAYONA 2004 – Los Poblados con Menhires del Algarbe Occidental: Nuevas Perspectivas para la Explicación del Origen y Implicaciones Históricas Sobre las Primeras Sociedades Sedentarias en el Suroeste Peninsular. *Jornadas Temáticas Andaluzas de Arqueología. Sociedades Recolectoras y los Primeros Productores. Libro de Resúmenes de Actas*, Sevilla, 2004:195-204.

CALADO, M. 2002 – Standing Stones and Natural Outcrops. The Role of Ritual Monuments in the Neolithic Transition in Central Alentejo. In C. Scarre (ed.), *Monuments and Landscape in Atlantic Europe*, Routledge, London, 2002:17-35.

CALADO, M. 2003/2004 – Entre o Céu e a Terra. Menires e Arte Rupestre no Alentejo Central. In M. Calado (ed.), *Sinais de Pedra. Actas do 1º Colóquio Internacional Sobre Megalitismo e Arte Rupestre na Europa Atlântica*. Évora, 2003/2004.

CARDOSO, J.L., and M.V. GOMES 1997 – Caracterização do Machado Mirense. Os Materiais do Monte dos Amantes (Vila do Bispo, Algarve). Setúbal Arqueológica XI-XII, 1997:121-46.

CAVALLI-SFORZA, L. 1996 – *Genes, Povos e Línguas*. Instituto Piaget, Lisboa.

CERRILLO, E. 2005 – Bases Económicas y Ambientales para el Estudio de las Comunidades Neolíticas del Centro-Oeste Peninsular: Perspectivas Desde el Yacimiento de Los Barruecos. In R. Ontañón Peredo, C. García-Moncó Piñeiro and P. Arias Cabal, *Actas del III Congreso del Neolítico en la Península Ibérica: Santander, 5 a 8 de octubre de 2003*, Universidad de Cantabria, Servicio de Publicaciones, Santander, 2005:93-102.

CHANG, K.C. 1981 – The Affluent Foragers in the Coastal Areas of China: Extrapolation from Evidence on Transition to Agriculture. In S. Koyama, and D.H. Thomas (ed.), *Affluent Foragers. Pacific Coast East and West. SENRI Ethnological Studies* 9, 1981:177-86.

DRISCOL, S.T. 1988 – Power and Authority in Early Historic Scotland: Pictish Symbol Stones. In J. Gledhil, B. Bender and M. Larsen (eds.), *State and Society. The emergence and development of Social Hierarchy and Political Centralization.* One World Archaeology. Thomson Publishing, London, 1988:215-236.

DUPRE, G., and P. REY 1973 – Reflections on the Pertinence of a Theory of Exchange. *Economy and Society* 2/2, 1973:131-163.

FERNÁNDEZ LÓPEZ De PABLO, J.M. GÓMEZ PUCHE 2009 – Climate Change and Population Dynamics During the Late Mesolithic and the Neo-lithic Transition in Iberia. Documenta Praehistorica XXXVI, 2009:67-96.

FORTEA, J. 1973 – *Los Complejos Microlaminares y Geometricos del Epipaleolitico Mediterraneo Español.* Universidad de Salamanca, Slamanca.

FRIEDMAN, J. 1977 – Tribus, Estados i Transformaciones. In M. Bloch (ed.), *Análisis Marxistas y Antropología Social.* Barcelona, 1977:191-239.

GAILEY, C. H., and T. PATTERSON 1988 – State Formation and Uneven Development. In J. Gledhill, B. Bender, and M.T. Larson, (ed.), *State and Society: The Emergence and Development of Social Hierarchy and Political Centralization.* Unwin and Hyman/ Routledge, London 1988/1995:77-90.

GODELIER, M. 1981 – Infrastructures, Societies and History. Reprinted in *Institutt for Sosialantropologi Occasional Papers* 4, Universitetet I Oslo 1981:206-240. Originally published 1978 *Current Anthropology* 19/4.

GOMES, M.V., and L.M. CABRITA 1997 – Dois Novos Povoados Neolíticos, com Menires, no Barlavento Algarvio. XI – XII *Setúbal Arqueológica,* 1997:191-98.

GONZÁLEZ-VILA, F., O. POLVILLO, T. BOSKI, D. MOURA and J.R. de ANDRES 2003 – Biomarker Patterns in a Time-Resolved Holocene/terminal Pleistocene Sedimentary Sequence from the Guadiana River Estuarine Area (SW Portugal/Spain border). *Organic geochemistry* 34, 2003:1601-1613.

JOHNSON, A.W., and T. EARLE 1987 – *The Evolution of Human Societies: From Foraging Group to Agrarian State.* Stanford University Press, Palo Alto, California.

KENT, S. 1989 – Cross Cultural Perceptions of Farmers as Hunters and the Value of Meat. In S. Kent, (ed.), *Farmers as Hunters,* Cambridge University Press, Cambridge, 1989:1-17.

LYONS, W.H., M.D. GLASCOCK and P.J. MEHINGER 2003 – Silica from Sources to Site: Ultraviolet Fluorescence and Trace Elements Identify Chert from Lost Dune, Southeastern Oregon. *Journal of Archaeological Science* 30, 2003:1139-59.

MARTÍN-SOCAS, D., M.D. CÀMALICH, *et al.* 2004 – La Cueva del Toro (Antequera, Málaga). Una Comunidad Ganadera en Andalucía, Entre el VI y el III Milenio ANE. *Jornadas Temáticas Andaluzas de Arqueología. Sociedades Recolectoras y los Primeros Productores. Libro de Resúmenes de Actas,* Sevilla.

MARTÍN-SOCAS, D., M.D. CÀMALICH and P. GONZÁLEZ (ed.) 2004 – *Cueva de El Toro (Antequera, Málaga). Una Comunidad Ganadera en Andalucía, Entre el VI y el III Milenios A.N.E.* Monografías de Arqueología 20, Sevilla.

MATEUS, J.L. 1992 – *Holocene and Present-Day Ecosystems of the Carvalhal Region, Southwest Portugal.* Unpublished Thesis, University of Utrecht, Utrecht.

MEILLASSOUX, C. 1977 – *Mujeres, Graneros y Capitales: Economía Doméstica y Capitalismo.* Siglo XXI, México.

NOCETE, F. 1984 – Jefaturas y Territorio. *Cuadernos de Prehistoria de la Universidad de Granada* 9, 1984:289-304.

NOCETE, F. 2001 – *Tercer Milenio Antes de Nuestra Era. Relaciones y Contradicciones Centro/Periferia en el Valle del Guadalquivir.* Bellaterra Arqueología, Marzo 2001, Barcelona.

NOCETE, F., A. ORIHUELA, R. OTERO, J.A. LINARES, J.C. ROMERO, P. ESCALERA and R. SÁEZ 1996 – Refutaciones al mundo arquograficamente organizado de los neolíticos del SW. *Actas del I Congrés del Neolitic a la Península Ibérica* II, 1996:853-862, Bellaterra, Gavá.

OKADA, Y. 2003 – Jomon Culture of Northeastern Japan and the Sannai Maruyama Site. In J. Habu, Savelle, J., Koyama, S. and Hongo, H. (ed.), *Hunter-Gatherers of the North Pacific Rim.* CHAGS 8. Aomori and Osaka, SENRI Ethnological Studies 63, 2003:173-86.

OLIVEIRA, J. de 2000 – O Megalitismo de Xisto na Bacia do Sever (Montalvão – Cedillo). In V.S. Gonçalves, (ed.), *Muitas Antas, Pouca Gente.* Actas do I Colóquio Internacional sobre Megalitismo, 2000:135-58, Lisboa.

ORIHUELA, A. 1999 – *Historia de la Prehistoria: El Suroeste de la Peninsula Ibérica.* Huelva.

PLOG, S. 1990 – Agriculture, Sedentism and Environment in The Evolution of Political Systems. In S.

Upham, (ed.), *The Evolution of Political Systems. Sociopolitics in Small-scale Sedentary Societies,* Cambridge University Press, Cambridge, 1990:172-202.

PRICE, T.D., and J.A. BROWN 1985 – Aspects of Hunter-Gatherer Complexity. In T.D. Price and J. Brown (ed.), *Prehistoric Hunter Gatherers, Gatherers: The Emergence of Cultural Complexity.* Academic Press New York, 1985:3-20.

RAMOS, J. 2000 – El Problema Historiográfico de la Diferenciación Epipaleolítico – Neolítico Como Debate Conceptual. *SPAL* 9, 2000:279-92.

LAZARÍCH, M., GONZÁLEZ, M. GARCÍA, V. CASTAÑEDA, N. HERRERO, M. PÉREZ, and J. RAMOS 2002 – Contribución de "El Retamar" al conocimiento del modo de producción, de los modos de vida y valoración socioeconómica de la formación social tribal. In J. Ramos, and M. Lazarích, (ed.), *El asentamiento de "El Retamar" (Puerto Real, Cádiz). Contribución al Estudio de la Formación Social Tribal y a los inicios de la economía de producción en la Bahía de Cádiz.* Universidad de Cádiz, Cádiz, 2002:425-432.

SAXE, A.A., and P.L. GALL 1977 – Ecological Determinants of Mortuary Practices: the Temuan of Malaysia. In W. Wood, (ed.), *Cultural Ecological Perspectives on SW Asia.* Ohio University Centre for International Studies 41Papers in International Southeast Asia Studies, Ohio University, Athens, Ohio, 1977:74-82.

SCOTT, J.C. 1985 – *Weapons of the Weak: Everyday Forms of Peasant Resistance.* Yale University Press, New Haven, CT.

SHARP, L. 1964 – Technical Innovation and Culture Change: An Australian Case. In P.B. Hammond, (ed.), *Cultural and Social Anthropology: Selected Readings.* MacMillan, New York.

SILVA, C.T. da and J. SOARES 1981 – *Pré-história da Área de Sines.* Gabinete da Área de Sines, Lisboa.

SMITH, B. 2001 – Low-Level Food Production. *Journal of Archaeological Research* 9/1, 2001:1-43.

SOARES, J. 1996 – Padrões de Povoamento e Subsistência no Mesolítico da Costa Sudoeste Portuguesa. *Zephyrus* 49, 1996:109-24.

SOUTHALL, A. 1999 – The segmentary state and the ritual phase in political economy. In S.K. McIntosh (ed.), *Beyond Chiefdoms: Pathways to complexity in Africa.* Cambridge University Press, Cambridge, 1999:31-38.

STRAUS, L.G., D. FORD, L. MARAMBAT *et al.* 1992 – Early Farming in the Algarve (Southern Portugal): A Preliminary View from two Cave Excavations Near Faro. *Trabalhos de Antropologia e Etnologia* 32, 1992:141-62.

THOMAS, D.H. 1981 – Complexity Among the Great Basin Shoshoneans. The World's Least Affluent Hunter-Gatherers?. In S. Koyama and D.H. Thomas (ed.), *Affluent Foragers. Pacific Coast East and West.* SENRI Ethnological Studies 9, 1981:19-52.

TURÓN, J., A. LÉZINE and M. DENÈFLE 2003 – Land-Sea Correlations for the Last Glacial Inferred from a Pollen and Dynocist Record from the Portuguese Margin. *Quaternary Research* 59, 2003:88-96.

WALLACE, A. 1966 – *Religion: an Anthropologic View.* Random House, New York.

WOBST, M. 1977 – Stylistic Behavior and Information Exchange. In C.E. Cleland (ed.), *For the Director: Research Essays in Honor of James B. Griffin.* Anthropological Papers 61, Museum of Anthropology, University of Michigan Museum of Anthropology, Ann Arbor, 1997:317-342.

WOLF, E.R. 1990 – Distinguished Lecture: Facing Power: Old Insights, New Questions. *American Anthropologist* 92/3, 1990:586-96.

ZILHÃO, J. 2000 – From the Mesolithic to the Neolithic in the Iberian Peninsula. In T.D. Price, (ed.), *Europe's First Farmers,* Cambridge University Press, Cambridge and New York., 2000:19-56.

Chapter 10
PRE-NEOLITHIZATION: RECONSTRUCTING THE ENVIRONMENTAL BACKGROUND TO LIFE WAY CHANGES IN THE LATE MESOLITHIC OF THE CARPATHIAN BASIN

Pál SÜMEGI
Department of Geology and Paleontology, University of Szeged and the Archeological Institute of the Hungarian Academy of Sciences, Budapest, Hungary

Abstract: *The very recent and important Hungarian Holocene cores from peat bogs of Bátorliget and Nagy Mohos of Kelemér are the primary source for the analysis of interaction between climate, environment, and socio-cultural change during the Mesolithic in the Carpathian Basin. The Early Holocene climate changes at 9000 BC seem to have triggered significant social and technological changes in the Pre-Pottery Neolithic of the Near East, as well as in the Central European Mesolithic, enabling parallel developments that ultimately led to the Neolithic. However, during the Carpathian Basin's Late Mesolithic when climate oscillations are less severe, Holocene radiocarbon-dated core profiles indicate several human induced vegetation changes between 7000-6000 BC. This is supported by the latest data from a site in the Jászság region of the Northern Great Hungarian Plain, as well as from re-examination of older data from other Mesolithic sites. In general, the development appears to include a more permanent settlement strategy, and possibly the early adaptation or imitation of some Neolithic techniques and methods, as well as internal social changes. Application of natural science techniques appear to support archaeologists' assumptions that around 7000 BC (i.e. right before the appearance of the Early Neolithic in the Carpathian Basin) an independent Pre-neolithic phase developed. This phase survived in the foothills of the Carpathians during the expansion of the Early Neolithic Körös culture.*

INTRODUCTION

This paper presents the results of an environmental-historical analysis of the Late Mesolithic in the Carpathian Basin. Analyses are limited to the most recent and important Hungarian Holocene profiles (Fig. 10.1), primarily those of Bátorliget Marsh (Sümegi 1996), the Nagy Mohos peat bog of Kelemér (Magyari *et al.* 2002), and the Csaroda peat bog (Sümegi 1999). These results are complemented by analyses of horizon profiles from earlier cores of the same age (Fig. 10.2).

Before the start of the Neolithic, several radical changes took place in the material culture of Europe's Mesolithic groups. Archaeologists regard the uniform transformation of stone tools as an important component of Pre-neolithization in the Balkans and Central Europe (Clark 1958, Kozłowski 1987, Kertész 1993). Consequently, this study asks whether there are any observable signs of such Pre-neolithic influence in Holocene period Hungarian soil profiles, and in the process clarify what can be considered as a Pre-neolithic influence from an environment-historical perspective.[1]

THE MESOLITHIC PHASES IN THE CARPATHIAN BASIN

Based on results from a site in the Jászság area of the northern Great Hungarian Plain, as well as reanalysis of older data from other Mesolithic sites (Kertész 1993, Bánffy 2006), two major chronological units are established for the Mesolithic period in the Carpathian Basin. These are the older Early Mesolithic and younger Late Mesolithic. According to Kertész (1993, 1996), the Early Mesolithic sites are Szekszárd-Palánk (Vértes 1962) and Sződliget (Gábori 1956, 1968).

The Late Mesolithic is further divided into two phases by Kertész (1993) based on data from several sites in the Jászság. The older Jászberény phase is thought to be coeval with the Boreal climatic regime, while the younger Jásztelek phase is dated to the Early Atlantic climatic regime (Kertész, 1994a, 1994b). Kertész (1993) correlates the sites of Barca I (Prošek 1959), Sered (Bárta 1957), Mostová (Bárta 1980), Tomášikovo (Bárta 1955), and the Czech and Austrian sites of Smolín (Valoch 1985), Přibice (Valoch 1975), Kamegg, and Limberg-Mühlberg with the Jászberény phase (Kertész 1993, 1996a, 1996b). The more recent Jásztelek phase is named after the Jásztelek I site in Hungary. Based on the tool kit, Kertész (1993, 1996a, 1996b) correlates Jásztelek I with the sites of Tarpa-Márki at Tiszahát (Dobosi 1969, 1983, Szathmáry 1978), Kaposhomok in Transdanubia (Pusztai 1957), Csomaköz, Ciumeşti II (Păunescu, 1964), and Kamenitsa I in the Sub-Carpathian highlands (Matskevoi 1987).

EARLY HOLOCENE CLIMATE AND ENVIRONMENTAL CHANGE

During the second half of the Mesolithic, starting ca. 8400 cal BC (9200 bp), lime gradually declines accompanied by the advent of oak (*Quercus*) within the research area. Palynological analysis shows a closed canopy oak woodland developed under more balanced,

[1] Our work has enjoyed support from the grants OTKA T-034 392 and NKFP 5/0063/2002.

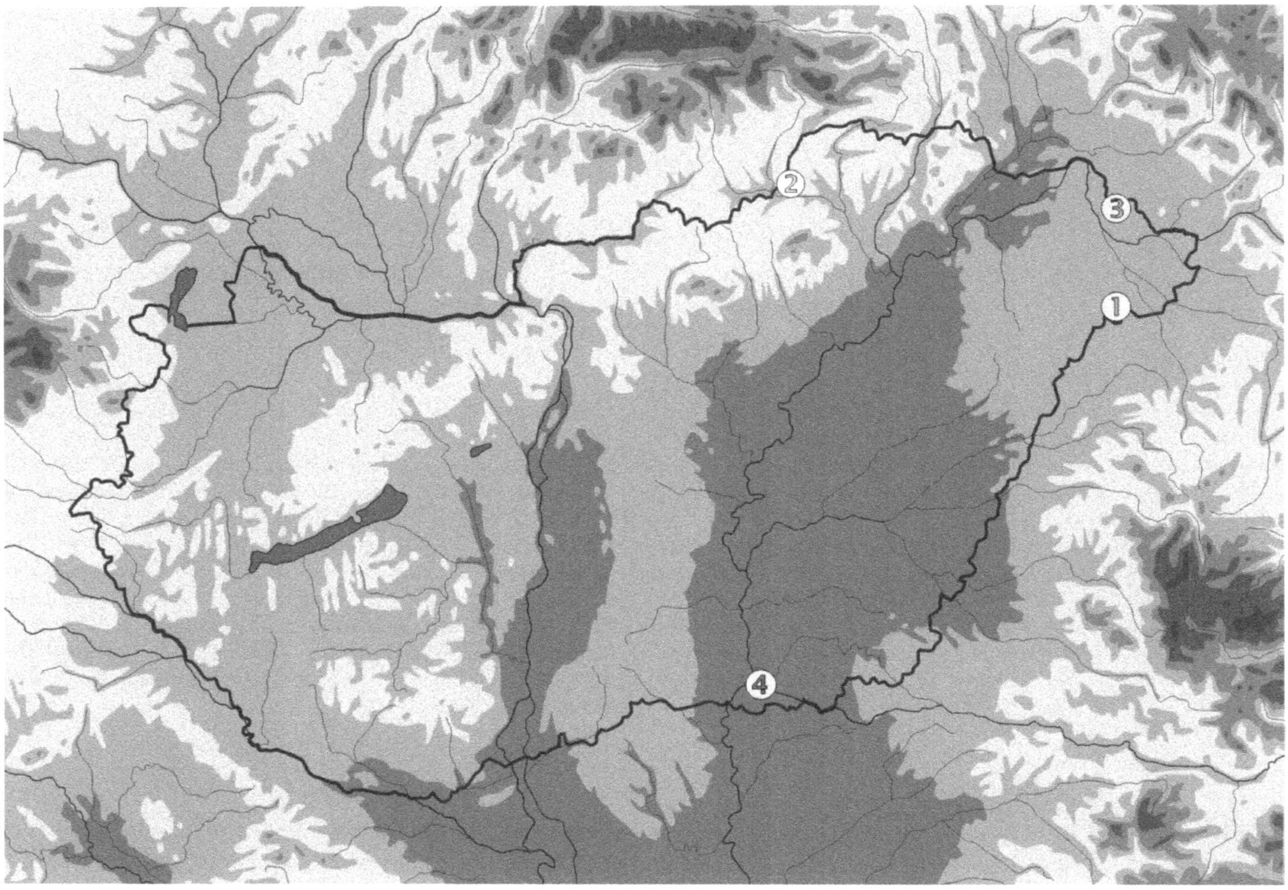

Figure 10.1. Location of the analyzed areas in Hungary. 1. Bártoliget Marsh; 2. Kis-Mohos Lake, 3. Kelemér; Nyíres Lake, Csaroda; 4. Batida-Oxbow, Szeged-Gorzsa

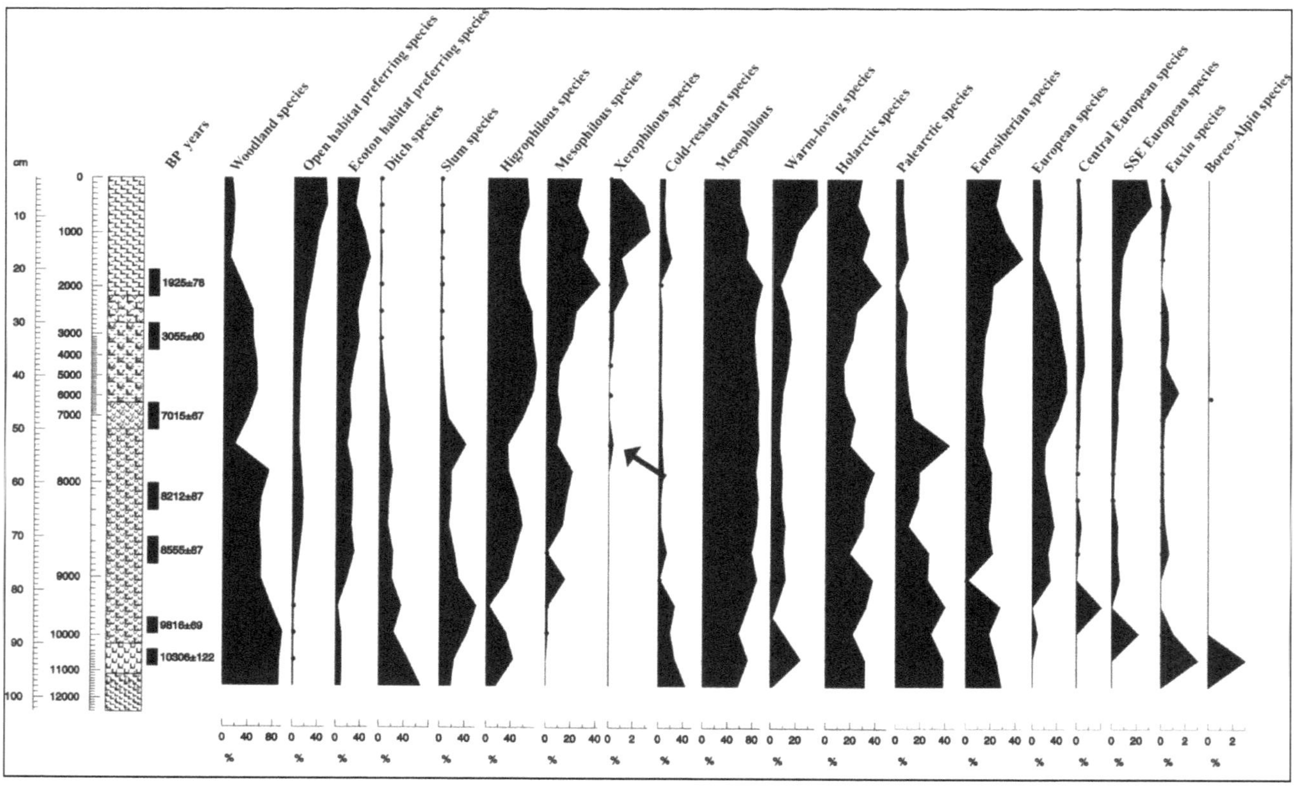

Figure 10.2. Probable Pre-neolithic effect based on changes in the component of the Early Holocene mollusk fauna in the Bátorliget Marsh catchment basin

milder, and wetter climatic conditions. The composition of the malacofauna, the dominance of species-rich forest dwelling classes, and the emergence of *Anisus spirorbis-Ruthenica filograna-Discus perspectivus* indicate the formation of forest vegetation. Similar changes are observed at this time in pollen profiles from Nyíres Lake at Csaroda (Sümegi 1999) and from Kis-Mohos Lake at Kelemér (Willis *et al.* 1998). According to these research results, Mesolithic hunter, fisher, and gatherer communities must have populated the closed canopy oak forests (with sufficient water supplies) located in the northeastern parts of the Great Hungarian Plain and in the foothills of the Carpathians. Palynological (Kertész *et al.* 1994) and radiocarbon analyses (Sümegi 2002) of deposits from an oxbow lake situated near the Mesolithic site of Jászberény also indicate the development of closed canopy oak hardwood gallery forests at 8400 BC.

According to the malacological data, the mean July paleo-temperatures reach 22°C, exceeding even today's values (Sümegi *et al.* 1996). Using the method developed by Davis for recent pollen data (Davis *et al.* 2001), a mean July paleo-temperature of 20-21° C is inferred from the pollen composition of the cores taken from the central part of the Bátorliget Marsh (Willis *et al.* 1995). These data seem to indicate the start of a warm phase. However, they are significantly lower than the 24-25° C for the whole Great Hungarian Plain (Járainé-Komlódi 1966, 1969).

This paleoenvironmental transformation, noted at the Pleistocene/Holocene boundary in the Bátorliget and other core profiles from the Northern Great Hungarian Plain, correlates with the environmental transformations of the Mesolithic of Franchthi Cave in Greece. The transformation at Franchthi Cave is observable in composition changes of the gathered malacofauna such as *Cyclope neritea*. This species prefers significant water levels and functions as a marker for flooding in coastal areas (Shackleton and Andel 1980, Andel *et al.* 1980). It coincides with development of the maximum water level in the eastern Mediterranean (Williams *et al.* 1978). These changes must have been triggered by intensive melting of the gradually retreating continental ice sheets and glaciers, consequent increased freshwater supply, and global sea level rise causing flooding of the sea shelves and river estuaries (Adamson *et al.* 1980).

Parallel with these climate changes, the Natufian culture in the Near East pursued an intensive gathering economy and associated way of life, resulting in the first aceramic Neolithic farming groups (Sherratt 1980). The relationship between the paleoenvironmental and cultural changes are so strong that all researchers tend to emphasize the role of climate change, although different aspects are emphasized in the emergence of a Neolithic in the Near East (Binford and Binford 1968, Childe 1936). Thus, the Early Holocene climate changes around 9000 BC seem to have triggered significant social and technological changes in the Pre-Pottery Neolithic (PPN) of the Near East as well as the Mesolithic of Central Europe.

THE MESOLITHIC ENVIRONMENT AND SOCIO-CULTURAL CHANGE

The paleoenvironmental data (Sümegi 2008, Sümegi *et al.* 2002) and the transformations observed in the lithic technology of the Carpathian Basin (Kertész 1996a, Kertész *et al.* 1994) indicate that the time around 9000 BC was an important watershed in the development of the Mesolithic. At this time some stone tool techno-cultural components, generally characteristic of the second half of the Mesolithic, began to appear (Kertész 1996b:144). However, other technical innovations, such as trapezoidal blades, come into use only in the youngest Mesolithic horizon. This is shown by the lack of trapezoidal blades in samples from Jászberény I. The samples came from an archaeological deposit with a radiocarbon date of (8030±250 bp) that yields a calibrated date of 7051 BC[2] (Kertész *et al.* 1994). Thus, the appearance of these tools must have taken place later, probably around 7000 BC.

Starting around 9000 BC, Mesolithic sites are located in the closed canopy oak forests of the alluvial plains and sand hummocks of the Northern Great Hungarian Plain. The plains are characterized by closed canopy pine-birch mixed taiga forests in the Sub-Carpathian and Sub-Alpine highland zones (Willis *et al.* 1997, Sümegi 1998, Gardner 1999, Magyari *et al.* 2001a). People at Mesolithic sites, such as Smolín, Kamegg, Sered I, and Barca I (Bárta 1980, 1981, Prosek 1959) use different material sources for their artifacts, but their lithic industries are closely linked to those seen in the Northern Great Hungarian Plain (Kertész 1996a, 1996b).

The sites of the Northern Great Hungarian Plain and the Sub-Carpathian and Sub-Alpine highland regions are located north of the distribution area of the Balkan Tardigravettien, a regional variant of the local lithic industry. This lithic complex is followed by the Epigravettian, where the application of western stone production techniques and the Sauveterrian and Beuronian cultural components are of primary importance (Kertész 1996a, 1996b). The Mesolithic sites of the Sub-Carpathian highland region belong to the Mesolithic of the Tisza valley (Bárta 1981), while those on the northern alluvial fans belong to the Mesolithic of the Northern Great Hungarian Plain (Kertész 1994).

The Mesolithic Sub-Carpathian sites of the Tisza valley are interpreted as special local subgroups of the Epigravettien derived from the Gravettien and characterized by a culture that applied western lithic production techniques to their local tool kit. These groups did not occupy the continental forest steppes, but rather the closed canopy woodlands of oak and pine during to the 9th – 8th millennium BC in contrast to the paleoenvironmental reconstructions of Kertész (1996a, 1996b). Our research shows that the flora, fauna, and soils occurring in the distribution areas of the two groups are fundamentally distinct (Willis *et al.* 1997, Sümegi

[2] Editor's note: Recalculation of 8030±250 bp with OxCal v3.10 (Bronk Ramsey 2005), using atmospheric data from Reimer *et al.* (2004) yields 7300 BC (68.2%) 6600 BC and 7600 BC (95.4%) 6400 BC.

1998, Gardner 1999, Magyari *et al.* 2001). Consequently, the differing natural environments must have played a role in the creation of group level cultural differences during the Mesolithic.

At the end of the 8th millennium BC and the beginning of the 7th millennium BC, the vegetation differences between the areas of the Northern Great Hungarian Plain, the Sub-Carpathian highlands, and the Sub-Alpine regions dissolved. Differences in the development of zonal montane vegetation during the Early Holocene in the Carpathian Basin are merely due to elevation. The emergence of a relatively uniform vegetation of closed canopy deciduous woodlands, rich in species, and characterized by a dominance of oak and hazel is found up to 600 m asl. From 600 to 1000/1200 m asl mixed deciduous woodlands developed, characterized by a dominance of lime, spruce, and Scotch pine (Farcaş *et al.* 1999, Willis *et al.* 1997, 1998, Gardner 1999, Magyari *et al.* 1999, 2000, Stieber 1967). In spite of the general homogeneity, the steppe and forest steppe vegetation from the end of the Pleistocene survived in those areas characterized by lower ground water levels or alkaline soils. Thus, even during expansion of the closed canopy oak woodlands there is evidence of mosaic-like vegetation complexity at both the local and regional scale within the Carpathian Basin (Sümegi *et al.* 1998a, 2002, Gal *et al.* 2006).

Important technological and cultural transformations take place paralleling the homogenization of vegetation at the beginning of the 7th millennium BC. These occur right before the appearance of the Neolithic (Andel and Runnel 1995). The Mesolithic transformation is evident in the lithic industries from various sites, such as Jásztelek I (Kertész 1993), Tarpa-Márki tanya, Ciumeşti II, and Kamenitsa I, in the Northern Great Hungarian Plain. These developments includes cone-shaped cores, micro-burin technology, and notched projectile points, as well as trapezoidal projectile points, blades, and retouched truncations (Kertész 1996a, 1996b). Surface finds of sickle blade-like lithics, probably serving as sickle inserts, are found at Jásztelek I, suggesting intensified gathering (Kertész 1993, 1996a, 1996b).

During this period, the soil pollen profile at Bátorliget (see Fig. 10.2) shows absolute dominance of closed canopy forest-dwelling species. Yet, some xero-thermophilous steppe and forest steppe dwelling mollusks (*Cepaea vindobonensis*, *Granaria frumentum*) are identified in littoral areas, indicating minor steppe-like patches within the woodlands. These patches indicate either a natural climate-related mosaic-like development, a climate-related transformation of the environment and the vegetation, or a minor human induced impact between 6900-6500 BC (8000-7500 bp). If the change is due to a minor human impact, it can be interpreted as 'Pre-neolithic human influence' resulting in the emergence of open vegetation areas, and extension of marginal vegetation in several other localities (Bánffy 2006, Sümegi 1998, 1999a, 2001, Sümegi-Kertész 2000, Sümegi *et al.* 2002a, b). The emergence of more open vegetation is directly linked to Mesolithic sites during active growth periods in a woodland setting.

MESOLITHIC HUMAN ENVIRONMENTAL IMPACT

The initially spontaneous, and later on intentional, spreading of light-preferring plants, such as hazel, occurring in the marginal vegetation, must have triggered similar vegetation transformations (Kalicz and Makkay 1976). The artificial creation of 'hunting trails' for the pursuit of forest-dwelling games (aurochs, deer, wild boar, and bison), or foliage feeding, must have contributed to the appearance and expansion of steppe-like vegetation spots (Sümegi 1998, 1999a, 2001, Sümegi-Kertész 2000, Sümegi *et al.* 2002). This transformation may be considered a Pre-neolithization, since radiocarbon dates for the core profile from the margin of Bátorliget Marsh indicate that malacological changes occur right before the earliest appearance of the Early Neolithic Körös and Starčevo cultures in the Carpathian Basin (Hertelendi *et al.* 1996).

Forrest fires are frequently cited as evidence of human impact, and the notion seems to apply in the research area as well. This is indicated by the possibility of spontaneous forest fires being greatly reduced in the area of the Northern highlands following the Early Holocene retreat of coniferous woodlands and the expansion of less flammable trees in the area (Willis *et al.* 1997). Forest fires did not fully cease, but their intensity and frequency changed as indicated by the mutual appearance of closed canopy forest-dwelling and thermoxerophilous mollusk species in the Bátorliget core profile. Furthermore, minor ash peaks display a strong correlation (Willis *et al.* 1997:745) with decreases of oak (*Quercus*) pollens accompanied by an increase of hazel (*Corylus*) pollen between 7000-6000 BC. This evidence is not confined to Bátorliget as similar transformations are also observed at other sites in the Northern Great Hungarian Plain and the highlands (Sümegi 1998, 1999, 2000). This evidence suggests the beginning of woodland management.

Unfortunately, paleoenvironmental research of Mesolithic sites is still in its infancy (Kertész *et al.* 1994a, 1994b, 1997), limiting detailed evidence for woodland management in Hungary. However, there are several analogies for the use of fire in America, Australia (Mellars 1976), and western Europe (Evans 1975). In these areas, hunter-gatherers used forest burning for several purposes including hunting, expanding marginal vegetation zones, and creating trails or pathways for herds of hunted animals (Bennett *et al.* 1990, Clark 1972, 1988, Clark *et al.* 1989, Smith 1970).

The intentional human-induced expansion of marginal vegetation zones, and the creation of complex mosaic-like woodland environments initiated at the end of the Mesolithic can be regarded as one of the most important artificial environmental transformations enhancing large-scale expansion of hazel (Smith 1970, Behre 1988). Thus

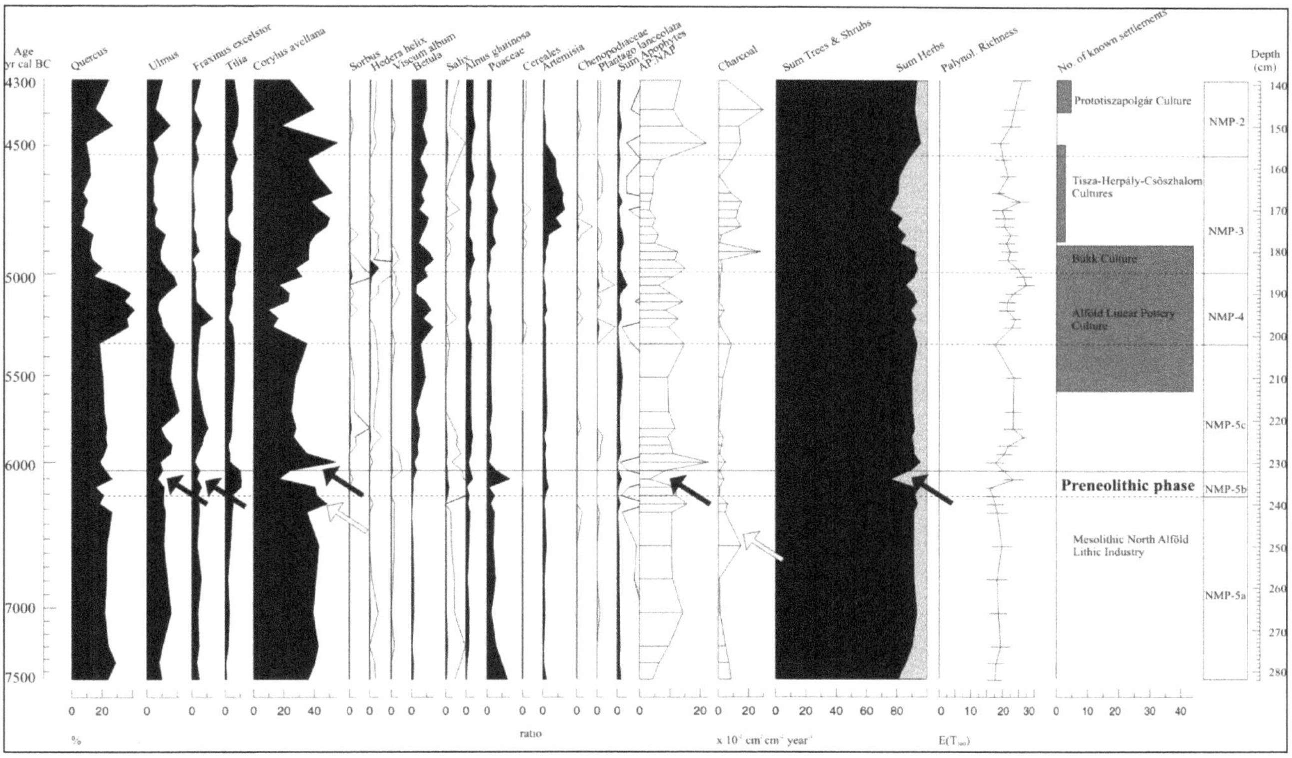

Figure 10.3. Hypothetical Mesolithic and Pre-neolithic effects based on changes in the pollen sequence of the Negy-Mohos Peat Bogs, Kelemér. Mesolithic: open arrow, Pre-neolithic: filled arrow

far, no archaeobotanical signs demonstrating the collection of hazelnuts are known from Hungarian Mesolithic sites. However, there seems to be a strong correlation between increases of hazel pollen, human-induced forest burning, and strategies applied during gathering noted at west European Mesolithic sites (Smith 1970). Similar processes have been inferred by Juhász (2002) in palynological investigations of profiles from western Transdanubia.

DISCUSSION

The intentional Late Mesolithic human-induced vegetation transformation creates a complex mosaic-like woodland environment. This is crucial evidence that the hunter-fisher-gatherers in the Carpathian Basin reach the 'substitution phase' (Zvelebil and Rowley-Conwy 1986). The experiences and knowledge gained through intentional interferences and activities enables the adaptation of methods and practices required for food production (Zvelebil 1986). Increases in the ratio of hazel pollen between 7000-6000 BC follow minor ash peaks in the neighborhood of the Mesolithic sites in the northern parts of the Great Hungarian Plain and the northern highlands of Hungary. This evidence indicates that the majority of the Mesolithic population, residing in the northern parts of the Carpathian Basin, may have reached the 'substitution phase' during this period (Sümegi 1998, 1999a, Sümegi *et al.* 2002).

The assumption that Mesolithic populations reached the 'substitution phase' is strengthened by evidence from the palynological analysis of the Mohos peat bogs of Kelemér. Here a cyclical decrease in the amount of elm and ash is observed during the Late Mesolithic (at 5960 BC) (Fig. 10.3) (Magyari *et al.* 2002). It implies selective gathering of leaves for animal feed (Heybroek 1963, Kertész-Sümegi 1999b).

Furthermore, there seems to be a chronological link between the appearance of open vegetation and developments in lithic technology. This is evidenced by intensive gathering of open vegetation-dwelling gastropods and steppe plants, the occurrence of minor ash peaks, the emergence of complex mosaic-like vegetation, the spread of hazel, and the likely use of litter for feed. Innovations in lithic technologies include trapezoidal projectile points and the appearance of probable sickle inserts. All these transformations occur in the archaeological record in the neighborhood of the Bátorliget Marsh during the 7th millennium BC. This implies intensive changes in Mesolithic communities and suggests the possible emergence of a Pre-neolithic phase (Fig. 10.4).

The observed environmental changes in the sedimentary basins, including the basin of the Bátorliget Marsh, strongly imply an emergence of an initially independent Pre-neolithic phase among autochthonous Mesolithic populations (Sümegi *et al.* 2008, Sümegi and Kertész 2001), even though no truly definitive Pre-neolithic sites have so far been found in the Carpathian Basin. Later social changes, resulting in more intensive utilization of the landscape, must have been triggered by cultural impacts affecting Mesolithic communities of the

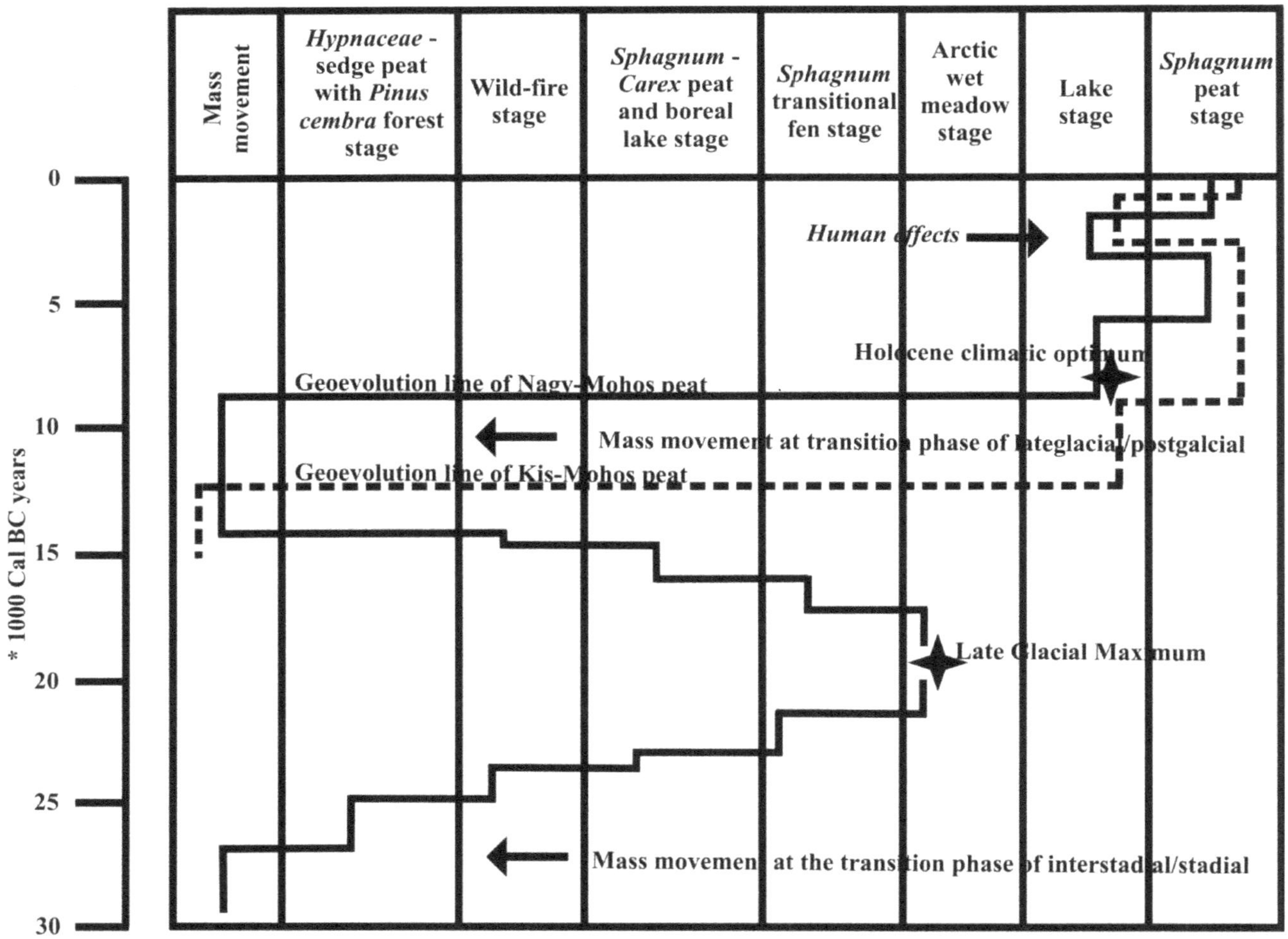

Figure 10.4. Geo-evolutionary stages in the Negy-Mohos and Kis-Mahos Peat Bogs (after Sümegi 2002)

Carpathian Basin following settlement of Neolithic groups with cultural roots in the Near East (Ammerman and Cavalli-Sforza 1971).

SUMMARY OF RESULTS

The scientific results of various Holocene radiocarbon-dated core profiles indicate several minor but important human-induced changes in the vegetation between 7000-6000 BC. These alterations seem to display direct connections with a more massive gathering of hazelnuts, and the utilization of shoots and foliage of younger ashes and linden, possibly for animal feed. They also include the formation of hunting trails, camp sites, and, on the whole, a more permanent settlement strategy. There is also the possibility of an early adaptation of techniques and methods for food production. Finally, there is the archaeological evidence of an internal Mesolithic development indicated by changes in the Mesolithic tool kit. Thus, the information gained through the application of tools and methods from the hard sciences seem to lend credence to the assumptions formulated by archaeologists. Accordingly, around 7000 BC, a probably independent Pre-neolithic phase must have developed that comes into being just before the appearance of Early Neolithic cultures in the Carpathian Basin. It continues to survive in the foothills of the Carpathians during expansion of the Neolithic Körös culture.

References

ADAMSON, D.A., F. GASSE, F.A. STREET and M.A.J. WILLIAMS 1980 – Late Quaternary History of the Nile. *Nature* 287, 1980:50-55.

AMMERMAN, A.J., and L.L. CAVALLI-SFORZA 1971 – Measuring the Rate of Spread of Early Farming in Europe. *Man* 6, 1971:674-688.

ANDEL, T.H., and C.N. van RUNNELS 1995 – The Earliest Farmers in Europe. *Antiquity* 69, 1995:481-500.

ANDEL, T.H. van, T.W. JACOBSEN, J.B. JOLLY and B. LIANOS 1980 – Late Quaternary History of the Coastal Zone near Franchthi-cave, Southern Argolid, Greece. *Journal of Field Archeology* 7, 1980:389-402.

BÁNFFY, E. 2006 – Eastern, Central and Western Hungary: Variations of Neolithisation Models. *Documenta Praehistorica* XXXIII, 2006:125-142.

BÁRTA, J. 1957 – Pleistocénne Piesocné duny pri Seredi a ich Paleolitické a Mezolitické Osídlenie. *Slovenská Archeológia* 5, 1957:5-72.

BÁRTA, J. 1959 – Mezolitické a Neolitické Kamenné Nástroje z dún "Vrsky" pri Dolnej Strede. *Slovenská Archeológia* 7, 1959:241-259.

BÁRTA, J. 1980 – Paleolit a Mezolit. *Slovenská Archeológia* 28, 1980:119-136.

BÁRTA, J. 1981 – Das Mesolithikum im nordwestlichen Teil des Karpatenbeckens. *Veröffentlichungen des Museums für Ur- und Frühgeschichte Potsdam*, Potsdam 14-15, 1981:295-300.

BEHRE, K.E. 1988 – The Role of Man in European Vegetation History. In B. Huntley and T. Webb III. (eds.), *Vegetation History*. Kluwer Academic Publishers, Amsterdam, 1988:633-672.

BENNETT, K.D., S. BOREHAM, M.J. SHARP and V.R. SWITSUR 1992 – Holocene History of Environment, Vegetation and Human Settlement on Catta Ness, Lunnansting, Shetland. *Journal of Ecology* 80, 1992:241-273.

BINFORD, S.R., and L.R. BINFORD 1968 – *New Perspectives in Archeology*. Aldine Press, New York.

CHILDE, G. 1936 – *Man Makes Himself*. Watts Press, London.

CLARK, J.G.D. 1958 – Blade and Trapeze Industries of the European Stone Age. *Proceedings of the Prehistory Society* 24, 1958:24-42.

CLARK, J.G.D. 1972 – *Star Carr: a Case Study in Bioarcheology*. Addison-Wesley Modular Publication, Wellesley, Ma., USA.

CLARK, J.G.D. 1988 – Particle Motion and the Theory of Charcoal Analysis, Source Area, Transport, Deposition and Sampling. *Quaternary Research* 30, 1988:67-80.

CLARK, J.D.G., J. MERKT and H. MULLER 1989 – Post-glacial Fire, Vegetation and Human History on the Northern Alpine Forelands, South Western Germany. *Journal of Ecology* 77, 1989:897-925.

CLARK, J.R. 1988 – Effect of Climatic Change on Fire Regimes on Northwest Minnesota. *Nature* 334, 1988:233-235.

DAVIS, B.A., S. BREWER, T. STEVENSON and S. JUGGINGS 2001 – High Resolution from Low Resolution, A new Method for Time Series Pollen-Climate Reconstruction And its Application in Investigating Abrupt Holocene Climate Change in Southern Europe/Mediterranean. *Abstracts of PAGES Conference*, Aix an Provance, France.

DOBOSI, T.V. 1969 – Tarpa. *Régészeti Füzetek*, Ser. I. No. 22, 1969:22.

DOBOSI, T.V. 1983 – Ásatás Tarpa-Márki Tanyán. *Communicationes Archaeologicae Hungariae* 12, 1983:5-18.

EVANS, J.G. 1975 – *The Environment of Early Man in the British Isles*. Paul Elek Press, London.

FARCAŞ, S., J.L. de BEAULIEU, M. REILLE, G. COLDEA, B. DIACONEAŞA, T. GOSLAR and T. JULL 1999 – First ^{14}C Datings of Late Glacial and Holocene Pollen Sequences from the Rumanian Carpathians. Comptes Rendues de l'Academie des Sciences de Paris, *Science de la Vie* 322, 1999:799-807.

GAL, E., I. JUHÁSZ and P. SÜMEGI. (eds.) 2006 – *Environmental Archaeology in North-Eastern Hungary*. Varia Archaeologica Hungarica 19, Archaeological Institute, Budapest.

GARDNER, A.R. 1999 – The ecology of Neolithic Environmental Impacts – Re-Evaluation of Existing Theory Using Case Studies from Hungary. *Dokumenta Prehistorica* 26, 1999:163-183.

GÁBORI, M. 1956 – Mezolitikus Leletek Sződligetről. *Archeológiai Értesítő* 83, 1956:177-182.

GÁBORI, M. 1968 – Mesolithischer Zeltgrundriss in Sződliget. *Acta Archeologica Hungarica* 20, 1968:33-36.

HERTELENDI, E., N. KALICZ, P. RACZKY, F. HORVÁTH, É. VERES, I. SVINGOR, I. FUTÓ and L. BARTOSIEWICZ 1996 – Re-evolution of the Neolithic in Eastern Hungary Based on Calibrated Radiocarbon Dates. *Radiocarbon* 37, 1996:239-241.

HEYBROEK, H.M. 1963 – Diseases and Lopping for Fodder as Possible Causes of a Prehistoric Decline of Ulmus. *Acta Botanica Neederlandica* 12, 1963:2-22.

JÁRAINÉ-KOMLÓDI, M. 1966 – Adatok az Alföld Klíma- és Vegetációtörténetéhez I. *Botanikai Közlemények* 53, 1966:191-201.

JÁRAINÉ-KOMLÓDI, M. 1969 – Adatok az Alföld Klíma- és Vegetációtörténetéhez II. *Botanikai Közlemények* 56, 1969:43-55.

JUHÁSZ, E.I. 2002 – *A Délnyugat Dunántúl negyedkori vegetációtörténetének palinológiai rekonstrukciója. (Reconstitution palynologique de la végétation depuis le Tardiglaciaire dans la région de Zala, sud-ouest de la Hongrie)*, Unpublished Ph. D. dissertation 215, University of Marseille-Pécs, Marseille.

KALICZ N., and J. MAKKAY 1976 – Frühneolithische Siedlung in Méhtelek-Nádas. *Mitteilungen des Archäologischen Instituts der Ungarischen Akademie der Wissenschaften*, 6, 1976:13-24.

KERTÉSZ, R. 1993 – Data to the Mesolithic of the Great Hungarian Plain. *Tisicum* 8, 1993:81-104.

KERTÉSZ, R. 1994a – Late Mesolithic Chipped Stone Industry from the Site Jásztelek I (Hungary). In G. Lőrinczy (ed.), *A Kőkortól a Középkorig: Von der Steinzeit bis zum Mittelalter*, Szeged, 1994:23-44.

KERTÉSZ, R. 1994b – The Present State of the Research of the Mesolithic in the Great Hungarian Plain. *Jász-Nagykun-Szolnok Megyei Múzeumok* 49, 1994:9-33.

KERTÉSZ, R. 1996a – *Az Alföld Őstörténete a Kora-Holocénban. Az Észak-Alföldi Mezolit ipar.* Unpublished Ph. D. dissertation, University of Szeged, Szeged.

KERTÉSZ, R. 1996b – The Mesolithic in the Great Hungarian Plain: A Survey of the Evidence. In L. Tálas (ed.), *At the Fringes of Three Worlds: Hunter-Gatherers and Farmers in the Middle Tisza Valley.*

Damjanich Múzeum Kiadványa, Szolnok, 1996:5-34.

KERTÉSZ, R., and P. SÜMEGI 1999a – Teóriák, Kritika és egy Modell: Miért állt meg a Körös-Starcevo Kultúra Terjedése a Kárpát-medence Centrumában? *Tisicum* 10, 1999:9-22.

KERTÉSZ, R., and P. SÜMEGI 1999b – Az Északi-középhegység Negyedidőszak végi Őstörténete. *Nógrád Megyei Múzeumok Évkönyve* 23, 1999:66-93.

KERTÉSZ, R., P. SÜMEGI, M. KOZÁK, M. BRAUN, E. FÉLEGYHÁZI and E. HERTELENDI 1994 – Archaeological and Palaeoecological Study of an Early Holocene Settlement in the Jászság Area. *Acta Geographica, Geologica et Meteorologica, Debrecina* 32, 1994:5-49.

KOZLOWSKI, S.K. 1987 – The Pre-Neolithic Base of the Early Neolithic Stone Industries in Europe. *Archaeologia Interregionalis* 5, 1987:9-18.

MAGYARI, E., P. SÜMEGI, M. BRAUN and G. JAKAB 2001 – Retarded Hydrosere: Anthropogenic and Climatic Signals in a Holocene Raised Bog Profile from the NE Carpathian Basin. *Journal of Ecology*, 89, 2001:1019-1032.

MATSKEVOI, L.G. 1987 – Mezolit Predkarpatya, zapadnogo Podolya i Zakarpatya. Arkheologiya Prikarpatya, *Volyni i Zakarpatya* (Kamennyi Vek), Kiev, 1987:75-89.

MELLARS, P. 1976 – Fire, Ecology, Animal Populations and Man: A Study of some Ecological Relationship in Prehistory. *Proceedings of the Prehistory Society* 42, 1976:15-45.

PĂUNESCU, A. 1964 – Cu Privire la Perioda de Sfîrsit a Epipaleoliticului în Nord-vestul si Nord-estul Romîniei si Unele Persistente ale lui în Neoliticul vechi. *Studii si Cercetari de Istorie Veche* 15, 1964:321-336.

PROSEK, F. 1959 – Mesolitická Obsidiánova Industrie ze Stanice Barca I. *Archeologické Rozhledy* 11, 1959:145-148.

PUSZTAI, R. 1957 – Mezolitikus Leletek Somogyból. *Janus Pannonius Múzeum Évkönyve* 6, 1957:96-104.

SHACKLETON, J.C., and T.H. van ANDEL 1980 – Prehistoric Shell Assemblages from Franchthi Cave and Evolution of the Adjacent Coastal Zone. *Nature* 288, 1980:357-359.

SMITH, A.G. 1970 – The Influence of Mesolithic and Neolithic Man on British Vegetation. In D. Walker and R.G. West, (eds.), *Studies in the Vegetational History of the British Isles*. Cambridge University Press, Cambridge, 1970:81-96.

SÜMEGI, P. 1989 – *Hajdúság Felső-pleisztocén Fejlődéstörténete Finomrétegtani (Üledékföldtani, Őslénytani, Geokémiai) Vizsgálatok Alapján.* Unpublished Ph. D. dissertation 96, University of Debrecen, Debrecen.

SÜMEGI, P. 1996 – *Az ÉK-magyarországi Löszterületek Összehasonlító Őskörnyezeti Rekonstrukciója és Rétegtani Értékelése.* Unpublished Ph. D. Dissertation 120, University of Debrecen, Debrecen.

SÜMEGI, P. 1998 – Az Utolsó 15000 év Környezeti Változásai és Hatásuk az Emberi Kultúrákra Magyarországon. In G. Ilon (ed.), *A Régésztechnikusok Kézikönyve.* Savaria Kiadványa, Szombathely, 1998:367-397.

SÜMEGI, P. 1999 – Reconstruction of Flora, Soil and Landscape Evolution, and Human Impact on the Bereg Plain from Late-Glacial up to the Present, Based on Palaeoecological Analysis. In J. Hamar and A. Sárkány-Kiss (eds.), *The Upper Tisa Valley*. Tiscia Monograph Series, Szeged, 1999, 173-204.

SÜMEGI, P. 2002 – Jászsági Adatok a Magyarországi Holocén Rétegtani és Őskörnyezettani Vitás Kérdéseihez. *Magyar Őslénytani Vándorgyűlés Kiadványa* 5, Pásztó, Abstract and Poster.

SÜMEGI, P. 2008 – Palaeogeographical Background of the Mesolithic and Early Neolithic Settlement in the Carpathian Basin. In J.K. Kozlowski and M. Nowak (eds.), *Mesolithic/Neolithic Interactions in the Balkans and in the Middle Danube Basin: Proceedings of the XV World Congress UISPP, Lisbon, 4-9 September 2006.* BAR International Series, BAR Publishing, Oxford, 2008:47-51.

SÜMEGI, P., S. GULYÁS and G. PERSAITS 2008 – Holocene Paleoclimatic and Paleohydrological Changes in the Sárrét Basin, NW Hungary, *Documenta Praehistorica XXXV*, 2008:25-31.

SÜMEGI, P., and R. KERTÉSZ 2001 – Palaeogeographic Characteristic of the Carpathian Basin: An Ecological Trap During the Early Neolithic? In R. Kertész and J. Makkay (eds.), *From the Mesolithic to the Neolithic.* Archaeolingua Press, Budapest, 2001:405-416.

STIEBER, J. 1967 – A Magyarországi Felső-Pleisztocén Vegetációtörténete az Anthrakotómiai Eredmények (1957-ig) Tükrében. *Földtani Közlöny* 97, 1967:308-316.

SZATHMÁRY, L. 1978 – A Déri Múzeum mezolitikus leleteiről. *Magyar Kurír* 26, 1978:3-6.

VÉRTES, L. 1962 – Die Ausgrabungen in Szekszárd-Palánk und die archäologischen Funde. *Swiatowit* 24, 1962:159-202.

VALOCH, K. 1975 – Eine Endpaläolithische Industrie von Pribice (Bez. Breclav) in Südmähren. *Casopis Moravského Musea v Brne* 60, 1975:45-78.

VALOCH, K. 1985 – The Mesolithic Site of Smolín, South Moravia. In C. Bonsall (ed.), *The Mesolithic in Europe*: Papers Presented at the Third International Symposium. John Donald Publishers, Edinburgh, 1985:461-470.

WILLIAMS, D.F., R.C., THUNELL and J.P. KENNETT 1978 – Periodic Freshwater Flooding and Stagnation of the Eastern Mediterranean Sea During the Late Quaternary. *Science* 201, 1978:252-254.

WILLIS, K.J., P. SÜMEGI, M. BRAUN and A. TÓTH 1995 – The Late Quaternary Environmental History of Bátorliget, N.E. Hungary. *Palaeogeography, Palaeoclimatology, Palaeoecology* 118, 1995:25-47.

WILLIS, K.J., M. BRAUN, P. SÜMEGI and A. TÓTH 1997 – Does Soil Change Cause Vegetation Change or Vice-Versa? A Temporal Perspective From Hungary. *Ecology* 78, 1997:740-750.

WILLIS, K.J., P. SÜMEGI, M. BRAUN, K.D. BENNETT, and A. TÓTH 1998 – Prehistoric Land Degradation in Hungary: Who, How and Why? *Antiquity* 72, 1998:101-113.

ZVELEBIL, M. 1986 – Mesolithic Prelude and Neolithic Revolution. In M. Zvelebil (ed.), *Hunters in Transition.* Cambridge University Press, Cambridge, 1986:5-16.

ZVELEBIL, M., and P. ROWLEY-CONWY 1986 – Foragers and Farmers in Atlantic Europe. In M. Zvelebil (ed.), *Hunters in Transition.* Cambridge University Press, Cambridge, 1986:67-93.

Chapter 11
MESOLITHIC-NEOLITHIC TRANSITION IN THE CARPATHIAN BASIN: WAS THERE AN ECOLOGICAL TRAP DURING THE NEOLITHIC?

Pál SÜMEGI and Sándor GULYÁS
Dept. of Geology and Paleontology, University of Szeged, Hungary

Róbert KERTÉSZ
János Damjanich Museum, Szolnok, Hungary

Gábor TIMÁR
Dept. of Geophysics, University of Budapest, Hungary

Abstract: *The central Carpathian Basin forms the northern boundary of the Körös-Starčevo culture. Various archaeological and scientific interpretations have been advanced to explain the presence of this boundary, which stopped the progress of the Neolithic expansion on its northern frontier. The traditional explanations for the boundary are hampered by the fact that there are no marked geomorphological boundaries and the Mesolithic-Neolithic transition has been confounded by a gap in concomitant Mesolithic settlements. Recent archaeological evidence and modern scientific environmental analysis indicates that climate, bedrock, soil, and the evolution of the environmental mosaics, together with the agricultural economy and cultural tradition of the Early Neolithic, resulted in a boundary at the northern expansion limit of the Körös-Starčevo culture. This Central European-Balkan Agro-ecological Barrier (CEB AEB) may have functioned as an ecological trap.*

INTRODUCTION

Archaeologist point out that the Carpathian Basin (Fig. 11.1) is one of the most important areas involved in the process of European neolithization. Its central part forms the northern boundary of expansion of the Anatolian-Balkan Neolithic during the Early Neolithic, represented by the Körös-Starčevo culture (Kutzián 1947, Kalicz and Makkay 1977, Kalicz *et al.* 1998, Makkay 1982, Raczky 1983, Bánffy 1996, 2000a, 2000b, Otté and Noiret. 2001). Various archaeological and scientific interpretations have been advanced to explain the presence of the Neolithic barrier, which stopped the Neolithic expansion on the northern frontier of the Körös-Starčevo culture (Fig. 11.2). Previous scientific explanations have concentrated on two main areas: models of river channel shifts and the interpretation of the paleoenvironmental record (Szathmáry 1978, Dobosi 1983, Makkay 1982, Raczky 1989). Such research also focused on the environmental and social factors influencing the expansion of the Early Neolithic communities, whose culture and agricultural production experience comes from the Balkans and the Mediterranean (Ammerman and Cavalli-Sforza 1971).

Traditional explanations for the barrier to Neolithic expansion are hampered by the fact that there are no marked geomorphological boundaries, such as mountains or large rivers. Furthermore, the more recent search of the mechanisms which drove the Mesolithic-Neolithic transition has been confounded by a gap in the archaeological record due to a dearth of concomitant Late Mesolithic settlements (Fig. 11.3).

In this chapter we seek an answer to the question of why the northern expansion of the earliest Neolithic ceased within central parts of the Carpathian Basin. We also deal with the complexities of the Mesolithic-Neolithic transition. In the process, we present new paleoclimatic and environmental data.

Based on paleoecological, climatological, and pedological research (Sümegi *et al.* 1998), as well as recent archaeological data (Kertész *et al.* 1994), a new agro-ecological model (Sümegi and Kertész 2001) is proposed for the whole of the Carpathian Basin. The model provides a fundamentally different approach to understanding the archaeological and paleoenvironmental records by explaining past relationships between man and environment in the Carpathian Basin, (Kertész and Sümegi 2001, Sümegi 2008). It systematically disproves the theories of the Proto-Tisza River boundary and several other paleoenvironmentally based concepts (ibid.).

THE FOUNDATIONS OF THE NEW AGRO-ECOLOGICAL MODEL

According to detailed geological and paleontological analyses on Pleistocene deposits of the Carpathian Basin, this 300,000 km^2 area lies at the interface of the woodlands of Central and Central-Eastern Europe, as well as the arboraceous grasslands or steppes of Eastern Europe. This area is characterized by highly complex and diverse developments during the past two to two and a half million years (Sümegi *et al.* 1998). From the start of

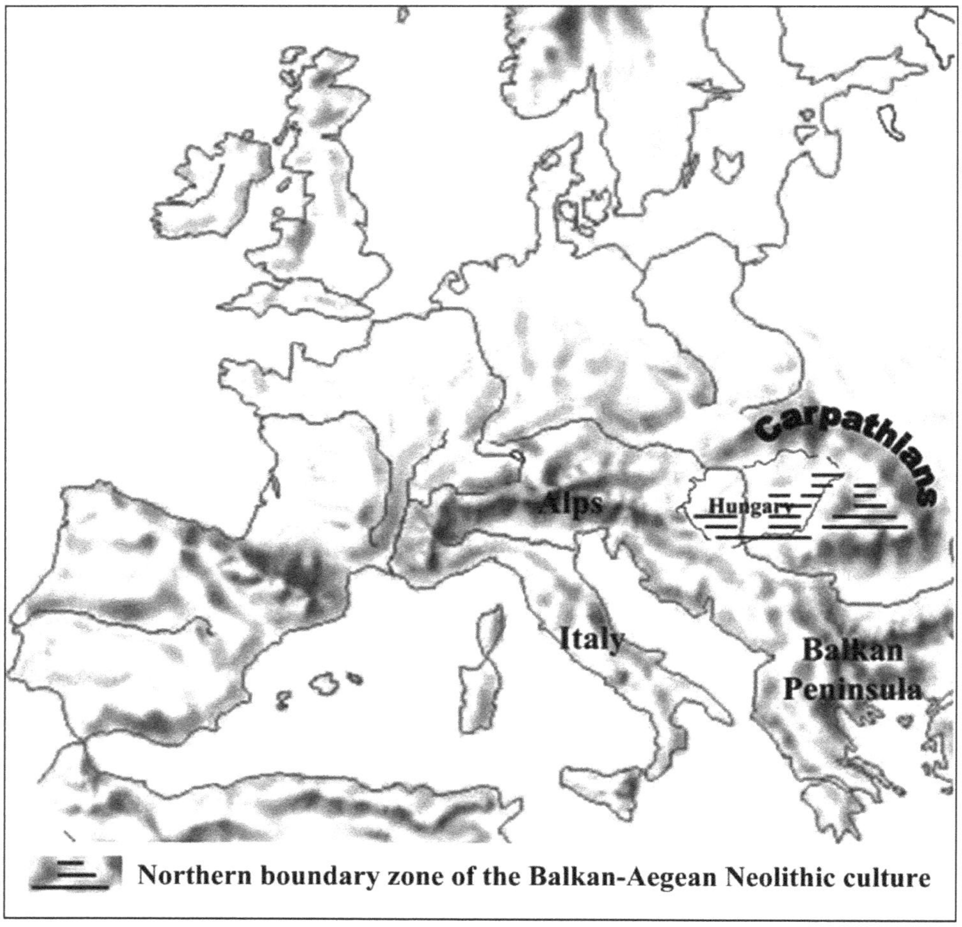

Figure 11.1. The location of the Carpathian Basin and Hungary

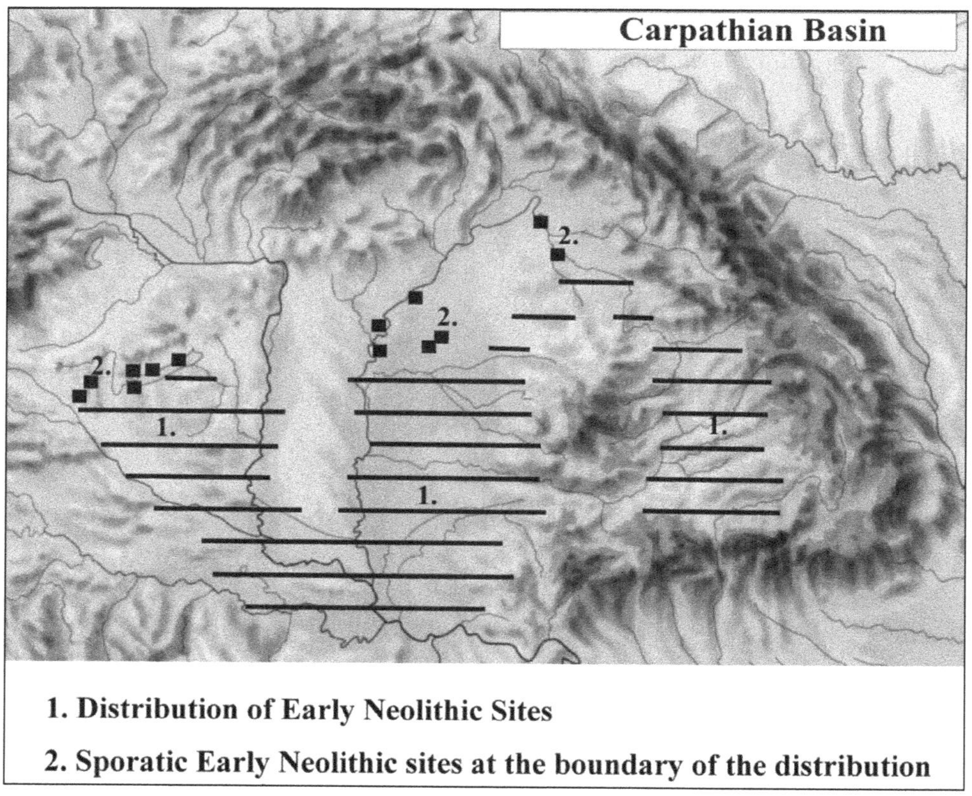

Figure 11.2. Early Neolithic sites in the Carpathian Basin

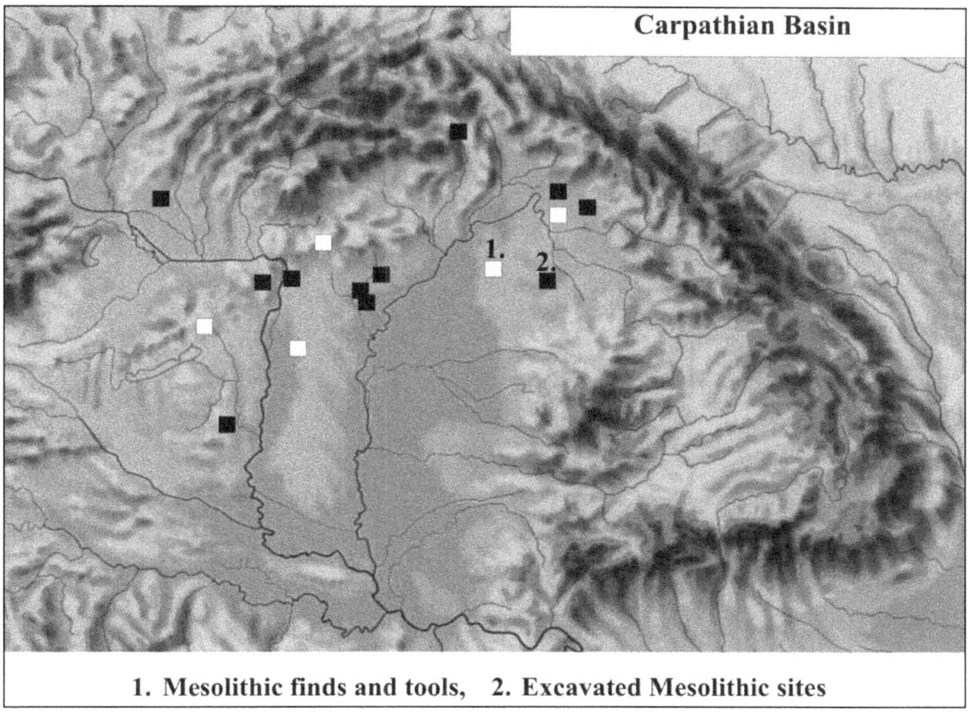

Figure 11.3. Mesolithic sites and finds in the Carpathian Basin

the Ice Age, three major areas developed in the Carpathian Basin region displaying large-scale complexity and mosaic-like structure at micro-, meso-, and macro- scales. The formation of a complex macro-scale mosaic is due to the presence of overlaps in four major climatic zones. This situation is further complicated by a gently decreasing Continental and increased Oceanic effect from east to west. Furthermore, an increasing Sub-Mediterranean influence from north to south and the presence of Sub-Carpathian climatic influence in the hilly areas and mountains are observed.

Due to these climatic zones, mosaic-like vegetation developed in the Carpathian Basin as early as the end of the Pleistocene, and this mosaic-like complexity (mosaicity) is reflected in the present composition and distribution of the vegetation as well (Sümegi and Hertelendi 1998). The regional and local morphological and hydrological conditions further intensified through time and influenced the effects of the major overlapping climatic zones. As a consequence of this mosaicity and development of climatic and vegetation zones, soil conditions tend to also display large-scale mosaic-like complexity within the Carpathian Basin region. This complexity is further intensified by the high variability of the bedrock.

The observable mosaicity of climatic, faunal, floral, and soil endowments developed during the Quaternary and fluctuated cyclically in time and space. The environmental complexity strongly affected the immigrating Neolithic communities, forcing them to adapt to the mosaic conditions. It prevented the expansion of the gathered, hunted, or even domestically grown plants and animals. Thus, the climate and environment produced the economic foundations of these societies throughout the entire Carpathian Basin. Therefore, the immigrant groups, coming from various climatic and environmental areas, could occupy only those parts of the Carpathian Basin that corresponded to their prior economic experience (Sümegi and Kertész 2001).

According to previous paleoecological studies, these paleoenvironmental differences have been continuously present since the beginning of Holocene 10,000 years ago, and they determined the possibilities of expanding settlements for individual cultures – especially those using domesticates (Sümegi 1999). However, our research indicates that a very complex interaction must have evolved between the Neolithic and Mesolithic groups in the central areas of the Carpathian Basin and their surrounding environments.

Only a new, complex geoarchaeological model can explain the neolithization process, as well as the complex socio-cultural interactions. This new model is based on the following archaeological and environmental evidence (Sümegi and Kertész 2001):

1. Early Neolithic groups of the Körös-Starčevo culture must have migrated to the southern parts of the Carpathian Basin from the Mediterranean, because their cultural background as well as the primary genetic centers of their domesticated crops and animals was in Asia Minor, the ancient Near East, and South-Eastern Europe (Vavilov, 1951, Bökönyi, 1974). Representatives of this Early Neolithic culture cultivated Mediterranean crops, such as einkorn, emmer, spelt, and barley (Füzes, 1990, Hartyányi *et al.* 1968) and raised animals with a dominance of sheep

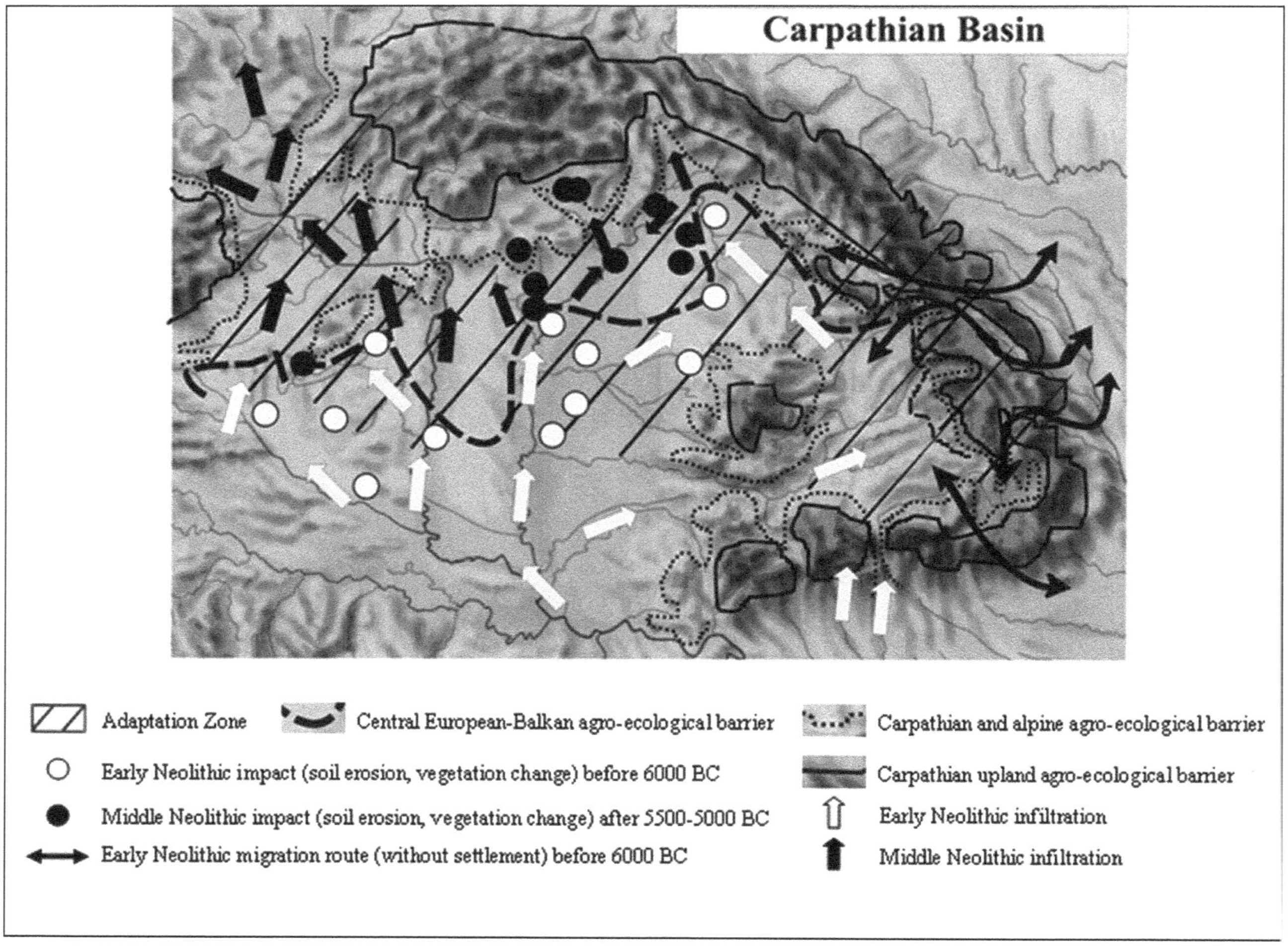

Figure 11.4. Central European Agro-ecological Barrier

and goat (Bökönyi, 1974). Besides crop cultivation and animal husbandry, hunting, fishing, and gathering continued to play an important role among the agriculturists (Bökönyi 1974). These Neolithic communities were characterized by sedentism, village-like settlements, and relatively high population densities.

2. Mesolithic groups in the northern parts of the Carpathian Basin can be culturally linked to Western and Central Europe (Kertész *et al.* 1994). Their subsistence was based exclusively on hunting and gathering, with a dominance of the former. Hunting of solitary prey, such as wild horse, red deer stag, and wild boar, called for different hunting strategies than the hunting of herd animals, such as bison, aurochs, red deer, hind, and roe deer. Due to their mobile lifestyles, these small hunter-gatherer groups established seasonal settlements, with summer and winter camps. The less expansive and thin occupation layers observed at these camps indicate a low population density and brief, seasonal occupations.

3. Every creature, including the plants and animals domesticated by the Early Neolithic groups, has certain habitat and environmental preferences. Climatic and environmental factors, such as temperature, precipitation, solar radiation during the growth season, the length of the vegetation period, etc., influenced the crops that were cultivated and the animals that were bred by the Earliest Neolithic communities, as well as the prey animals of the hunter-gatherer communities (Sümegi *et al.* 1998). This phenomenon is called the minimum limiting factor (Liebig 1840).

When factors such as climate, bedrock, soil, and the evolution of the environmental mosaics are taken together with the agricultural economy and cultural tradition, a boundary is defined by the northern limit of the Körös-Starčevo culture's expansion. This theoretical line is called the Central European-Balkan Agro-ecological Barrier (abbreviated as CEB AEB). It limits the northward spread of the Balkan type neolithization during the Early Neolithic of the Carpathian Basin (Fig. 11.4) and strongly influences the agricultural economy. This implies that Early Neolithic communities with a Mediterranean cultural background found themselves in an ecological trap (Sümegi and Kertész 2001) on the margins of a Balkan climate and environment. As a result, their advance into the Carpathian Basin first slowed down and eventually halted along the CEB AEB.

What was the effect of the above on the Mesolithic communities, living north of this barrier, and their gathering, fishing, and hunting lifestyles? In order to find an answer to this question two important points must be considered.

First, Mesolithic hunter territories came close to that of the Neolithic farmers. This enabled the transmission of Neolithic culture from one community to the other (Zvelebil and Rowley-Conwy 1986, Zvelebil 1986). However, due to different socio-cultural developments and adaptation to environmental factors, there must have been significant quantitative population differences between the Neolithic and Mesolithic populations. However, the differences could have been easily mitigated in the river valleys, which acted as a kind of infiltration zone. These relatively narrow, channel-like areas forced the smaller Mesolithic communities into circumscribed regions, resulting in increasing population densities (Sümegi 2000). This leads to the assumption that Early Neolithic and Late Mesolithic communities met in these zones. Thus, the parts of the river valleys lying on the CEB AEB must have led to the first stage of the neolithization process in the Carpathian Basin.

Second, it is important to emphasize that due to the lack of conditions necessary for the formation of Balkan type agricultural production north of the CEB AEB, Early Neolithic communities were unable to settle these northern areas. This provides time then for the Mesolithic communities to adopt the Neolithic technical and production innovations without actually assimilating culturally, economically, or demographically into Balkan Neolithic communities. However, most Mesolithic communities south of the CEB AEB fully assimilate, undergoing the entire neolithization process. The only exceptions are those places where possibilities for isolation were present, e.g., the Iron Gate, where the Mesolithic population around Lepenski Vir lived near the Danube gorges area. Thus, the CEB AEB played a crucial role in the formation of a very different, new Neolithic culture, fully adapted and assimilated to the local conditions north of the boundary.

Our agro-ecological model explains the expansion of neolithization at the macro-scale of several thousand square kilometers, but how can we account for the paleoenvironmental effects in the process of neolithization at a regional scale?

NEW ASPECTS OF THE AGRO-ECOLOGICAL BARRIER MODEL

The appearance, settlement, and dissemination of advanced Neolithic cultures from Asia Minor and the Balkans was a very important process in the evolution of the Carpathian Basin environment during the Early Neolithic. It started an anthropogenic process that led to the transformation of the natural diversity of the Great Hungarian Plain (Willis *et al.* 1995, 1998, Sümegi 1999, Sümegi and Bodor 2000). At the same time, settlements of the Körös culture were clearly connected to alluvial flats and alluvial plains by the banks of the rivers (Kalicz and Makkay 1977, Sherratt 1982, 1983, Whittle 1996). Several scholars who examined this factor at the macro-scale only arrived at the conclusion that hydromorphic soils of the alluvial flats played a very important part in the economies of Körös culture (Nandris 1972, Kosse 1979, Sherratt 1982). However, examinations of this macro-scale process at local and regional scales yields rather surprising results (Sümegi 2000).

The groups within the Körös culture developed two settlement types, which can easily be distinguished by their location within the environment. One type of Körös site is situated on Holocene alluvia beside the riverbeds. Thus far, this type has not been researched extensively, and we have only very scarce knowledge concerning a few aspects of these settlements. They might have been used for hunting, fishing, and gathering activities. Thus, they could have been the outgrowth of adaptation by local agricultural groups or formed through interaction with nearby Mesolithic communities.

The other type is *normally* found on the natural levees that developed toward the end of the Pleistocene. These levees are in most cases covered by infusion Aeolian loess deposits. Morphologically they constitute the highest points of the plain depression areas. Therefore, they are above flood level. Due to their water absorbing capacity, grain composition, bedrock and structural features, closely related to the black soils, no alluvial and hydromorphic soils developed.

If we examine the Pleistocene relict surface thoroughly in relation to the Early Neolithic settlement and economy, we can formulate a model that demonstrates the differences in settlement strategies resulting from the regional and micro-scale mosaicity. This model suggests that during the course of Early Neolithic expansion, groups moved from dominantly alluvial areas towards elevated loess-covered surfaces (natural levees) as part of their landscape use and settlement strategies. In this adaptation process, a special part is played by these isolated surfaces on the Holocene alluvia of the Great Hungarian Plains. Levees are covered with drier, loessic rocks and black earth soils (Fig. 11.5). In other words, they are covered mostly with infusion loess, because they represent a transitional area between alluvia and dry elevated loess-covered surfaces (levees) (Sümegi 2000). These micro-mosaics of river alluvia and loess-covered Pleistocene elevated relict surfaces must have played an important role in the neolithization process of the Carpathian Basin (Fig. 11.6). The island-like loessic alluvial heights were primarily chosen by the Neolithic communities for settling (Sümegi *et al.* 2000), because they provided an excellent opportunity for the extension of previously acquired farming economies to loess-covered elevations.

Animal husbandry and agricultural production were the dominant subsistence activities on the elevated flood-free Pleistocene lag surfaces (natural levees). This relates to one type of the Körös settlement strategy. The other type of Körös site is situated on Holocene alluvia adjacent to

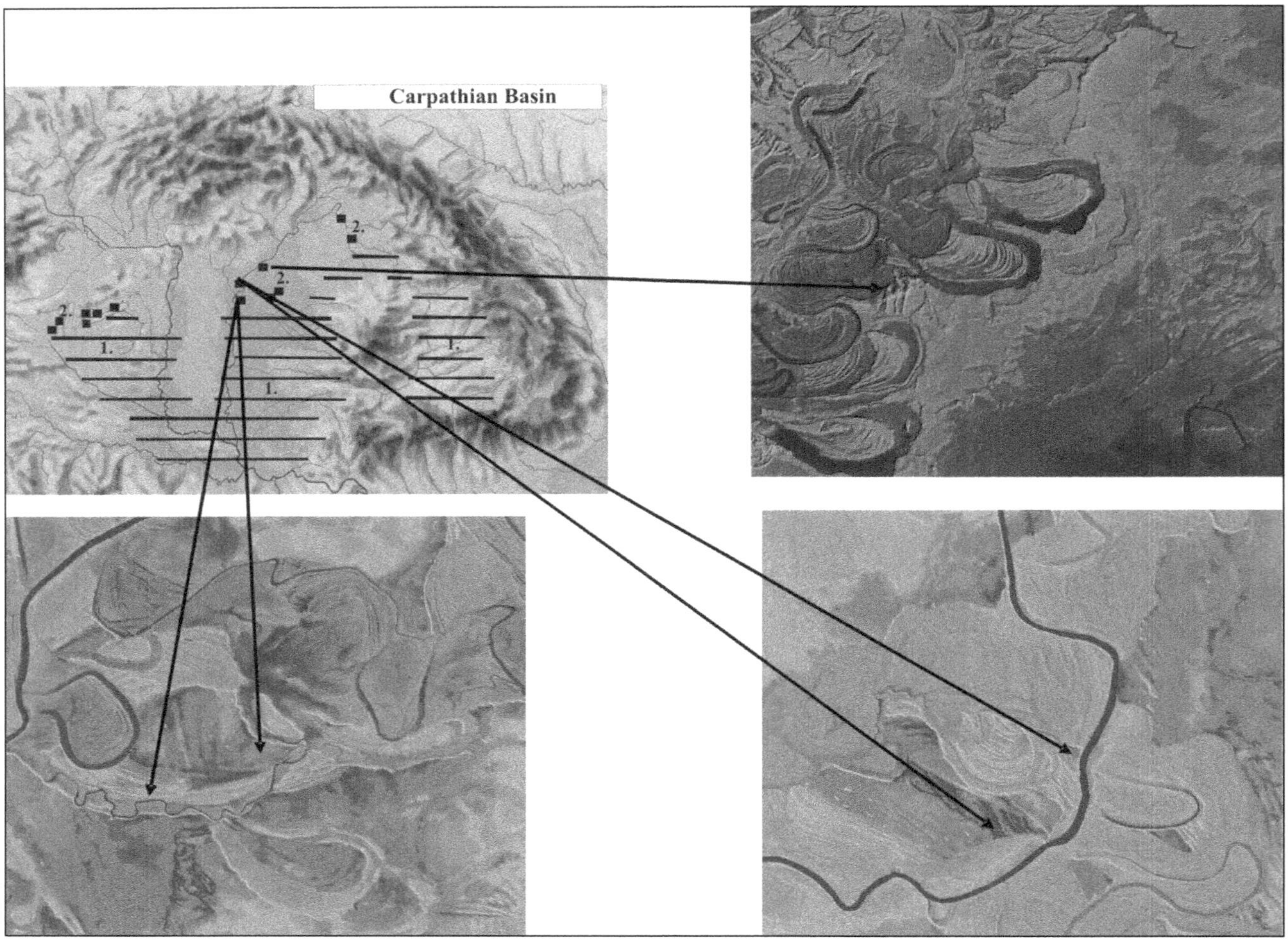

Figure 11.5. Sporadic Early Neolithic sites on the loess-covered Pleistocene lag surfaces (loessy islands in alluvia) on the northern boundary of the Aegean-Anatolian culture

the riverbeds. However, on the plain (next to the contemporary active river branch) the Holocene alluvial surfaces were covered with difficult to cultivate clay-rich soils and gallery forests. Here hunting, fishing, and gathering were the dominant subsistence activities. Since the archeologically identified Mesolithic communities inhabited the closed forest-covered alluvia in the central parts of the Carpathian Basin and travelled through them in search for sustenance (Kertész *et al.* 1994), they must have occasionally encountered Early Neolithic farmers. These chance encounters occurred in spite of the fact that they lived on and used different habitat portions of the riparian environment than Neolithic population.

SUMMARY

The view we have presented above differs significantly from earlier notions, even at the macro-scale. Environmental differences already present during the Holocene determined the life of Mesolithic and Neolithic groups. The Central European-Balkan agro-ecological barrier (CEB AEB) constrained the settlement and expansion possibilities of the Early Neolithic Körös culture, whose roots lie in the Balkan Peninsula and Asia Minor.

The same barrier also must have played an essential part in the neolithization process of Late Mesolithic communities in the Carpathian Basin, and in the establishment of autochthonous Neolithic groups that did not have Mediterranean cultural roots, including the development of the Linear Pottery or *Linienbandkeramik* (LBK) complex that developed between ca. 5500-5000 BC. Moreover, infiltration zones in the analyzed area along the main river valleys acted as important media, where Late Mesolithic and Early Neolithic communities could maintain contact with one another. These areas, transitional in morphology, climate, vegetation, and soil conditions, represent an environmental shift due to a sudden increase in elevations and the effects of micro-scale mosaicity. The Pleistocene lag surfaces and alluvial plains that formed because of different geological histories are characterized by different subsoil and morphological conditions. These differing micro-environmental conditions must have significantly altered and modified settlement strategies of the Körös culture and led to the establishment of settlements permitting eventual occupation of the higher loess-covered surfaces. Therefore, central parts of the Carpathian Basin constitute a very important transitional region between environments of the Balkan Peninsula and those of the western part of Europe.

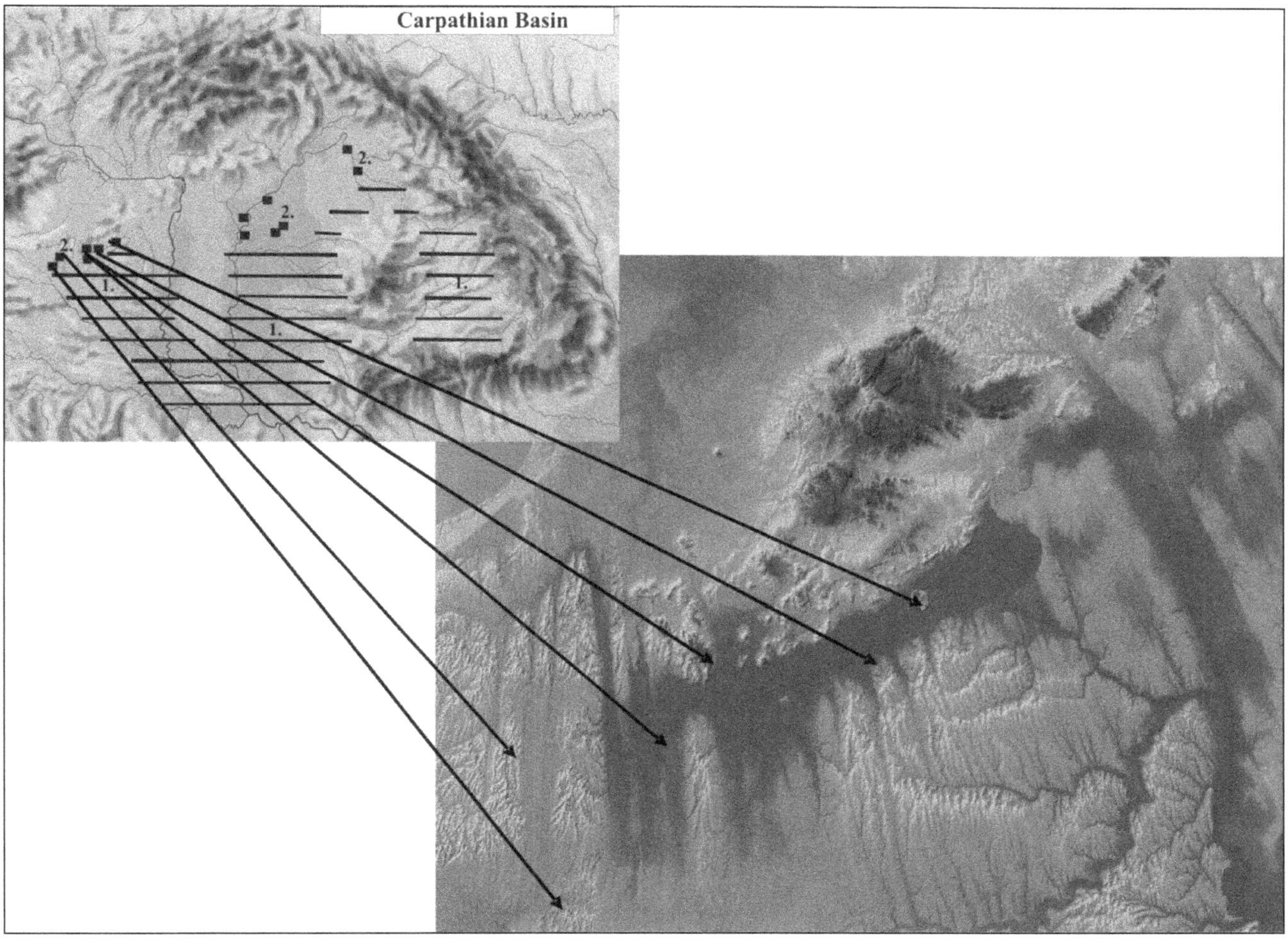

Figure 11.6. Sporadic Early Neolithic (Starčevo) sites on the loess-covered surfaces in the northern boundary of Balkan-Aegean culture

References

BÁNFFY, E. 1996 – Neolithic and Copper Age Settlements at Hahót and Zalaszentbalázs. Archaeology And Settlement History. In The Hahót Basin SW Hungary. *Antaeus* 22, 1996:35-50.

BÁNFFY, E. 2000a – Starčevo und/oder LBK? *Varia Neolithica I. Beiträge zur Ur- und Frühgeschichte Mitteleuropas* 22, 2000:47-60.

BÖKÖNYI, S. 1974 – *History of Domestic Mammals in Central and Eastern Europe*. Akadémiai Kiadó, Budapest.

DOBOSI, T.V. 1983 – Ausgrabung von Tarpa-Gehöft Márki. *Communicationes Archaeologicae Hungariae* 12, 1983:5-18.

FÜZES, M. 1990 – Die Pflanzenfunden in Ungarn der Anfänglichen Entwicklungsphase des Ackerbaues (Neolithikum und Kupferzeit). *Tapolcai Városi Múzeum Közleményei* 1, 1990:139-238.

HARTYÁNYI, B.P., G.Y. NOVÁKI and Á. PATAY 1968 – Samen- und Fruchtfunde in Ungarn von der Jungsteinzeit bis zum 18. Jahrhundert. *Magyar Mezőgazdasági Múzeum Közleményei*, 1968:5-84.

KALICZ, N., and J. MAKKAY 1977 – Die Linienbandkeramik in der Großen Ungarischen Tiefebene. *Studia Archeologica* 7, 1977:25-32.

KALICZ, N.M., Zs.T. VIRÁGH and K. BÍRÓ 1998 – The Northern Periphery of the Early Neolithic Starčevo Culture in Southwestern Hungary: A Case Study of an Excavation at Lake Balaton. *Documenta Praehistorica* 25, 1998:151-181.

KERTÉSZ, R., P. SÜMEGI, M. KOZÁK, M. BRAUN, E. FÉLEGYHÁZI and E. HERTELENDI 1994 – Archeological and Paleoecological Study of an Early Holocene Settlement in the Jászság Area (Jászberény I). *Acta Geographica Debrecina* 32, 1994:5-49.

KERTÉSZ, R., and P. SÜMEGI 2001 – Theories, Critiques and a Model: Why did the Expansion Of The Körös – Starčevo Culture Stop In The Centre Of The Carpathian Basin? In R. Kertész and J. Makkay (eds.), *From the Mesolithic to the Neolithic*. Archaeolinqua Press, Budapest. 2001:225-246.

KUTZIÁN, I. 1947 – The Körös Culture. *Dissertationes Pannonicae*, Series II/23.

LIEBIG, J. 1840 – *Organic Chemistry and its Application to Agriculture and Physiology*. Taylor and Walton Press, London.

MAKKAY, J. 1982 – *A magyarországi neolitikum kutatásának új eredményei.* (*The latest results of Hungarian Neolithic Research*), Akadémiai Kiadó, Budapest.

KOSSE, K. 1979 – *Settlement Ecology of the Early and Middle Neolithic Körös and Linear Pottery Cultures in Hungary.* British Archaeological Reports 64, 1979:35-46.

NANDRIS, J. 1972 – Relation Between the Mesolithic, the First Temperate Neolithic and the Bandkeramik: The Nature of the Problem. Alba Regia, *Annales Musei Stefani Regis* 12, 1972:61-70.

OTTE, M., and P. NOIRET 2001 – La Mésolithique du Bassin Pannonien et la Formation du Rubané. *L'Anthropologi* 105, 2001:409-419.

RACZKY, P. 1983 – Questions of Transition Between the Early and Middle Neolithic in the Middle and Upper Tisza Region. *Archeologiai Értesítő* 110, 1983:161-194.

RACZKY, P. 1989 – Chronological Framework of the Early and Middle Neolithic in the Tisza Region. *Varia Archaeologica Hungarica* II, 1989:233-251.

SHERRATT, A. 1982 – The development of Neolithic and Copper Age Settlement in the Great Hungarian Plain. Part 1: The regional setting. *Oxford Journal of Archaeology* 1, 1982:287-316.

SHERRATT, A. 1983 – The development of Neolithic and Copper Age settlement in the Great Hungarian Plain. Part 2: Site Surveys and Settlements Dynamics. *Oxford Journal of Archaeology* 2, 1983:13-41.

SÜMEGI, P. 1999 – Reconstruction of Flora, Soil and Landscape Evolution, and Human Impact on the Bereg Plain from Late-Glacial up to the Present, Based on Palaeoecological Analysis. In J. Hamar and A. Sárkány-Kiss (eds.), *The Upper Tisa Valley.* Tiscia Monograph Series, Szeged, 1999:173-204.

SÜMEGI, P. 2000 – An Environmental Archaeological Analysis of "Bihar-area". In L. Selmeczi (ed.), *Neolithic of "Bihar-area".* Déri Múzeum, Debrecen, Debrexen, 2000:7-18.

SÜMEGI, P. 2008 – Palaeogeographical Background of the Mesolithic and Early Neolithic Settlement in the Carpathian Basin. In J.K. Kozlowski and M. Nowak (eds.), *Mesolithic/Neolithic Interactions in the Balkans and in the Middle Danube Basin: Proceedings of the XV World Congress UISPP, Lisbon, 4-9 September 2006.* BAR Publishing, 2008:53-82.

SÜMEGI, P., and E. BODOR 2000 – Sedimentological, Pollen and Geoarcheological Analysis of Core Sequence at Tököl. In I. Poroszlai and M. Vicze (eds.), *Százhalombatta Archaeological Expedition.* Archeolinqua Press, Budapest, 2000:83-96.

SÜMEGI, P., and E. HERTELENDI 1998 – Reconstruction of Microenvironmental Changes in Kopasz Hill Loess Area at Tokaj (Hungary) Between 15000 – 70000 BP Years. *Radiocarbon* 40, 1998:855-863.

SÜMEGI, P., E. HERTELENDI, E. MAGYARI and M. MOLNÁR 1998 – Evolution of the Environment in the Carpathian Basin During the last 30000 BP Years and its Effects on the Ancient habits of the Different Cultures. In L. Költő and L. Bartosiewicz (eds.), *Archimetrical Research in Hungary* II. Budapest, 1998:183-197.

SÜMEGI, P., and R. KERTÉSZ 2001 – Palaeogeographic Characteristic of The Carpathian Basin: An Ecological Trap During The Early Neolithic? In R. Kertész and J. Makkay (eds.), *From the Mesolithic to the Neolithic.* Archaeolinqua Press, Budapest. 2001:405-416.

SZATHMÁRY, L. 1978 – About Mesolithic Finds of the Déri Museum. *Múzeumi Kurír* 26, 1978:3-6.

VAVILOV, N.I. 1951 – The Origin, Variation, Immunity and Breeding of Cultivated Plants. *Chronica Botanica* 13, 1951:25-32.

WHITTLE, A. 1996 – *Europe in the Neolithic.* Cambridge University Press, Cambridge.

WILLIS, K.J., P. SÜMEGI, M. BRAUN and A. TÓTH 1995 – The Late Quaternary Environmental History of Bátorliget, N.E. Hungary. *Palaeogeography, Palaeoclimatology, Palaeoecology* 118, 1995:25-47.

WILLIS, K.J., P. SÜMEGI, M. BRAUN, K.D. BENNETT and A. TÓTH 1998 – Prehistoric Land Degradation in Hungary: Who, How and Why? *Antiquity* 72, 1998:101-113.

ZVELEBIL, M. 1986 – Mesolithic Prelude and Neolithic Revolution. In M. Zvelebil (ed.), *Hunters in Transition.* Cambridge University Press, Cambridge. 1986:5-16.

ZVELEBIL, M., and P. ROWLEY-CONWY 1986 – Foragers and Farmers in Atlantic Europe. In Zvelebil, M. ed. *Hunters in Transition.* Cambridge University Press, Cambridge. 1986:67-93.

Chapter 12
NEW DATA CONCERNING THE DETECTION AND NATURE OF HUMAN IMPACT ON THE MOHOS LAKES, NORTHEAST HUNGARY

Imola E. JUHÁSZ
Institute of Archeology, Hungarian Academy of Sciences, Budapest, Hungary

Abstract: *Detailed palynological analysis of samples from the borehole in the Kis-Mohos marshland yields new information regarding vegetation changes at the Pleistocene/Holocene boundary in the Kelemér region of North Hungary. According to these data, the advent of deciduous woodlands lagged 1000 years behind succession in the Great Hungarian Plains, which started approximately 9500 years ago. Comprehensive evaluation of pollen and radiocarbon data for the Nagy-Mohos area enables a more accurate chronological delineation of Holocene vegetation development phases. The most significant results are the identification of a pre-neolithization process in the ecological record, the deforestation activities of the Neolithic population, the environmental impact of Copper and Bronze Age communities, and the reconstruction of environmental conditions at the time of the Kelts (Celts) and the cultures of the Migration Period (including the immigrating Hungarians). The observed environmental changes are largely attributed to human impact. However, in some instances climatic change does coincide with human impact, suggesting a human adjustment to such a change.*

INTRODUCTION

The Mohos Lakes of Kelemér region, Northeast Hungary, are among the most precisely studied paleoecological sites in the Carpathian Basin (Fig. 12.1 and 12.2). The earliest study on vegetation development was published in 1915 by geologists from the Hungarian National Geological Institute (László and Emszt 1915). Later, one of the most important personalities in Hungarian botanical research, B. Zólyomi carried out palynological studies in order to date the peat bogs and reconstruct their vegetation history.

Radiocarbon dates place the formation of two peat bogs (Kis-Mohos and Nagy Mohos) at 25000 and 15000 years ago (Willis *et al.* 1997, Magyari *et al.* 2001) contrary to the previously supposed 10000 and 5000 years by Zólyomi (1931). Three new cores reveal a clay-rich layer nearly devoid of macrofossils developed in the southern basin of the Nagy Mohos at the time when the Kis-Mohos formed.

The Kis-Mohos basin layer was created by a landslide, of which deposits appear between 270 and 300 cm in the NM2b sediment section. Below this is another 260 cm thick Pleistocene peat sediment (Juhász 2002). The upper meter of which consists of a solid (difficult to break through) deposit rich in charcoal fragments and pebbles. Zólyomi (1931) supposed it to be the base of the sediment. However, underneath Willis *et al.* (1997, 1998) and Sümegi (1998) found an additional 400 cm thick moss and meadow peat layer overlying a 200 cm deep Late Glacial lacustrian sediment. Although Zólyomi previously supposed this layer to be the base, it actually slid into the basin due to erosion resulting from heavy fire-aided forest clearance by the Kelts (Celts). The rise in the water level during medieval times was probably caused by artificial increase of water in the bog (Sümegi *et al.* 2008). The latest data, presented here, allows more accurate chronological delineation of the Holocene vegetation development phases.

RADIOCARBON AGES AND CHRONOLOGICAL PROBLEMS CONCERNING CORE NM-2b

The vegetation history of the last 27000-25000 years is represented in the pollen sequence from Nagy-Mohos (NM-2b). Included is a longer hiatus during which pollen deposition was inhibited. Here I concentrate on reconstruction of vegetation development during the Holocene (approximately the last 10000 years) when the influence of human populations living near the site are detectable.

The ^{14}C date of 16700±950 bp (19000/16900 BC)[1] obtained from a depth of 310-315 cm in the NM-2b core must be rejected based on comparison with pollen sequences from the same area of the Nagy-Mohos (Magyari *et al.* 2000) as well as from results obtained from the Kis-Mohos located about 400 m distance (Fig. 12.2; Willis *et al.* 1997, Willis *et al.* 1998). This date by itself would indicate the pollen assemblage from zone NM2b-f was deposited during the Pleniglacial. However, the pollen content of this layer shows a likely Early Holocene vegetation pattern, which is in total disaccord with the Pleniglacial Period.

Our diagram from NM2b-f correlates well with the pollen sequence from NM I (Magyari *et al.* 2000, Magyari 2002). Both of the cores show signs of a hiatus between

[1] Editors' note: Uncalibrated dates are listed as bp and calibrated dates are given as BC. The uncalibrated bp ^{14}C dates have been calibrated by the senior editor, using atmospheric data from Reimer *et al.* (2004) with OxCal version 3.10 (Bronk Ramsey 2005). Only the tabulated "singe range" at ±0 is given.

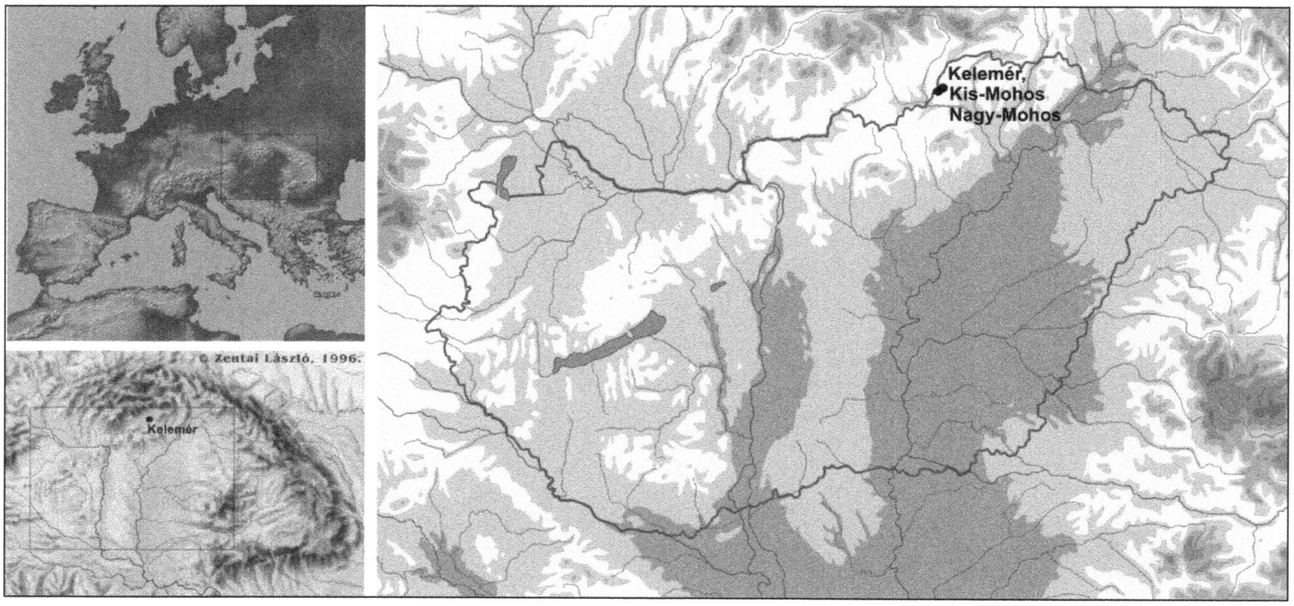

Figure 12.1. Location of the Nagy-Mohos and Kis-Mohos Lakes at Kelemér in the Carpathian Basin

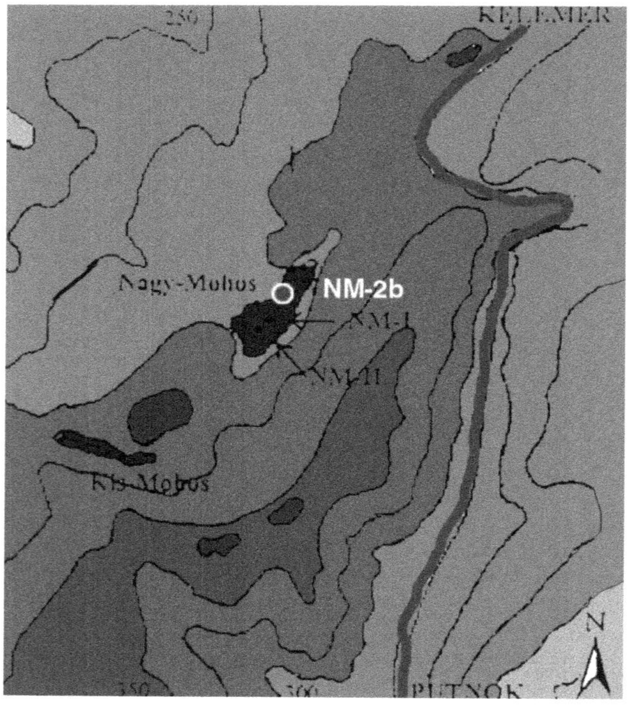

Figure 12.2. Geographical position of Nagy-Mohos and Kis-Mohos Lakes at Kelemér, North Hungary

12150/12065-7600/7590 BC (12200-8600 bp) at NM I (Magyari *et al.* 2000), and between ca. 13520/13250-10670/10440 BC (13000-10500 bp) at NM-2b (Juhász 2002). This time period covers the end of Dryas I and the terminal Late-Glacial/Early Preboreal. A silty, sterile lake sediment layer representing the end of the Preboreal and the beginning of the Boreal (ca. 9300-8300 bp or 8610/8540-7450/7330 BC) was deposited above the Early Holocene organic-rich peat layers, at a depth between 300 and 275 cm.

DESCRIPTION AND THE INTERPRETATION OF THE LOCAL POLLEN ASSEMBLAGE ZONES (LPZ) OF NAGY-MOHOS (NM2b) CONCERNING THE BEGINNING OF THE HOLOCENE

NM2b-f (320-300 cm) ca. 10500-9300 bp (10670/10440-8610/8540 BC)

Samples from 320 cm represent the end of the Late-Glacial and the beginning of the Holocene (Fig. 12.3). The Preboreal is characterized by an amelioration of climate with warmer temperatures favorable to the appearance and development of broad-leaved termophyllous tree taxa. Temperate oak forests develop with typical elements, such as hazel (*Corylus*), elm (*Ulmus*) and linden (*Tilia*). Birch (*Betula pendula*), a wind-pollinated taxon, shade intolerant, pioneer species, has a regional presence. *Betula* is usually represented at the forest periphery or in small stands close to it. *Pinus sylvestris*, though still represented at 20%, decreases towards the end of the zone. The vegetation pattern is mosaic-like with herbaceous vegetation, elements of cold steppe (*Artemisia*-mugwoth, *Caryophyllaceae*, *Chenopodiaceae*-goosefoot) and typical taxa of wet meadows (*Ranunculaceae*, *Potentilla* sp., *Rubiaceae*) represented. Along the edges of the lake, *Typha/Sparganium* and sedges (*Cyperaceae*) suggest lower water levels. There is a sporadic appearance of *Myriophyllum* sp. along with Sphagnum moss spores. Pollen proportions of steppe species are much higher than usually detected during the Early Holocene despite elevated values of temperate tree taxa. From this pollen assemblage we may conclude that temperate tree-taxa are present during the Pleistocene/Holocene transition along with the steppe vegetation, or the younger Early Holocene pollen grains are mixed with grains of older sediment layers.

As mentioned earlier, sterile pollen layers at depths of 300 and 275 cm probably represent the end of the

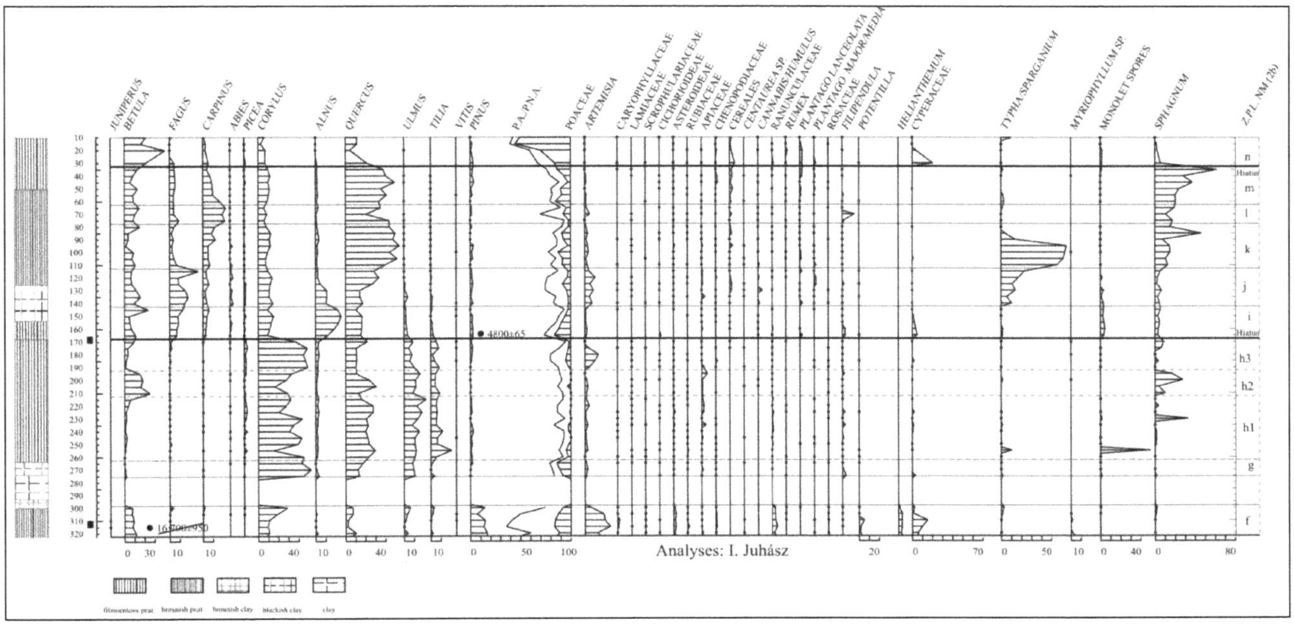

Figure 12.3. Simplified percentage pollen diagram of the Holocene part of Nagy-Mohos NM2b pollen sequence (Juhász 2002)

Preboreal and the beginning of the Boreal (ca. 9300-8300 bp or 8610/8540-7450/7330 BC). Sedimentation continues in our Nagy-Mohos NM2b sequence from a depth of 275 cm onwards. This differs from the Nagy-Mohos NMI and II samples (Magyari *et al.* 2000), where the sedimentation rate slowed down around 5515 bp (4355/4340 BC). There is little information concerning vegetation development for this period until the upper 115 cm are reached (Magyari 2002:25 and Appendix 4.1.2.). This general lack of information may represent an unsuccessful attempt to drain the mire in the early 20th century (Gyulai 1995). Correlations can be made only with the pollen cores from the Kis-Mohos Lake (Willis *et al.* 1998), lying about 400 m south-west of Nagy-Mohos next to Vár-domb or Castle Hill (Fig. 12.2).

DESCRIPTION AND THE INTERPRETATION OF THE LOCAL POLLEN-ASSEMBLAGE ZONES OF NAGY-MOHOS (NM2b) ENCOMPASSING THE EARLY AND MIDDLE HOLOCENE

NM2b-g (275-260 cm) 8300-7900 bp (7450/7330-6755/6685 BC)

This local pollen assemblage zone (LPZ) was most probably deposited during the Boreal period (the Mesolithic). Very high percentages of arboreal (A.P.) species, close to 80%, dominated by *Corylus* are represented, and the proportion of the pollen grains for other temperate tree taxa including *Quercus* (oak), *Ulmus* and *Tilia* increase. The low but continuous presence of ivy (*Hedera*) indicates the favorable climate during the Boreal period. The temperate oak-hazelnut dominated forest is characterized by a robust forest canopy. Earlier well represented Scots pine (*Pinus sylvestris*) and *Betula* become sporadically distributed. Herbaceous vegetation is dominated by grasses (*Poaceae*). Mugworth (*Artemisia*), *Chenopodiaceae*, and *Asteraceae* (composites) are represented by relatively low values. Vegetation indicative of humid zones (*Apiaceae, Rubiaceae, Thalictrum, Ranunculaceae* and *Filipendula*) is also represented by sporadic pollen grains. Alder (*Alnus*), indicating a rise in the water levels, increases. While *Cyperaceae* is found very sporadically, peat mosses and the Pteridophytes along with Monolete pollen grains (probably *Thelypteris palustris*) are present.

NM2b-h (260-170 cm) 7900-6000 bp (6755/6685-4930/4845 BC)

A strong species-rich mixed oak forest dominated by hazel forest cover is present during this period. It is most important for documenting the beginning of the neolithization process. This zone can be divided into the following three sub-zones:

a) NM2b-h1 (260-210 cm) 7890-6800 bp (6750/6680-5715/5670 BC)

The termophyllous, mixed deciduous forest elements increase with proportions of arbor pollen (A.P.) rising to a maximum of 80-90%. Three distinguishable peaks of *Corylus* pollen percentages, which are preceded by less well expressed increases of *Ulmus*, can be seen. The first and smaller peak is indicated at a depth of 250 cm, and dates to ca. 7600 bp (5475/5380 BC). The second peak is indicated at a depth of 230 cm and dates to ca. 7200 bp (6070/6030 BC). The third peak is indicated at a depth of 215 cm, and dates to ca. 7000 bp (5970/5840 BC). All of these peaks are followed by a sudden increase in the pollen percentage (from 10% to 40% and later from 30% to 40%) of ash (*Fraxinus*), which implies deliberate *Corylus avellana* collecting.

Most oak wood taxa are present with high frequencies. Several steppe elements, such as *Artemisia*, *Chenopodiaceae, Asteraceae, Apiaceae, Rubiaceae*, are also present, as are pollen grains of *Thalictrum* and *Ranunculaceae*. Increased values for *Picea abies* (spruce) and *Pinus sylvestris* occur. This suggests either a mosaic-like environment following the exposures, or a real mixture, of the different taxa such as that recently found in the region. Beech (*Fagus sylvatica)* is present regionally as represented by sporadic pollen occurrences in the sequence. Herbaceous vegetation and non-arboreal pollen content (N.A.P.) are responsible for 10-20% of the total pollen. Grasses are represented by smaller peaks of 2-3%, and the rise of anthropogenic taxa is concentrated around the depth where hazel peaks are indicated. We can detect a modest progression of *Alnus* in the humid zones, which is probably associated with gallery forests (ripisylvae) dominated by the Pteridophytes and *Typha/Sparganium* in the marginal zones of the mire. *Sphagnum* spore values also show a transitional peak.

b) NM2b-h2 (210-190 cm) 6800-6400 bp (5715/5670-5470/5340 BC)

The composition of the deciduous forest changed. A transitional peak of 20% for birch is indicated within this sub-zone, and *Picea abies* and *Abies alba* (fir) are only sporadically represented. *Corylus* decrease to 20%, concurrent with an increase in *Ulmus*, *Quercus*, and *Betula* pollen. *Filipendula*, *Rubiaceae* and *Asteraceae* are typical herbaceous species found during this period in the areas surrounding the site. An important increase in *Chenopodiaceae* pollen grains is detected at the depth of 225 cm, with its highest values occurring at the end of the h2 sub-zone. Some anthropogenic taxa are also represented; however, arboreal vegetation predominates.

c) NM2b-h3 (190-170 cm) 6400-6000 bp (5470/5340-5470/5340 BC)

During this time *Betula* decreases from 30% to 2%, and *Quercus* diminishes from 20% to 10%. The sudden increase of *Corylus* now makes it the dominant taxa of the deciduous forest. *Picea* and *Alnus* also increase slightly. *Fagus* is gradually integrated into the forest. Among the herbaceous taxa the *Poacea* dominate, but the presence of some anthropogenic plant species are also noteworthy. *Artemisia* increases suddenly and becomes dominant at 25-30%. Cereals indicate agricultural activity by human population living close to the site. Although tree species (A.P.) decrease at the beginning of this zone, they return to their maximum values at the very end of this sub-zone. The rise of different arboreal and herbaceous species peaks, when Neolithic populations are probably settled in this region (also seen in Magyari *et al.* 2000, Magyari 2002).

NM2b-i (170-140 cm) 4800-4300 bp (3640/3540-2910/2895 BC)

A hiatus in sedimentation evidenced by sharp changes in several pollen grain values is present at 170 cm, and corresponds to the time between ca. 6000 and 5100 bp (5470/5340-3960/3820 BC) (Fig. 12.3). There is a sharp decrease in *Corylus* (from 40 to 10%), an increase of *Alnus* (from 1-2% to 25%), and the beginning of the hornbeam (*Carpinus*) curve, to mention just the most important taxa. From the beginning of the following zone, there is a major change in the composition of the forest. Prior to deposits from this zone (NM2b-i), *Quercus*, *Tilia*, (linden) and *Ulmus* predominated, but within this zone *Fagus* and *Carpinus* become the dominant elements of the temperate oak woodland. Coniferous trees (*Picea*, *Abies*, *Pinus)* are also present. In the humid zones, close to the site, *Alnus* stands are widespread. A lake with a floating carpet of *Sphagnum* has developed, and the shore is surrounded by reeds and *Typha/Sparganium*.

These changes could be due to human impact on the environment. Use of fire or other woodland management techniques between 5000-3000 BC (Willis 1996) has already been described for several places in the Balkans, corresponding well with our chronology. The Nyírség and Hatvan cultures are present in the region at this time (Sümegi 1998).

NM2b-j (140-110 cm) 4300-3800 bp (2910/2895-2280/2200 BC)

This period is characterized by the decrease of tree taxa and opening of the forest canopy indicating extensive forest clearance for agricultural activity. The Hatvan culture was under strong influence from the Mediterranean (Sümegi 1998) during the Middle Bronze Age (2000-1800 BC). Agricultural activities become increasingly important and cereals are regularly cultivated. Agricultural satellite taxa of plantain (*Plantago lanceolata* and *Plantago major*) and several taxa of the *Chenopodiaceae* family are found in the pollen assemblage. Agricultural activities also affected the gallery forests in the humid zones. *Alnus* decreases becoming sporadic at the end of this zone. Although *Fagus* and *Carpinus* are among the most important forest taxa, human impact affected these taxa as well (see a transitional decrease of beech). Sedges disappear; Pteridophytes become sporadic and Typha/Sparganium increases up to 30% along marginal zone of the lake. *Sphagnum*-mosses remain at constant levels.

NM2b-k (110-75 cm) 3800-3100 bp (2280/2200-1415/1385 BC)

Changes in the forest composition at this time differ from previous periods. The agricultural activities of the flourishing Hatvan and Füzesabony cultures (2000-1500 BC) of the Middle-Bronze Age, and that of the Piliny culture (1450-1250 BC) at the end of Bronze Age, are evident in the profile. The production of wine and metal required large quantities of wood (Sümegi 1998), as evidenced by the decrease in *Fagus* pollen from 25% at the end of the previous zone to 1-2%. Concurrent with this change, *Carpinus* increases to 20-25% of the total pollen percentage as a result of the opening of the beech forest. Meanwhile *Quercus* increases from 30% to 50% and dominates tree taxon in the forest. The decrease of *Artemisia* pollen and fluctuations in anthropogenic taxa suggests that agricultural activity may have shifted to more distant areas. Grasses fluctuate slightly, but overall their values remain constant, and cereals are still present. The high peak values of *Typha/Sparganium* indicate an increase of the floating *Sphagnum* swamp cover, suggesting higher water levels.

NM2b-l (75-60 cm) 3100-2900 bp (1415/1385-1115/1050 BC)

The Piliny and Kyjatice cultures exited the Kelemér region (Sümegi 1998) at the transition between the Bronze and Iron Ages (1200-800 BC). The Kyjatice culture is known for its system of fortified settlements on the borderline of the Great-Hungarian Plain and the Bükk Mountain Range (Kemenczei 1970). Beech, already in low proportions, is not able to increase in the oak woods. Hornbeam increases up to 20% and becomes one of the dominant taxa as oak percentages decrease from 45 to 25%. Concurrent with the opening forest cover, birch increases. Among the herbaceous taxa, *Artemisia* and the *Poaceae* increase, and pollen grains of cereals and plantain (*Plantago lanceolata*) are present. Agricultural activities may have included fire for clearing forests and increasing arable land, as indicated by the increase in the charcoal content at this level (Willis *et al.* 1998). *Typha/Sparganium* decreases, becoming sporadic, concurrent with a peak in *Filipendula*. *Sphagnum* remains constant and dominates the lake.

DESCRIPTION AND THE INTERPRETATION OF THE LOCAL POLLEN-ASSEMBLAGE ZONES OF NAGY-MOHOS (NM2b) CONCERNING THE LATE- HOLOCENE

NM2b-m (60-30 cm) 2900-2500 bp (1115/1050-760/560 BC)

Increased agricultural activity attributable to the Pre-Scythians and Scythians dated between ca. 900-300 BC and then the Kelts around 400-0 BC (Sümegi 1998) is evident in this zone. Due to human impact, there are important changes in the composition of the forest. Elevated values of *Carpinus*, seen during the previous phase, decline concurrent with a rise in *Quercus*, which once again becomes the dominating taxon of the forest. *Betula* maintains constant values with some transitional increases, and *Fagus* and *Corylus* remain stable. Herbaceous vegetation increases as the forest cover gradually opens. Towards the end of the zone, anthropogenic taxa are present with peak values of cereals, *Chenopodiaceae* and *Plantago lanceolata*. A change in the sedimentation can be seen around a depth of 40 cm. *Sphagnum*-moss increases to peak values of 60% at the very end of this zone. The first of only two traces of turf cutting in the Kelemér region occur in this zone (Magyari 2002). Around 2500-2100 bp (ca. 760/560-165/100 BC) the Kelts (Kalicz 1970) cut into the *Sphagnum* peat bog probably to gain potable water (Willis *et al.* 1998). This may explain changes in the sedimentation rates and the important decrease of the *Sphagnum* values in the next zone.

NM2b-n (40-10 cm) 1345 bp- ? (655/670-? AD)

Sedimentation resumes following the cut-off of the *Sphagnum* peat bog. There are several sharp increases in the pollen values of different taxa in this zone, supporting the idea of a hiatus caused by the above mentioned peat exploitation. Further escalation of agricultural activity is also evident. Anthropogenic taxa remains stable (cereals, *Plantago*, and *Rumex*), and *Poaceae* increase to the 60% of the total pollen. A sudden decrease in the arboreal pollen from 60% to 30% of the total pollen content resulted from human impacts. *Quercus* decreases from 40% to 10%, and *Fagus* becomes almost extinct. All tree taxa have low values except *Betula,* which increases to its maximum value during this period. *Betula* appears in the openings of the forest cover where there is enough light. Following the cutting of the *Sphagnum* layers, the peat bog continues to develop. Pollen values for butreeds (*Cyperaceae*) increase and peak at the beginning of the zone. *Typha/Sparganium* and Pteridophytes with Monolete spores are again present, and reached levels similar to earlier peak values. The second indication of turf cutting in the Kelemér region happens around the 1920s based on written sources. The owner of the site cut the peat of the Nagy-Mohos Lake to allow swimming and to use the peat for soil amelioration (Fig. 12.3 and 12.7; Magyari 2002, Gyulai 1995).

DISCUSSION

The vegetation history of the Mohos Lakes during the Mesolithic and Neolithic period

There is a strong correlation between the increase of hazel pollen grains and the strategies applied during gathering. Mesolithic field camps occur within a 70-100 km radius of the site (Willis *et al.* 1998). Mobile human groups could thus have caused the small-scale disturbance (Kertész and Sümegi 1999) seen in the NM2b-h1 zone. Palynological research on various Holocene radiocarbon dated profiles (Fig. 12.4) suggests that these small magnitude human-induced changes took place between the years 7000-6000 BC (Sümegi 2004). Alterations seem to be directly connected to several adjustments made by hunter-gatherer populations:

- A more massive gathering of hazelnuts (Smith 1970)
- The utilization of shoots and foliage of younger *Fraxinus* and *Tilia* for animal feed (Heybroek 1963)
- The formation of hunting trails and campsites,
- A generally more permanent settlement, and
- The possibly of an early adaptation or imitation of certain Neolithic techniques and methods.

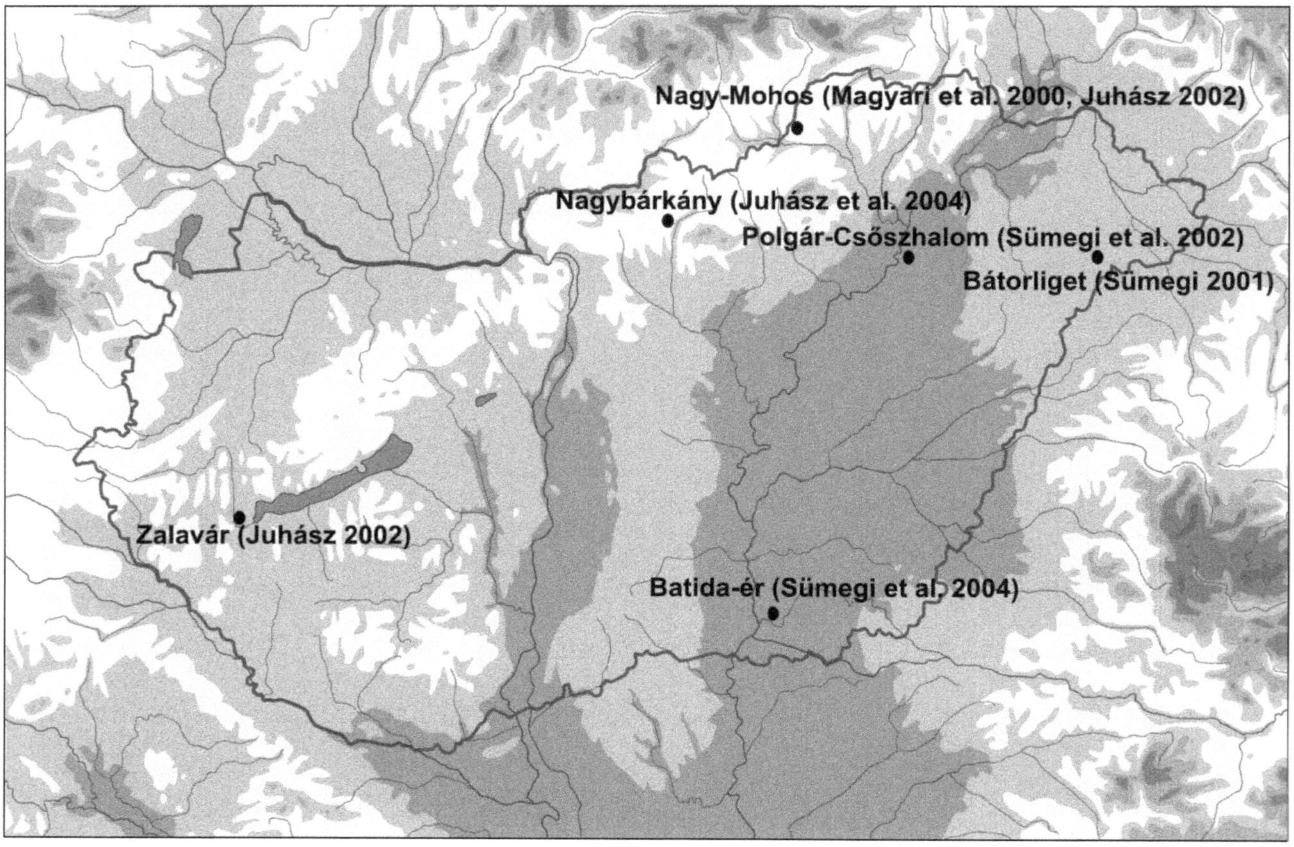

Figure 12.4. Location of other pollen sequences with Pre-Neolithic human impact (after Juhász 2004 modified)

Data gained through the application of natural historical scientific tools seem to justify assumptions made by archaeologists who previously proposed that just before the appearance of Early Neolithic in the Carpathian Basin, an independent Pre-Neolithic phase may have developed around 7000 BC (Bánffy *et al.* 2008, Sümegi 2008, Sümegi and Kertész 2001). A similar argument is made for Transdanubia (Bánffy 2004, Juhász 2004, Sümegi 2004). This phase probably survived in the foothill areas of the Carpathians when the Körös culture settled in the southern part of the Carpathian Basin (Fig. 12.5 and 12.6).

During the sub-zone h2 and sub-zone h3, dating between 6800-6000 bp (5715/5670-4930/4845 BC), anthropogenic taxa are present in only small amounts and decreases of tree taxa pollen are at a minimum levels suggesting wood usage during the Neolithic was minimal (Heybroek 1963). Archaeological evidence further suggests that the settlements of the Bükk culture existed within a 50 km radius of the Kelemér region (Sümegi 1998). This population apparently fed animals with shoots and foliage from *Fraxinus* and *Tilia* (Heybroek 1963). According to Favre and Jacomet (1998) several mountain-adapted Neolithic groups in Germany and Switzerland used hazel and elm twigs, catkins and branches as livestock fodder during wintertime. Gardner (1999, 2002) has also suggested that the Neolithic Alföld *Linienbandkeramik* or Linear Pottery culture (AVK) in the Mátra Mountains coppiced hazel and hornbeam trees. It is likely that the members of the Bükk culture did the same. Thus, the less well expressed rise in oak pollen in Kis-Mohos (Willis *et al.* 1998) may have anthropogenic origins. The Holocene pollen diagram of Kis-Mohos (Willis *et al.* 1998) also indicates a decrease in the proportion of hazel, but because of its lower resolution this process is far less evident (11.7 and 11.8).

Leaf fodder favors understory taxa (e.g. Rosaceae), increase of *Vitis* (fox-grape) and *Hedera*, arguing for the opening up of the forest canopy with the pioneer species *Betula* probably appearing in the clearings. However, there are no signs of extensive forest clearances and cereals appear only at the very end of this zone. In the third sub-zone dating to between 6400-6000 bp (5470/5340-4930/4845 BC), *Corylus* again becomes the dominant taxon. The presences of the anthropogenic marker taxa indicate human impact; probably by the Tisza-Herpály-Csőszhalom culture (Fig. 12.7 and 12.8) on the vegetation (Magyari 2002). Cereal pollen indicates the beginning of agricultural activities close to the site. Arable fields were probably cleared of forest possibly through the use of fire, a notion further justified by peaks of macro-charcoal in these samples.

Similar changes are detected and described in the other NM I-II pollen sequence of Nagy-Mohos (Fig. 12.8; Magyari *et al.* 2000, Magyari 2000). There are four

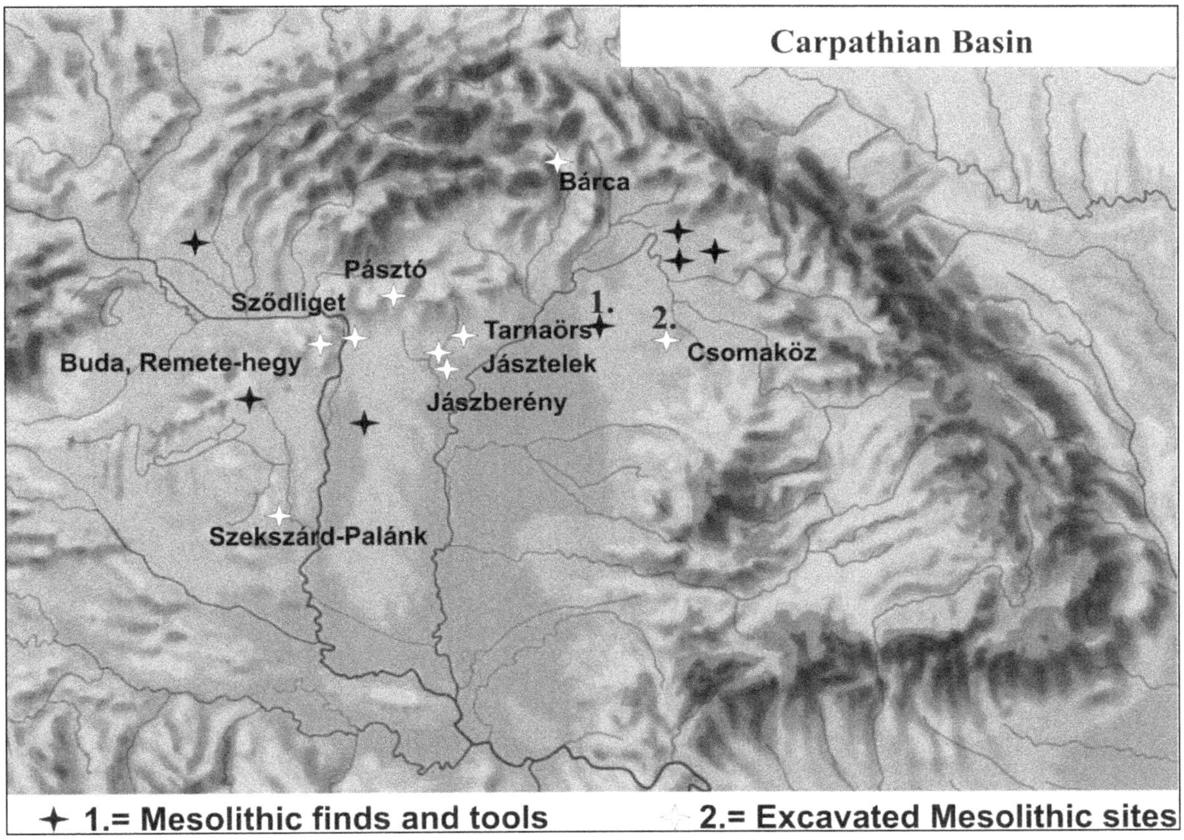

Figure 12.5. Mesolithic sites in the Carpathian Basin (after Sümegi 2003)

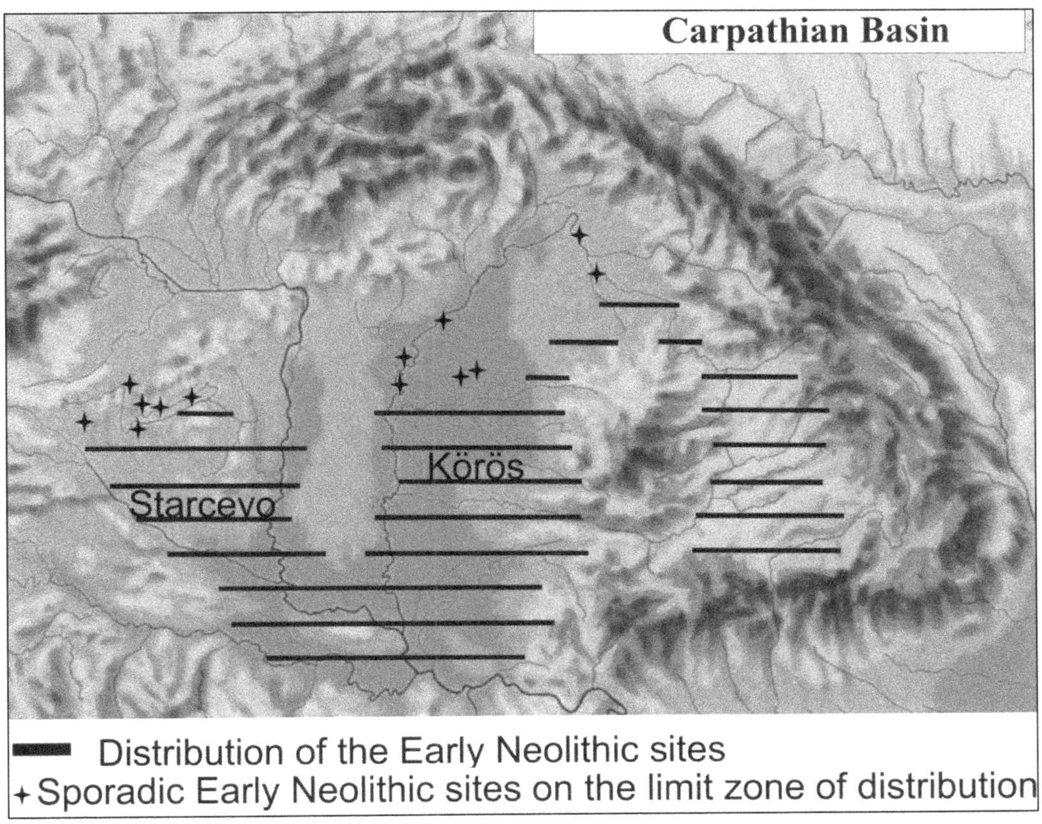

Figure 12.6. Neolithic sites in the Carpathian Basin north of the Neolithic barrier (after Sümegi 2003)

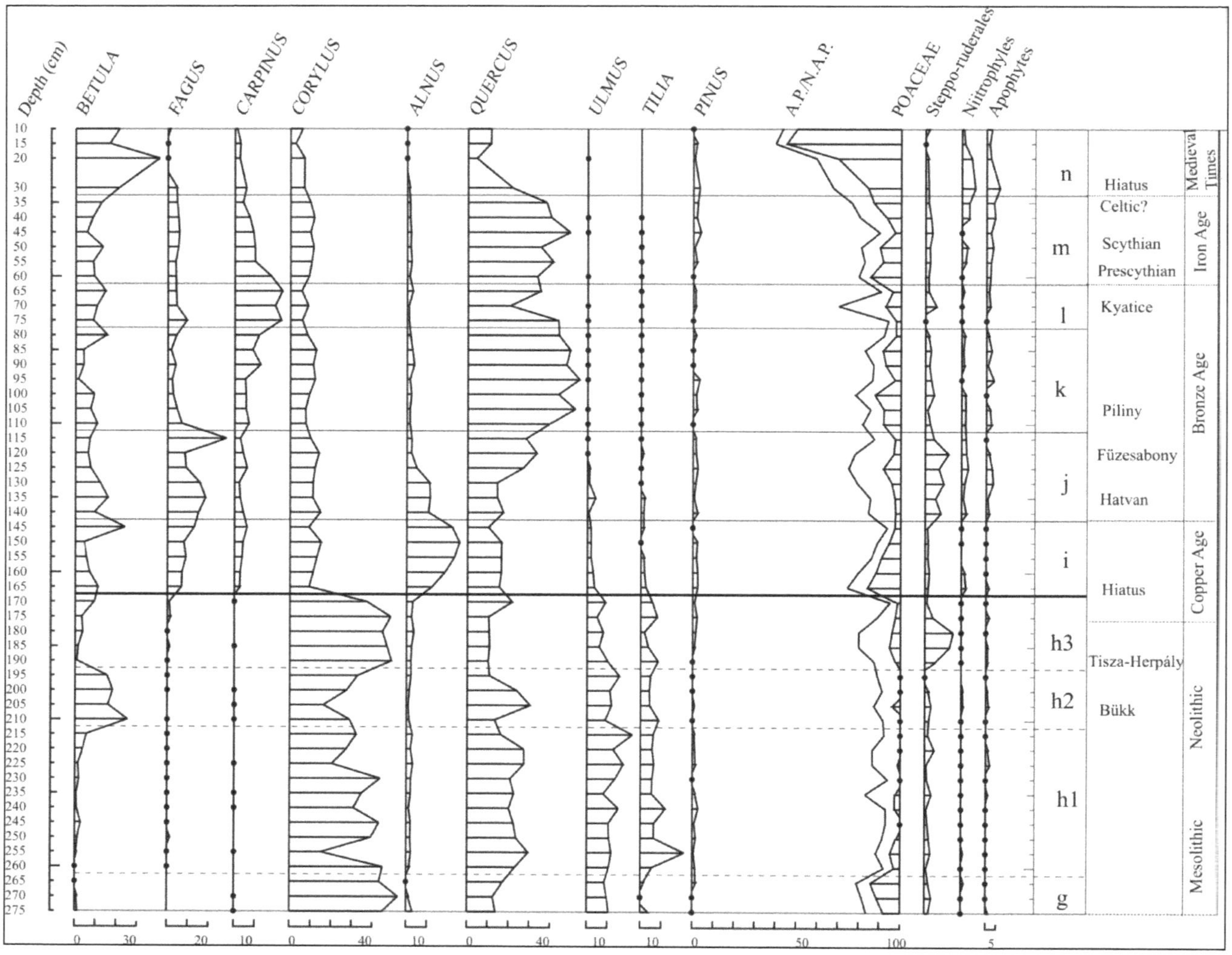

Figure 12.7. Simplified pollen diagram of selected anthropogenic taxa and the archaeological record

distinguishable decreases in the *Ulmus* pollen curve followed by increases in ash (*Fraxinus*) pollen percentages around 5850, 5250, 5000 and 4900 BC. The time periods when the Mesolithic and Neolithic populations lived near the Mohos Lakes at Kelemér are well expressed in this sequence. The first peak at 5850 BC is thought to have been caused by Mesolithic populations (Sümegi 1999, 2000). Late-Mesolithic hunter-gatherers may have lived next to new arrivals from the Balkans in the northern part of the Carpathian Basin during this time. These new arrivals practiced an Early Neolithic agrarian economy (Kalicz 1970), and could have engaged in cultural and technological exchange with the hunter-gatherers (Kertész and Sümegi 1999, 2001).

At ca. 5250 BC the decrease of *Ulmus* pollen percentages parallels the decline of *Corylus avellana*. This change in the pollen assemblage may be connected to the specialized life form of the AVK or the related Bükk culture and the Tiszadob Group in the North Mountain Range (Kalicz-Makkay 1977). Around 5000-4900 BC *Poaceae* and *Artemisia* increase, and the ratio of the arboreal taxa decrease, suggesting small-scale clearances. The appearance of cereal pollen grains suggests formation of arable fields. The anthropogenic impact coincides with an expansion of the Lengyel culture from Transdanubia (Bánffy 1994, Regenye 2005). In contrast to archaeological assumption of a lower settlement density and the emptying of this area (Sherratt 1982; Kalicz, 1970), this event signals human impacts affecting the woodlands around Kelemér.

The vegetation history of the Mohos Lakes during the Copper and Bronze Age

Peat layers younger than 5515 bp (4355/4340 BC) are missing due to peat exploitation in the southern Nagy-Mohos Basin, where the NMI-II sequence was cored (Fig. 12.2). Thus, only the pollen core of Kis-Mohos (Willis *et al.* 1998) can be used to correlate our sequence (NM2b) for the later periods. The proportion of *Fagus sylvatica* increases, while *Corylus avellana*, *Ulmus* and *Quercus* retreat into the woodlands around the mire. Maximum levels of copper and charcoal are detected, and the increase of grasses suggests new human impacts.

Different suggestions have been advanced concerning the increase in *Fagus* during this period, including climatic change (Beug 1992, Gardner and Willis 1999) and soil

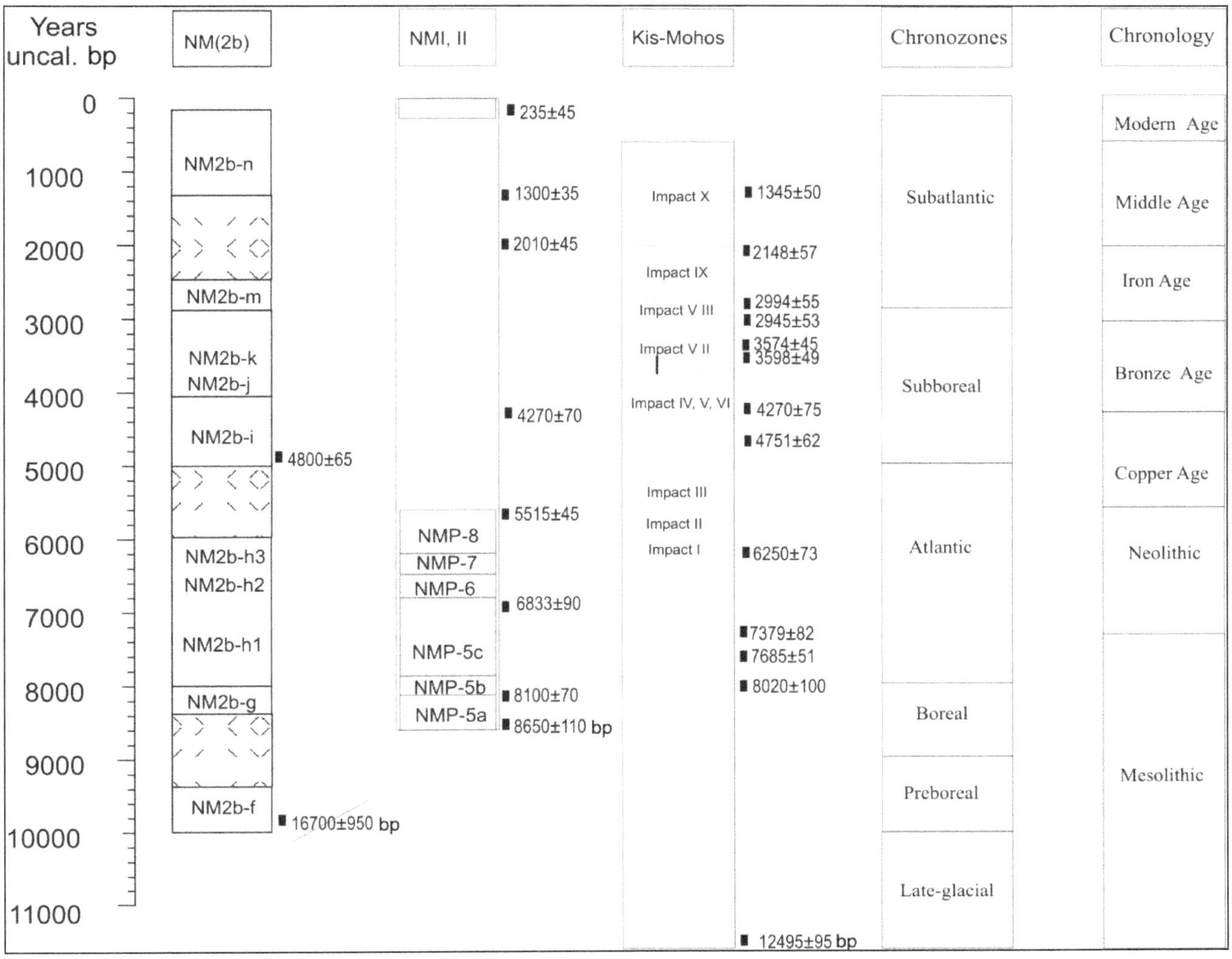

Figure 12.8. Correlation of the pollen sequences from Kis-Mohos (Willis *et al.* 1998), Nagy-Mohos NMI-II (Magyari 2002) and Nagy-Mohos NM2b (Juhász 2002)

change (Willis 1994), but anthropogenic influence appears more likely (Behre 1988, Jahns 1993, Willis 1996). The dispersion of beech around the Mohos Lakes can be reconstructed as a result of human impact attributable to the Late Copper age Baden culture (3500 to 2800 BC), similar to the process investigated in present-day Scandinavia (Björkmann 1999). It is likely that beech populations benefited from climate change and from the anthropogenic woodland clearances that accelerated species change in the closed canopy oak woodland (see also Magyari 2002). The role of *Carpinus betulus*) also becomes increasingly important in the woodland because of human impact between 5000-3900 bp (3795/3765-2460/2340 BC).

The vegetation history of the Mohos Lakes during the Iron Age

Dominance of the hornbeam-beech-oak woodland was typical at the Mohos Lakes until ca. 3000 bp (1290/1210 BC), when they were replaced again by *Quercus* (Willis *et al.* 1998). However, this oak forest composition differs from that of the Early Holocene oak forest. *Corylus* and *Ulmus* are at much lower proportions, and *Fagus* and *Carpinus* also occur. It is likely that disturbances by the Kyatice culture between 1250 and 900 bc (Kemenczei 1970) played a role in the composition of this vegetation change, as indicated by the high proportion of *Betula*, Filicales and the charcoal maximums. Construction of a chain of wooden fortresses on the borderline of the Bükk Mountains and the Great Hungarian Plain resulted in woodland clearance (Sümegi 1998).

Soil erosion was continuous from about 2800 bp (975/920 BC) as the area of woodland (AP) decreased. This period correlates with the presence of the Scythians and Kelts in the region. The high levels of charcoal during this period at Kis-Mohos (Willis *et al.* 1998), as well as the geochemical results, suggest that the Kelts engaged in considerable construction activities. Use of large quantities of wood is illustrated at the medieval castle of Vár Hill, where Sümegi (1998) suggests that the foundation was built during the Keltic period.

The presence of a hiatus, as peat was probably removed in order to create an open-water surface for potable water, is evident in our Nagy-Mohos pollen sequence (NM2b) during the Keltic Period (at the end of NM2b-m),. Sedimentological studies carried out on Kis-Mohos and Nagy-Mohos NMI-II suggests that water levels for the mire were artificially raised during this period. Development of a shallow lake environment is associated with hemp retting (Willis *et al.* 1998, Sümegi 1998), and the climate may have become more humid. Nevertheless, the connection between eroded soil layers, charcoal content, and subsequent higher water levels implicate strong human impact on the mire environment. Kalicz (1970) suggests that the Kelts were living in the area surrounding the site and affected the woodland around Vár Hill. Forest clearance may have caused the first erosion event observable in the inspected peat layers. The Kelts, who already had intricate iron tools, caused considerable environmental change in the Carpathian Basin and other areas of Europe (Jerem *et al.* 1984, 1985, Sümegi, 1999). Additionally, climate change towards a more humid environment is suggested on geoarchaeological grounds from Fertő Lake, Hanság, Northwest Transdanubia (Jerem *et al.* 1984, 1985).

The vegetation history of the Mohos Lakes from the Migration period until Middle Ages

Sedimentation resumed around 1300 bp (670/765 AD), as represented in the NM2b-n zone. The pollen record indicates that the region was not depopulated during the Migration Period and human impact is detectable for the Hungarian Conquest Period (9th to 10th century). Elevated values for cereal and nitrophyllous taxa correlate with population increases observed in the archaeological data (Sümegi 1998).

During the Middle Ages (1000-1600 AD) the Kis-Mohos Lake was used for hemp retting as indicated by the elevated values for *Cannabis* (hemp) pollen in the sediment layers. During Medieval times the Nagy-Mohos was damned, probably for defensive reasons, and the peat-layers exploited. The castle that was probably populated during the Middle Ages and Turkish era (16-17th century) gradually fell into disuse by the 18th century, resulting in a decrease of human disturbances in the vicinity of the Mohos-bogs. The slopes were newly forested and stabilized, resulting in reduced erosion. These changes decreased the rate of in-wash of organic and inorganic material, reinvigorating the swamp environment and development of a *Sphagnum* peat-bog.

The southern basin of Nagy-Mohos Lake (NMI and II, Magyari 2002) was most probably exploited twice (Fig. 12.8; Magyari 2002) first during the Keltic Period (100-50 BC) and later, during the 1920s by the land owner Diószeghy's family when peat was used as a soil amendment, as suggested by historical sources (Gyulai 1995).

CONCLUSIONS

Palynological analyses of a 600 cm deep sedimentary sequence taken from Kelemér-Nagy-Mohos peat bog, Northeast Hungary, provide a paleoenvironmental record for roughly the last 25300 years. The region is of special interest, because it contains preserved a continuous peat deposit encompassing the period from ca. 25300-13500 bp (ca. 25300-14280/13910 BC) (Magyari *et al.* 2000, 2001). At the Upper Pleniglacial/Late Glacial boundary, the sequence is interrupted by approximately 20 cm thick intercalation of inorganic sediment that can be traced throughout the entire basin. Holocene sedimentation above this clay horizon commences around 10500 bp (10670/10440 BC), and the different phases of Holocene vegetational change since are presented in this chapter.

The basin was occupied by a shallow open lake with substantial input of inorganic material until ca. 8120 bp (6170 BC), when floating reed-swamp vegetation encroached onto the lake surface, turning into a *Carex* fen within a century. A *Sphagnum*-bog became established by ca. 7200 bp (6070/6030 BC). Species composition of the peat bog underwent further changes when *Eriophorum vaginatum* prevailed over *Sphagna*. Increases in the local mite values during the *Sphagnum*-bog phases appeared synchronously with upland vegetation changes, suggesting selective exploitation of *Ulmus* and *Corylus* coupled with burning and soil erosion.

Comprehensive evaluation of pollen and radiocarbon dates for the Nagy-Mohos area (Fig. 12.8; Magyari *et al.* 2000, 2001; Juhász 2002) enable a more accurate chronological delineation of the Holocene vegetation development in this part of Hungary. The scientific data were obtained to ascertain the landscape use and the transformative effects of human communities through detailed paleoenvironmental analyses of the boreholes from the marshlands. The analyses resulted in the identification of a pre-neolithization period, the human deforestation activities of Neolithic, Bronze, and Copper Age, as well as the reconstruction of the environmental background conditions during the occupation by the Kelts and later cultures of the Great Migration Period, including the immigrating Hungarians. The observed results appear to imply that environmental changes are largely attributable to human impact. However, in some instances climatic change does coincide with human impact, suggesting a human adjustment to such a change.

Acknowledgments

The palynological research was carried out with a Ph.D. grant by the French Government and the Institut Français of Budapest. I express my gratitude to Dr. Hervé Richard and his colleagues (Besançon CNRS Laboratoire de Chrono-écologie, UFR des Science et Techniques) for providing the sediment core and the radiocarbon dates. During the analyses and interpretation of the sediment cores my colleagues in the Institut Mediterranéen d'Écologie et de Palynology (IMEP), Laboratoire de Botanique et de Botanique Historique in Marseille provided great help. I could always turn to Anne Vaillant with problems during the sample treatments. My deepest respect and all my gratitude go to my tutor, Prof. Jacques-Louis de Beaulieu for his patience, endurance and firmness. I could always count on his thorough knowledge about the vegetation history of Europe. His confidence in me and in my work led me to the accomplishment of my research. The work would not have reached its fruition without him. I am also very grateful to Dr. Maximilian Baldia and to Dr. Christel Baldia for their kind reception in the US and their help in revising my English text.

References

BÁNFFY, E. 1994 – Anthropomorphic Figurines in Cult Corners of Neolithic Houses. In A.C. Ambrosi (ed.), *La Statuaria Antropomorfa in Europa dal Neolitico alla Romanizzazione. Atti del Congresso La Spezia-Pontremoli 1988*. Istituto Internazionale di Studi Liguri, Sarzana, 1994:41-46.

BÁNFFY, E. 2004 – *The 6th Millennium BC Boundary in Western Transdanubia and its Role in the Central European Neolithic Transition*, Varia Archeologica Hungarica XV, Archaeological Institute, Hungarian Academy of Sciences, Budapest, 2004.

BÁNFFY, E., W.J. EICHMANN and T. MARTON 2008 – Mesolithic foragers and the spread of agriculture in Western Hungary. In J.K. Kozlowski and M. Nowak (eds.) *Mesolithic/Neolithic Interactions in the Balkans and in the Middle Danube Basin: Proceedings of the XV World Congress UISPP, Lisbon, 4-9 September 2006.* BAR International Series, BAR Publishing, Oxford, 2008:53-82.

BEHRE, K. 1986 – *Anthropogenic Indicators in Pollen Diagrams*. Balkema, Rotterdam, 1986.

BEUG, H.J. 1982 – Vegetation History and Climate Changes in Central and Southeastern Europe. In A.F. Harding (ed.) *Climate Change in Later Prehistory*, Edinburgh University Press, Edinburgh, 1982:85-102.

BJÖRKMAN, L. 1999 – The Establishment of *Fagus sylvatica* at the Stand-Scale in Southern Sweden. *The Holocene* 9/2, 1999:237-245.

FAVRE, P., and S. JACOMET 1998 – Branch Wood from the Lake Shore Settlements of Horgen Scheller, Schwitzerland: Evidence for Economic Specialization in the Late Neolithic Period. *Vegetation History and Archaeobotany* 7, 1998:167-178.

GARDNER, A.R. 1999 – The impact of Neolithic Agriculture on the Environments of South-East Europe. Unpublished Ph. D. Dissertation, University of Cambridge, Cambridge, 1999.

GARDNER, A.R. 2002 – Neolithic and Copper Age Woodland Impacts in Northeast Hungary? Evidence from the Pollen and Sediment Chemistry Records. *The Holocene* 12/5, 2002:541-553.

GARDNER, A.R., and K.J. WILLIS 1999 – Prehistoric Farming and the Postglacial Expansion of Beech and Hornbeam: A Comment on Küster. *The Holocene* 9/1, 1999:119-122.

HEYBROEK, H.M. 1963 – Diseases and Lopping for Fodder as Possible Causes of a Prehistoric Decline of *Ulmus*. *Acta Botanica Neerlandica* 12, 1963:1-11.

GYULAI, I. 1995 – A Keleméri Mohos-Tavak. [The Mohos Lakes at Kelemér.] *Természet Világa* 126, 1995:137-138.

JAHNS, S. 1993 – On the Holocene Vegetation History of the Argive Plain' *Vegetation History and Archeobotany* 2, 1993:187-203.

JEREM, E., G. FACSAR, L. KORDOS, E. KROLOPP and I. VÖRÖS 1984 – A Sopron-Krautackeren Feltárt Vaskori Telep Régészeti és Környezetrekonstrukciós Vizsgálata. [The Archaeological and Environmental Investigation of the Iron Age Settlement Discovered at Sopron-Krautacker.] *Archeológiai Értesítő* 111, 1984:141-169.

JEREM, E., G. FACSAR, L. KORDOS, E. KROLOPP and I. VÖRÖS 1985 – A Sopron-Krautackeren Feltárt Vaskori Telep Régészeti és Környezetrekonstrukciós Vizsgálata. [The archaeological and Environmental Investigation of the Iron Age Settlement Discovered at Sopron-Krautacker.] *Archeológiai Értesítő* 112, 1985:3-24.

JUHÁSZ, I.E. 2002 – A Délnyugat Dunántúl Negyedkori Vegetációtörténetének Palinológiai Rekonstrukciója. [Reconstitution Palynologique de la Végétation Depuis le Tardiglaciaire Dans la Région De Zala, Sud-Ouest de la Hongrie.] Unpublished Ph. D. Dissertation, University of Pécs and University Aix-Marseille III, 2002.

JUHÁSZ, I.E. 2004 – Palynological Evidences of Preneolithization in South-Western Transdanubia. *Antaeus* 27, 2004:213-225.

KERTÉSZ, R., and P. SÜMEGI 1999 – Teóriák, Kritika és egy Modell: Miért állt meg a Körös-Starcevo Kultúra Terjedése a Kárpát-Medence Centrumában? [Theories, Critiques and a Model: Why did the Expansion of The Körös-Starčevo Culture stop in the Centre of the Carpathian Basin?] *Tisicum*, 10, 1999:9-22.

KERTÉSZ, R., and P. SÜMEGI 2001 – Theories, Critiques and a Model: Why did the Expansion of the Körös-Starčevo Culture stop in the Centre of the Carpathian Basin? In R. Kertész and J. Makkay (eds.), *From the Mesolithic to the Neolithic*. Proceedings of the International Archeological Conference held in the Damjanich Museum of Szolnok, 22-27 September, 1996, Archeolingua Foundation, Budapest, 2001:225-246.

KALICZ, N. 1970 – *Agyagistenek. A Neolitikum és Rézkor Emlékei Magyarországon.* [Clay Gods: The Memories of the Neolithic and Copper Age in Hungary.] Corvina Kiadó, Budapest, 1970.

KALICZ, N., and J. MAKKAY 1977 – *Die Linienbandkeramik in der Großen Ungarischen Tiefebene*. Akadémiai Kiadó, Budapest.

KEMENCZEI, T. 1970 – A Kyjatice Kultúra Észak Magyarországon. [The Kyjatice culture in Hungary.] *Herman Ottó Múzeum Évkönyve* 9, 1970:17-78.

KÜSTER, H. 1997 – The Role of Farming in the Postglacial Expansion of Beech and Hornbeam in the Oak Woodlands of Central Europe. *The Holocene* 7/2, 1997:239-242.

LÁSZLÓ, G., and K. EMSZT 1915 – *A Tőzeglápok és Előfordulásuk Magyarországon.* [*The Peat-Bogs and Their Occurrences in Hungary.*] Fritz Ármin Nyomdája, Budapest, 1915.

MAGYARI, E. 2002 – *Climatic Versus Human Modification of the Late-Quaternary Vegetation in Eastern Hungary.* Unpublished Ph. D. Dissertation, University of Debrecen, Debrecen.

MAGYARI, E., G. JAKAB, P. SÜMEGI, E. RUDNER and M. MOLNÁR 2000 – Paleoökológiai Vizsgálatok a Keleméri Mohos Tavakon. [Paleoecological Investigations at Kelemér Mohos Lakes.] In E. Szurdoki (ed.), *Tőzegmohás Élőhelyek Magyarországon: Kutatás, Kezelés, Védelem.* [*Peat-Bog Habitats in Hungary: Research, Conservation and Management.*] CEEWEB Workshop, Miskolc. 2000:101-131.

MAGYARI, E., P. SÜMEGI, M. BRAUN and G. JAKAB 2001 – Retarded Hydrosere: Anthropogenic and Climatic Signals in a Holocene Raised Bog Profile from the NE Carpathian Basin. *Journal of Ecology* 18, 2001:113-128.

REGENYE, J. 2005 – A Vizsgálandó Közép-Dunántúli Újkőkori Kerámia és Környezete. *Archeometriai Műhely* 2. 2005:16-23.

SHERRATT, A. 1982 – The Development of Neolithic and Copper Age Settlements in the Great Hungarian Plain. Part I: The Regional Setting. *Oxford Journal of Archaeology* 1, 1982:287-316.

SMITH, A.G. 1970 – The Influence of Mesolithic and Neolithic Man on British Vegetation. In D. Walker and R.G. West (eds.), *Studies in the Vegetational History of the British Isles.* Cambridge University Press, Cambridge, 1970:81-96.

SÜMEGI, P. 1998 – Ember és Környezet Kapcsolata a Kárpát-Medencében az Elmúlt 15.000 év Során. *Panniculus* 3, 1998:367-395.

SÜMEGI, P. 1999 – Az Utolsó 15000 év Környezeti Változásai és Hatásuk az Emberi Kultúrákra Magyarországon. [The Environmental Changes of the last 15,000 Years and Their Effect on the Human Cultures in Hungary.] In G. Ilon (ed.), *A régésztechnikusok kézikönyve.* Savaria Kiadványa, Szombathely, 1999:367-397.

SÜMEGI, P. 2003 – Early Neolithic Man and Riparian Environment in the Carpathian Basin. In E. Jerem and P. Raczky (eds.), *Morgenrot der Kulturen. Frühe Etappen der Menschheitsgeschichte in Mittel- und Südosteuropa.* Festschrift für Nándor Kalicz zum 75. Geburtstag. Archaeolingua Main Series 15, Budapest 2003:39-53.

SÜMEGI, P. 2004 – Preneolitizáció: Egy Kárpát-Medencei, Késő-Mezolítikum Során Bekövetkezett Életmódbeli Változás Környezettörténeti Rekonstrukciója. [Pre-Neolithization: The Environmental Historical Reconstruction of a Change in Lifestyle Occurring During the Late-Mesolithic in the Carpathian Basin.] *Mumosz II. Őskoros kutatók II. Összejövetelének konferenciakötete, 6-8 November 2000*, Debrecen, 2004:21-32.

SÜMEGI, P. 2008 – Palaeogeographical Background of the Mesolithic and Early Neolithic Settlement in the Carpathian Basin. In J.K. Kozlowski and M. Nowak (eds.), *Mesolithic/Neolithic Interactions in the Balkans and in the Middle Danube Basin: Proceedings of the XV World Congress UISPP, Lisbon, 4-9 September 2006.* BAR Publishing, Oxford, 2008:47-51.

SÜMEGI, P., and R. KERTÉSZ 1998 – A Kárpát-Medence Őskörnyezeti Sajátosságai: Egy Ökológiai Csapda az Újkőkorban? [Palaeogeographic Characteristics of the Carpathian Basin: An Ecological Trap During the Early Neolithic?] *Jászkunság* 44, 1998:144-157.

SÜMEGI, P., and R. KERTÉSZ 2001 – Palaeogeographic Characteristics of the Carpathian Basin: An Ecological Trap During the Early Neolithic? In R. Kertész and J. Makkay (eds.), *From the Mesolithic to the Neolithic.* Proceedings of the International Archeological Conference held in the Damjanich Museum of Szolnok, 22-27 September, 1996, Archeolingua Foundation, Budapest 2001:405-415.

SÜMEGI, P., I. JUHÁSZ, E. MAGYARI, G. JAKAB, E. RUDNER, ZS. SZÁNTÓ and M. MOLNÁR 2008 – A Keleméri Mohos Tavak Fejlődéstörténetének Rekonstrukciója Paleobotanikai Vizsgálatok Alapján. [The Environmental Reconstruction of the Mohos Lakes at Kelemér with the help of Paleobotanical Investigations]. In S. Boldogh and G.T. Farkas (eds.), *A Keleméri Mohos-tavak: Kutatás, Kezelés, Védelem*, ANP füzetek 4, Aggteleki Nemzeti Park, Igazgatóság, 2008:59-63.

WILLIS, K.J. 1994 – The Vegetational History of the Balkans. *Quaternary Science Reviews* 13, 1994:769-788.

WILLIS, K.J. 1996 – Where did the Flowers go? The Fate of Temperate European Flora During Glacial Periods. *Endeavour* 20/3, 1996:110-114.

WILLIS, K.J., P. SÜMEGI, M. BRAUN and A. TÓTH 1997 – Does Soil Change Cause Vegetation Change or vice versa? A Temporal Perspective from Hungary. *Ecology* 78, 1997:740-750.

WILLIS, K.J., P. SÜMEGI, M. BRAUN, K.D. BENNETT and A. TÓTH 1998 – Prehistoric Land Degradation in Hungary: Who, how and why? *Antiquity* 72, 1998:101-113.

ZÓLYOMI, B. 1931 – A Bükkhegység Környékének *Sphagnum*-Lápjai. [The *Sphagnum* Peat-Bogs of the Bükk Mountains.] *Botanikai Közlemények* 28/5, 1931:89-121.

Chapter 13
LATE NEOLITHIC MAN AND ENVIRONMENT IN THE CARPATHIAN BASIN: A PRELIMINARY GEOARCHEOLOGICAL REPORT FROM CSŐSZHALOM AT POLGÁR

Pál SÜMEGI, Sándor MOLNÁR, Katalin HERBICH, Mariann IMRE, Gabriella SZEGVÁRI, Sándor GULYÁS
Dept. of Geology and Paleontology, University of Szeged, Hungary

Pál SÜMEGI, Gábor TIMÁR
Dept. of Geophysics, University of L. Eötvös, Budapest, Hungary

Abstract: *The Late Neolithic tell of Csőszhalom near Polgár (ca. 5000-4600 cal BC) is an unusual site, situated in the northeastern corner of the Great Hungarian Plains. It is surrounded by five concentric ditches with four causeways, similar to those found in the Lengyel culture to the west. However, it belongs to the Tisza culture of eastern Hungary. Adjacent to the tell is a village with hundreds of wattle and daub houses. This paper establishes a Late Neolithic paleogeography for the area, using sedimentological and micromorphological analysis. Human activity, including grazing and cultivation, are concentrated near the tell and settlement as well as the loess-covered levee and isthmus-like remnant surface near the site. Minimal activity can be identified in the alluvial plain area of the Tisza River. Results from carpological analysis also point to relatively intensive agricultural activity in the neighborhood. Pollen and pedological analyses imply that the highest flood protected areas of Csőszhalom, originally covered with loess-steppe or forest-steppe, was used for the cultivation of crops. However, macrofossil research indicates the extension of the croplands to the seasonally flooded areas. The micro-scale zonality and mosaic complexity in the area was enhanced by human activities, completely altering the original vegetation cover of the loess levees. Yet, the composition of flora in marsh and forest areas were only marginally modified.*

INTRODUCTION

The prehistoric site of Csőszhalom near the town of Polgár belongs to the Tisza culture of the Hungarian Late Neolithic (Bánffy and Bognár-Kutzián 2007, Kalicz-Raczky 1987, Raczky 1989, 1998, Raczky *et al.* 1994). It is occupied from ca. 5000-4600 BC. Csőszhalom is an unusual site consisting of two different kinds of concomitant occupation sites: a tell and adjacent village. Tells occur in the Near East, the Balkan Peninsula, and the Tisza River drainage of Hungary. Their mound-like appearance is the result of numerous successive building phases in a confined space. However, in this case, the building space is confined by five concentric ditches with four causeways. Such enclosures are commonly found in the Lengyel culture to the west of the Tisza culture, where tells do not exist. It is perhaps even more surprising that archaeologists discovered a huge village adjacent to the tell. This *horizontal settlement* contains the remnants of hundreds of wattle and daub houses. Pottery is often painted in polychrome, exhibiting geometric designs. Burials are richly furnished.

Csőszhalom is situated within the Carpathian Basin in the northeastern corner of the Great Hungarian Plains. The site is at the northern interface of two main regions: the Hajdúság and the Hortobágy. It is bordered by the alluvial plain of the Tisza (Fig. 13.1). The geological history of these regions is fundamentally determined by bedrock and soil conditions, as well as the composition of the vegetation within the surroundings of Csőszhalom.

According to various scientists representing different branches of 20th century earth sciences, the Tisza River (after an east-west migration at the end of the Ice Age) reached its final course only in the second half of the Holocene, about 6000 years ago (Cholnoky 1907, Borsy 1995, Borsy *et al.* 1969). However, recent geological and paleontological results indicate that the river established its present course by 20,000 BP (Nyilas and Sümegi, 1991, Szöőr *et al.* 1991). Thus, a wide alluvial plain, bearing similar hydrological and geomorphologic conditions as today (Fig. 13.2 and Fig. 13.3), developed by the end of the Pleistocene (Sümegi *et al.* 2000).

Nascent environmental conditions strongly determine human settlement strategies and agricultural production. Rivers deriving from the northern parts of the Carpathians (the Sajó River and its tributaries) formed an alluvial fan in the area surrounding Polgár. At the end of the Pleistocene, the Tisza cut into this fan complex following a northeast-southwest course and finally beheaded it. Thus, the varied composition alluvial fan complex became somewhat elevated, reducing the level of saturation. This enabled air-dust particles to settle out on wet floodplains allowing formation of infusional loess deposits on natural levees. Former tributaries of the Sajó River gradually caused infilling in the wind gaps. As a result, the Tisza discharged water only during floods.

The alluvial fan complex at Csőszhalom was dissected into islands and peninsulas by the alternating and formerly active wind gaps at the end of the Pleistocene.

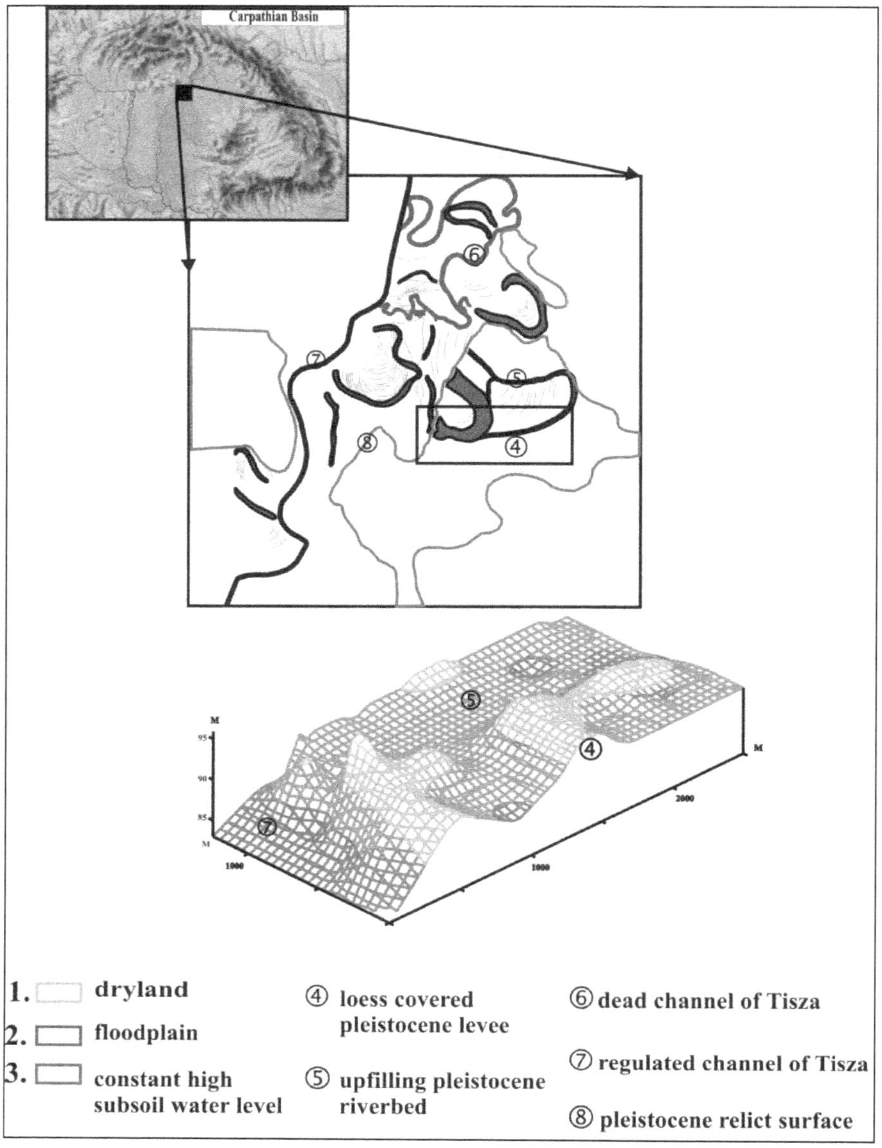

Figure 13.1. Geological and morphological map of Csőszhalom at Polgár (after Gillings, 1995)

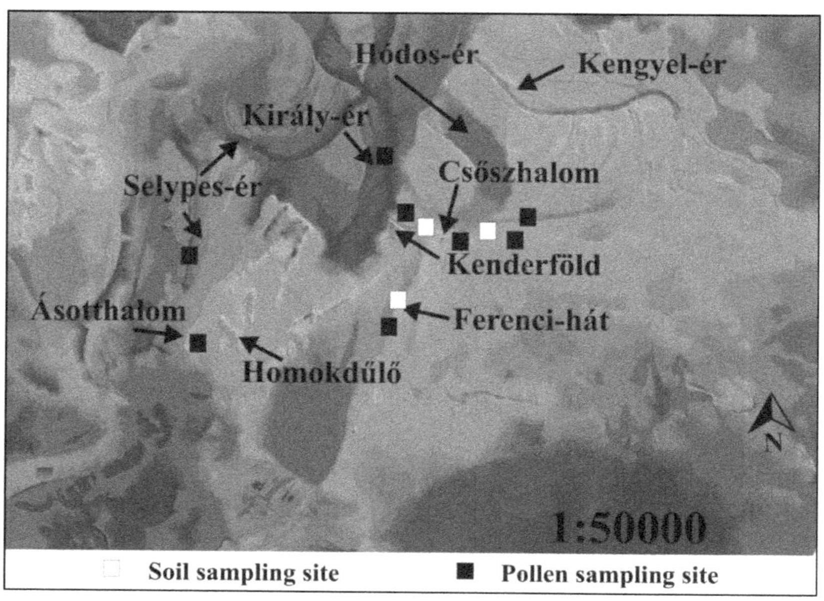

Figure 13.2. Digital field map of Polgár-Csőszhalom, and its surroundings (Timár-Rácz, 2002)

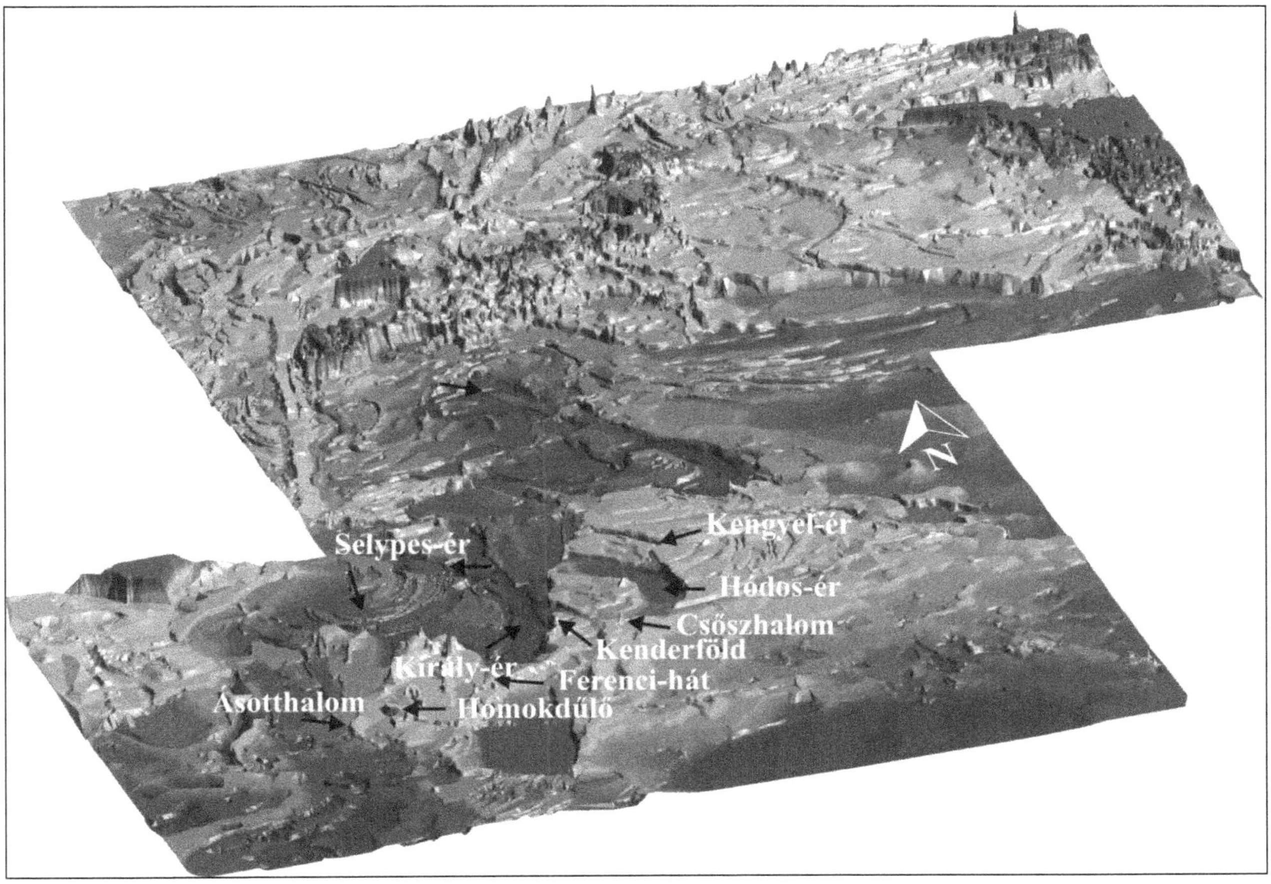

Figure 13.3. Three-dimensional digital field map of Polgár-Csőszhalom and its surroundings (Timár-Rácz, 2002)

Later, in the Neolithic, flooding on the dry levees occurred only during major floods. The deep and widening alluvial plain of the Tisza bordered the whole area to the northwest. Thus, on a micro-scale, the Late Neolithic community chose to settle a morphologically extreme mosaic area bearing an alluvial loess cover. Local morphological and hydrological conditions greatly influenced natural soil conditions and vegetation in this habitat.

ANALYSIS OF THE SOIL CORES

Coring results show that the Csőszhalom tell and the settlement are situated on a wide levee formed by a large sandy area covered with alluvial loess. From the top down to the surrounding river beds, the hydro series is greatly dependent on local morphology responsible for the mosaic of soil conditions and vegetation (Fig. 13.4). The following Late Neolithic paleogeography is established for the area based on sedimentological, micro-morphological (Fig. 13.5), pollen analytical, and carpological research (Raczky *et al.* 2002, Gyulai 2000). The analyzed samples come from the surrounding ditches, river beds, and archeological sites.

Micro-morphologic analysis

Micro-morphologic analysis of sediments from ditches surrounding the Late Neolithic tell, along with coring samples, indicate relatively thick vegetation cover (mostly soft-bodied plants) with fine, red, amorphous, jelly-like iron-oxihydroxid (limonite-goethite) bands around roots. Carbonate grains and bands of reddish iron concretions (pea structures) with a diameter of 0.5-0.6 mm are identified between 92-92.5 m and maximally up to the 93 m above sea level (asl). This indicates that floods with surficial groundwater arteries reached relatively high zones of saturation between 92–93 m asl prior to river regulation. This implies that only areas higher than 93 msl were suitable for the establishment of settlements and farming in the investigated area.

Given the morphological and hydrological conditions of the area and the altering levels of saturation around Csőszhalom, the processes of soil formation are highly varied. In areas lying below 92 m asl, floods and stagnant waters caused dark, water-influenced polihedric meadows and paludal, hydromorf soils. These soils contain iron humate as well as reed, sedge, and bulrush remains. Such is the case with Kender brook, surrounding the Csőszhalom area to the northeast. Around 92 m asl, a special type of water-influenced woodland soil formed along the infilling channels. In these regions, a thick cover of gallery forests composed of softwood (*Salix, Populus*) and hardwood trees (*Quercus, Ulmus, Fraxinus*) developed. The autochthonous, carbonized bark and leaf prints of oaks and alders from the 92 m level of archaeological site No. 6 at Csőszhalom, on the banks of

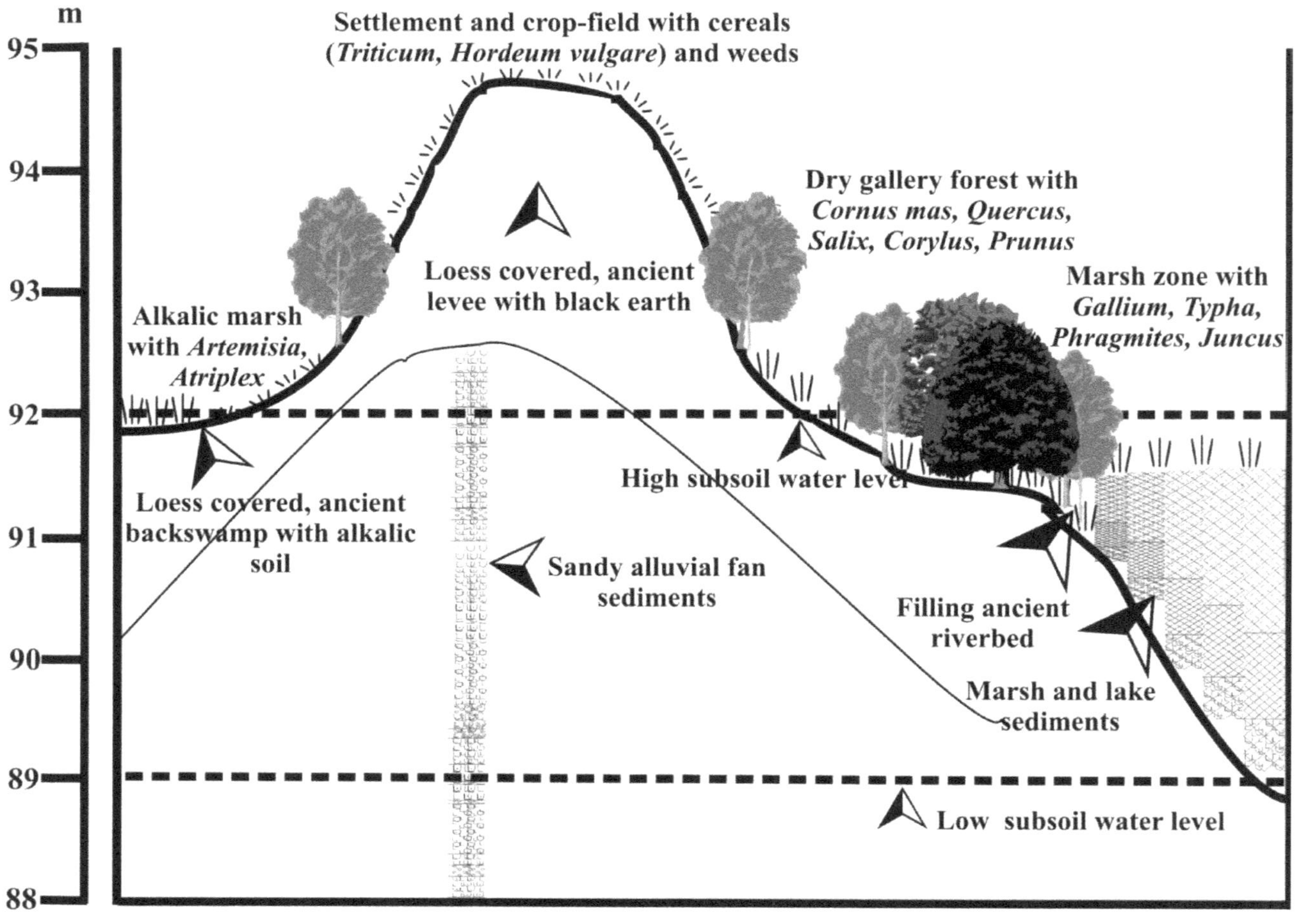

Figure 13.4. Soil conditions and vegetation series of the Polgár-Csőszhalom area in the Neolithic (Raczky *et al.* 2002)

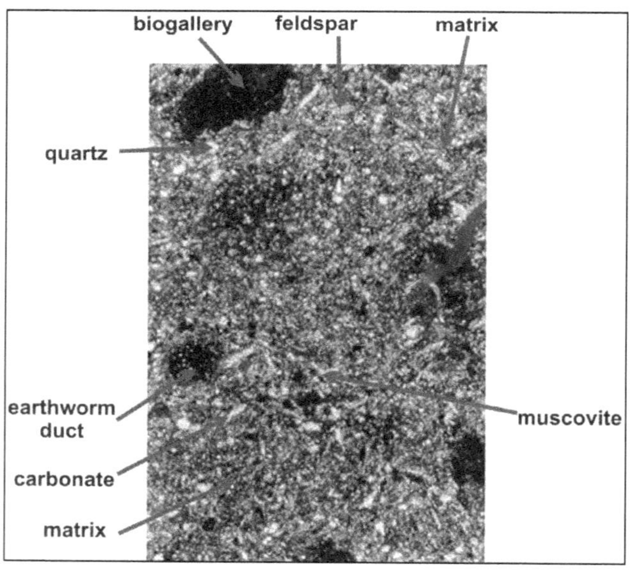

Figure 13.5. Micromorphological section from the basal loess layer at Polgár-Csőszhalom

Kengyel brook, also indicate gallery forests. On the other hand, patchy salt accumulation levels and an alkali soil layer with columnar B horizon are observed between levels of 92–93 m asl. These geological signs indicate a process of alkalization in the former back swamp, located opposite Kengyel-brook on the southeastern side of a spit-like natural levee at Csőszhalom. Similar signs were found during the archaeological investigations of the northeastern margin of the Late Neolithic settlement.

Determining original soil cover for levels above 93 m asl is fraught with difficulties. According to historical maps, the area was under cultivation at least since the Middle Ages, so the soils are greatly and deeply disturbed. However, from the ditches surrounding the Csőszhalom tell and during the archaeological investigations of the Neolithic settlement, we managed to take soil samples, representing the original soil cover of the Neolithic. Micro-morphological analysis of these soils indicates dominance of soft-bodied plants in this horizon. High quality black soils developed having adequate carbonate and organic content along with very good porosity and water balance conditions. Signs of a patchy forest cover are found only near a former bed of Kengyel brook. However, in certain places the black soil horizon shows signs of salt accumulation indicating some sort of alkalization process. Thus, several subtypes or transitional forms of the black soil may have developed in the area, although these cannot be precisely determined today.

Sedimentological and micro-morphological analyses

Sedimentological and micro-morphological analyses indicate the development of paludal soils in the infilling channels and hydromorf woodland soils on the banks. A gradual formation of black soils is observed towards the top of the alluvial loess-covered levees. Meanwhile, the former back swamp area of the levee, opposite the riverbed, shows seasonal water coverage resulting in alkali soils. These belts and the soil mosaics significantly influenced the economics of the Neolithic community. The black soil-covered areas would have been suitable for cultivation, while the saline parts and the areas with hydromorphic soils could be used for grazing and harvesting fodder. However, animals would be exposed to significant danger in the marshy areas due to the presence of *Lymnea truncatula*, which is responsible for spreading liver rot. The wood from gallery-trees, which developed on the alluvial plain of the Tisza and the banks of brooks in the former alluvial fan, would be used for heating. Leaves could be used as fodder, and the environment would have supported the collection of various fruits, berries, and plants. Large forests may have played a significant role in the lives of Neolithic community as indicated by the high number of hunted game.

Palaeobotanical analysis

Palaeobotanical results reinforce the interpretations of human activities from the pedological analyses. The pollen diagrams show a gradual increase in crop pollen (*Triticum*) and weeds as one moves towards the Csőszhalom ditch system from the alluvial plains of the Tisza through the Kengyel brook. The weeds consist of *Chenopodiaceae, Compositae, Poaceae, Plantago lanceolata*, and *Polygonum*. They were probably spread through plowing agricultural fields as well as walking. In the same area, pollen of *Cyperaceae Gramineae*, indicative of open vegetation, show a similar pattern. In the alluvial plain of the Tisza, pollen rates for woody plants (AP) are over 60% with a decrease to below 40% in the surroundings of Kengyel brook. The rates are below 30% in the ditch next to the tell. These pollen diagram trends clearly point to the gradual decrease of woody plants towards the Csőszhalom tell and settlement. An increase in steppe elements and cultivated plants and weeds are observed.

According to these data, human activities (grazing, walking, and cultivation) are concentrated mainly in the areas of the tell, the settlement, as well as the loess-covered levee and the isthmus-like remnant surface lying above the 93 m asl horizon. On the one hand, signs of human activities in the alluvial plain area of the Tisza during the Late Neolithic are minimal. On the other, considerable activity is observed in the area surrounding Kengyel brook. A more open gallery-forest (oak, ash, alder, willow, and poplar) with undergrowth of thick fern-cover and various shrubs of hazel, blackthorn, elder, and cornel are reconstructed for this area.

Large amounts of fruits, berries and their pits, along with fruit and pip remains of crab-apples are characteristic of open gallery-forests. Acorns and pieces of oak charcoal are found in the area of the Csőszhalom tell (Gyulai 2000). In addition, a significant amount of emmer (*Triticum turgidum ssp dicoccum*), spelt (*T. aestivum ssp. spelta*) einkorn (*T. monococcum*), rivet type wheat (*T. turgidum ssp. turgidum*), club wheat (*T. aestivum ssp. compactum*), bread wheat (*T. aestivum spp. aestivum*), barley (*Hordeum vulgare ssp. hexastichum*), and even millet (*Panicum miliaceum*) are recovered archaeologically (Gyular 2000). Seeds of papilionaceous plants, such as lentil (*Lens culinaris*), green pea (*Pisum sativum*), and vetch (*Vicia cf. sativa*) (Gyulai 2000) verify that Late Neolithic people living at Csőszhalom reached a peak of early agricultural production. A wide array of weeds from various habitats (alkalic marshes, loess-steppes, wetland meadows, floodplain forests, fields, pastures, and dirty walkways are identified through pollen analysis (Gyulai 2000). The weeds associated with these areas included *Polygonum arenastrum, P. persicaria, Chenopodium album, C. hybridum, Gallium spurium, G. palustre, Carex hirta, Rumex hydrolapathum, Scoenoplectus lacustris, Hordeum bulbosum, Fallopia*, etc.

Results of the carpological analysis also point to the presence of large-scale, versatile agricultural production, mainly the growth of crops in the neighborhoods of the Csőszhalom tell during the Late Neolithic. The crops were complemented with arid loving crop weeds (*Secalietea* species) (Küster 1985) and seeds of wetland and swamp plants.

CONCLUSION

The nascent environmental conditions strongly determined human settlement strategies and agricultural production. Palaeobotanical data along with the results of pollen analysis and pedological research imply that the highest flood protected areas of Csőszhalom, originally covered with loess-steppe or forest-steppe, were chosen for the cultivation of crops as evidenced by the presence of *Stipa* seeds. However, results of macrofossil analysis indicate the extension of the croplands to the seasonally flooded areas. Furthermore, seeds of plants preferring wetlands, swamps, forests, and saline marshes suggest development of plant assemblages in the low lying areas befitting local morphological, hydrological, and soil conditions (Fig. 13.6), creating micro-scale zonality and a complex mosaic that was further enhanced by human activities. In this mosaic environment where Late Neolithic people settled, human activities completely altered the original vegetation cover of the loess levees while the composition of marshy flora and forest areas was only mildly modified.

References

BÁNFFY, E., and I. BOGNÁR-KUTZIÁN 2007 – *The Late Neolithic Tell Settlement at Polgár-Csőszhalom,*

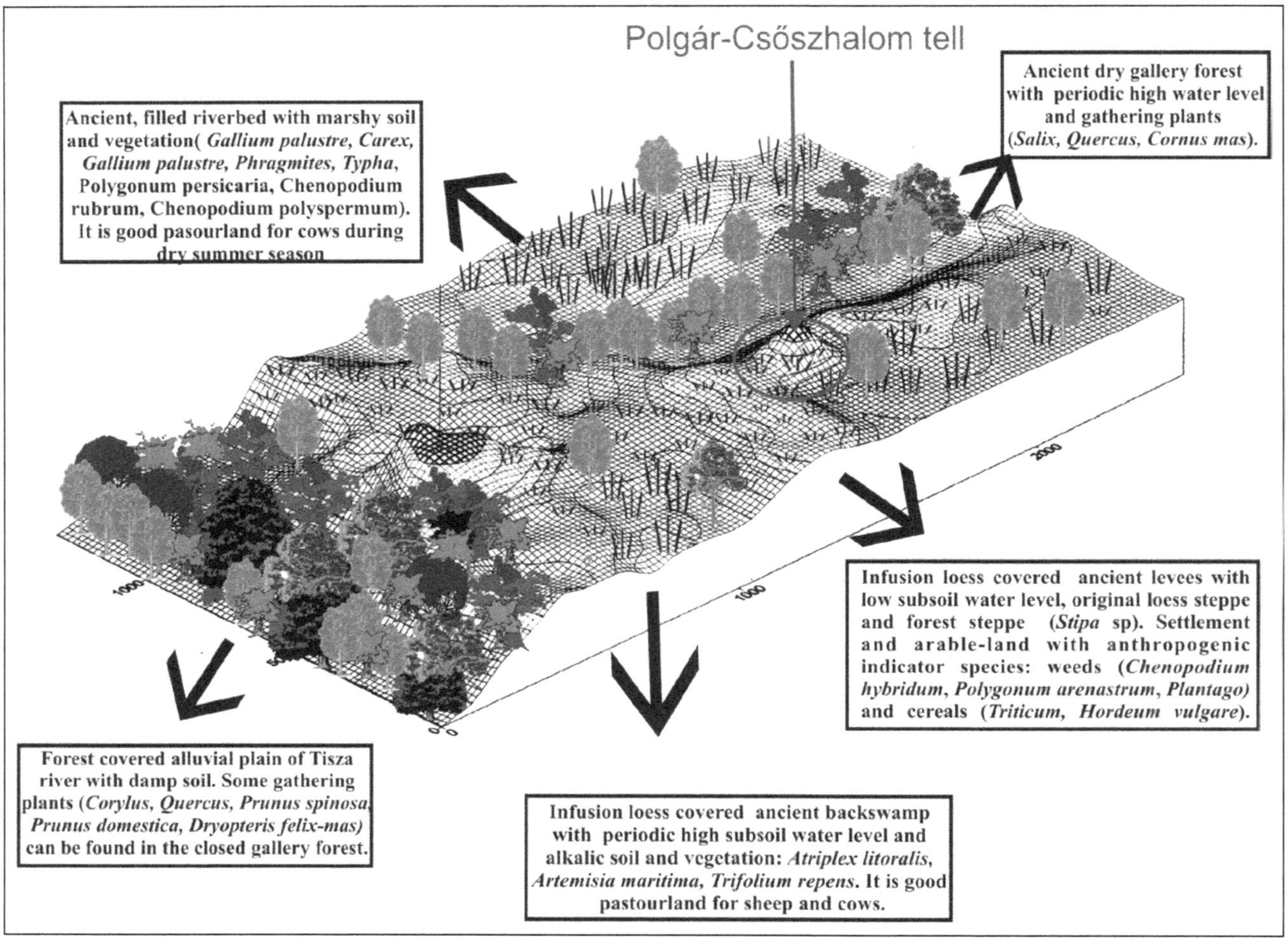

Figure 13.6. Reconstruction of the local morphological, hydrological and soil condition of Polgár-Csőszhalom in the Neolithic (Raczky *et al.* 2002)

Hungary: The 1957 Excavation. BAR S1730, Archaeolingua, BAR Publishing, Oxford, 2007).

BORSY, Z. 1995 – Evolution of the North-eastern part of the Great Hungarian Plain in the past 50,000 years. *Quaestiones Geographicae*, Special Issue 4. 1995:65-71.

BORSY, Z., B. MOLNÁR and S. SOMOGYI 1969 – Az Alluviális Medencesíkságok Morfológiai Fejlődéstörténete Magyarországon. *Földrajzi Közlemények* 93, 1969:237-254.

CHOLNOKY, J. 1907 – A Tiszameder Helyváltozásai. *Földrajzi Közlemények* 28, 1907:381-405.

GILLINGS, M. 1995 – Flood Dynamics and Settlement in the Tisza Valley of North-East Hungary: GIS and the Upper Tisza Project. In G. Lock and Z. Stancic (eds.), *Archaeology and Geographical Information Systems*. Taylor and Francis, London, 1995:67-84.

GYULAI, F. 2000 – *Archeobotanika*. Jószöveg kézikönyvek, Debrecen.

KALICZ, N., and P. RACZKY 1987 – The Late Neolithic of the Tisza Region. A Survey of Recent Archaeological Research. In P. Raczky (ed.), *The Late Neolithic of the Tisza Region*. Szolnok, Budapest, 1987:11-30.

KÜSTER, H. 1985 – Herkunft und Ausbreitungsgeschichte einiger Seclietea-Arten. Tüxenia. *Mitteilungen der Floristisch-soziologischen Arbeitsgemeinschaft* 5, 1985:89-97.

NYILAS, I., and SÜMEGI, P. 1991 – The Mollusc Fauna of Hortobágy at the end of the Pleistocene (Würm 3) and in the Holocene. In C. Meier Brook (ed.), *Proceeding of the 10 International Malacological Congress*, 27 August-2 September 1989. University of Tübingen, Claus Meier-Brook, Tübingen, 1991:481-486.

RACZKY, P. 1989 – Chronological Framework of the Early and Middle Neolithic in the Tisza Region. In S. Bökönyi (ed.), Neolithic of Southeastern Europe and its Near Eastern Connections. *Varia Archaeologica Hungarica* II, Budapest, 1989:233-251.

RACZKY, P. 1998 – The Late Neolithic Tell of Polgár-Csőszhalom and its Relationship to the External Horizontal Settlement in Light of Recent Archaeological Data. In P. Anreiter, L. Bartosiewicz, E. Jerem, and W. Meier-Arendt (eds.), *Man and the animal World*. Archeolingua Press, Budapest, 1998:481-489.

RACZKY, P., W. MEIER-ARENDT, K., KURUCZ, Zs. HAJDÚ, and Á. SZIKORA 1994 – *Polgár-*

Csőszhalom. A Late Neolithic Settlement in the Upper Tisza Region and its Cultural Connections (Preliminary Report). Jósa András Múzeum Évkönyve, 36, 1994:231-240.

SÜMEGI, P, A. MOLNÁR and G. SZILÁGYI 2000 – Szikesedés a Hortobágyon. Alkalization process in the Hortobágy region. *Természet Világa* 131, 2000:213-216.

RACZKY, P., W. MEIR-ARENDT, A. ANDERS, Zs. HAJDÜ *et al.* 2002 – Polgár-Csőszhalom (1989-2000): Summary of the Hungarian-German Excavation on a Neolithic Settlement in Eastern Hungary. In R. Aslan, S. Blum, G. Kastl, F. Schweizer and D. Thumm (eds.), *Mauerschau II, Festschrift für Manfred Korfmann.* Verlag Bernard Albert Greiner, Remshalden-Grunbach, 2002:838-840.

SZÖŐR, GY., P. SÜMEGI and É. BALÁZS 1991 – Sedimentological and Geochemical Facies Analysis Upper Pleistocene Fossil Soil Zones Discovered in the Hajdúság Region, NE Hungary. In M. Pécsi and F. Schweitzer, (eds.), *Quaternary Environment in Hungary. Studies in Geography in Hungary* 26. Akadémiai Kiadó, Budapest, 1991:47-59.

TIMÁR, G., and T. RÁCZ 2002 – The effects of neotectonic and hydrological processes on the flood hazard of the Tisza region (East Hungary). In S.A.P.L. Cloetingh, F. Horváth, G. Bada and A. Lankreijer (eds.), *Neotectonics and Surface processes: the Pannonian Basin and Alpine/Carpathian System.* European Geosciences Union, Stephan Mueller Special Publication Series 3, 2002:267-275 (Open access: http://www.stephan-mueller-spec-publ-ser.net/3/267/2002/smsps-3-267-2002.pdf).

Chapter 14
FRESHWATER MUSSELS AND LIFE IN THE LATE NEOLITHIC TELL OF HÓDMEZŐVÁSÁRHELY-GORZSA, SOUTHEAST HUNGARY

Sándor GULYÁS and Pál SÜMEGI
Dept. of Geology and Paleontology University of Szeged, Hungary

Abstract: *Freshwater shellfish were retrieved from Profile 18 (ca. 4900-4530 BC) of the Late Neolithic tell of Hódmezővásárhely-Gorzsa, in the southeastern Hungary are analyzed. Three Unionid taxa are identified. U. pictorum and U. tumidus dominate throughout the whole profile. The calculated proportions of mussel meat and corresponding nutrition point to the use of shellfish only as a secondary food source. Preliminary analysis of the valves shows that the mussels are harvested during the late spring, early summer, either after a harsh winter or during the establishment of the settlement. Shellfish harvest intensity varies throughout the profile. Consequently, the intensified harvest periods may be associated with new settlement stages and/or indicate changes in climatic and environmental conditions.*

INTRODUCTION

The vast majority of zooarcheological research on mussels is strictly focused on marine forms (Bailey 1975a, 1975b, 1978, 1994, Bailey and Milner in Press, Deith 1983a, 1983b, 1984, 1985, Andersen 1989, Enghoff 1989, Milner 2001, Rowley-Conwy 1983, Swadling 1976, Waselkov 1987, Fieller *et al.* 1985, Petersen 1986, Meehan 1982, Jones and Fisher 1990, Claassen 1998, Luby and Gruber 1999, Henderson *et al.* 2002, etc.) in the international literature. However, archaeological evidence indicates that throughout the Holocene the exploitation of freshwater mussels was just as important to prehistoric and historic people living inland (Gulyás *et al.* 2007, Gulyás 2011, Parmalee 1956, Parmalee and Klippel 1974, 1986, Matteson 1959, Warren 1975, Peacock 1996, 1997, 2000, 2002).

The earliest studies on freshwater mussels come from archaeological sites of Pre-Columbian aboriginal cultures occupying the fluvial plains of the central and southeastern United States (Baker 1923, 1930, 1936, 1941). In such research, shellfish is primarily used for the reconstruction of the former fauna and environment of the shell-fishing sites (Matteson 1958, 1959, 1960, Parmalee 1956, Parmalee *et al.* 1972). The meat of these bivalves must have represented an important food supplement in the daily subsistence of the Native American cultures (Parmalee and Klippel 1974). Moreover, the shells themselves were often modified into tools such as scrapers, hoes, bowls, spoons, and ornaments (Picha and Swenson 2000, Lippincott and Davis 2000, Warren 1975, 2000, Myers and Perkins 2000, Hirst 2000, Dorsey 2000, Parmalee 1956, 1988, 1994, 1998, Parmalee and Bogan 1986, Parmalee *et al.* 1980, 1982, Theler 1987, 1990, 1991, Blitz 1993, Bogan 1981, Casey 1987, Claassen 1994, 1998, Klippel *et al.* 1978, Muller 1986, Murphy 1971, Peacock 1996, 1997, 2000, 2002, Peacock and James 2002, Robinson 1983, Scott 1982, Taylor 1989, Taylor and Spurlock 1982).

In contrast to North America, the general lack of European archaeological studies on freshwater riverine mussels must be the result of there being only a few taxa of freshwater mussels (Badino *et al.* 1991, Bogan 1993, Nagel and Badino 2001, Williams *et al.* 1993, Turgeon *et al.* 1998). Consequently, exploitation of freshwater mollusks could never have been as important as marine bivalves in the lives of European communities except in limited areas and at certain periods during the course of prehistory. Nonetheless, the archaeological record implies that these mussels were readily available in the Neolithic, Copper, and Bronze Ages in Hungary (Czógler 1934, Horváth 1982, 1987 Domokos 1980, Sümegi *et al.* 1996, Sümegi 1999, Sümegi 2003, Tóth 2003, Whittle 1999, 2000, Whittle *et al.* 2001). In fact, several hundred kilograms of shell have so far been excavated in Hungary as part of archaeological excavations at Neolithic and Bronze Age sites. Yet their analysis has been rare. Gulyás *et al.* (2007), Gulyás (2011), Sümegi *et al.* (1996), Sümegi (2003), Tóth (2003) discuss the results of detailed zooarcheological and paleoecological analyses.

The present research focuses on various aspects of life in the Late Neolithic community inhabiting the tell settlement at Hódmezővásárhely-Gorzsa, southeastern Hungary. Scientific analysis studies dietary habits and gathering modes of freshwater bivalve shells retrieved from the site, and follows the aims and approaches previously applied to shellfish from a Hungarian Early Neolithic Körös culture site. The long-term aim is to shed light on the interrelationship of various subsistence activities within this community, as well as the wider social and economic relations.

MATERIAL AND METHODS

Initial archaeological excavations were implemented in the 1980s by Ferenc Horváth at Gorzsa-Földvár Halom (Gorzsa Earth Mound Hill), located a few kilometers

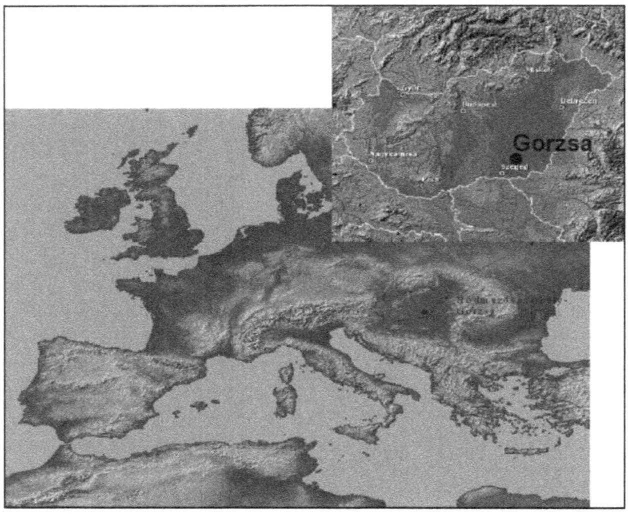

Figure 14.1. Location of the Late Neolithic site of Hódmezővásárhely-Gorzsa

from the city of Hódmezővásárhely, in southeastern Hungary (Fig. 14.1). These yielded numerous finds from the Tisza and Proto-Tiszapolgár culture, including remains of wattle and daub houses, burials, pottery and other artifacts, as well as a large quantity of shellfish (Horváth 1982). Profile 18, which has been divided into 24 micro-horizons (Fig. 14.2) dates from ca. 4900-4530 BC. Shell material from each of the individual micro-horizons is being investigated separately. The shells from the lowest micro-horizon (Horizon 24), a shallow pit at the bottom of Profile 18, were chosen for detailed analysis. The pit is dated between 4898-4874 BC and must have functioned as a kitchen midden.

A total of 1213 shells are examined in detail. The shells suffered cracking due to stress from compaction by weight of the overlying layers. Thus, only a small fraction of valves and fragments are suitable for further statistical analysis. Additionally, about 1500 shells, scattered within individual horizons of the site are examined. The shells are analyzed with respect to the following issues:

TAXONOMIC IDENTIFICATION OF INDIVIDUAL TAXA

Taxonomic identification is most important, because all subsequent analyses depend on it. It is also an indispensable asset for the reconstruction of shell-fishing near the site. Shells with completely intact beaks, or at least the umbo, are most useful for accurate identification of species. Publications on modern freshwater species of Hungary are used to distinguish between the individual taxa based on the shape of the shells, the characteristics observable on the hinge area, and the beak or umbo (Richnovszky and Pintér 1979, Soós 1943). The subsequent biometric analysis of complete shells corroborates the accuracy of these determinations.

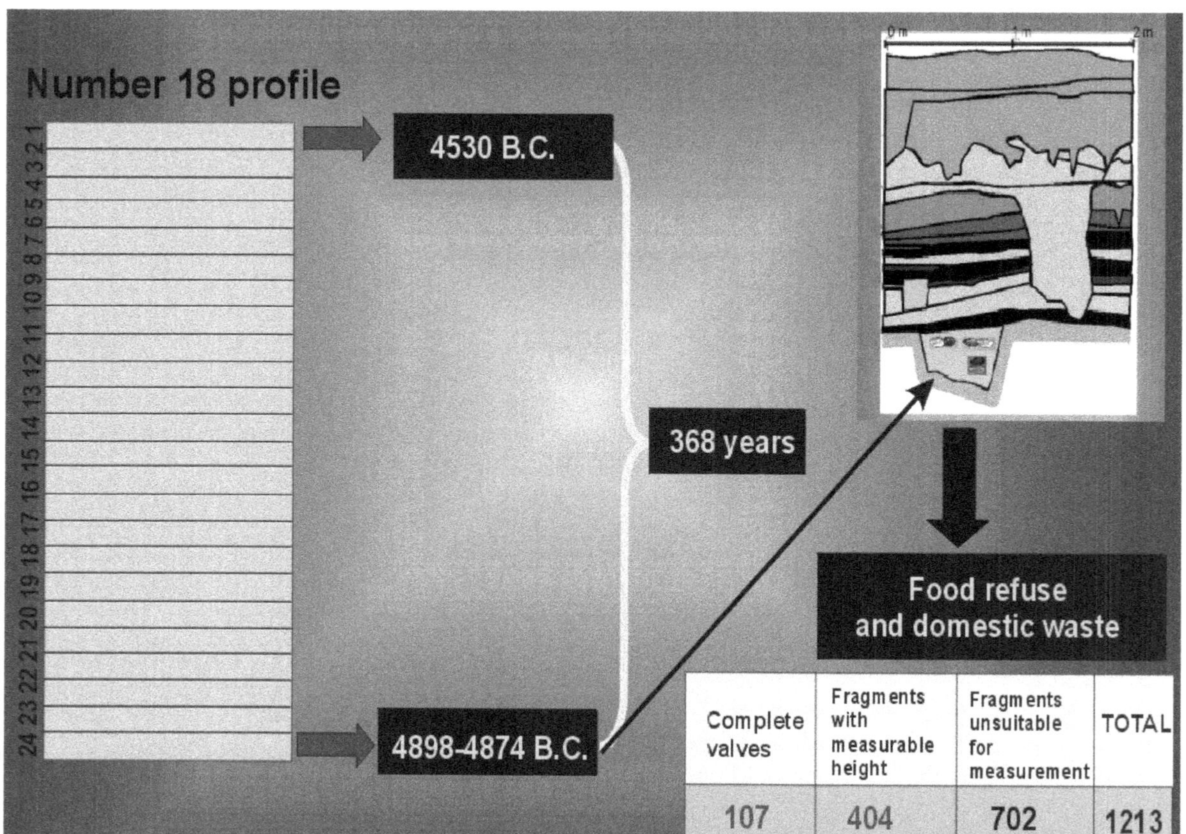

Figure 14.2. The origin and distribution of the studied shell material

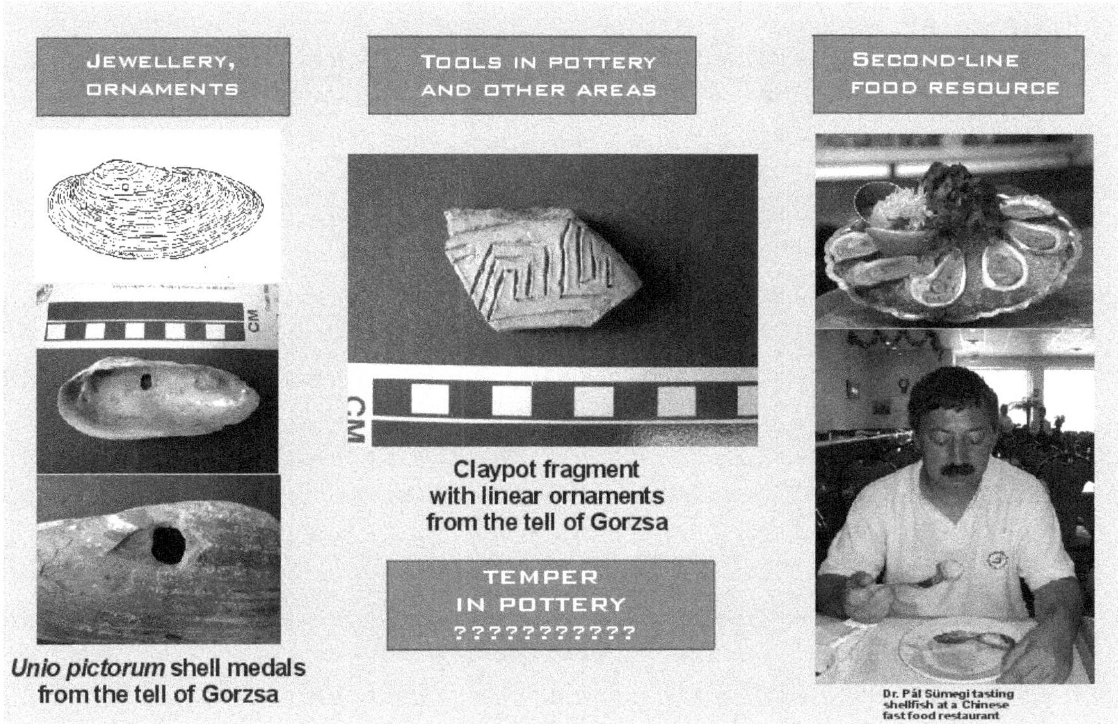

Figure 14.3. The inferred use of shellfish during the Neolithic in Hungary

Minimum Number of Individuals and the Proportional Abundance of Species

Determination of the minimum number of individuals (MNI) serves as a starting point for the calculation of specimen numbers and the proportional abundance or dominance values of the various mollusk species. The environmental characteristics of the site are determined from the ecological needs, substrate, and habitat preferences of the individual taxa. The values of proportional abundance or dominance in site catchment analysis (SCA) are used for this analysis, allowing determination of whether the mussels are gathered from nearby oxbow lakes and backwater areas on the floodplain or from the channel of the Tisza River located some 10 km distant from the site. For this purpose, all complete shells or shell fragments with a beak and/or hinge are sorted into left and right valves. Two valves represent an individual for bivalves. However, both valves are very rarely preserved together in one place (paired valves) due to the selected nature of the material and disposal of empty shells by humans. Additionally, either valve can represent a separate individual. For these reasons, the largest number of valve half (left or right) is used in calculation of the MNI. Thus, the ratio of the left and right valves for individual species (i.e., the number of the paired valves) can yield information on whether or not the shells are discarded in connection with a single relatively short event.

Insight into Ancient Culinary Habits

The dark brownish hue on the shells and ash remains covering the surfaces of the shell may indicate that the mussels were roasted in a fire before consumption. The lack of such signs and fractures along the margins suggests raw consumption, just as oysters are eaten today (Czógler 1934) (Fig. 14.3).

Calculating the Quantities and Nutrition of Meat Gained from Mussels

Parmalee and Klippel (1974) provide the first reliable qualitative and quantitative assessment of food potential for various American indigenous freshwater mussel species harvested by the natives of North America. A similar approach is applied in the development of an acceptable method that predicts the actual amount of edible meat and nutritive value of local Hungarian shellfish fauna. The resulting values are used to determine whether the mussels served as a primary or secondary food source. This information can also provide the approximate number of individuals for whom the total meat could have provided.

In order to determine the live weight of animals, the weight of their soft material or meat, and the derivable energy content, the main biometric parameters of the mussels must be determined. These are shell height (H), shell width (W), the index of flatness (H/W), and weight. Where possible, these parameters are measured to a 0.5 mm accuracy using a caliper. The weight of the shells is recorded using laboratory scales. The resulting size and shape variants are used as input for further analyses. Where we can measure shell widths, the species dependent empirical formulas by Kiss (2000) are used to calculate living weight and derived meat (Table 14.1). Regression analysis is used by Kiss (2000) in detailed

Table 14.1. Equations used to calculate of live weight and meat weight of the mussels: Calculated parameters of three Unionid species (after Kiss 2000)

Species	X	Y	Regression	Type	r2	N
U.crassus	L(mm)	LW(g)	Y=5.49E-04*X2.630	exp	0.798	32
U.crassus	L(mm)	WW(g)	Y=8.29E-05*X^2.816	exp	0.766	32
U.pictorum	L(mm)	LW(g)	Y=2.45E-04*X^2.784	exp	0.948	27
U.pictorum	L(mm)	WW(g)	Y=1.35E-04*X^2.658	exp	0.915	27
U.tumidus	L(mm)	LW(g)	Y=2.03652-04*X^2.83052	exp	0.955	28
U.tumidus	L(mm)	WW(g)	Y=0.430706*X^0.912775	exp	0.932	28

Calculated parameters with X= shell height for the three Unionid species using the unpublished data of Kiss (2000)

Species	X	Y	Regression	Type	r2	N
U.crassus	H (mm)	LW(g)	Y=4.48E-04*X^3.22774	exp	0.848	77
U.crassus	H (mm)	WW(g)	Y=6.90762E-04*X^2.74039	exp	0.766	77
U.pictorum	H (mm)	LW(g)	Y=1.42073E-03*X^2.94331	exp	0.891	27
U.pictorum	H (mm)	WW(g)	Y=8.7871E-04*X^2.79654	exp	0.908	27
U.tumidus	H (mm)	LW(g)	Y=4.39812E-04*X^3.19665	exp	0.946	28
U.tumidus	H (mm)	WW(g)	Y=3.30106E-04*X^2.95259	exp	0.903	28

L= length of the shell; H= height of the shell, LW= live weight; WW=meat weight

morphometric and population ecological research on living *Unionidae* from the Tisza River in Hungary. He finds a strong correlation between the width and the live weight of the shells, as well as the weight of the soft material.

The approximate energy and number of people for which the food served is estimated from these values. According to Tudorancea and Florescu (1968), 1 g dry weight of *Unionidae* meat equals 4488.1 calories (cal). Since the ratio between the wet and dry weight of the soft tissues of *Unio* is 5.38, 1 g of wet weight is equivalent to 834.22 cal (1 kg = 834 kcal) (Ravera and Sprocati 1997). This means that freshwater mussels yield more protein and nutrition (Parmalee and Klippel 1974) than their marine counterparts do. The calorie content of marine bivalves is estimated to be 600 kcal per kg (Bailey and Milner in press, Clark 1972 [1954]). Unfortunately, for the majority of shells only the height is recorded. Since there is a strong correlation between height and width (r = 0.87) during growth, similar empirical formulas have been set up for individual species, using the original data by Kiss (2000) (Table 14.1).

Selectivity of Shell-Fishing and the Shell Gathering Strategies

People may exercise preference for particular species and particular sizes of mollusks. This may be due to nutritional factors, labor requirements, ease of access, taste, social practices, or cultural idiosyncrasies. Foragers preying upon freshwater mollusks, like rats, muskrats, and humans, generally show a preference for larger, older specimens to the limit of their handling capacities (Convey *et al.* 1989, Hanson *et al.* 1989, Zahner-Meike and Hanson 2001, Richardson and Yokley 1996, Ravera *et al.* 2003). In some cases, however, such as those of aboriginal cultures of North America, the largest, oldest specimens seem to have been avoided. Perhaps their meat was not palatable or they simply did not occur in the habitat collected by humans (Parmalee 1956, Peacock and Chapman 2001, Peacock and James 2002, Peacock 2000, 2002).

To determine collection methods and preferences, statistical analyses are applied. Shell height or shell length displays a strong correlation with age of the mussel, because of accretionary shell growth. Thus, frequency histograms, plotting the size of the shells, together with univariate statistical parameters (standard deviation, mean, skewness, and kurtosis), help determine whether the gatherers of the Neolithic community collected shells selectively. They can also test for preferences of certain size or age groups within the fauna, as well as intentional planning for shell-fishing. The type of distribution (unimodal, bimodal, or multi-modal) provides information from which we can ascertain whether the shells come from the same population as well as determine the approximate time of harvesting. Additional information on harvesting practices comes from modern analogies. These include sampling experiments on freshwater mollusks carried out by manual collection (Hanson *et al.* 1989, Richardson and Yokley 1996). The authors found that visual and tactile harvesting are quite inefficient for mussels less than 35 mm in length. These smaller mussels are generally buried too deep in the substrate for humans to detect by hand. In the analyzed archaeological material, the dominance of larger size classes and lack of smaller ones implies that they were gathered by hand. Unfortunately, there is no direct evidence for such a practice in the literature. For additional verification, the amount of smaller gastropod shells dwelling in the same habitat and on the same substrate as the large mussels should also be taken into consideration. In addition, correlation of prehistoric data with data from modern ethnographic research may be useful in reconstructing the method of shell-fishing.

Finally, density values for the three Hungarian *Unio* species in modern rivers, ponds, and oxbow lakes can be used to determine whether the mussels originate from a single population. However, they must come from environments with similar substrate conditions and faunal composition as the prehistoric system under investigation (Kiss 1996, B. Tóth and Bába 1980, Horváth 1955). Such data provide a relatively good estimate of the size of the collection site (given the MNI) once a unimodal distribution is established.

The Time or Season of Collection

Freshwater mussels in rivers, ponds, and oxbow lakes contrast with marine forms (Bailey and Milner in press, Deith 1983a, 1983b, 1985, Rowley-Cowny 1984, Petersen 1986, etc.). They are generally available for collection during seasons of active growth. Collection is usually possible between April and November in temperate continental climates when the waters are free of ice. In general, this is the case in the Carpathian Basin, but its environment is formed by the mutual interaction between several climate zones, exuberant vegetation, and various soil types. This yields a mosaic-like environmental pattern (Sümegi and Kertész 1998, 2001, Sümegi *et al.* 1996) suggesting some variability in the harvesting season. Therefore, the actual collection season may shed light on factors that may have influenced timing of shellfish collection.

There are several factors that influence the timing and need of mussel gathering. Of special interest to archaeologists is the possible need for an easily available food source during construction of new settlements when other forms of subsistence activities are out of the question. Another factor is the need for supplementary food resources in times of harsh environmental conditions. A third one relates to social and economic conditions, which foster harvesting (Bailey and Milner in press, Claassen 1998). To provide archaeologists with information related to these factors, it is necessary to establish a reliable sclerochronology.

Establishing a reliable sclerochronology suitable for use in archaeological studies, such as those developed for several marine forms (Deith 1983a, 1983b, 1985, Quitmeyer *et al.* 1997, Rowley-Cowny 1983, Petersen 1986, Bailey and Milner in press, Claassen 1998) is complicated (Neves and Moyer 1988). It requires that the growth of the given bivalve species is accurately documented at both the macro- and the micro-scale. Interpreting seasonality based on the alternating dark and light growth increments is relatively easy for marine forms. They have relatively stable habitat conditions, enabling a year-long growth season. For freshwater mussels, frequent environmental changes introduce potential variability into the formation of the micro-growth increments. This hampers the possibility of easy high resolution age and seasonality determinations.

Figure 14.4. Thin-sections and acetate peels of the shells

Table 14.2. The proportion of taxa present in the analyzed material and the calculated MNI (minimum number of individuals) represented by the valves

SPECIES	Complete valves (No).	Fragments with measurable height (No.)	Fragments unsuitable for measurement (No.)	Minimum number of Individuals (MNI)
U. tumidus right valve	34	114	106	
U. tumidus left valve	35	146	40	287
U. pictorum right valve	27	63	233	
U. pictorum left valve	10	77	322	426
U. crassus right valve	0	2	0	
U. crassus left valve	1	2	1	4
MNI	63	225	429	717

Most growth-related studies for the Hungarian *Unionidae* are based on counting and measuring the adjacent growth rings on surfaces of the shells (Kiss and Pekli 1987, 1988a, 1988b, Kiss and Petró 1992, Kiss *et al.* 1988, Entz 1932, Pónyi *et al.* 1981, Pónyi 1990, Sebestyén 1939). However, analysis of micro-growth increments providing a suitable analogue for age determination beyond the seasonal scale is not available. Fortunately, extensive studies done outside Hungary are available for the same Unionid species (Checa 2000, Dunca and Mutvei 1996, Mutvei and Westermark 2001, Mutvei *et al.* 1994, Timm and Mutvei 1993, Westermark *et al.* 1996, Dettmann *et al.* 1999). These serve as acceptable analogues in interpretation of micro-growth patterns retrieved from Gorzsa. Unfortunately, in the past, differences in climatic and environmental condition in the Carpathian Basin were complicated by the mosaic-like complexity and diversity of the environment. This is present at macro-, meso-, and micro-scales and extends to prehistoric Gorzsa (Sümegi and Kertész 1998, 2001). Therefore, our interpretations beyond the seasonal scale are only preliminary and necessitate further research.

The following methods are applied for the purpose of understanding growth and seasonal studies of mollusks. A part of the shell is sectioned from the umbo to the growing shell margin. The polished thin-sections are analyzed under a stereomicroscope at a magnification of 25 times to observe the last annual band and the last growth increment. This allows establishment of the possible season (spring, summer, autumn) of death (Fig. 14.4). In order to corroborate these rough and low-resolution predictions, thin-sections are prepared of selected shells. They are analyzed under a petrographic microscope at a magnification of 50x. Additionally, acetate peels are made of both the transversally sectioned shell and the whole shell surface, following the method of Sato (1999). The acetate peels are examined under a microscope at a magnification of 100x. This approach is most promising for the life-history analysis of marine bivalves (Richardson *et al.* 1993, Sato 1999).

The aim is to correlate observable growth patterns on the shell surfaces with the micro-growth increments of the thin-sections and peels taken from the transversally sectioned shells. This methodology is more rapid, less expensive, and more sophisticated than the preparation of thin sections. In addition, it allows a quick, reliable mass evaluation of seasonality on a larger number of shells. However, we also find that the resolution of this approach is not as good as the thin-section analysis. This is corroborated by other researchers (Neves and Moyer 1988).

RESULTS

The lowest horizon of the waste pit in the Late Neolithic tell of Gorsza is examined. Out of the 1213 shells analyzed, 107 are complete valves, and 404 are suitable for taking height measurements, yielding 511 valves suitable for statistical analysis. The remaining 702 specimens are shell fragments. For these, only taxonomy can be determined, precluding their use in statistical analysis and the exact prediction of biomass. Accordingly, three major taxa are present in the shell material. These are *Unio pictorum* (Linné 1758), *Unio crassus* (Retzius 1788), and *Unio tumidus* (ibid.). All three are also present in modern fauna of Hungarian rivers and ponds, rendering them ideal analogues for the analysis of archaeological shell (Richnovszky and Pintér 1979, Soós 1943).

Since either valve can represent an individual, the valves present in larger numbers are taken to signify a single specimen for the calculation of the MNI. The 1213 valves must have corresponded to 717 specimens of the three taxa (MNI) (Table 14.2.). From the MNI of the individual taxa the following proportional abundance values are determined (Fig. 14.5): *Unio pictorum* = 59%, *Unio tumidus* = 40%, *Unio crassus* = 1%.

Based on the ecological needs and habitat preferences of individual species, the following conclusions can be made: The species *Unio crassus* generally prefers moving water habitats. Its modern representatives are present in the more distant and recent Tisza riverbed, where they dwell on the sandy substrate close to the main channel

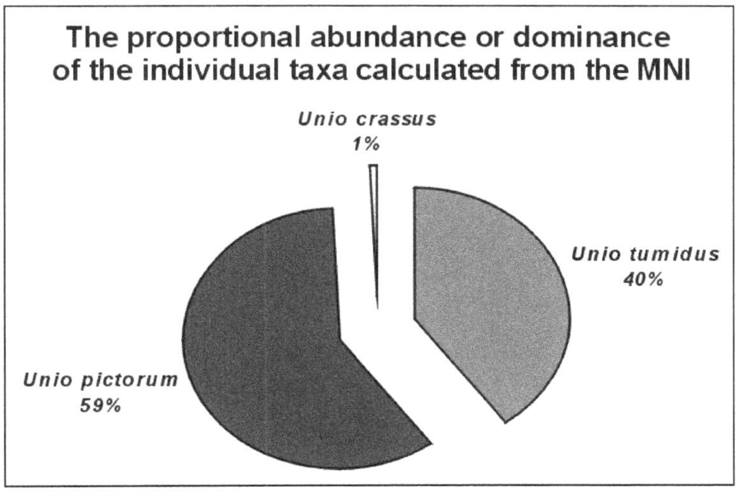

Figure 14.5. The distribution of the different species

line. *Unio pictorum* prefers stagnant water habitats. Its modern representatives populate ponds, oxbow lakes, the littoral, and protected parts of rivers, where they dwell on a muddy substrate. *Unio tumidus* also prefers stagnant waters. They dwell on the muddy substrate in similar habitats as the previous species (Richnovszky and Pintér 1979, Soós 1943). The dominance of *Unio pictorum* (Retzius 1788), seems to indicate that these shells originate exclusively in the backwaters of the floodplain surrounding the site (Fig. 14.6).

Disposal of the shells must have been linked to a single harvest. This can be concluded from the large number of paired valves in the studied material from the pit. This notion is supported by the analytical results from complete valves, i.e., those with measurable heights, as well as the shell fragments (Fig. 14.7).

According to the literature on mussel shells, average population densities for the three unionid species are between 50-70 specimen per m^2 in well-oxygenated oxbow lakes (Kiss 1996). Assuming that gathering was done in a single place, it must have been restricted to a relatively small area of a few m^2. As such, it would have required the cooperative work of three or four people – perhaps a small nuclear family (Fig. 14.6).

The excavations located several damaged shell pendants, which eventually ended up in the prehistoric waste ditch (Horváth 1982). They indicate the possible use of shells as adornments. Furthermore, the empty shells might have functioned as tools used to create pottery ornamentation (Fig. 14.3).

The likelihood that bivalve shells were used to temper pottery is unlikely. Shells of unionid bivalves are relatively thick and not easily crushed into a fine and relatively homogenous powder. Furthermore, the thick periostracum (uppermost organic layer) yields large amounts of impurities to temper. The successful removal of this layer seems unlikely, given available Neolithic technologies.

Material recovered from the pit shows signs of burning. Roasting could be identified on some specimens. Development of roasting or cooking before consumption contrasts with the Early Neolithic site of Ecsegfalva (Whittle 2007), where such evidence is absent.

If mussels served primarily as food, it becomes imperative to determine the amount of meat available for consumption and the possible energy content it provides. Furthermore, one should estimate the number of individuals fed by the total amount of meat, knowing that their gathering and disposal might be linked to a single event within one household. In order to insure reliable calculations, measurements of the valves present in larger numbers in the sample are utilized uniformly for all species. Result shows no significant deviation from the values of dominance for calculation of the live weight of mussels gathered and the weight of meat procured. Of 717 specimens present in the sample, only 288 have complete valves and a measurable height as indicator of size. Methods presented above are used for these estimates. Although *Unio pictorum* is prevalent in the excavated material, only 87 are suitable for calculation of total live weight. It is calculated that 2.59 kg of mussels yield 991.6 g of meat. The 181 specimens of the second largest group, *Unio tumidus*, yield a total live weight of 4.7 kg. This corresponds to 1484 g of meat. The 20 specimen of *Unio crassus* have a total live weight of about 13 g, yielding 4.28 g of meat (Fig. 14.8). Clearly, *Unio pictorum* was the preferred target of shell-fishing, because it yields more edible meat with less live weight to be carried home than does *Unio tumidus* (Fig. 14.8).

Total live weight of the 288 analyzed shells is around 7 to 7.3 kg, corresponding to 2 to 2.5 kg of meat. In terms of energy content, this equals 2085 kcal, which hardly meets the daily need of a young male or an adult female. The proportions of identified taxa are almost the same for complete valves (fragments with measurable heights) as for the 429 fragments unsuitable for measurement (Fig. 14.7). Combining the data, one can calculate a rough total meat weight of 6.2 kg for the whole shell

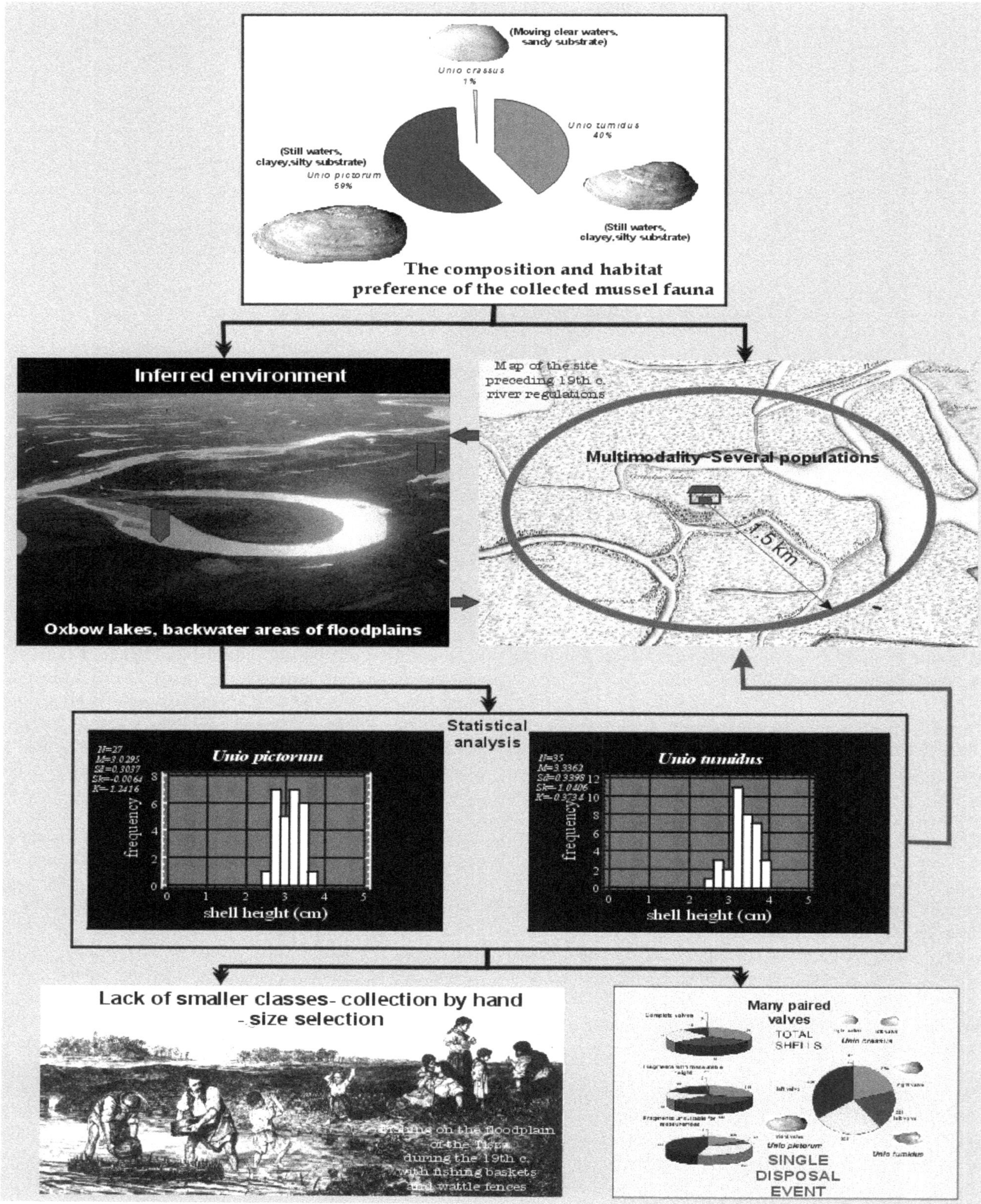

Figure 14.6. Steps in the site catchment analysis (SCA)

material (717 MNI) retrieved from the pit. Assuming an exclusive reliance on shellfish in the diet, this would provide a total nutrition of 5170 kcal. This is about the daily need of a family of two adults and two small children. Consequently, the mussels must have served only as a supplementary food source. Perhaps this occurred when other subsistence forms were not available, that is, during harsh climatic conditions or during the construction of new settlements. The amount of area needed to collect the 717 specimens must have been about 10 m^2 based on observations of the oxbow lakes of Hungarian Tisza, Sajó, and Takta rivers. These

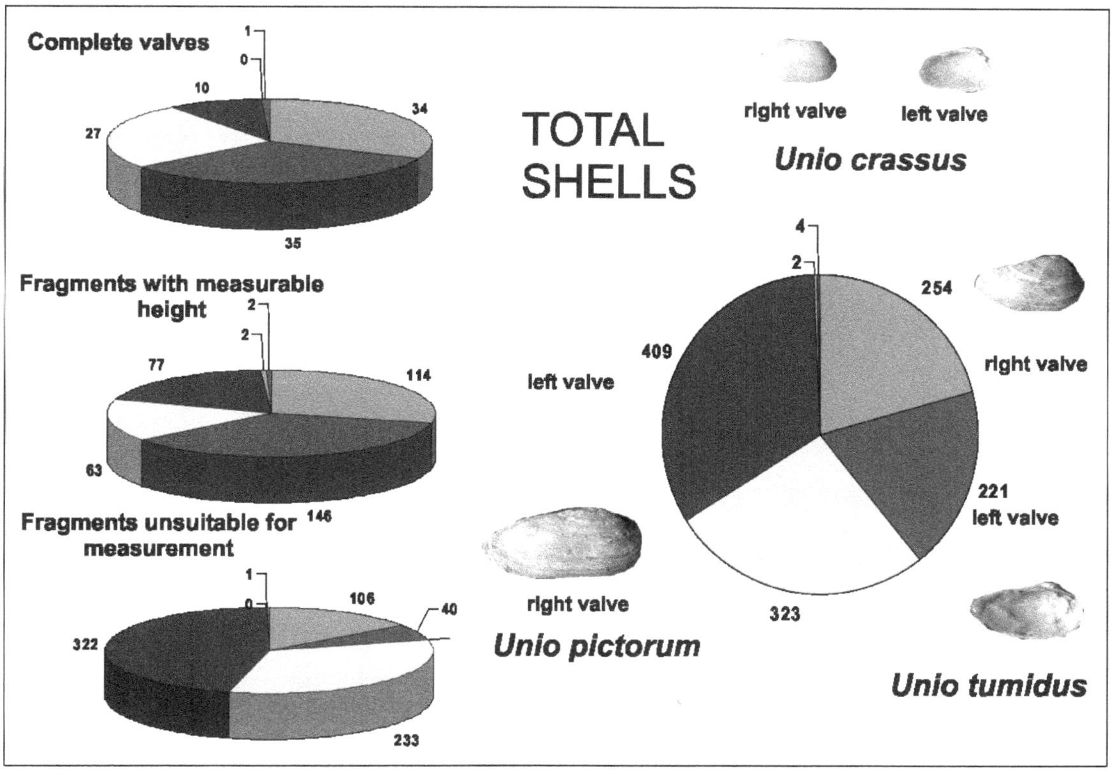

Figure 14.7. The proportion of paired valves

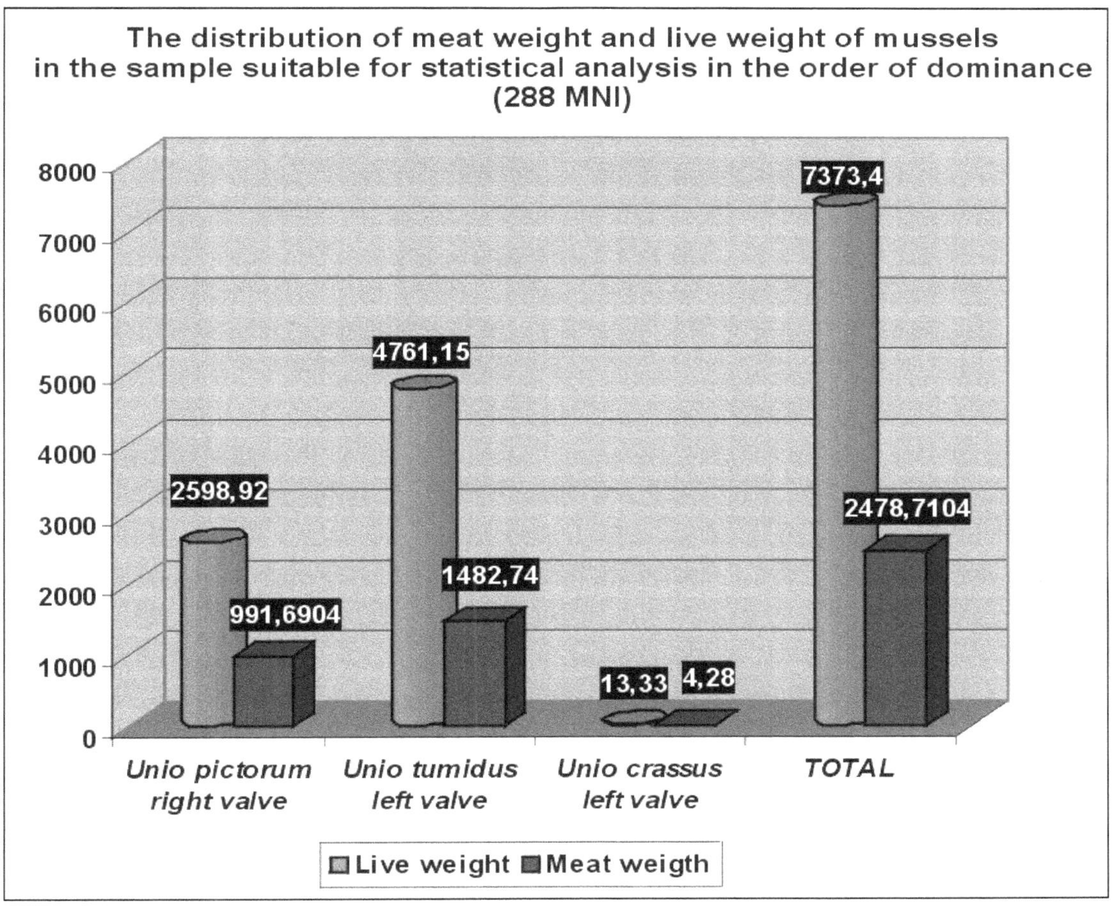

Figure 14.8. The distribution of calculated living weight and meat weight between the individual taxa and the whole sample in the order of taxon dominance

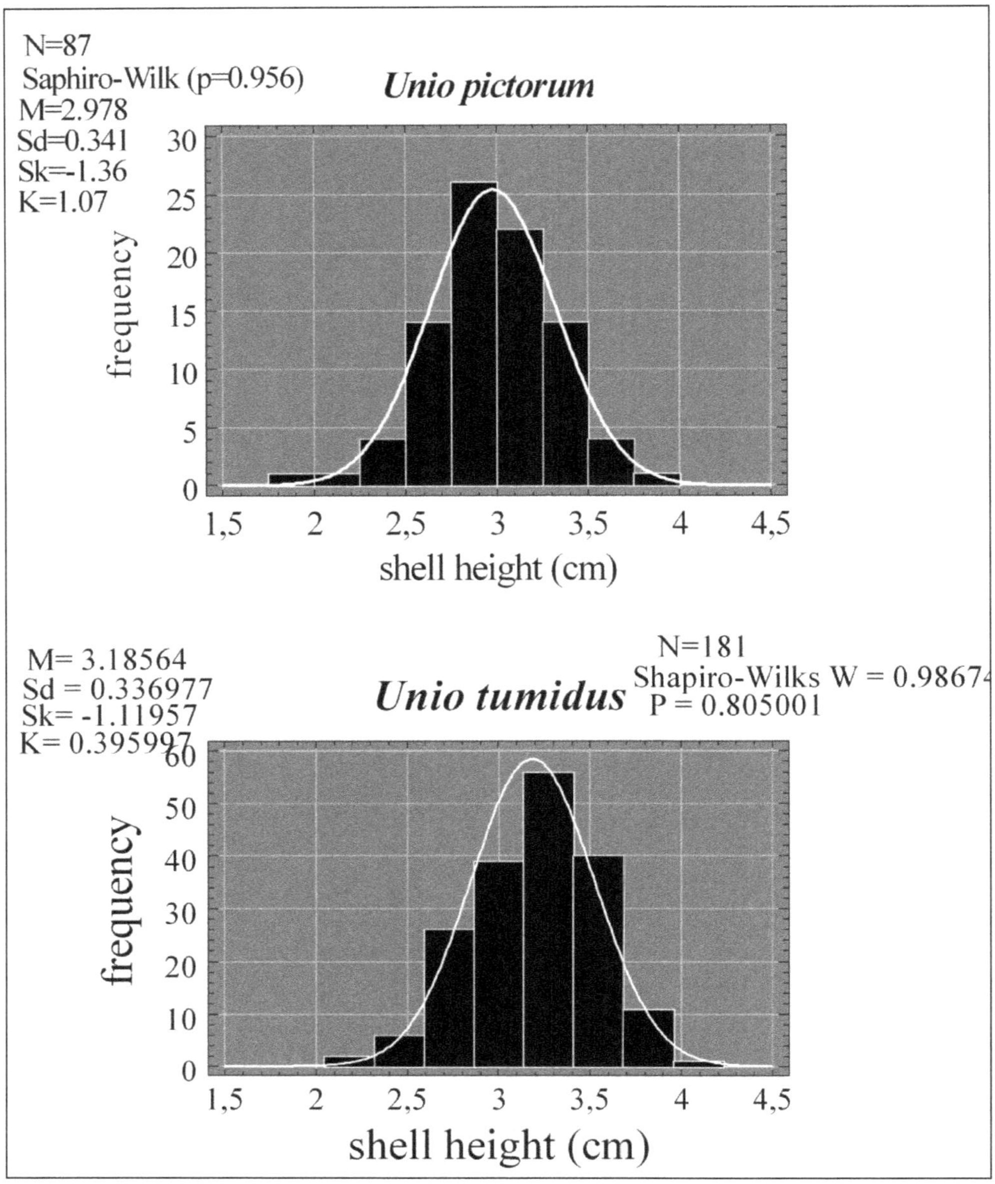

Figure 14.9. The size distribution of the two dominant species in the order of dominance within the sample suitable for statistical analysis

places have similar benthic conditions as our study area. Mussel densities range between 50-70 specimen/m^2 (Kiss 1996). Despite the small sample size, it is clear that smaller classes (< 1.9 cm height) are generally missing in the data from the dominant species (Fig. 14.9). Standard deviation values further corroborate the assumptions that primarily larger and older mussels were collected. This contrast with the specimens harvested by Native North Americas. Here, the unimodal distribution of size values for the two dominant specimens corroborates that the mussels must have come from a single population. The general lack of juvenile forms implies that collection was done by hand. Nonetheless, use of ethnographically known fishing kits and open-bottom fishing baskets cannot be ruled out entirely (Fig. 14.6).

Preliminary results suggest that mussels were collected in late spring to early summer, and may represent harvest after a harsh winter, or during construction of a new settlement. However, corroboration of these assumptions requires further investigations.

The following provides our findings for the whole profile. Shellfish comes from all sampled horizons, although the quantity varies. Foraging of mussels must have been continuous throughout the whole life of the tell (Fig. 14.10). Figure 14.10 shows the original proportional abundance of individual taxa from Horizon 24 up through Horizon 1. *Unio pictorum* is the prevalent form. This implies that the location or environment of shell-fishing remained relatively stable within a radius of 1.5-2 km around the settlement.

Four major and six minor peaks in the diagrams are marked by stars (Fig. 14.10 and Fig. 14.11). The dark stars depict shell-fishing events, when more than 50

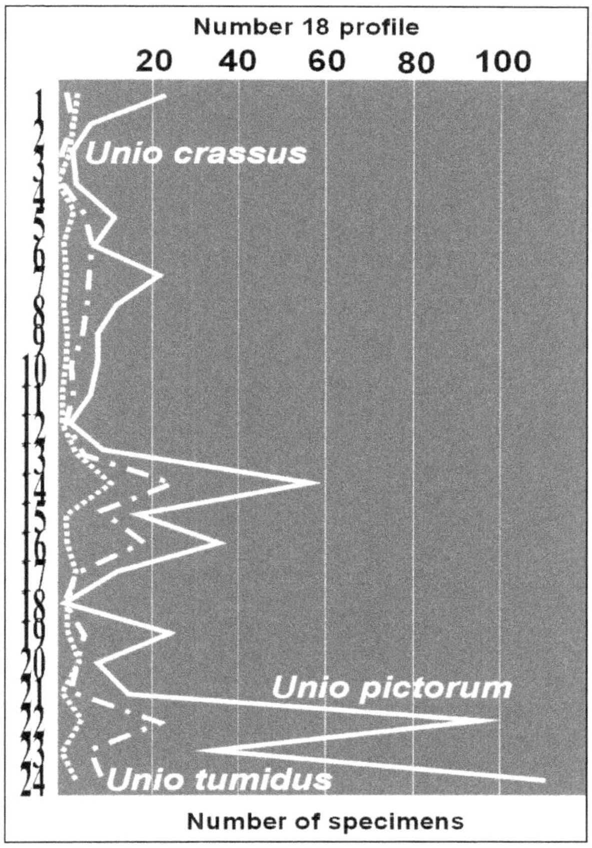

Figure 14.10. The number of collected shells according to taxa within the whole of Profile No. 18

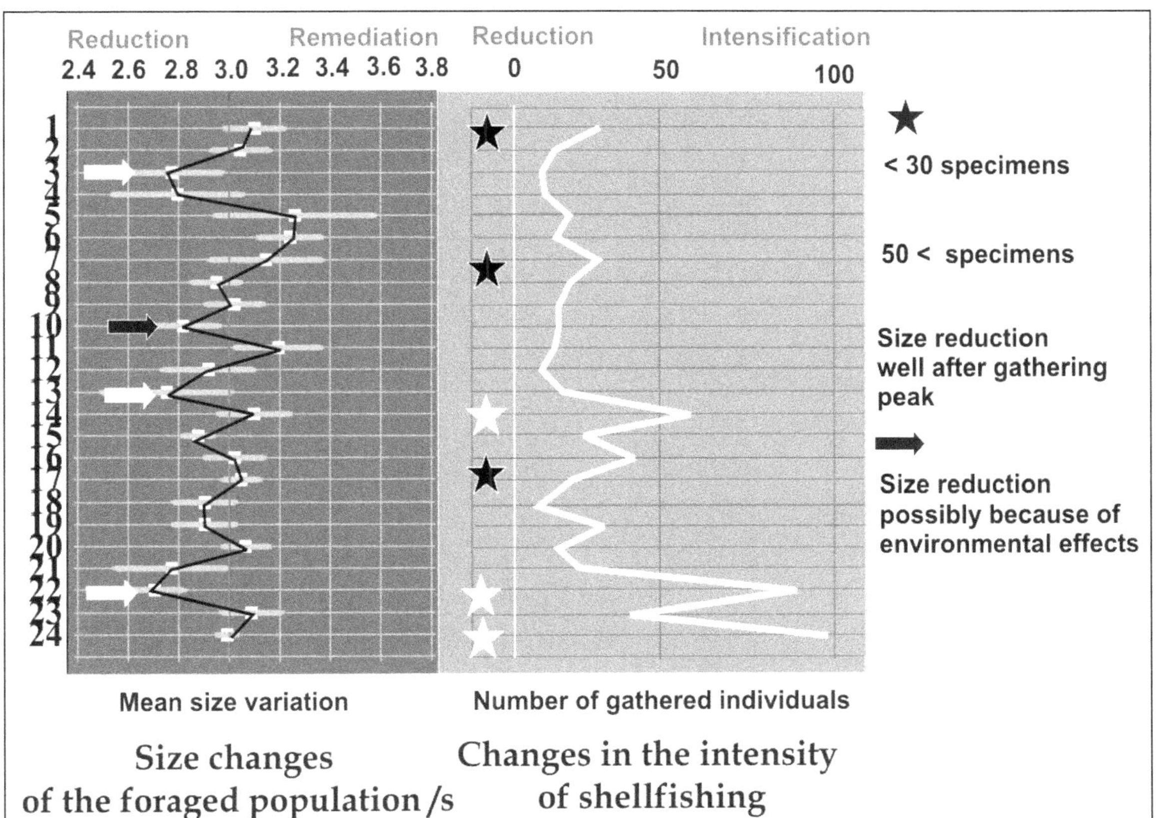

Figure 14.11. Fluctuations in the intensity of shell-fishing and concomitant alterations observable in the foraged population of the prevalent *Unio pictorum*

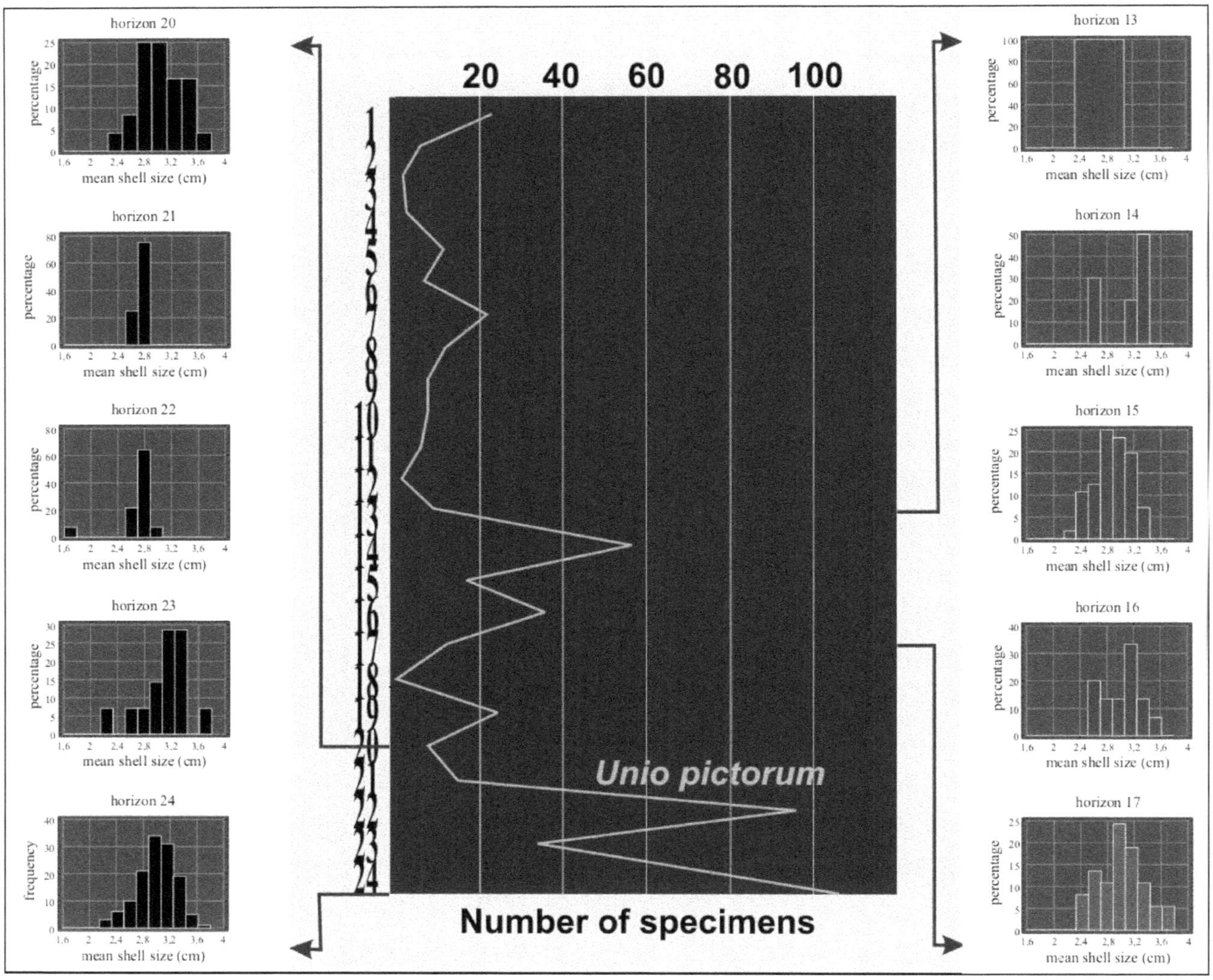

Figure 14.12. Changes in the number of exploited populations with the intensification of shell-fishing

specimen of *Unio pictorum* were harvested. Three of these major and one of the minor collection peaks were followed by a significant drop in the average size of the population (marked by dark arrows). These size reductions were followed by remediation of the foraged populations as intensity of shell-fishing declined. This appears to be the aftermath of human over-harvesting, with a delayed response of the natural population or populations (Horizons 22, 13, 3). There is a minor reduction of the mean size in Horizon 10, which is preceded by remediation of the foraged mussel population after a size reduction in Level 13. As the intensity of harvest remained relatively stable within the horizons above Level 13, this minor drop in mean size may be attributed to unfavorable growth conditions that may mark alterations in climate and/or water quality.

Another phenomenon related to the intensification of shellfish harvesting can be observed in the graphs and frequency histograms (Fig. 14.12). Here arrows indicate horizons where intensified exploitation finally led to a delayed drop in mean size of the foraged population(s). Within the lowermost, most heavily exploited horizon (Horizon 24), harvesting seems to have been originally restricted to a single population or single location as reflected by the unimodal distribution on the frequency histogram. Continuation of shell-fishing at a reduced intensity (Horizon 23) is followed by another peak at Horizon 22. This finally caused a delayed reduction of shell size in Horizons 22 and 21. This seems to have been extended to at least two populations as implied by the pattern observable in the frequency histograms for the two horizons. The unimodal distribution gradually becomes bimodal and multimodal. Taken together, this means that after the first heavy exploitation (Horizon 24), followed by another exploitation peak (Horizon 22), humans were forced to exploit new harvesting sites or populations, if they wanted to continue and/or intensify their foraging of mussels. Later exploitation intensity decreases significantly (Horizon 21 and 20), bringing about a slow remediation of the foraged populations. At that point, shell-fishing again was restricted to a single population.

When the intensity of shell-fishing is low, there is a unimodal distribution in the population (Horizons 17, 15, 13), and in the frequency histograms displaying size alterations. This means that a single location must have been in use for collecting. Conversely, with intensification of exploitation, distributions become bimodal and multimodal. Following the initial shell-fishing peak (Horizon 24), and then every other major collection peak (Horizon 16, 14, 7), humans had to rely more intensively on two or more harvesting sites simultaneously to meet their original demands.

CONCLUSION

Shells from the lowest horizon of the tell are analyzed in detail. They appear to be from a kitchen midden associated with the initial establishment of the tell (4898-4874 BC) by the Tisza culture. This analysis together with analysis of the later 23 micro-horizons leads to the conclusion that shellfish was only a secondary food source in the lives of the tell inhabitants. Therefore, we believe that intensified exploitation periods may correspond to new settlement stages, when there was an urgent need for quickly available alternative food sources, in addition to the normal agricultural products.

Acknowledgements

We are greatly indebted to Max Baldia for his kind revision and corrections to the original MS. We thank Árpád Kiss for sharing his findings and data on recent *Unionidae* with us. We also thank Félix Schubert at the University of Szeged, Department of Mineralogy, Petrology and Geochemistry for his kind assistance in the preparation of the thin-sections.

References

ANDERSEN, S.H. 1989 – Norsminde: A "Kokkenmodding" with Late Mesolithic and Early Neolithic Occupation. *Journal of Danish Archeology* 8, 1989:13-40.

BADINO, G., G. CELEBRANO and K.O. NAGEL 1991 – *Unio Elongatulus* and *Unio pictorum*: Molecular Genetics and Relationship of Italian and Central European Populations, Bulletino del Museo Regionale di Scienze Naturali, Torino 9/2, 1991:261-274.

BAKER, F.C. 1923 – The use of Molluscan Shells by the Cahokia Mound Builders. *Transactions of the Illinois State Academy of Science* 16, 1923:328-334.

BAKER, F.C. 1930 – The use of Animal Life by the Mound-Building Indians of Illinois. *Transactions of the Illinois State Academy of Science* 22, 1930:41-64.

BAKER, F.C. 1936 – Remains of Animal Life from the Kingston Kitchen Midden site near Peoria, Illinois. *Transactions of the Illinois State Academy of Science* 29, 1936:243-246.

BAKER, F.C. 1941 – A Study in Ethnozoology of the Prehistoric Indians of Illinois. *Transactions of the American Philosophical Society* 3, 1941:141-64.

BAILEY, G. 1975a – *The role of Shell Middens in Prehistoric Economies*. Unpublished Ph. D. Thesis, Cambridge University, Cambridge, 1975.

BAILEY, G. 1975b – The Role of Mollusks in Coastal Economies: The Results of Midden Analysis in Australia. *Journal of Archeological Science* 2, 1975:45-62.

BAILEY, G. 1978 – Shell Middens as Indicators of Postglacial Economies: A Territorial Perspective. In P.A. Mellars (ed.), *The Early Postglacial Settlement of Northern Europe*, Duckworth, London, 1978:37-64.

BAILEY, G. 1994 – The Weipa Shell Mounds: Natural or Cultural? In M. Sullivan, S. Brockwell and A. Webb (eds.), *Archeology in the North: Proceedings of the 1993 Australian Archeological Association Conference.* North Australia Research Unit, The Australian National University, Darwin, 1994:107-29.

BAILEY, G., and N. MILNER In Press. – The Marine Molluscs from the Norsminde Shell Midden. In S. Andersen (ed.), *Stone Age Settlement in the Coastal Fjord of Norsminde, Jutland, Denmark.*

BLITZ, J.H. 1993 – Big Pots for Big Shots: Feasting and Storage in a Mississippian Community. *American Antiquity* 58/1, 1993:80-96.

BOGAN, A.E. 1981 – A Reconstruction of the Freshwater Molluscan Fauna of the Little Tennessee River, East Tennessee. *The Bulletin of the American Malacological Union for 1981*, 1981:33-34.

BOGAN, A.E. 1993 – Freshwater Bivalve Extinctions: A Search for Causes. *American Zoology* 33, 1993:599-609.

CASEY, J.L. 1987 – Aboriginal and Modern Mussel Assemblages of the Lower Cumberland River. *Southeastern Archaeology* 6/2, 1987:115-124.

CHECA, A. 2000 – A New Model for Periostracum and Shell Formation in *Unionidae* (Bivalvia, Mollusca). *Tissue and Cell* 32/5, 405-416.

CLAASSEN, C. 1994 – Washboards, Pigtoes and Muckets: Historic Musseling in the Mississippi Watershed. *Historical Archaeology* 28/2, 2000:45-67.

CLAASSEN, C. 1998 – *Shells*. Cambridge University Press, Cambridge, 1998.

CLARK, J.G.D. 1972[1954] – *Starr Carr: A Case Study in Bioarchaeology*. Reprint. Addison-Wesley Module Publications 10. Reading, Massachusetts.

CONVEY, L.E., J.M. HANSON and W.C. MACKAY 1989 – Size Selective Predation on Unionid Clams by Muskrats. *Journal of Wildlife Management* 53, 1989:654-657.

CROWLEY, T.E. 1957 – Age Determination in Anodonta. *Journal of Conchology* 24, 1957:201-207.

CZÓGLER, K. 1934 – Édesvízi Kagylók Szegedvidéki Régészeti Leletekben. [Freshwater Mussels in the Archaeology near Szeged.] Dolgozatok a M. Kir. Ferenc József. *Tudományegyetem Archeológiai Intézetéből* IX-X, 1934:298-302.

DEITH, M.R. 1983a – Seasonality of shell collecting determined by oxygen isotope analysis of marine shells from Asturian sites in Cantabria. In C. Grigson and, J. Clutton-Brock (eds.), *Animals and Archeology 2. Shell Middens, Fishes and Birds*, British Archeological Reports, Oxford, 1983:67-76.

DEITH, M.R. 1983b – Molluscan Calendars: The use of Growth-Line Analysis to Establish Seasonality of Shellfish Collection at the Mesolithic Site of Morton, Fife. *Journal of Archeological Science* 10, 1983:423-440.

DEITH, M.R. 1984 – The Role of Shellfish in the Economy of Neolithic Sites in Apulia. *Origini*, 1984:23-44.

DEITH, M.R. 1984 – Seasonality from Shells: An Evaluation of two techniques for Seasonal Dating of Marine Molluscs. In N.R.J. Fieller, D. Gilbertson and N.G.A. Ralph (eds.), *Palaeoenvironmental Investigations: Research Design, Methods and Data Analysis, Symposia of the Association for Environmental Archeology* 5A. British Archeological Reports 258, 1984:119-130.

DETTMAN, D.L., A.K. REISCHE and K.C. LOHMANN 1999 – Controls on the Stable Isotope Composition of Seasonal Growth Bands in Aragonite Freshwater Bivalves (*unionidae*). *Geochimica et Chosmochimica Acta* 63/7-8, 1999:1049-1057.

DOMOKOS, T. 1980 – A bélmegyeri holocén (rézkori) Unio-félék statisztikai összehasonlítása recens anyaggal. [Statistical comparison of the Holocene Copper Age Unio shellfish material of Bélmegyer with that of the modern Hungarian fauna.] *Békés megyei Múzeumok Közleménye* 6, 1980:103-114.

DORSEY, R. 2000 – Archeological Interpretation of Freshwater Mussel Assemblages near the Solomon River, Kansas. *Central Plains Archeology* 8/1, 2000:25-30.

DUNCA, E., and H. MUTVEI 1996 – Periodic Microgrowth Patterns in Shells of Freshwater Unionid Bivalves. *Bulletin de l'Institut Oceanographique, Monaco*, Special Issue 14, 1996:127-131.

ENGHOFF, I.B. 1989 – Fishing from the Stone Age *settlement* Norsminde. *Journal of Danish Archeology* 8, 1989:41-50.

ENTZ, G. 1932 – A Balaton Békateknö Kagylóiról (Unionodae). [On the Naiads of Lake Balaton.] *A Magyar Biológiai Intézet I. Osztályának munkáiból, Tihany V.*, 1932:1-14.

FIELLER, N.R.J., D.D. GILBERTSON, and N.G.A. RALPH (eds.) 1985 – Palaeoenvironmental investigations: research design, methods and data analysis, Symposia of the Association for Environmental Archeology No. 5A. British Archaeological Research, Oxford 258, 1985.

GULYÁS, S. 2011 – Freshwater Shellfish as Archives in Reconstructing Subsistence Strategies of Neolithic Communities and Local, Regional Endowments of The Riparian Environment in The Carpathian Basin. Doctoral Dissertation. University Of Szeged, Szeged, Hungary, 2011.

GULYÁS, S., A. TÓTH and P. SÜMEGI 2007 – The Zooarcheological Analysis of Freshwater Bivalve Shells and their Relevance Regarding the Life of a Neolithic Community at the Körös Site of Ecsegfalva In A. Whittle (eds): *The Early Neolithic on the Great Hungarian Plain: Investigations of the Körös Culture Site of Ecsegfalva 23, Békés County*, Varia Archeologica Hungarica 21/II 2007:395-411

HANSON, J.M., W.C. MACKAY, and E.E. PREPAS 1989 – Effect of Size-Selective Mackay Predation by Muskrats on a Population of Unionid Clams. *Journal of Animal Ecology* 58, 1989:15-28.

HENDERSON, W.G., L.C. ANDERSON, and C.R. McGIMSEY 2002 – Distinguishing Natural and Archeological Deposits: Stratigraphy, Taxonomy, and Taphonomy of Holocene shell-Rich Accumulations from the Louisiana Chenier Plain, PALAIOS 17/2), 2002:192-205.

HIRST, K. 2000 – Freshwater Mussels from Nebraska Phase Sites Along the Missouri River Drainage in Southwestern Iowa *Central Plains Archeology* 8/1, 2000:38-45.

HORVÁTH, A. 1955 – Die Mollusken Fauna der Theiss. *Acta Biologica Szegediensis* I., 1955:174-180.

HORVÁTH, A. 1982 – A Gorzsai Halom késő Neolit Rétege. [The Neolithic beds of Gorzsa hill.] *Archeológiai Értesítő* 109, 1982:201-220.

HORVÁTH, F. 1987 – Hódmezõvásárhely-Gorzsa. In P. Raczky (ed.), *The Late Neolithic of the Tisza Region.* Szolnoki Múzeum, Szolnok, 1987:31-46.

JONES, J.R., and J.J. FISHER 1990 – Environmental Factors Affecting Prehistoric Shellfish Utilization, Grape Island, Boston Harbor, MA. In N.P. Lasca and J. Donohue (eds.), *Archeological Geology of North America*, GSA Centennial Special 4, 1990:137-47.

KISS, Á. 1995 – *The propagation, growth and biomass of the Chinese Huge Mussel Anodonta woodiana woodiana LEA, 1834 in Hungary.* Private Ms based on the PhD thesis, second edition, University of Agricultural Sciences of Gödöllő, Tropical and Subtropical Department, 1995:1-29.

KISS, Á. 1996 – Jelentés a Keresztényi TK, a Tiszadobi TTK Vizeinek Malakológiai Vizsgálatáról. [Report on the Mollusc Fauna of the Waters of the Natural Reserves at Keresztényi, Tiszadob.] Unpublished Masters Theses. GATE, Gödöllő, 1996.

KISS, Á. 2000 – Unonid Mussels from the River Tisza at Szeged. Unpublished Article, 2000.

KISS, Á., and J. PEKLI 1987 – A tavi Kagyló (*Anodonta cygnea* L.) és az Amuri Kagyló *Anodonta woodiana woodiana* LEA, 1834 Szaporodásának Feltételei. [The Prerequisites for the Recruitment of *Anodonta cygnea* L. and *Anodonta woodiana woodiana* LEA, 1834.] *Georgikon Napok XXI. Proceedings*, Keszthely, 1987:257-261.

KISS, Á., and J. PEKLI 1988a – Eltérések a Gyulai és a Szarvasi Amuri Kagyló *Anodonta woodiana woodiana* LEA, 1834 Méretadatai Között. [Deviations in the Metrics of *Anodonta woodiana woodiana* LEA, 1834 Deriving from the Sites at Gyula and Szarvas SE Hungary.] *SOOSIANA* 16, 1988:15-18.

KISS, Á., and J. PEKLI 1988b – On the Growth Rate of *Anodonta woodiana woodiana* LEA, 1834 (Bivalvia: *Unionacea*). *Bulletin of the University of Agricultural Science* 1, Gödöllő, 1988:119-124.

KISS Á., and E. PETRO 1992 – Distribution and Biomass of some Chinese Mussel Population (*Anodonta woodiana woodiana* LEA, 1834; Bivalvia *Unionacea*) in Hungary. *11th International Malacological Congress at Siena ITALY, Abstract* 245, 1992.

KISS Á., J. PEKLI and S. HÉJJA 1988 – Takarmányozási célú Kagylótenyésztés. [Shellfish Propagation for Fodder.] *Halászati Tudományos Tanácskozás XII*, Szarvas, Abstract 13, 1988.

KLIPPEL, W.E., G. CELMER and J.R. PURDUE 1978 – The Holocene Naiad Record at Rodgers Shelter in the Western Ozark Highlands of Missouri. *Plains Anthropologist* 23/82, Part 1, 1978:257-271.

LIPPINCOTT, K., and L.B. DAVIS 2000 – A Prehistoric Freshwater Mussel Collection from the Schmitt Chert Mine Site (24BW559) near Three Forks, Montana. *Central Plains Archeology* 8/1, 2000:70-79.

LUBY, E., and M.F. GRUBER 1999 – The Dead must be Fed: Symbolic Meanings of the Shellmounds of the San Francisco Bay Area. *Cambridge Archeological Journal* 9, 1999:95-108.

MATTESON, M.R. 1958 – Analysis of an Environment as Suggested by Shells of Freshwater Mussels Discarded by Indians of Illinois. *Transactions of the Illinois State Academy of Sciences* 51, 1958:8-13.

MATTESON, M.R. 1959 – An Analysis of the Shells of Fresh-Water Mussels Gathered by Indians in Southwestern Illinois. *Transactions of the Illinois Academy of Science* 52, 1959:52-58.

MATTESON, M.R. 1960 – Reconstruction of Prehistoric Environments Through Analysis of Molluscan Collections from Shell Middens. *American Antiquity* 26, 1960:117-120.

MEEHAN, B. 1982 – *Shell Bed to Shell Midden.* Australian Institute of Aboriginal Studies, Canberra, 1982.

MILNER, N. 2001 – At the Cutting Edge: Using Thin-Sectioning to Determine Season of Death of the European Oyster, *Ostrea edulis*. *Journal of Archeological Science* 28, 2001:54-68.

MULLER, J. 1986 – *Archaeology of the Lower Ohio River Valley*. Academic Press, Orlando.

MURPHY, J.L. 1971 – Molluscan Remains from four Archaeological Sites in Northeastern Ohio. *Sterkiana* 43, 21-25.

MUTVEI, H., and T. WESTERMARK 2001 – How Environmental Information can be Obtained from Naiad Shells? In G. Bauer and K. Wachtler (eds.), *Ecology and Evolution of the Freshwater Mussels Unionidae*. Ecological Studies 145, Springer, Berlin, 2001:367-379.

MUTVEI, H., T. WESTERMARK, E. DUNCA, B. CARELL, S. FORBERG and A. BIGNERT 1994 – Methods for Study of Environmental Changes Using Structural and Chemical Information in Molluscan Shells. *Bulletin de l'Institut Oceanographique, Monaco*, Special Issue 13, 1994:163-191.

MYERS, T.P., and K. PERKINS 2000 – Mussels and Marginal Utility. *Central Plains Archeology* 8/1, 2000:52-59.

NAGEL, K.O., and G. BADINO 2001 – Population Genetics and Systematics of European *Unionidae*. In G. Bauer and K. Wachtler (eds.), *Ecology and Evolution of the Freshwater Mussels Unionidae*. Ecological Studies 145, Berlin, Springer, 2001:51-82.

NEVES, R.J., and S.N. MOYER 1988 – Evaluation of Techniques for Age-Determination of Freshwater Mussels (*Unionidae*). *American Malacological Bulletin* 6/2, 1988:179-188.

PARMALEE, P.W. 1956 – A Comparison of Past and Present Populations of fresh-Water Mussels in Southern Illinois. *Illinois Academy of Science Transactions* 49, 1956:184-192.

PARMALEE, P.W. 1988 – A Comparative Study of late Prehistoric and Modern Molluscan Faunas of the Little Pigeon River System, Tenessee. *American Malacological Bulletin* 6/2, 1988:165-179.

PARMALEE, P.W. 1994 – Freshwater mussels from Dust and Smith Bottom Caves, Alabama. In N.S. Goldman-Finn and B.N. Driskell (eds.), Preliminary Archaeological Papers on Dust Cave, Northwest Alabama. *Journal of Alabama Archaeology* 40 (1 and 2):135-162.

PARMALEE, P.W. 1998 – *The Freshwater Mussels of Tennessee*, University of Tennessee Press, Knoxville, 1998.

PARMALEE, P.W., and A.E. BOGAN 1986 – Molluscan Remains from Aboriginal Middens at the Clinch River Breeder Reactor Plant Site, Roane County, Tennessee. *American Malacological Bulletin* 4/1, 1986:25-37.

PARMALEE, P.W., and W.E. KLIPPEL 1974 – Freshwater Mussels as a Prehistoric Food Resource. *American Antiquity* 39/3, 1974:421-434.

PARMALEE, P.W., and W.E. KLIPPEL 1986 – A Prehistoric Aboriginal Freshwater Mussel Assemblage from the Duck River in Middle Tennessee. *The Nautilus* 100/4, 1986:134-140.

PARMALEE, P.W., A.P. ANDREAS and N. WILSON 1972 – *Animals Utilized by Woodland Peoples Occupying Apple Creek Site, Illinois.* Illinois State Museum, Reports of Investigations 23, Springfield, 1972.

PARMALEE, P.W., W.E. KLIPPEL, and A.E. BOGAN 1980 – Notes on the Prehistoric and Present Status of the Naiad Fauna of the Middle Cumberland River, Smith County, Tennessee. *The Nautilus* 94/3, 1980:93-105.

PARMALEE, P.W., W.E. KLIPPEL, and A.E. BOGAN 1982 – Aboriginal and Modern Freshwater Mussel Assemblages (Pelecypoda: *Unionidae*) from the Chickamauga Reservoir, Tennessee. *Brimleyana* 8, 1982:75-90.

PEACOCK, E. 1996 – Future Directions in the Analysis of Freshwater Bivalves in Archaeology. *Assemblage* 1, http://www.assemblage.group.shef.ac.uk/1/peacock.html (Last accessed July 7, 2009).

PEACOCK, E. 1997 – Current and Future Directions in the Analysis of Freshwater Bivalves in Archaeology. In C.M. McNutt (ed.), Results of Recent Archaeological Investigations in the Mid-South: Proceedings of the 17th Annual Mid-South Conference, Memphis, Tennessee: June 29-30, 1996, The University of Memphis, Anthropological Research Center, *Occasional Paper* 18, 1997:71-93.

PEACOCK, E. 2000 – Assessing Bias in Archaeological Shell Assemblages. *Journal of Field Archaeology* 27/2, 2000:183-196.

PEACOCK, E. 2002 – Shellfish Use During the Woodland Period in the Middle South. In D.G. Anderson and, R. Mainfort (eds.), *The Woodland Southeast*, The University of Alabama Press, Tuscaloosa, 2002:444-460.

PEACOCK, E., and S. CHAPMAN 2001 – Taphonomic and Biogeographic Data from a Plaquemine Shell Midden on the Quachita River, Louisiana. *Southeastern Archeology* 20, 2001:44-55.

PEACOCK, E., and T.R. JAMES 2002 – A Prehistoric Unionid Assemblage from the Big Black River Drainage in Hinds County, Mississippi. *Journal of the Mississippi Academy of Sciences* 47/2, 2002:119-123.

PETERSEN, K.S. 1986 – The Ertebølle Køkkenmøddig and the Marine Development of the Limfjord, with Particular Reference to the Molluscan Fauna. *Journal of Danish Archeology* 5, 1986:77-84.

PICHA, P.R., and F.E. SWENSON 2000 – Freshwater Shell Tool/Ornament Production and Resource Use in the Middle Missouri Subarea of North Dakota. *Central Plains Archeology* 8/1, 2000:85-94.

PONYI, J. 1990 – Az *Unionidae család* (Mollusca: Bivalvia) elterjedése és tömege a Balatonban. [The distribution and biomass of *Unionidae* in Lake Balaton.] *Állattani Közlemények* 76, 1990:91-97.

PONYI, J., M. REHÁK, and L. GELENCSÉR 1981 – Három balatoni Unio-faj (U. crassus Retzius, *U.tumidus* Retzius, *U. pictorum* Linne) Héjméreteinek és Testsúlyának Viszonya. [The Relationship Between shell Parameters and wet Weight in Three Unionid Taxa from Lake Balaton.] *Állattani Közlemények* 68, 1981:129-130.

QUITMYER, I.R., D.S. JONES and W.S. ARNOLD 1997 – The Sclerochronology of hard Clams, *Mercenaria* spp., from the South-Eastern USA: A Method of Elucidating the Zooarchaeological Records of Seasonal Resource Procurement and seasonality in Prehistoric Shell Middens. *Journal of Archaeological Science* 24/9, 1997:825-840.

RAVERA, O., and A.R. SPROCATI 1997 – Population Dynamics, Production, Assimilation and Respiration of Two Fresh Water Mussels: *Unio mancus Zhadin* and *Anodonta cygnea Lam. Memorie dell'Istituto Italiano di Idrobiologia* 56, 1997:113-130.

RAVERA, O., A.R. SPROCATI and L. VIDO 2003 – Metal Concentrations of *Unio pictorum mancus* from 12 N Italian Lakes in Relation to their Trophic Level. *Journal of Limnology* 62/2, 2003:121-138.

RICHARDSON, C.A. 2001 – Molluscs as Archives of Environmental Change. *Oceanography and Marine Biology* 39, 2001:103-64.

RICHARDSON, C.A., S.A. COLLIS, K. EKARATNE, P. DARE and D. KEY 1993 – The Age Determination and Growth Rate of the European Flat Oyster, *Ostrea edulis* in British Waters Determined from Acetate Peels of Umbo Growth Lines. *ICES Journal of Marine Science* 50/4, 1993:493-500.

RICHARDSON, T.D., and P. YOKELY 1996 – A Note on Sample Technique and Evidence of Recruitment in Freshwater Mussels (*Unionidae*). *Archiv für Hydrobiologie* 137, 1996:135-140.

RICHNOVSZKY, A., and L. PINTÉR 1979 – A vízi csigák és kagylók (mollusca) kishatározója. [Handbook of freshwater mussels and snails.] Vízügyi Dokumentáció és Továbbképző Intézet. *Vízügyi Hidrológia* 6, 1979:135-144.

ROBISON, N.D. 1983 – Archeological Records of Naiad Mussels along the Tennessee-Tombigbee Waterway. In A.C. Miller (compiler), *Report of Freshwater Mussels Workshop*, October 26-27 1982. U.S. Army Corps of Engineers, Waterways Experiment Station, Environmental Laboratory, Vicksburg, Mississippi, 1983:115-129.

ROWLEY-CONWY, P. 1983 – Sedentary Hunters: The Ertebølle Example. In G.N. Bailey (ed.), *Hunter-Gatherer Economy in Prehistory*. Cambridge University Press, Cambridge, 1983:113-126.

SATO, S. 1982 – Yarborough Site Faunal Remains. In C. Solis and R. Walling Archaeological Investigations at the Yarborough Site (22Cl814), Clay County, Mississippi. University of Alabama, Office of Archaeological Research, *Report of Investigations* 30, 1982:140-152.

SATO, S. 1999 – Temporal Change Of Life-History Traits in Fossil Bivalves: On the Example of *Phacosoma japonicum* from the Pleistocene of Japan. *Paleogeography, Paleoclimatology, Paleoecology* 154, 1999:313-323.

SEBESTYÉN, O. 1939 – A Balatoni Najádok Növekedéséröl. [On the Growth of the Naiads of Lake Balaton.] *Tihany* 11, 1939:259-271.

SOÓS, L. 1943 – *A Kárpát-Medence Mollusca-Faunája*. [The mollusc fauna of the Carpathian Basin.]

Magyarország Természetrajza, Magyar Tudományos Akadémia, Budapest.

SÜMEGI, P. 1999 – Csigák és Kagylók a Régészeti Kutatásokban I. [Mussels and Snails in Archaeology I.] *Természet Világa* 130/10, 1999:454-457.

SÜMEGI, P. 2003 – *A Régészeti Geológia és a Történeti Ökológia Alapjai.* [Introduction to Geoarcheology and Historical Ecology], JatePress, Szeged.

SÜMEGI, P., and R. KERTÉSZ 1998 – A Kárpát-medence Õskörnyezeti Sajátosságai: Egy Ökológiai Csapda az Újkõkorban? [The Paleoenvironmental Characteristics of the Carpathian Basin- an Ecological Trap During the Neolithic?] *Jászkunság* 44, 1998:144-157.

SÜMEGI, P., and R. KERTÉSZ 2001 – Palaeogeographic Characteristic of the Carpathian Basin: An Ecological Trap During the Early Neolithic? In R. Kertész and J. Makkay (eds.), *From the Mesolithic to the Neolithic.* Archaeolinqua Press, Budapest, 2001:405-416.

SÜMEGI, P., J. KOZÁK and C.S. TÓTH 1996 – Jelentés a Polgár: Kenderföld Bronzkori Tell Hulladékgödrébõl Származó Kagylók Archeozoológiai Feldolgozásáról. [Report on the Findings of the Shellfish Material Retrieved from the Refuse Pit of a Bronze Age Tell at Polgár-Kenderföld.] *Report to the Hajdú-Bihar Megyei Múzeumok Igazgatósága, Debrecen,* 23, Debrecen, 1996.

SWADLING, P. 1976 – Changes Induced by Human Exploitation in Prehistoric Shellfish Populations. *Mankind* 10, 1976:156-62.

TAYLOR, R.W. 1989 – Changes in Freshwater Mussel Populations of the Ohio River: 1,000 BP to Recent Times. *Ohio Journal of Science* 89/5, 1989:188-191.

TAYLOR, R.W., and B.D. SPURLOCK 1982 – The Changing Ohio River Naiad Fauna: A Comparison of Early Indian Middens with Today. *The Nautilus* 96/2, 1982:49-51.

THELER, J.L. 1987 – Prehistoric Freshwater Mussels (Naiads) from Brogley Rockshelter in SW Wisconsin. *American Malacological Bulletin* 5/2, 1987:165-173.

THELER, J.L. 1990 – Prehistoric Freshwater Mussel (Naiad) Assemblages from SW Iowa. *American Malacological Bulletin* 7/2, 1990:127-131.

THELER, J.L. 1991 – Aboriginal Utilization of Freshwater Mussels at the Aztalan Site, Wisconsin. In J.R. Purdue, W. Klippel E. and B.W. Styles (eds.), *Beamers, Bobwhites, and Blue-Points: Tributes to the Career of Paul W. Parmalee,* University of Tennessee, Department of Anthropology, *Report of Investigations* 52, 1991:315-332.

TIMM, H., and H. MUTVEI 1993 – Shell Growth of the Freshwater Unionid *Unio crassus* from Estonian Rivers. *Proceedings of the Estonian Academy of Sciences: Biology* 42, 1993:55-67.

TÓTH, A. 2003 – *Hódmezõvásárhely-Gorzsa késõ Neolit Tell Kagylóanyagának Archeozoológiai Szerepe.* [*On the Importance of Shellfish in the Life of the Tell Community of Hódmezõvásárhely-Gorzsa.*] Unpublished Masters of Science Thesis, University of Szeged, Department of Geology and Paleontology. Szeged, 2003.

TÓTH, B. M., and K. BÁBA 1980 – The Molluscan Fauna in the Bed of Tisza and its Tributaries. *Tiscia (Szeged)* XV, 1980:143-149.

TUDORANCEA, C., and M. FLORESCU 1968 – Considerations Concerning the Production and Energetics of *Unio tumidus Philipsson* Population from the Crapina Marsh. *Travaux du Muséum d'Histoire Naturelle "Grigore Antipa"* VIII, 1968:395-409.

TURGEON, D.D., A.E. BOGAN, E.V. COAN, W.K. EMERSON, *et al.* 1998 – Common and Scientific Names of Aquatic Invertebrates from the United States and Canada: Mollusks, 2nd edition. *American Fisheries Society Special Publication.* Bethesda, Maryland, 1998.

WASELKOV, G.A. 1987 – Shellfish Gathering and Shell Midden Archeology. In M.B. Schiffer (ed.), *Advances in Archeological Method and Theory* 10. Academic Press, New York, 1987:93-171.

WILLIAMS, J.D., M.L. WARREN, K.S. CUMMINGS, J.L. HARRIS *et al.* 1993 – Conservation Status of Freshwater Mussels of the United States and Canada. *Fisheries* 1/9, 1993:6-22.

WARREN, R.E. 1975 – *Prehistoric Uniacean Utilization at the Widows Creek site, Northeastern Alabama.* Unpublished M.A. Thesis. University of Nebraska, Lincoln, 1975.

WARREN, R.E. 2000 – Prehistoric Procurement and Use of Freshwater Mussels Along the Missouri River in the Northern Great Plains. *Central Plains Archeology* 8/1, 2000:60-69.

WESTERMARK, T., B. CARREL, S. FORBERG, H. MUTVEI and E. KULAKOWSKI 1996 – Freshwater Unionid Shells as Environmental Archives: Methodology and Observations. *Bulletin de l'Institut Oceanographique, Monaco, Special Issue* 14, 1996:73-81.

WHITTLE, A. 1999 – *The Early Neolithic of Hungary.* http://www.cf.ac.uk/hisar/archaeology/projects/hungary/ (Last accessed July 7, 2009).

WHITTLE, A. 2000 – New Research on the Hungarian Early Neolithic. *Antiquity* 74, 2000:13-14.

WHITTLE, A. (ed.) 2007 – *The Early Neolithic on the Great Hungarian Plain: investigations of the Körös culture site of Ecsegfalva 23, County Békés.* Varia Archaeologica Hungarica XXI, Archaeological Institute of the Hungarian Academy of Sciences and the School of History and Archaeology, University of Cardiff. Budapest and Cardiff.

WHITTLE, A., I. ZALAI-GAÁL and P. SÜMEGI 2001 – *Körös Culture Environment, Settlement and Subsistence-Preliminary Report on the Second Season of an Interdisciplinary Project at Ecsegfalva, County*

Békés, Hungary. http://www.cardiff.ac.uk/hisar/archaeology/reports/koros/ (Last accessed July 07, 2009).

ZAHNER-MEIKE, E., and J.M. HANSON 2001 – Effect of Muskrat Predation on Naiads. In G. Bauer and K. Wachtler (eds.), *Ecology and Evolution of the Freshwater Mussels Unionidae, Ecological Studies* 145, Springer, Berlin, 2001:163-184.

Chapter 15
IMPRINTS OF THE ANTHROPOGENIC INFLUENCES IN A PEAT BOG FROM TRANSDANUBIA, HUNGARY

Imola E. JUHÁSZ
Institute of Archeology, Hungarian Academy of Sciences, Budapest, Hungary

Abstract: *The Holocene vegetation history is reconstructed for the Hahót Basin, a large swampy area of the Zala Hills region in southwest Hungary. The palynological data are correlated with this regional archaeological settlement information. The human impact on the vegetation dynamics are assessed, beginning with the Copper Age and ending with the Migration Period. The changes in vegetation during this time appear to be primarily due to human agency rather than climate change.*

INTRODUCTION

This chapter reconstructs the Holocene vegetation history of the Hahót Basin located west of the Danube (Transdanubia) in Hungary. The area is part of the Carpathian Basin. The basin's central part has a transitional environmental and climate zone wedged between the Balkan Peninsula and Western Europe. The Balkan and West European climate and environment were significantly different beginning during the early parts of the Holocene. The transition between the two regions is expressed in vegetation and soil conditions, resulting in a micro-scale mosaic environment. Pleistocene lag surfaces, meridional valleys and alluvial plains possess different subsoil and morphological conditions. These differences affected the lives of numerous cultures and groups of people that have lived in this part of Europe for millennia (Bánffy 2000, 2004, 2006, Sümegi and Kertész 2001). The different micro-environments promoted differing settlement strategies starting in the Neolithic (Sümegi 2004, Sümegi *et al.* 2008).

The paleoecological research presented blow was preceded by a large-scale archaeological investigation, which was carried out from 1986 to 1994. The settlement density was low in the Hahót Basin when compared to the southern Great Hungarian Plain with its excellent climatic and soil conditions. A total of 78 habitation sites were identified in a 120 km^2 area. Nevertheless, the number of sites and the fact that about two-thirds of them averaged three occupations (including cemeteries) and several were occupied in eight to 11 different periods suggests a continuous occupation (Szőke 1995). This and the paleoecological research results lead to the conclusion that the changes of the vegetation dynamics from the beginning with the Copper Age to the end of the Migration Period were primarily the result of human impact instead of climatic change.

SITE DESCRIPTION

The Pötréte Site is situated on the west side of Lake Balaton (Fig. 15.1), in the proximity of the Kis-Balaton Nature Preserve of the Hahót Basin. The western part of this region is characterized by the meridional valleys of the Zala Hills. One of these valleys is the Hahót Basin, which is adjacent to the Szévíz Valley where additional pollen profiles were analyzed (Juhász *et al.* 2001, Juhász 2002). This basin is surrounded by the rolling hills of Zala, which run north to south. The stream valleys run in a meridional direction following a joint basin, and include the Szévíz and the Principális canals constructed by the Romans. The Digital Landscape Model prepared by Gábor Timár (Fig. 15.2) is based on a satellite image from 1990, prior to inundation of the Kis-Balaton. Today, the entire western Balaton is covered with water.

The archeological history of the western part of Balaton is complicated by the terrestrial islands and smaller elevations in the valleys (Fig. 15.3). The archeological sites of Zalavár (Szőke and Vándor 1986, Juhász 2004) near Kis-Balaton and Vörs-Mária asszony-sziget (Kalicz *et al.* 1998) are also situated on a terrestrial island some meters above the present water level.

MATERIAL AND METHODS

Coring locations were plotted on a 1:2880 scale map and sampled by the National Peat Cadastrial Survey. Peat coring locations were based on suggestions made by J.L. de Beaulieu, M. Reille, and T. Frankovics during the summer of 1998. Palynological analyses were undertaken by Institut Mediterranéen d'École de Palynology (IMEP) in the Marseille Laboratory as part of my doctoral research. The boreholes were extended using a Russian peat sampler, which excludes younger pollen contamination and guarantees the acquisition of intact sediment cores. Chemical treatments followed the standard laboratory method of Erdtman (1936), and included use of Cd-I$_2$ treatment (Thoulet, Nakagawa *et al.* 1998). When necessary, additional sieving of organic material was conducted to remove coarse debris and to facilitate pollen counts.

With the cooperation of M. Pattern, five ^{14}C samples (Gif-100247 to Gif-100251) from a 4 m long sediment core were sent to Gif-sur-Yvette for AMS dating.

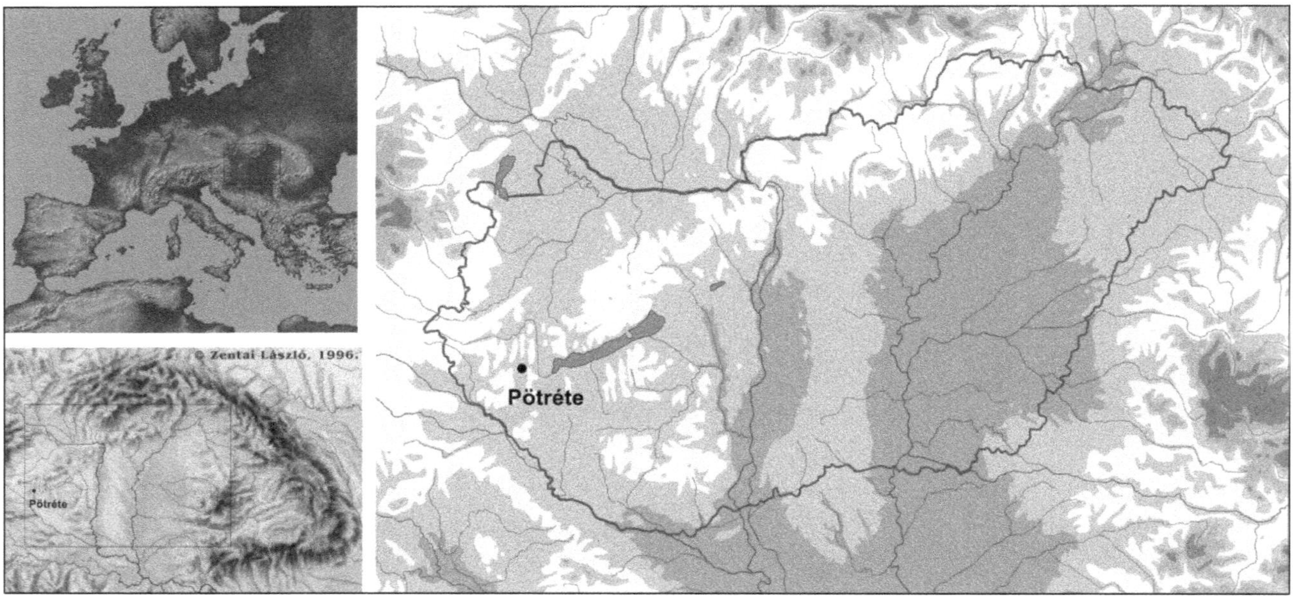

Figure 15.1. Location of the Pötréte marshland and the Carpathian Basin

Figure 15.2. Three-dimensional map of the Hahót Basin (prepared by Gábor Timár)

Unfortunately, all dates must be rejected due to an evident inversion of the dates. The base of the sequence (between 295 cm and 255 cm depth) is considered to have accumulated between 12000 bp and 11000 bp (11900-10960 BC).[1] Several indicators point to a large hiatus at the depth of 250 cm, coinciding with the middle Subboreal at ca. 5000 bp (3800 BC). Comparison of the pollen assemblage with other palynological results from the same region (Juhász *et al.* 2001, Juhász 2002, 2004 Medzihradszky 2001) indicates a decrease of linden and hazel (*Tilia* and *Corylus*) before the appearance of the continuous curve of hornbeam (*Carpinus)* at 230 cm depth. This layer is considered to have been deposited around 4800 bp (3550 BC). A decrease of beech (*Fagus*) and *Carpinus*, followed by the increase of oak (*Quercus*) at 110 cm, appears to occur at around 2700 bp (890/810 BC). The topmost part of the sequence is dated to 1200 bp (775-880 AD).

[1] Editors' note: Uncalibrated dates are listed as bp and calibrated dates are given as BC. The uncalibrated bp ^{14}C dates have been calibrated by the senior editor, using atmospheric data from Reimer *et al.* (2004) with OxCal version 3.10 (Bronk Ramsey 2005). Only the tabulated "singe range" at ±0 is given.

THE LOCAL POLLEN ASSEMBLAGE ZONES OF PÖTRÉTE (LPZ PTA)

PTA-a (295-250 cm) (12000-11000 bp/11900-10960 BC) *Poaceae-Pinus* phase

This pollen zone is characterized by a relatively high amount of Scots pine (*Pinus sylvestris*). Meanwhile the thermophilous tree taxa, except for *Corylus*, are only sporadically represented. This suggests the presence of coniferous woodland with birch (*Betula*). The relatively

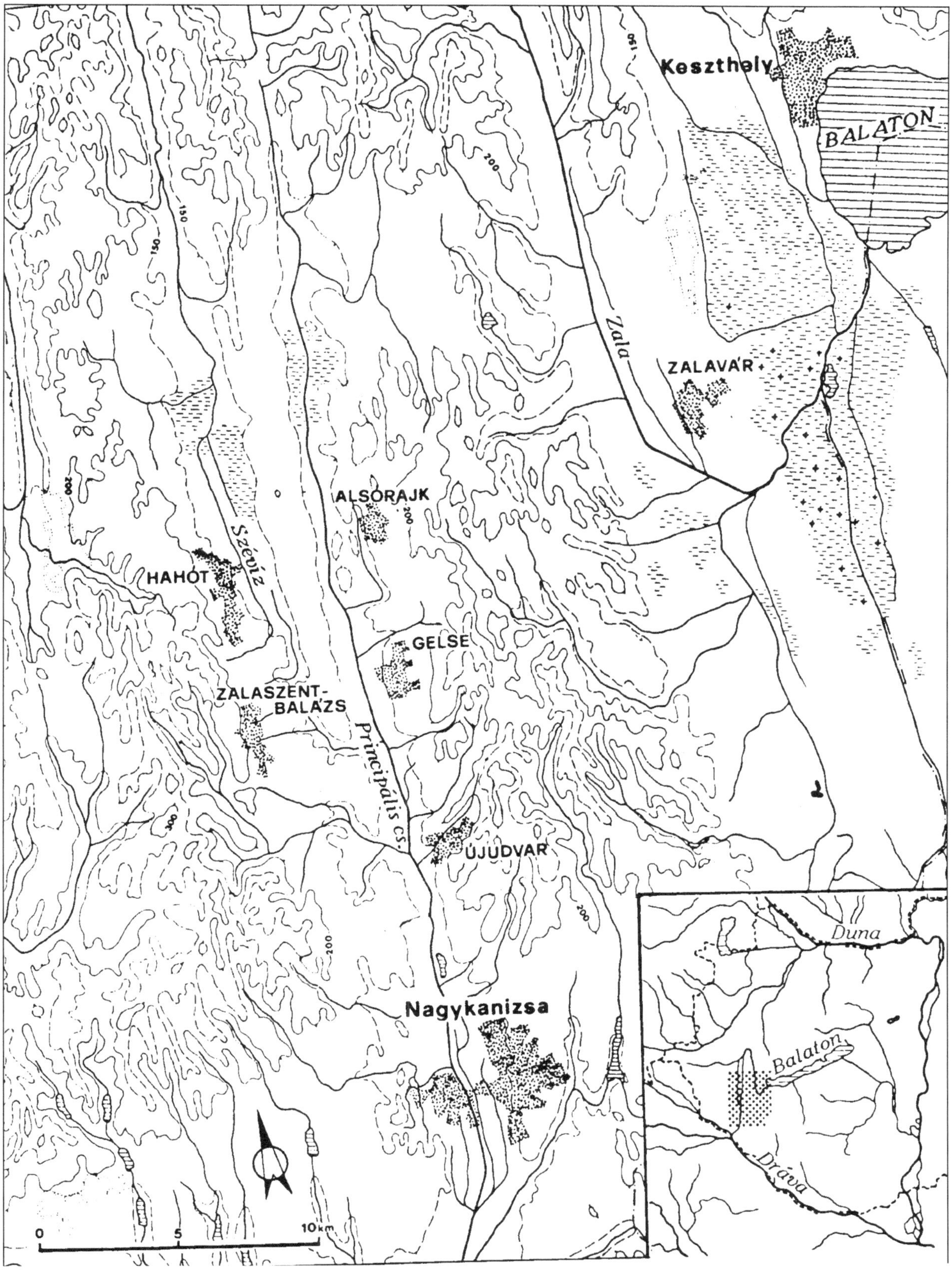

Figure 15.3. Contour map of the Hahót Basin

high presence of *Corylus*, the continuous presence of mugworth (*Artemisia*), and other herbaceous taxa suggest that these layers were deposited during the Allerød. A low number of taxa are represented in the samples. Palynomorphs, such as reeds (*Cyperaceae*), Pteridophytes with Monolete spores (*Thelypteris palustris*), and *Sparganium*/*Typha* are sporadically present, which implies deposition during a period of colder climate. The

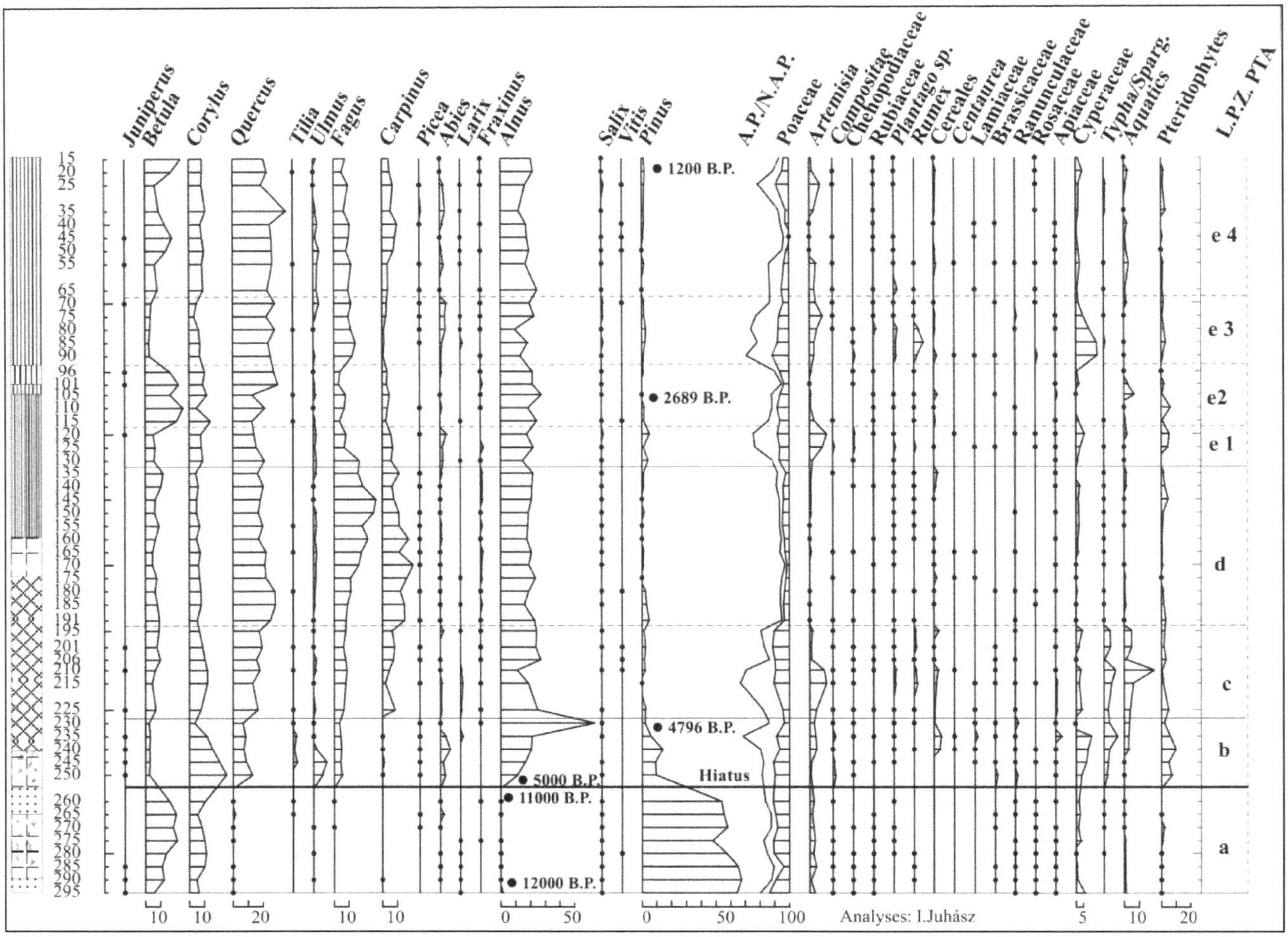

Figure 15.4. Simplified percentage pollen diagram of the Pötréte peat section and location of ^{14}C samples

sediment consists of a carbonated marne and gyttja with shells from a local open surface pond. This period is followed by a large, approximately 6000 year-long, gap in time, lacking sedimentation (Fig. 15.4).

PTA-b (250 – 225 cm) (5000-4700 bp or 3800-3500/3400 BC) Mixed Oak Woodland

This zone is characterized by a mature oak forest with all its usual taxa. The pollen assemblage for this zone is typical for the Middle Holocene, or the Subboreal in Central Europe. Under closed forest canopy arboreal species dominate. Continuous and stable values for pine (*Pinus sylvestris*) pollen grains decrease towards the end of the zone, where mixed oak woodland is suggested. *Quercus*, *Ulmus* (elm), and *Tilia* (Linden) appear at this point in the profile. Pollen grains from *Fagus* are also found, but *Carpinus* is present only regionally and sporadically in this part of the pollen diagram. Among the herbaceous species, grasses (*Poaceae*) dominate over other steppe elements that show up only in small quantities at the beginning of this period. Following the decrease of arboreal pollen (A.P.), *Poacea* becomes increasingly common in the vegetation. An increase of anthropogenic species, such as *Artemisia* (mugworth), *Chenopodiaceae* (goosefoot), and *Plantago* sp. (plantain), and the high level of the cereals found at the end of this zone indicate the presence of intensive agricultural activities. Since the heavy cereals pollen grains are disseminated only over short distances, the fields under cultivation would have been close to the site (Behre 1988). Chronologically, the deposit develops at the end of the Balaton-Lasinja culture and the beginning of *Furchenstich* culture.

The retreat of the tree taxa *Tilia, Ulmus, Corylus*, and *Abies* (fir), concurrent with the appearance of anthropogenic markers, is followed by a change in the sediment regime. Instead of the clays normal to the beginning of this zone, gyttja is detected in the bog profile. On the humid parts of the site, Pteridophytes with Monolete spores dominate over butreeds (*Cyperaceae*), and *Sparganium/Typha* species are also present, indicating low water levels. Concurrent with the sedimentation of gyttja layers, the numbers of aquatic and fen species increase, pointing to a rise in the previously low water table for these layers.

PTA-c (225-191 cm) (4700-4000 bp or 3500/3400-2550/2500 BC) Agricultural Period

Human activity is still detectable in the Subboreal layers near the site with the most important change being the introduction of *Carpinus* within the oak woodland. Pollen grains of *Tilia* and *Ulmus* are scarce, and the forest is dominated by *Quercus* and *Fagus*. *Corylus* and *Betula* are also present, next to the natural (primary) forest. The herbaceous taxa are dispersed, and the number of cereal pollen grains decrease compared to the previous zone. Typical fallow land taxa, ruderals, and nitrophyllous taxa (*Artemisia, Plantago, Rumex, Chenopodiaceae, Fabaceae* and partly *Poaceae*), and indicators of pasture (*Centaurea, Caryopyllaceae*) are increasing. *Alnus* (alder) dominates the vegetation of the wetlands. Local bog vegetation (*Cyperaceae, Sparganium/Typha latifolia, Filicales* and even *Pediastrum*, a genus of green algae) is decreasing. *Poaceae/Phragmites* and the helofites become more widespread. Water levels are increasing and the population that lived close to the site earlier has moved towards drier areas and the territories which were earlier used for agriculture now became pasturelands. This change in land use probably happened around 3500-3200 BC during the Middle Copper Age of the Proto-Boleráz, Boleráz culture.

PTA-d (191-130 cm) (4000-3000 bp or 2550/2500-1280/1220 BC) Forest Optimum

The vegetation reaches a forest maximum during this period, which is thought to have been deposited during the Subboreal. At first, the woodland is dominated by oak, but later *Carpinus* and *Fagus* become dominant. *Carpinus* is present with other tree taxa, and *Tilia* and *Ulmus* are sporadically present.

Proportions of herbaceous taxa do not change, but the number of cereals decreases further. Ruderals and nitrophilous taxa (*Artemisia* and *Poaceae*) are common. The presence of pasture indicators (*Plantago, Rumex*) indicate a survival of anthropogenic influences; however, their lower proportion suggests settlement more distant from the site. The presence of *Abies alba* and *Picea abies* (spruce) and the high level and dominance of *Alnus* shows that this is a relatively wet period. The oak forest has not yet suffered from human disturbance.

PTA-e (130-25 cm) (3000-1200 bp or 1280/1220-775/880 AD) Oak Forest Under Anthropogenic Influence

Several temporary decreases of tree taxa (A.P.) are observed with *Fagus* and *Carpinus* retreating from the forest canopy. *Quercus* stays at a stable level and is predominant in the forest. *Poaceae* and *Artemisia* have temporarily higher values, and anthropogenic taxa increase. *Cyperaceae, Typha/Sparganium,* and *Filicales*, or other species typical for swampy areas, are present in higher numbers during the first two phases (PTA-e1-e2) of this pollen zone, which can be divided into four separate phases:

a) PTA-e1 (130-115 cm) (3000-2800 bp or 1280/1220-950/920 BC) agricultural period

The previous phase, which had no obvious associated human settlement, is followed by this phase that represents renewed forest clearance and agricultural activities. A transitory peak in herbaceous taxa, indicated by the relatively low values of *Asteraceae* and *Chenopodiaceae*, suggests small-scale animal husbandry and temporary cultivation.

Human impact mostly affected *Carpinus* and *Fagus*, whose percentages significantly decrease, while *Quercus*, *Betula*, and *Corylus* remain stable. These layers are considered to have been deposited during the Urnfield culture of the Upper Bronze Age (1300-900 BC).

b) PTA-e2 (115-90 cm) (2800-2300 bp or 950/920-390 BC)

Betula reaches a temporary maximum during this period. As the forest recovers from previous human impacts, *Quercus* becomes the dominant taxa, with *Corylus* present at relatively stable levels. These results correspond to historical references to the Transdanubia during the Iron Age as "*immanes sylvae*" (Szabó 1971). *Fagus* and *Carpinus* do not return to previous high levels in Zone d. *Poaceae* remain at the same level as in the previous zone, but *Artemisia* and the other herbaceous taxa reach a temporary minimum. This period is characterized by abandonment of agricultural activities and by the closing of the forest canopy.

c) PTA-e3 (90-65 cm) (2300-1800 bp or 390 BC-220 AD)

A new period of agricultural activity can be seen in this zone. The forest clearance that started during the first phase (e3) affected mostly the secondary *Betula* and *Corylus*, while the natural forest recovers at the end of this period. Nitrofilous taxa, such as *Plantago lanceolata*, and *Rubiaceae*, increase at the beginning of the zone, and cereals, composites, and goosefoot (*Chenopodiaceae*) indicate a high level of human disturbance, as well as the presence of agricultural activities. This level probably represents the Roman Period.

d) PTA-e4 (65-15 cm) (1800-1211 bp or 220 -800 AD)

Cultivation ends and ruderals appear between the herbaceous vegetation, while the forest develops a dense canopy. This level is believed to have been deposited during the invasion of the Huns and Alans (4th-5th century), which is followed by more vigorous human activity. Forest clearance is evident at the very end of the sequence.

The results of the Principal Component Analyses of Pötréte (PCA PTA)

On the first axis of the Principal Component Analysis (Fig. 15.5), positive values represent the effect of climate change on the vegetation development. These samples show dominance of *Pinus* and *Betula*. Samples up to 250 cm having negative values represent a well-developed, mature oak forest. The second axis of the Principal Component Analysis shows the values related

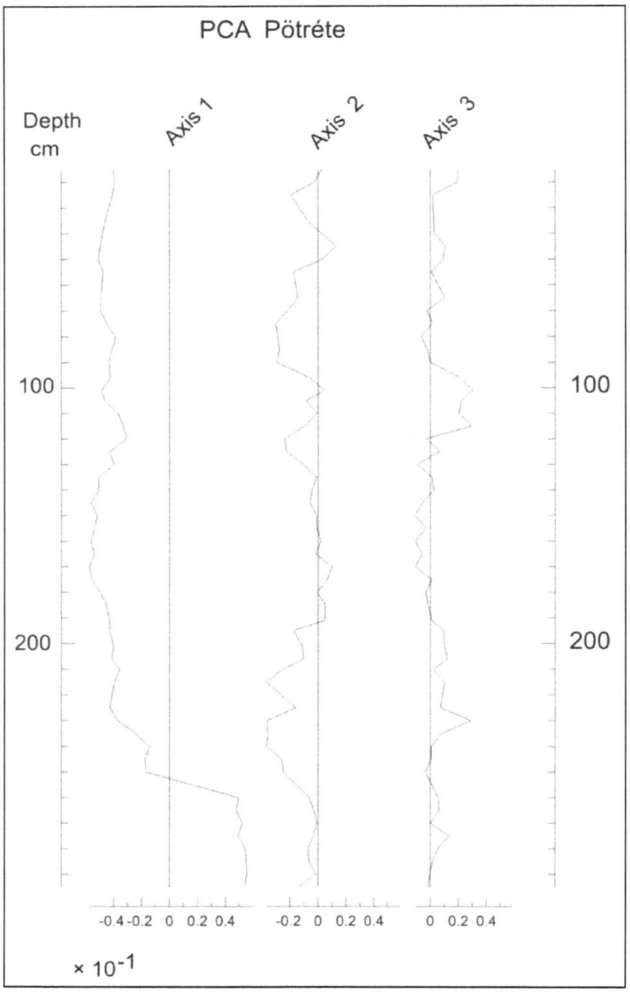

Figure 15.5. Principal Component Analysis of the Pötréte peat section

to human impacts. Most of the samples stay in the negative range. However, some samples with positive values close to 0 indicate abandonment of agricultural activities. During the first abandonment phase, *Carpinus* dominates the tree species. During the second phase, *Betula* increases. Samples with strong negative values indicate human influences and decreasing tree taxa number (Fig. 15.6).

HISTORY OF THE HUMAN IMPACT IN THE HAHÓT BASIN

The Copper Age (4500-2700 BC)

During the Copper Age settlement density reaches its highest level in the Lower-Zala Valley, at Kis-Balaton, and at the South Zala Plain, (Horváth 1994, Virág 1989). The increased importance of the region becomes evident during this period. However, at the end of the Copper Age the Hahót Basin is uninhabited. Remains of the Classical Baden culture (Site Nb. 74) can be found at only a single site. This settlement (Szőke 1995) lies very close to the Pötréte palaeoecological study site.

The PTA-b LPAZ Zone, dated to ca. 3800-3500 BC, is characterized by species-rich oak woodland, with beech present and hornbeam found only regionally (Fig. 15.4). Grasses dominate among the herbaceous vegetation, and other steppe elements have lower values during the beginning of the zone. Later these become more important as arboreal pollen percentages (A.P.) decrease. The rise of the anthropogenic taxa, especially the cereals, suggests relatively intensive agricultural activity. Arable fields are probably situated relatively close to the study site. Human impacts during this period (ca. 3800 BC) can be dated to the end of the Balaton-Lasinja and the beginning of the *Furchenstich* culture (Fig. 12.6).

The later PTA-c Zone, 3500/3400-2550/2500 BC (4700-4000 bp), is also characterized by a human presence. The most important change is the establishment of *Carpinus* in the *Fagus* and *Quercus* dominated woodland. *Betula* and *Corylus* are also important elements of the vegetation in the marginal zones of the forest. Herbaceous vegetation is very rich. However, the proportion of the cereals is lower than in the previous zone. Ruderals, nitrophilous taxa, and indicators of pastoral activity are increasingly widespread in the vegetation. There is probably an increase in water level during this time. People have moved to drier areas, and the territories that were in agricultural use have been transformed into pastureland. This period coincides with the Proto-Boleráz and Boleráz cultures (Fig. 15.6). Archaeological investigations have not found any signs of settlement from this period in the Hahót Basin. On the other hand, twenty-eight settlements were identified in the area around the Kis-Balaton region (Bondár 1989) and on the Zalai Plain where eleven sites have been excavated (Horváth 1994). North and west of the Hahót Basin there are few settlements dated to the end of the Late Copper Age. Within the research area there are only three sites from the Boleráz culture and a single site from the Baden culture. The unequal settlement density suggests that the western boundary of these cultures is located in this part of Transdanubia (Szőke 1995).

The Bronze Age

There are only a few dispersed settlements located outside the Hahót Marsh area at the beginning of the Bronze Age (2700-900 BC) (Horváth 1994). With the expansion of the Encrusted Pottery culture (Mészbetétes Kerámia) (Roberts *et al.* 2008), the broader environment of the Zala Hills, the Zala Plain, and the Hahót Basin are unpopulated during the second part of the early Bronze Age. The settlement boundary lay at the western edge of the Alsó-Zala Valley and the Kis-Balaton (Bondár 1989). At the beginning of the Early Bronze Age expansion of the pastoralist Early Tumulus culture from the northwest brought an end to the isolation of the Hahót Basin. The finds from the excavation of site Nb. 15 can be assigned to Phase C2 of the Tumulus culture. The ceramic assemblage from sites Nb. 33 and Nb. 57 (close to Pötréte) are dated to the Late Tumulus (1450-1300 BC) and the Early Urnfield culture periods (1300-900 BC) (Szőke and Vándor 1994). The Hahót Basin is more or

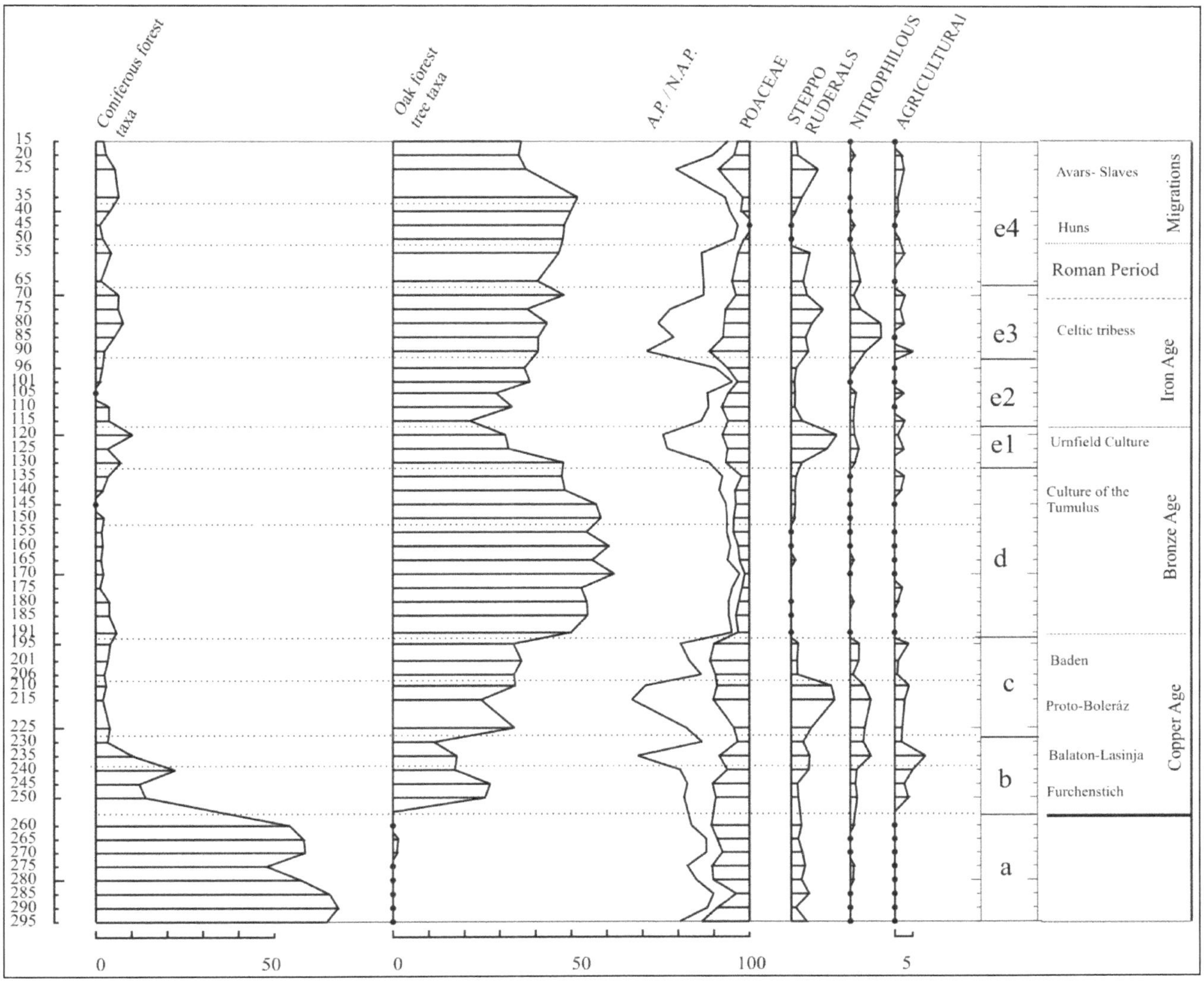

Figure 15.6. The chronology of the Pötréte pollen record with succession of cultures

less evenly settled during the Late Bronze Age, with a roughly similar settlement density as found in the southern Zala Plain and the Kis-Balaton region. The palynological results illustrate the human impact on the vegetation already proposed by archaeologists (Fig. 15.6).

The Iron Age

The Hahót Basin and its broader environment, the Zala Woodland, were again deserted in the Early Iron Age (9th-5th century BC). In contrast, a number of sites from this period have been identified and reported in the Keszthely area, to the northeast of the basin, close to Lake Balaton (Horváth 1994). The apparent lack of occupation can be attributed either to an internal border, most likely between tribal territories, or to an as yet little known economic phenomenon. The situation changed in the wake of the Celtic expansion of the 4th century BC. Following the advance of the Celtic tribes along the Danube, the Zala Uplands were drawn into the orbit of the Celtic world during the LT B phase (Szőke 1995).

Written sources and archaeological evidence suggest that the Zala region between the Rába River, the Balaton, and the Mura River was occupied by an unknown tribe, which seems to be socio-economically less advanced (Horváth 1989). The Hahót Basin was evenly settled with a density comparable to that of the southern Zala Plains (Horváth 1994) and the Kis-Balaton region (Horváth 1989, 1994).

Based on the pollen count from Zone PTA-e1 (Fig. 15.7), dating to ca. 1300-700 BC (3000-2600 bp), agricultural activity and clearing are evident. High amounts of herbaceous taxa suggest the presence of small-scale agriculture. Human impact, occurring most likely during the Urnfield Culture period (1300-900 BC), affected mostly *Fagus* and *Carpinus*. The above mentioned archaeological data suggests that the Hahót Basin was sparsely populated during this time, but a settlement was found close to the palaeoecological site, namely Site Nb. 57 (next to Pötréte). Later, at the beginning of the Iron Age, during the PTA-e2 Zone, the woodland canopy closes as *Betula* and *Quercus* become dominant. This observation is corroborated by written sources from the

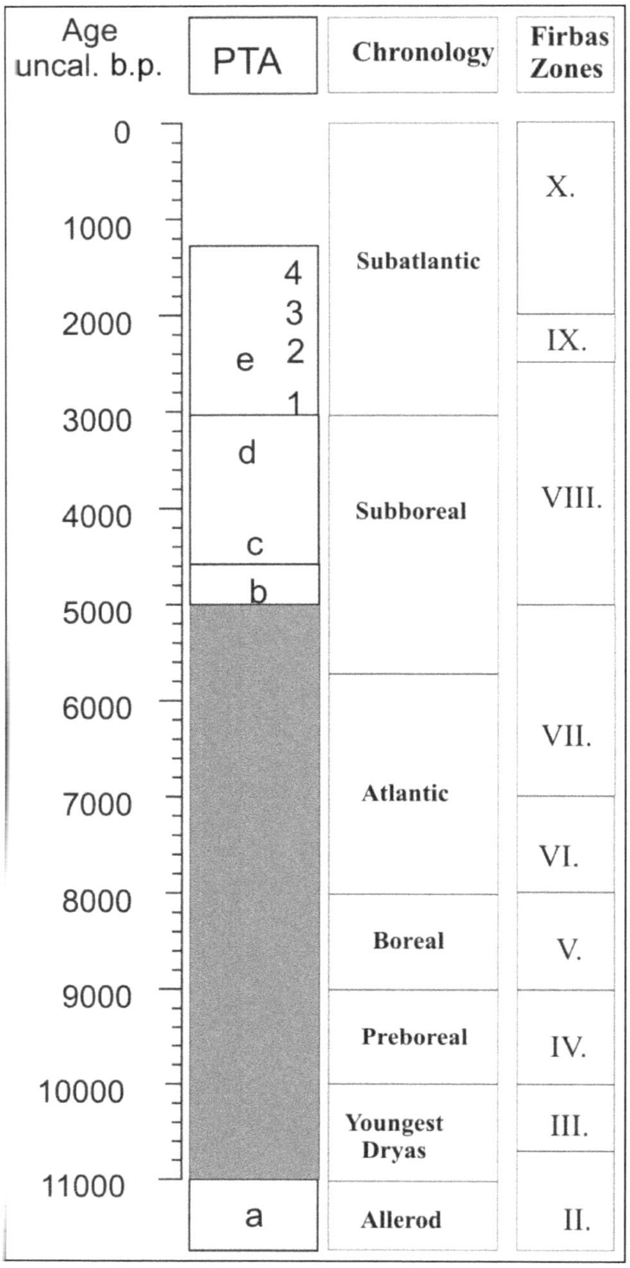

Figure 15.7. PTA, climatic periods, Firbas Zones and the Pötréte pollen record

end of the Iron Age, which characterize Transdanubia as "*immanes sylvae*", an immense woodland (Szabó 1971).

The Roman Period (AD 0-476 AD)

Roman conquest of Transdanubia was completed by the mid-first century, and the province of Pannonia was occupied by the Romans. Roman occupation of the Zala Uplands involved the influx and settlement of a northern Italian population as reflected by scattered urn interments and cremations burials from the mid-first to mid-third century AD at Magyarszerdahely-Homoki dűlő (Site Nb. 28, Szőke 1995). The true florescence of the Roman period in this region, however, fell into the late first and the early second century as the settlement of Salla (Zalalövő) prospered and attained municipal rank (Horváth 1979, 1994). The peaceful development of this settlement was interrupted by the Marcomannic-Sarmatian intrusions (160-180 AD), but the Hahót Basin, which lay beyond major communication routes, avoided this fate. This is demonstrated by the continuity of the burials in the Magyarszerdahely cemetery and by the absence of a destruction layer at Alsórajk- Kastélydomb Site Nb. 6 (Szőke 1995).

The new agricultural period is discernable in the PTA-e3 Zone (2200-1211 bp) deposited during the Roman period (0-476 AD). Clearing activities at the beginning of the zone primarily affected the secondary *Betula* and *Corylus*) woodland. The natural oak-hornbeam woodlands later recovered (Fig. 15.6, 15.7). The number of the nitrophilous taxa increased and the presence of cereals suggest heavy pastoral and agricultural activities.

The Migration Period (AD 300-700)

A new wave of Barbarian attacks swept through the province around AD 377. The Alsórajk-Kastélydombi villa was probably destroyed by groups of the Ostrogoth-Hunn-Alan alliance led by Alatheus and Saphrax. None of the successive Hunnic, Osthrogothic, Suebian, and Langobardic groups settled the Zala Upland. The area was virtually deserted during the first half of the 5th century, because it lay beyond the western settlement zone for these groups.

In the later 6th and the early 7th century the Hahót Basin was more or less a frontier zone of the Avar Kaganate that incorporated the entire Carpathian Basin (Horváth 1994). In the second part of the PTA-e3 Zone agricultural activity came to an end as ruderals appeared, and herbaceous vegetation appeared anew. This is probably the time when marauding tribes of Hunns and Alans invaded the area. As a result, people did not return to the territory until later in the 7th century, when signs of agricultural activity and the opening-up of the forest canopy are indicated in the pollen sequence (Fig. 15.6, 15.7). At that time the Zala Valley and the southern Zala Plains were settled by a mixed population of Avars and Slavs. Sporadic settlement finds suggest that the people may also have settled in the Hahót Basin. Their cemeteries remained in use until close to the end of the 7th century. In the late 7th and early 8th century the Hahót Basin and the Zala Upland were again deserted. Renewed settlement occurred only in the late 8th/early 9th century, after the dissolution of the Avar Kaganate (Szőke 1994). The new occupants were probably the descendants of a population that had migrated from elsewhere at the close of the 7th century.

CONCLUSION

Several of the 78 occupation and burial sites around the Pötréte marshland have been occupied for as many as eight to 11 different periods, reflecting an almost continuous human occupation. The paleoenvironmental data of Pötréte shows that the vegetation history starts around 3800-3550 BC (5000-4800 bp), during the Hungarian Late-Copper Age), when signs of human impact on the environment are already visible in the pollen sequence. From this time onward, the impact is almost continuously traceable up to the Migration Period (ca. 7th century AD). Therefore, the described environmental changes between ca. 3800 BC and the 8th century AD appear to be mostly due to human agency, rather than climate change.

Acknowledgments

The palynological analysis was carried out with help of a Ph.D. grant by the French Government and the Institut Français of Budapest. I would like to express my gratitude to Dr. Hervé Richard and his colleagues (Besançon CNRS Laboratoire de Chrono-écologie, UFR des Science et Techniques) for providing the sediment core and the radiocarbon dates. During the analyses and interpretation of the sediment cores my colleagues in the Institut Mediterranéen d'Écologie et de Palynology (IMEP), Laboratoire de Botanique et de Botanique Historique in Marseille provided great help. I could always turn to Anne Vaillant with problems during the sample treatments. My deepest respect and gratitude has to be given to my tutor, Prof. Jacques-Louis de Beaulieu for his patience, endurance, and firmness. I could always count on his thorough knowledge about the vegetation history of Europe, and his confidence in me and in my work led me to the accomplishment of my research. The work would not have reached its fruition without him. I am also grateful to Dr. Maximilian Baldia and to Dr. Christel Baldia for their help with the manuscript's English.

References

BÁNFFY, E. 2000 – The late Starčevo and the Earliest Linear Pottery Groups in Western Transdanubia. *Documenta Praehistorica* 27, 2000:173-185.

BÁNFFY, E. 2004 – Advances in the Research of the Neolithic Transition in the Carpathian Basin. In Lukes and Zvelebil (eds.) *LBK Dialogues: Studies in the Formation of the Linear Pottery Culture*. BAR S1304 2004:49-70.

BÁNFFY, E. 2006 – Eastern, Central and Western Hungary: Variations of Neolithisation Models. *Documenta Praehistorica* XXXIII, 2006:125-142.

BEHRE, K.E. 1988 – The role of Man in European Vegetation History. In B. Huntley and T. Webb III (eds.), *Vegetation History*. Kluwer Academic Publishers., Dordrecht, 1988:633-672.

BONDÁR, M. 1989 – Spätkupferzeit, Früh- und Mittelbrozezeit. In R. Müller (ed.), *Sieben Jahrtausende am Balaton von der Ur- und Frühgeschichte bis zum Ende der Türkenkriege*. Städtisches Reiss-Museum für Naturkunde und Vorgeschichte, Mannheim, 1989:26-36.

ERDTMAN, G. 1936 – New Methods in Pollen Analysis. *Svensk Botanisk Tidschrift* 30, 1936:154-164.

HORVÁTH, L. 1979 – A Magyarszerdahelyi Kelta és Római Temető. [The Magyarszerdahely Celtic and Roman Cemetery.] *Zalai Gyűjtemény* 14. Supplement A and B. 1979:95-98.

HORVÁTH, L. 1989 – *Späteisennzeit*. In Müller, R (ed.), *Kis-Balaton, Sieben Jahrtausende am Balaton von der Ur- und Frühgeschichte bis zum Ende der Türkenkriege*. Städtisches Reiss-Museum für Naturkunde und Vorgeschichte, Mannheim, Mannheim, 1989:47-50.

HORVÁTH, L. 1994 – Nagykanizsa és Környékének Története az Újkőkortól a Római kor Végéig. [The History of Nagykanizsa and its Surroundings Between the Neolithic and Roman Period.] In J. Béli and A. Rózsa-Lendvai (ed.), *Városi Monográfia 1*. Nagykanizsa, 1994:85-141.

JUHASZ, I.E. 2002 – Reconstitution Palynologique de la Végétation Depuis le Tardiglaciaire dans la Région de Zala, Sud-Ouest de la Hongrie [Palynological Reconstruction of the Late-Glacial and Holocene Vegetation of South-West Hungary.] Unpublished Ph. D. Thesis, University of Pécs and University Aix-Marseille III, Pécs, 2002.

JUHÁSZ, I.E. 2003 – *Some Evidences of Preneolitization in SW-Transdanubia, Hungary*. Unsettling the Neolithic Conference, 11-13 May 2003, poster, Cardiff, 2003.

JUHÁSZ, I.E. 2004 – Palynological Evidences of Preneolithization in South-Western Transdanubia. *Antaeus* 27, 2004:112-124.

JUHÁSZ I.E., R. DRESCHER-SCHNEIDER, V. ANDRIEU-PONEL and J.L. de BEAULIEU 2001 – Anthropogenic Indicators in a Palynological Records from Pölöske, Zala Region, Western Hungary. In P. Lippert (ed.), *Die Drau-, Mur- und Raab-Region im 1. vorchristlichen Jahrtausend, Akten des internationalen und interdisciplinären Symposiums von 26. bis 29. April 2000 in Bad Radkersburg*, Habelt, Bonn, 2001:29-38.

KALICZ, N., Zs.M.T. VIRÁGH and K. BÍRÓ 1998 – The northern periphery of the Early Neolithic Starčevo Culture in South-Western Hungary: A Case Study of an Excavation at Lake Balaton. *Documenta Praehistorica*, 25, 1998:151-187.

NAKAGAWA, T., E. BRUGLIAPALIA, G. DIGERFELD, M. REILLE, J.L. de BEAULIEU and Y. JASUDA 1998 – Dense Media Separation as a more Efficient Pollen Extraction Method for use with Organic Sediment Deposit Samples: Comparison with the Conventional Method. *Boreas* 27, 1998:15-24.

MEDZIHRADSZKY, Zs. 2001 – The reconstruction of the vegetation in the Kis-Balaton area during Lengyel period (Preliminary Report). In J. Regenye (ed.), *Sites and stones: Lengyel Culture in Western Hungary and Beyond*. Veszprém County Museum, Veszprém. 2001:143-148.

ROBERTS, S., J. SOFAER and V. KISS 2008 – Characterisation and Textural Analysis of Middle Bronze Age Transdanubian Inlaid Wares of the Encrusted Pottery Culture, Hungary: A Preliminary Study. *Journal of Archaeological Science*, 35/2, 2008:322-330.

SÜMEGI, P. 2004 – The Results of Paleo-Environmental Reconstruction and comparative Geoarchaeological Analysis for the Examined Area. In P. Sümegi and S. Gulyás (eds.), *The Geohistory of Bátorliget Marsh-land*. Archaeolingua Press, Budapest, 2004:301-348.

SÜMEGI, P., S. GULYÁS and G. PERSAITS 2008 – Holocene Paleoclimatic and Paleohydrological Changes in the Sárrét Basin, NW Hungary, *Documenta Praehistorica XXXV*, 2008:25-31.

SÜMEGI, P., and R. KERTÉSZ 2001 – Palaeogeographic Characteristic of the Carpathian Basin: An Ecological Trap During the Early Neolithic? In R. Kertész and J. Makkay (eds.), *From the Mesolithic to the Neolithic*. Archaeolingua Press, Budapest 2001:405-416.

SZABÓ, M. 1971 – *Celtic Heritage in Hungary*. Hereditas Kiadó, Corvina, Budapest, 1971.

SZŐKE, B.M. 1995 – Borderland of Cultures: Settlement History Research in the Hahót Basin (Aims, methods, results). In B.M. Szőke (ed.), *Archeology and Settlement History in the Hahót Basin, South-West Hungary, from the Neolithic to the Roman Age*. Communicationes de Instituto Archaeologico Academiae Scientarum Hungaricae 22. Antaeus, Budapest, 1995:13-34.

SZŐKE, B.M. 1994 – Népvándorláskor és koraközépkor története Nagykanizsán és környékén. [The History of the Migration Period and the early Middle Ages in Nagykanizsa and its surroundings.] *Nagykanizsa*, 1994:145-214.

SZŐKE, B.M., and L. VÁNDOR (eds.) 1994 – *Nagy Utazás.... Kultúrák Határán. Településtörténeti Kutatások a Szévíz és Principális Völgyében.* [The Great Journey... Where Cultures Meet.] Zalaegerszeg, 1994.

VIRÁG, Zs.M. 1989 – Jungsteinzeit und Frühkrupferzeit. In R. Müller (ed.), *Sieben Jahrtausende am Balaton von der Ur- und Frühgeschichte bis zum Ende der Türkenkriege.* Städtisches Reiss-Museum für Naturkunde und Vorgeschichte. Mannheim, 1989:17-18, Map 2.

Chapter 16
BREAKING UNNATURAL BARRIERS: COMPARATIVE ARCHAEOLOGY, CLIMATE, AND CULTURE CHANGE IN CENTRAL AND NORTHERN EUROPE (6100-2700 BC)

Maximilian O. BALDIA
The Comparative Archaeology WEB and ISEM, Southern Methodist University

Abstract: *Comparative archaeology aims to comprehend humanity's capability to cope with a continuously changing world. To peruse this aim archaeological and paleoclimate data are used to analyze socio-cultural responses to climate change across time and space. The barriers to the Mesolithic-Neolithic transition are followed from the Carpathian Basin to Scandinavia. The resulting research differs from environmental determinism and is at variance with prehistoric human environmental impact studies. It is concluded that the observable socio-cultural adjustments can be correlated with climatic change, but the kind and degree of human adjustment varies, depending on the preexisting cultural repertoire as well as the kind and intensity of climatic change.*

INTRODUCTION

Comparative Archaeology aims to understand humanity's ability to overcome barriers created by a changing world. Here, I further this aim by focusing on the human response to climate change in conjunction with the barriers to the expansion of farming in Central and Northern Europe. This requires data derived from multidisciplinary research, which is best integrated through the discipline of archaeology (van der Leeuw and Redman 2002).

Unfortunately, multidisciplinary research is hampered by the unnatural barriers of academia. These barriers artificially divide research into separate faculties. In crossing these academic barriers, archaeologists find that paleoclimate research is dominated by the physical sciences (e.g. van Buren 2001:131-132). This often inhibits attention to the human dimension. As a result, one encounters much technical data, which is frequently of little utility for the analysis of human behavior as expressed in socio-cultural adjustment to climate change. Even basic information, such as dating paleoclimatic events, can provide too coarse a time scale for a meaningful analysis of the range of human adjustments. Fortunately, recent achievements in paleoclimate research allow finer scales of analyses, in some cases providing detail beyond even the archaeological resolution. Given these advances, I analyze the socio-cultural response to climatic change on a cross-cultural, interregional, and diachronic level.

The research presented here differs from past studies of human adaptation, which are now deemed environmental determinism. It also differs from research, which links climate events with disasters and cultural collapse (e.g. van Buren 2001, Fagan 2004). Furthermore, it is at variance with the historical interest in prehistoric human environmental impact that has predominated Central and North European prehistoric environmental research since the 1940s (e.g. Iversen 1941, 1949, Kalis *et al.* 2003:40-41, Litt 2003).

Background

In opposition to traditional environmental determinism, many archaeologists have invoked human agency in detailing how Near Eastern domesticates or actual human migrations spread farming from the Fertile Crescent into Europe (Fig. 16.1). The influence of climatic change on this process is either refuted outright (Price 2000c:309) or not considered to be a major factor (e.g. Barnett 2000, Binder 2000, Price 2000b, Watkins 2006, Zilhão 2000, Zvelebil and Lille 2000). On the other hand, Richerson and Boyd (2001) and Richerson *et al.* (2001) propose that climatic conditions made the development of farming impossible before the Holocene and propose that the increase in CO_2 and precipitation levels of the early Holocene provide favorable conditions for cultivation, making agriculture compulsory in many parts of the world.

A survey of the literature shows that Late-Quaternary and Holocene climate research brought together paleoecologists and archaeologists for the purpose of studying human environmental impact by the late 1980s Chambers (1993). Chambers outlines the uses of numerous climate proxies for this kind of research. Paetzold (1992) hints at the correlation of temperature oscillations derived from the radiocarbon output of the sun in discussing the settlement of different cultures from the Neolithic to the LaTéne period in South Germany. For the adjacent Czech Republic Bouzek (1993, 2001, 2005) correlates 77 sites ranging from the earliest Neolithic to the early Iron Age with numerous climate proxies and concludes that climatic change usually causes culture change. Bosnall *et al.* 2002 propose that the important riverbank sites at the Danube Gorges area of Romania and Serbia were abandoned due to wet climate conditions, while the elevated site of Lepenski Vir

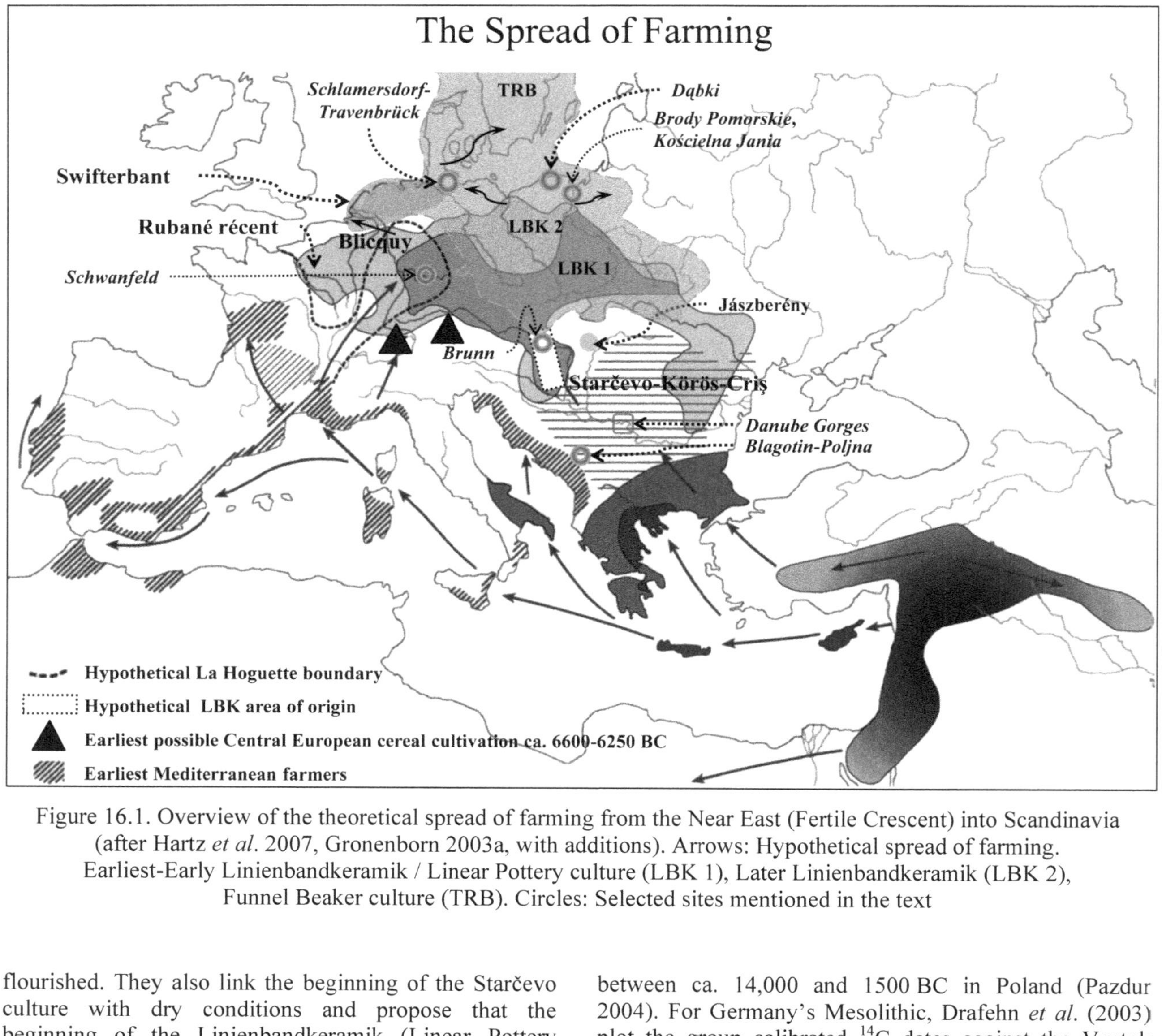

Figure 16.1. Overview of the theoretical spread of farming from the Near East (Fertile Crescent) into Scandinavia (after Hartz *et al.* 2007, Gronenborn 2003a, with additions). Arrows: Hypothetical spread of farming. Earliest-Early Linienbandkeramik / Linear Pottery culture (LBK 1), Later Linienbandkeramik (LBK 2), Funnel Beaker culture (TRB). Circles: Selected sites mentioned in the text

flourished. They also link the beginning of the Starčevo culture with dry conditions and propose that the beginning of the Linienbandkeramik (Linear Pottery culture), and possibly the Viča and Tisza cultures are linked to the start of a wet climate phase. Such conclusions are critically received by many archeologists. For instance Deslerova (2005) admits that climate may influence culture change in marginal regions, such as Switzerland, but not in the Czech Republic.

In spite of Deslerova's (2005) view that the Alpine region and its northern piedmont is a marginal area, the region's research provides the chronologically most fine-grained data for the period between 4300 and 2400 BC (e.g. Stöckli *et al.* 1995, Arbogast *et al.* 2006). Among the most pertinent early research is the study by Schibler *et al.* (1997:329-361), who describe economic changes observable in the tree-ring dated lake dwellings between 4000 and 3500 BC. In the same volume Maise provides an exemplary outline for the use of high-resolution climate proxies (Schibler *et al.* 1997:335-337).

A chronologically somewhat rough-grained study uses archaeological data, speleothems, peat formations, and cumulative ^{14}C dated probability density distributions to link climate oscillations with changes in cultural activity between ca. 14,000 and 1500 BC in Poland (Pazdur 2004). For Germany's Mesolithic, Drafehn *et al.* (2003) plot the group calibrated ^{14}C dates against the Vostok (Continental climate) and the GISP 2 (North Atlantic climate) ice core Δ^{18}Oxygen isotope proxies for temperature fluctuations.

Bogucki (1998) focuses on the differences between periods of high and low frequency climatic oscillations in relation to farming and cultural changes in Poland, starting around 5500 BC. He proposes that the apparent lack of high frequency oscillations allowed the Linienbandkeramik to introduce farming to this region. He further hypothesizes that the high frequency climatic oscillations just before 3800 BC probably affected the population of the North European Plain, resulting in the diversification of the Lengyel culture in Kujavia and the adaptation of farming by the foragers on the sandier soils.

Articles in the book edited by Gronenborn (2005) and Gronenborn (2009) consider numerous indicators of climate variability in conjunction with culture change during the Neolithic. They imply that climate oscillations may not only be linked to changes from one culture to another, but also to individual phases within a culture.

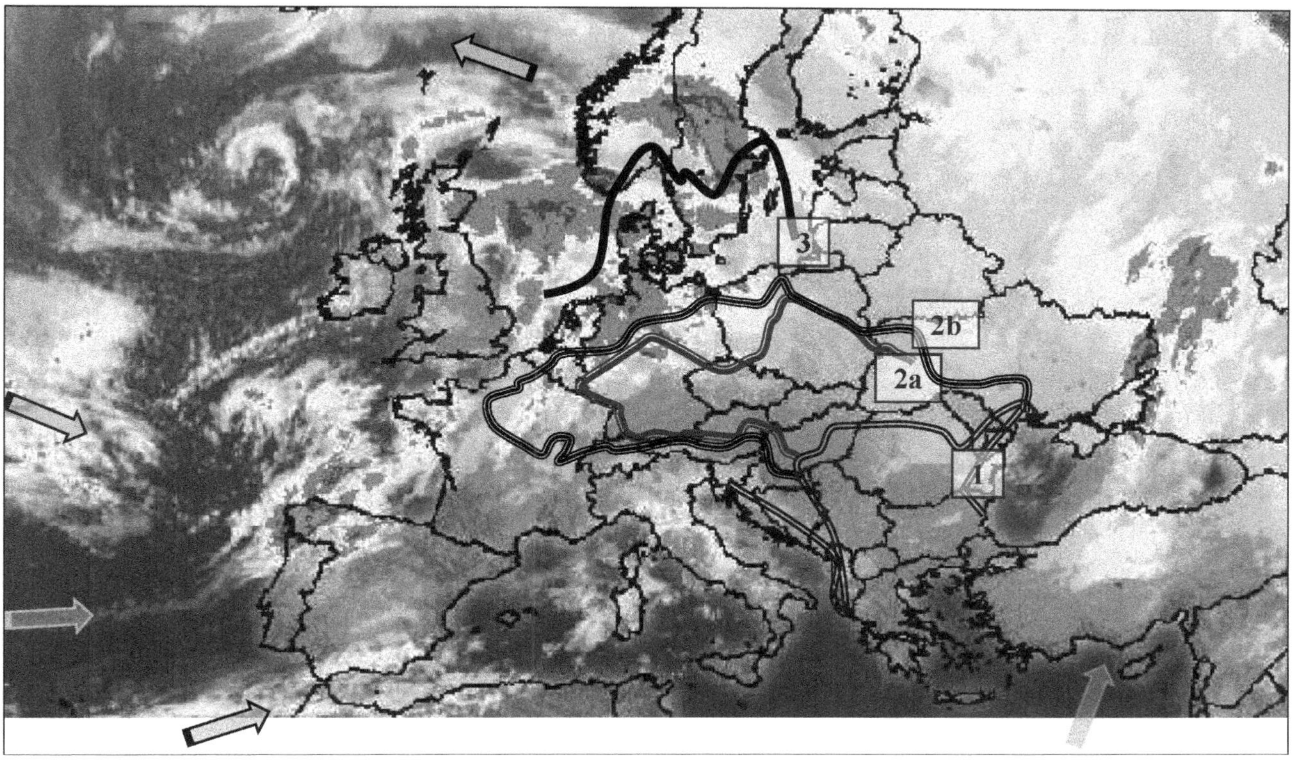

Figure 16.2. The Neolithic farming barriers superimposed over a satellite weather map, showing Europe's summer circulation pattern (weather map after AcuWeather 2002). Arrows: direction of circulation. 1: Barrier 1 (Starčevo, Körös, Criş cultures etc.), 2a: Barrier 2a (LBK 1 boundary), 2b: Barrier 2b (LBK 2 and Post-LBK cultures), 3: Barrier 3 (TRB North Group)

The study presented below stresses the interplay between socio-cultural adjustments and climatic oscillations. It begins with a section on terminology, dating and conventions used throughout this chapter. This is followed by a discussion of the most applicable paleoclimate proxies for this research. Thereafter, I discuss the climatic and archaeological data pertaining to each of the barriers to agricultural expansion. The presentation starts in Southeastern Central Europe and ends in Scandinavia. (Fig. 16.1). The barriers are presented in chronological order, that is, from the oldest to the most recent (Fig. 16.2).

Terminology, Dating and Conventions

In the Near East and Europe the development of farming (agriculture and animal husbandry) originally meant the simultaneous appearance and spread of cultivated plants, domesticated animals, ground stone tools, pottery and sedentism, i.e. the *Neolithic package*. Although, I use *Neolithic* and *farming* synonymously, this merely denotes one or more of these characteristics. It must also be noted that the terms farming and agriculture, or even horticulture, gloss over a considerable range of food production techniques (cf. B. Smith 2001) and a variety of differing farming approaches are plausible for the Neolithic of the Near East and Europe (Bogaard 2002, 2004, 2005). In addition to differing degrees of agricultural intensity, there must also have been varying degrees of hunting versus animal husbandry. Nonetheless, for the sake brevity I will continue to use the terms Neolithic, agriculture, animal husbandry, and pottery production interchangeably and only distinguish between foraging and farming.

To correlate human socio-cultural responses with climatic change, accurate archaeological and paleoclimatic chronologies are fundamental. Therefore, a large portion of this paper focuses on dating. Unfortunately, in paleoclimatology, different climate indicators do not pinpoint climate events with equal precision (e.g. Blaauw *et al.* 2007). Transitional archaeological (cultural) phases are also problematic. They have low archaeological visibility until they reach a *critical mass* of acceptance within the prehistoric society (Fokkens (2008:19). Climate also influences the quantity and quality of archaeological data (Schibler and Jacomet 2005). The same applies to indirect dating through stratigraphic association, since the speed with which a stratigraphic layer forms or is eroded depends on climatic and environmental factors. Even modern methods of direct (absolute) dating are influenced by climatic factors, such as variation of solar radiation. Finally, absolute dating methods use probability statistics, which result in a range of possible dates. Therefore, I focus on date ranges, when analyzing ^{14}C dates.

The ^{14}C dates come from the *14CEurope* database of more than 6100 dates collected from many sources, including the electronic databases of Furholt *et al.* (2002)

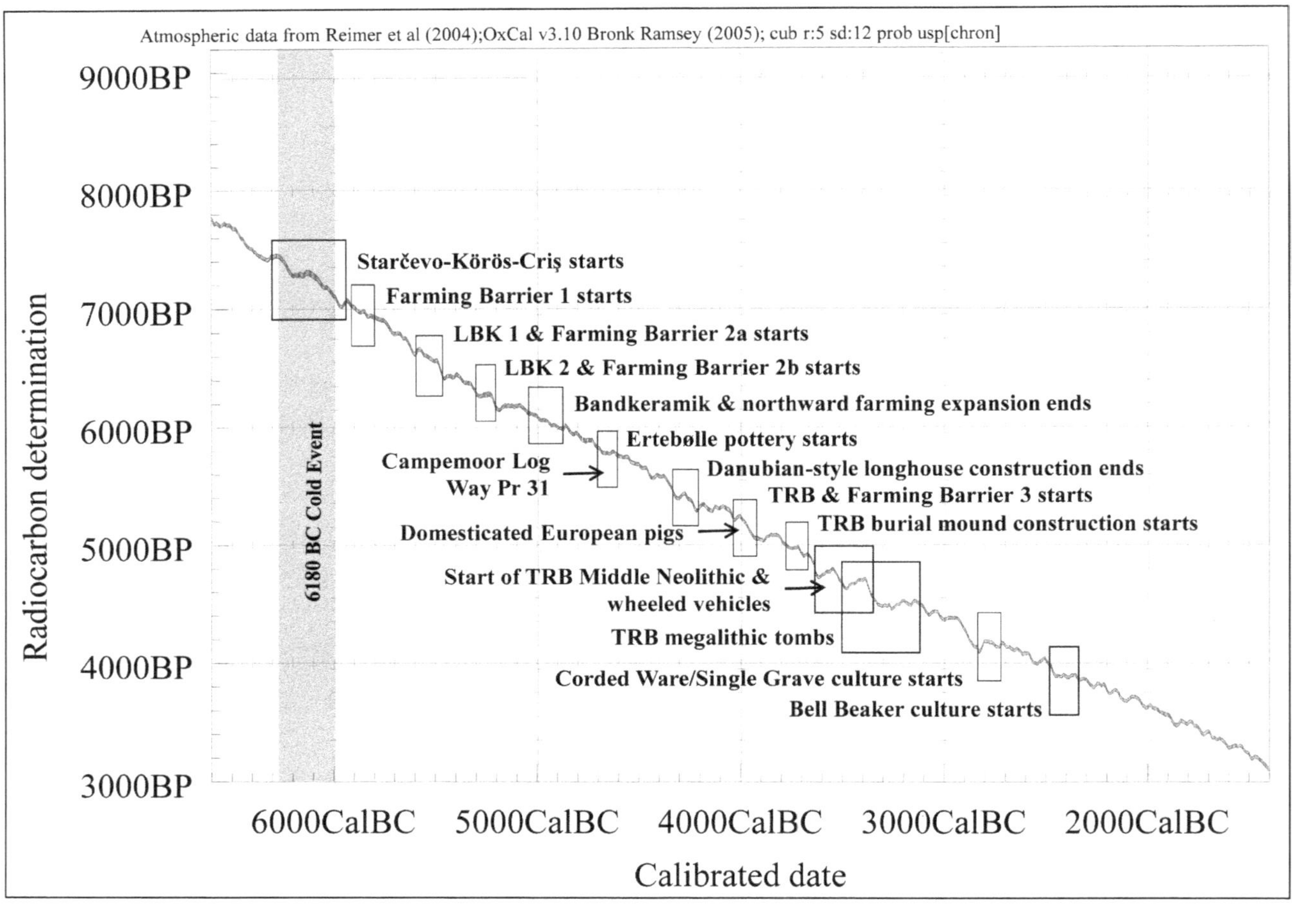

Figure 16.3. Radiocarbon curve and significant cultural events

and Shennan and Steele (2000 with updates). Additional sources are listed in the text or foot notes where pertinent. For the calibration of ^{14}C dates, OxCal 2005 (Version 3.10) developed by Bronk Ramsey with INCAL 04 atmospheric data (Reimer *et al.* 2004) is applied, using the default settings. Upper case *BC* denotes calibrated ^{14}C dates. The OxCal graphing tool is used to create individual and multiple plots of calibrated ^{14}C dates. However, the resulting OxCal graphs produce capital letters *BP* for uncalibrated ^{14}C dates. This has not been altered in the figures, but lower case *bp* is used in the text. Extreme ^{14}C dated samples (*outliers*) are generally omitted. Sigma (σ) refers to Standard Deviation, and delta (Δ or δ) is used for isotope ratios.

THE PALEOCLIMATE INDICATORS

Paleoclimate research indicates that the period starting around 8000 BC, and especially between 7000 to 5000 BC (the Mid-Holocene), coincides with a *particularly profound* climatic and environmental change (Steig 1999). During this period farming spreads throughout much of Europe (Fig. 16.2).

Satellite images show how the present European climate is dominated by the North Atlantic Oscillation (NAO). NAO correlates directly with Central European temperature variability (Bodri and Cermak 2003). Low Greenland winter temperatures, resulting from more northerly storm tracks, cause less precipitation in southern Central Europe and increase precipitation and temperatures in the north (Spurk *et al.* 2002:714). The reverse also occurs, so that wetter phases in the south are synchronous with dryer ones in the north (Spurk *et al.* 2002:713). Similarly, Magny *et al.* (2003), based on French data, propose wetter conditions for Europe's latitudes of ca. 50° to 43°, as opposed to those found farther north and south during the 6180 BC cold event.

Although, the ocean circulation pattern exists during the entire Holocene, detailed reconstruction of prehistoric climate change has to be inferred from various climate proxies. Radiocarbon,[1] tree-ring, and Greenland ice core proxies are of primary utility here, because they provide reasonable chronological control and interproxy verifiability.

The radiocarbon record relates to radiation intensity from the sun, which varies over time. The resulting wiggles are associated with climate (temperature) change, especially

[1] Surface ocean-atmosphere ^{14}C offsets for the later Holocene in the North Atlantic show modulation of the wider ΔR signal in some locations, but it appears that overall surface ocean ^{14}C responds to large-scale regional climatic fluctuations, and local variations are subordinate to that of the major oceanic forcing (Ascough *et al.* 2009).

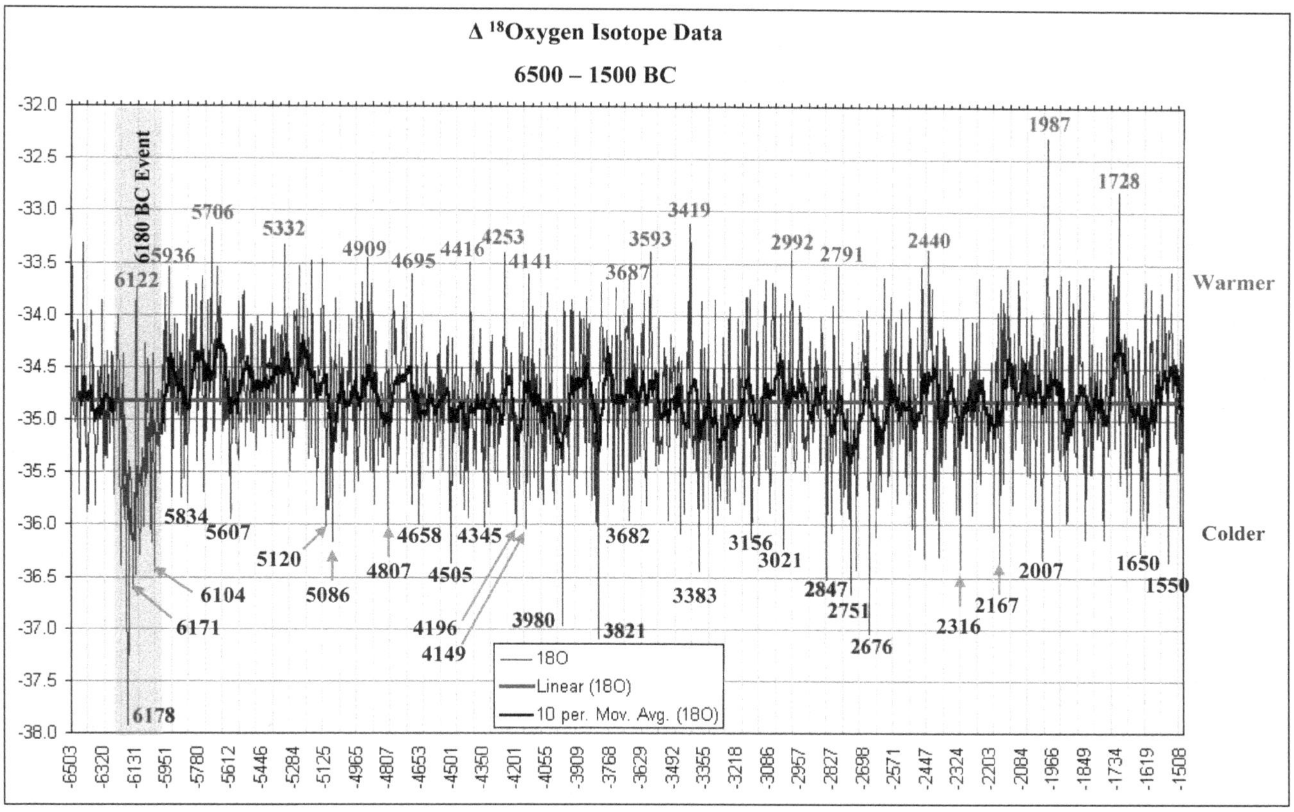

Figure 16.4. GRIP ice core Δ^{18}Oxigen isotope temperature proxy. Extreme warm and cold spikes are marked with their respective dates. Horizontal line: Linear log transformation of isotope data (normal temperature). Curve: Trend line (moving average)

when plotting the residuals, and the results roughly correlate with culture change (e.g. Maise in Schibler *et al.* 1997:335-337, Paetzold 1992, Strien and Gronenborn 2005). Unfortunately, the infamous wiggles also irritate archaeologists, because they raise havoc with radiocarbon dates of even the highest precision. This makes it difficult to isolate observable socio-cultural change in the archaeological record with the desired precision. Nonetheless, the thresholds of significant culture change roughly coincide with certain wiggles (Fig. 16.3). Moreover, the same wiggles, as well as segments of low oscillation, are connected with global climate events. In fact, the Δ^{14}C residuals (Stuiver *et al.* 1998) roughly correlate inversely with temperature and there is a correlation with climate anomalies observed in European fossil or bog oaks and other proxies. However, neither all wiggles nor their specific amplitudes seem to apply equally.

Complementary to Δ^{14}C is the more sensitive stable ^{18}Oxygen isotope data, calculated as Δ^{18}O.[2] The most valuable source for such data stems from the ice cores. The longest and most complete records come from the Greenland ice-core projects. Of six ice core bore areas (Johnsen *et al.* 2001), the most suitable for the present purpose are GRIP and GISP2. There is a strong correlation between Δ^{18}O records of these Greenland ice cores and isotopic as well as other proxies of the Northern Hemisphere paleoclimate sequences. However, an exact correlation is hampered by the existence of slightly different ice core chronologies (Southon 2004). Southon concludes that GRIP and GISP2 show good agreement from recent times up to 1350 BC, but thereafter GISP2 requires various offsets to match the data from the GRIP chronology (Southon 2004 Fig. 3). Furthermore, comparisons of the sun's energy output (^{10}Be – ^{14}C measurements) indicate that GISP2 is generally ca. 60 years too old after 1450 BC. To complicate matters, GRIP has three different counting chronologies. As a result, the anomaly often called the 8.2 kyr or 8200 BP cold event is dated 6180 BC, 6220 BC, or 6260 BC. Even though, the last date happens to agree with the GISP2 dating (Southon 2004 Table 1), statistical modeling, combing all data from four estimates, suggests a dating of 6297 to 6136 BC (Thomas *et al.* 2007). Nevertheless, the original and most complete counting method is applied here. It yields a date of 6177/6178 BC for the strongest cold spike of this phenomenon. Consequently, I choose to call the 8200 BP cold event the 6180 BC cold event throughout this chapter. The 6180 BC cold event is the most prominent cold event in the period of 6500-1600 BC (Fig. 16.4). Although, it is usually referred to as a single event, it may be seen as more than one, resulting from several factors, including changes in the sun's radiation, variation in North Atlantic deep-water contribution, and subtropical

[2] Δ^{18}O or δ^{18}O is the ratio of stable isotopes ^{18}O:^{16}O expressed as parts per thousand (‰).

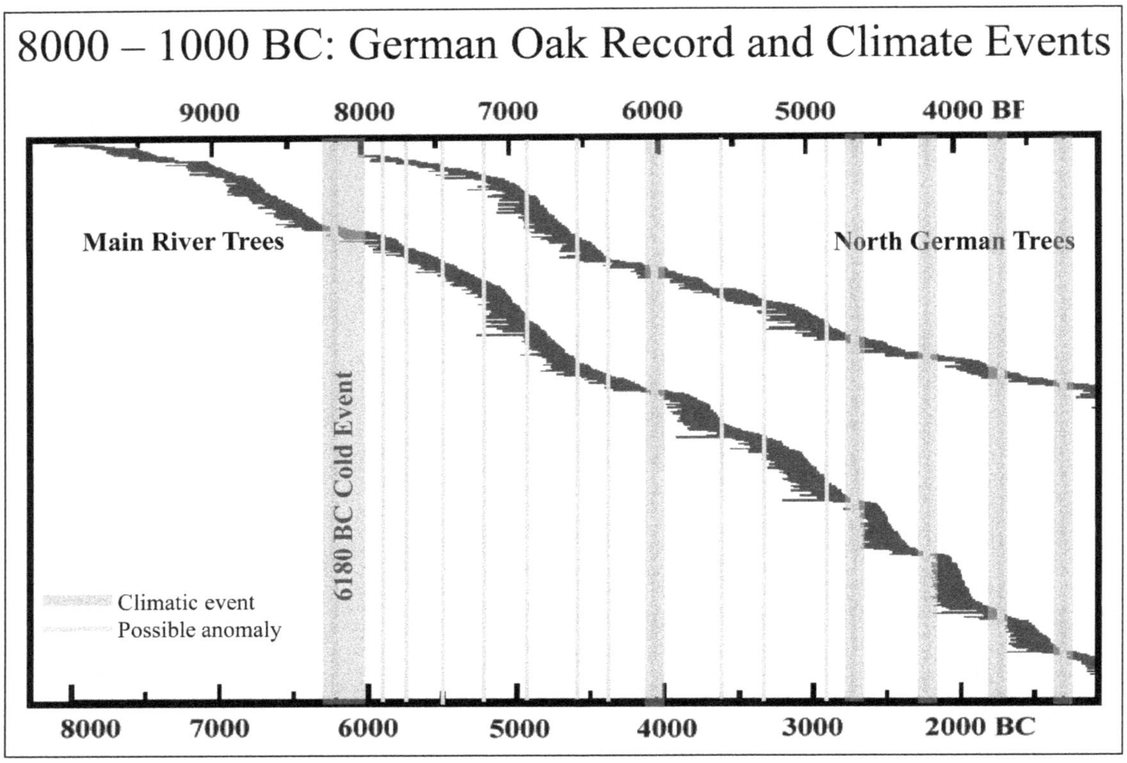

Figure 16.5. Depositional frequency of German oaks (after Spurk *et al.* 2002 Fig. 1 with modification). Shaded bar: Approximate maximum duration of the 6180 BC climate event between ca. 6270-6000 BC. Wide vertical line: climate anomaly. Narrow vertical line: Possible climate anomaly

sea surface temperature (Kofler *et al.* 2005 Fig. 4, Wiersma 2008).

Applying a linear log transformation to the GRIP $\Delta^{18}O$ isotope data results in a straight, horizontal line for the period of 6500 to 1500 BC (Fig. 16.4).[3] This base line may be considered the *normal temperature* for the research period. It is ideal for gauging the intensity of warm and cold spikes. Extreme warm and cold spikes are marked with their respective dates. The curve (trend line) is calculated to indicate the relative intensity and duration of warm periods above and cold periods below the normal temperature line.

Tree-ring data are probably the most useful high-resolution proxy. They provide climatological, environmental, and chronological information for archaeologists (Nash 2002). The Central European tree-ring data form part of the foundation of the IntCal98 radiocarbon calibration curve and have led to a revision in the IntCal04 curve (Reimer *et al.* 2004, Friedrich *et al.* 2004. The results correlate well with the GRIP ice core results (Spurk *et al.* 2002 Fig. 6 and 7). The Preboreal pine chronology of the Rhein (Rhine) and Danube Rivers in Germany have since been linked to the German oak chronology (Friedrich *et al.* 2004) and the Northwest German oak data have been combined with bog-pine data from 36 sites between the Elbe River and the Dutch-German border (Eckstein *et al.* 2009, Leuschner *et al.* 2007).

The oak chronology dates back to 8480 BC in the Rhein River valley and the Main River oak chronology reaches back to 8263 BC, providing data for the 6180 BC cold event (Friedrich *et al.* 2004:115). Analysis of the mean lifespan of these oaks shows two breaks during this event (Spurk *et al.* 2002 Fig. 4). One break occurs between ca. 6350 and 6300 BC, the other between roughly 6200 and 6000 BC. This again suggests that the 6180 BC event may be more than one oscillation. Combining the Main River and the North German oak record leads the authors to observe six climate events of varying length between 8000 and 1000 BC (Fig. 16.5). However, Leuschner *et al.* (2002) and Eckstein *et al.* (2009) isolate additional events.

The Main River, Northwest German, and Dutch bog oak records published by Spurk *et al.* (2002) and Eckstein *et al.* (2009) complement each other and suggest numerous climatic events. Figure 16.6 combines the anomalies noted by Spurk *et al.* (2002) and Leuschner *et al.* (2002), but also includes my additions estimated from the peaks (narrow vertical lines) and troughs (hatched lines) of the mean age chronology curves. These possible anomalies largely coincide with the major germination and/or dying-off (GDO) phases identified by Eckstein *et al.* (2009). Analysis of the wider West European tree-ring data by

[3] This contrasts with the period after 1500 BC, when the linear log transformation begins to curve downward toward a colder climate. A similar trajectory is seen in the tree growth homogeneity index of Germany, which trends toward wetter climatic conditions after 1500 BC (Schmidt and Gruhle 2003 Fig. 6).

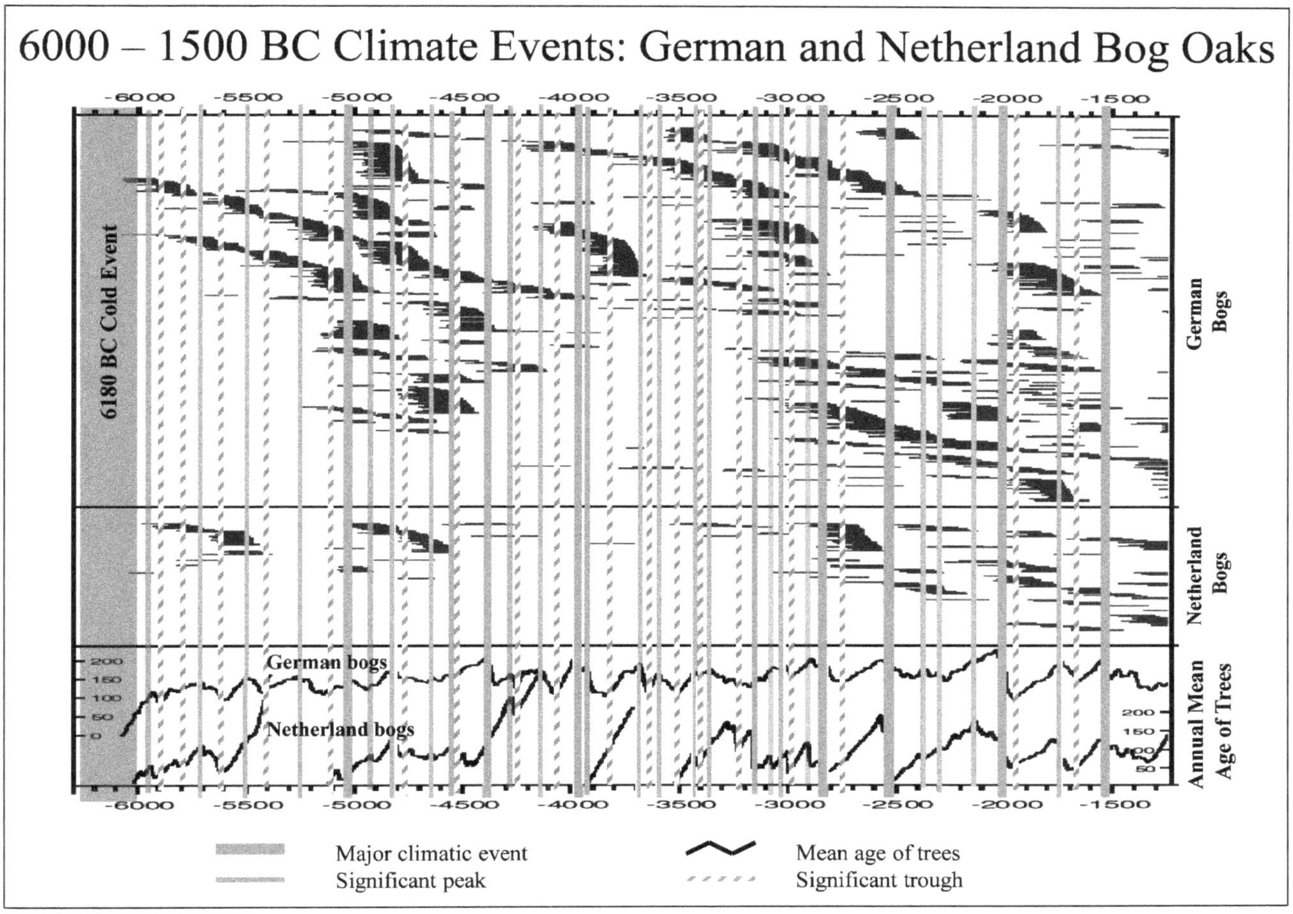

Figure 16.6. Lifespan of German and Netherland bog oaks by site with annual mean age graph (after Leuschner *et al.* 2002 Fig. 4 with modifications). Wide vertical line: Major climate events of varying lengths. Narrow shaded line: Possible climate anomaly deduced from peaks in the mean age curve. Hatched line: Significant trough

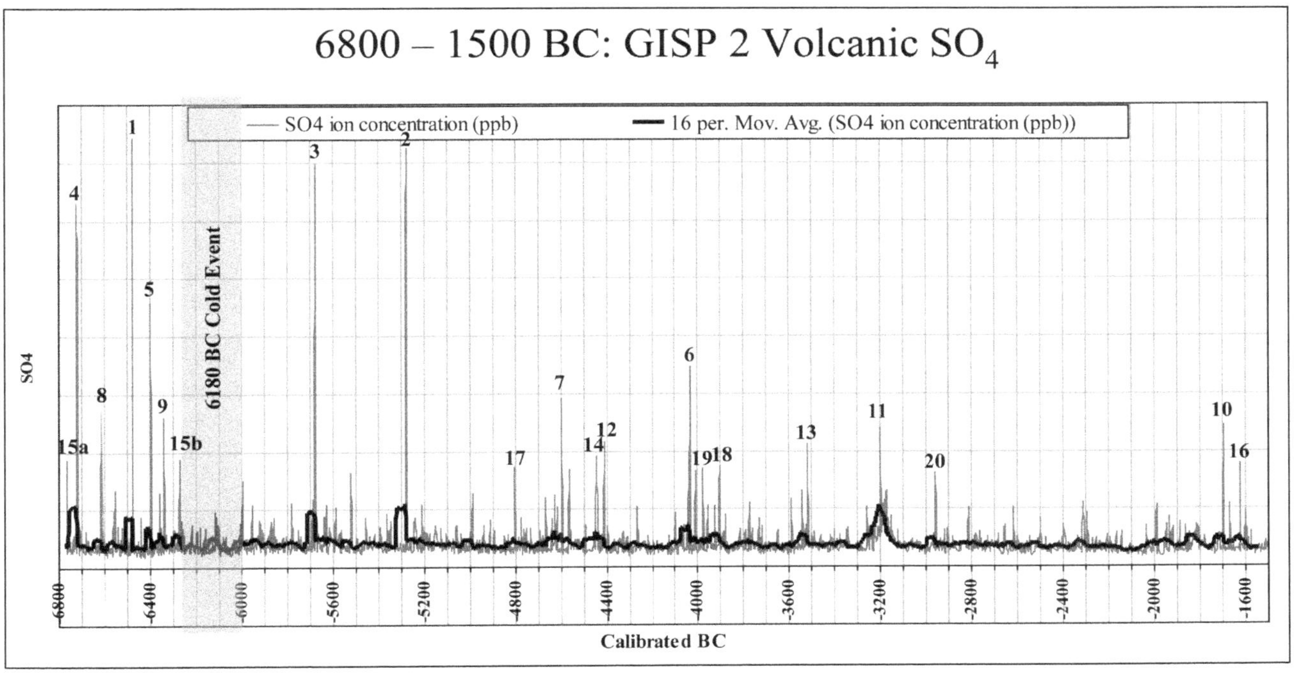

Figure 16.7. GISP 2 volcanic SO_4 concentrations (6800-1500 BC). Curve: Trend line (moving average). 1-20: Highest SO_4 levels, with 1 being the highest and the rest being successively lower. 15a/15B: Approximately the same SO_4 quantity. Lower quantities in the proximity of higher SO_4 levels are not labeled

Schmidt and Gruhle (2003) confirm these trends, especially during wet maritime climate episodes, when tree growth is most homogeneous.

Other proxies, such as eastern Alpine glacial advances or retreats, probable NAO events, and volcanic SO_4 accumulations in the GISP2 record complement the climatic events delineated here (Fig. 16.7). In addition to these climate proxies, others, such as those presented by Kalis *et al.* (2003), will occasionally be used in tracing the agricultural barriers of Central and Northern Europe and numerous related socio-cultural adjustments to prehistoric climate change.

FARMING BARRIER 1

Even though, reports of the earliest pollen and seeds (ca. 6600-6250 BC) from the Alpine region (Fig. 16.1) have resulted in a debate about the earliest cereal cultivation (e.g. Behre 2007, 2009, Nielsen 2009, Tinner *et al.* 2007), it is paradigmatic that the first farmers or at least farming came to Central and Northern Europe from the Near East via the Balkan Peninsula. Differences of opinion hinge largely on the degree to which invasion or migration of allochthonous farmers and the acculturation of autochthonous European foragers are involved (e.g. Ammerman and Cavalli-Sforza 1971, 1973, 1984, Budja 2001, 2004a, 2006, Breunig 1987, Chapman 2003, Davison *et al.* 2006, 2007, Dolukhanov *et al.* 2005, Gkiasta *et al.* 2003, Pavúk 2004, Perlés 2001, Radovanović 2006, Schubert 1999, Tringham 2000, Whittle *et al.* 2002, Andel and Runnels 1995, Zvelebil and Lillie 2000:57-72, Zvelebil 2004:183-187, Klopfstein *et al.* 2006, Zeder 2008). This section does not aim to settle these differences. Instead, its goal is to evaluate the archaeological data to ascertain correlations between human adjustments and climatic fluctuations during this Mesolithic-Neolithic transition in the Carpathian Basin.

The Mesolithic-Neolithic Transition

Much of the research has concentrated on the farmers. Excavated Late Mesolithic sites are still rare and their ^{14}C dates are even rarer. Thus, dating of the Mesolithic-Neolithic transition is highly inconsistent and stems essentially from Neolithic sites (cf. Zvelebil and Lillie 2000 Fig. 3.8, Schubert 1999:247 No. 70, Sherratt 1997 Table 11.6-11.7, Tringham 2000:27, Whittle 1996 Fig. 3.4). Recent increases in the number and quality of the ^{14}C dates has improved the chronology, but the new insights also illustrate the limits of our archaeological knowledge.

Dating the Mesolithic

The Mesolithic sites of the Danube gorges (Iron Gate area) are located on the southeastern threshold of the Carpathian Basin (Fig. 16.1). Their stratigraphy and absolute dating are rather problematic and under constant review (Borić and Dimitrijević 2007, Boroneanţ and Dinu 2006 Boroneanţ and Dinu 2006, Dinu *et al.* 2007, Tringham 2000, Whittle *et al.* 2002). Borić and Miracle (2004 Table 4) provide ^{14}C dates that appear to indicate repeated Mesolithic occupations starting at Padina around 9600 BC. The Mesolithic-Neolithic transition is placed around 6300 BC (Borić *et al.* 2004:236). At Lepenski Vir an early occupation with trapezoidal houses (Phase I/II) is calculated to start shortly after 6300 BC and its end is placed at 6000/5900 BC (Borić and Dimitrijević 2007 Fig. 3 and 4).

Close to Farming Barrier 1, near the confluence of the Tisza and Maros River, the Early Neolithic site of Marosele-Pana provides a ^{14}C date, which is also presumed to be of Mesolithic origin (Whittle *et al.* 2002:77-78). Recalibration of this date (OxA-9403 7765±55 bp) at 1σ yields 6650 (68.2%) 6510 BC. Located barely beyond Farming Barrier 1, two of four Mesolithic dates from Jászberény fall between ca. 6500 to 6000 BC (Fig. 16.8), indicating coexistence and suggesting the possibility of contact between foragers and farmers (cf. Chapter 10-11 this volume, Otte and Noiret 2001).

Dating the Early Neolithic

The earliest farming activity in the Carpathian Basin is customarily connected with the appearance of the Starčevo, Körös and Criş cultures, which are often combined into a single complex. Dating the complex varies and there is a debate about the relationship between monochrome and painted pottery. Without intending to engage in this debate, I will explore the main dating schemes in conjunction with dates from the *14CEurope* database.

In Hungary Horváth and Hertelendi (1994:118) propose 6300 to 5500 BC for Early and Middle Körös. Late Körös (Proto-Vinča 1) is dated 5500 to 5400 BC, while 5400-5300 BC is assigned to what is called the *surviving* Körös (Proto-Vinča 2). A year later this chronology is followed by a statistical analysis of the ^{14}C dates, which dates an Early Körös Phase to 5950-5400 BC and a Late Körös Phase to 5770-5230 BC (Hertelendi *et al.* 1995 Fig. 1). With minor revisions, this chronology has been applied in Hungary to environmental/climatic research focusing on marshlands (e.g. Sümegi 2004:318 Table 29).

Schubert (1999) starts with a Neolithic monochrome pottery horizon at the northern edge of the Balkans, for which he lists a start date of 6400/6200 BC from Maroslele-Pana (Schubert 1999:247, No. 70, Plate 68).[4] A similar date is reported by Horváth and Hertelendi (1994:122) from the same site (Deb-2733: 7497±56 bp). Its calibration yields 6440 (48.4%) 6350 BC and 6310 (19.8%) 6260 BC at 1σ, or 6460 (95.4%) 6240 BC at 2σ. Schubert (1999:245) also mentions a virtually identical date from the Poljanica Plateau, northeastern Bulgaria. The dates are within the cumulative 2σ probability curve

[4] Schubert only publishes the calibrate range of this date. He notes that it is from Körös II context and was originally reported by the Heidelberg Radiocarbon Laboratory in 1996.

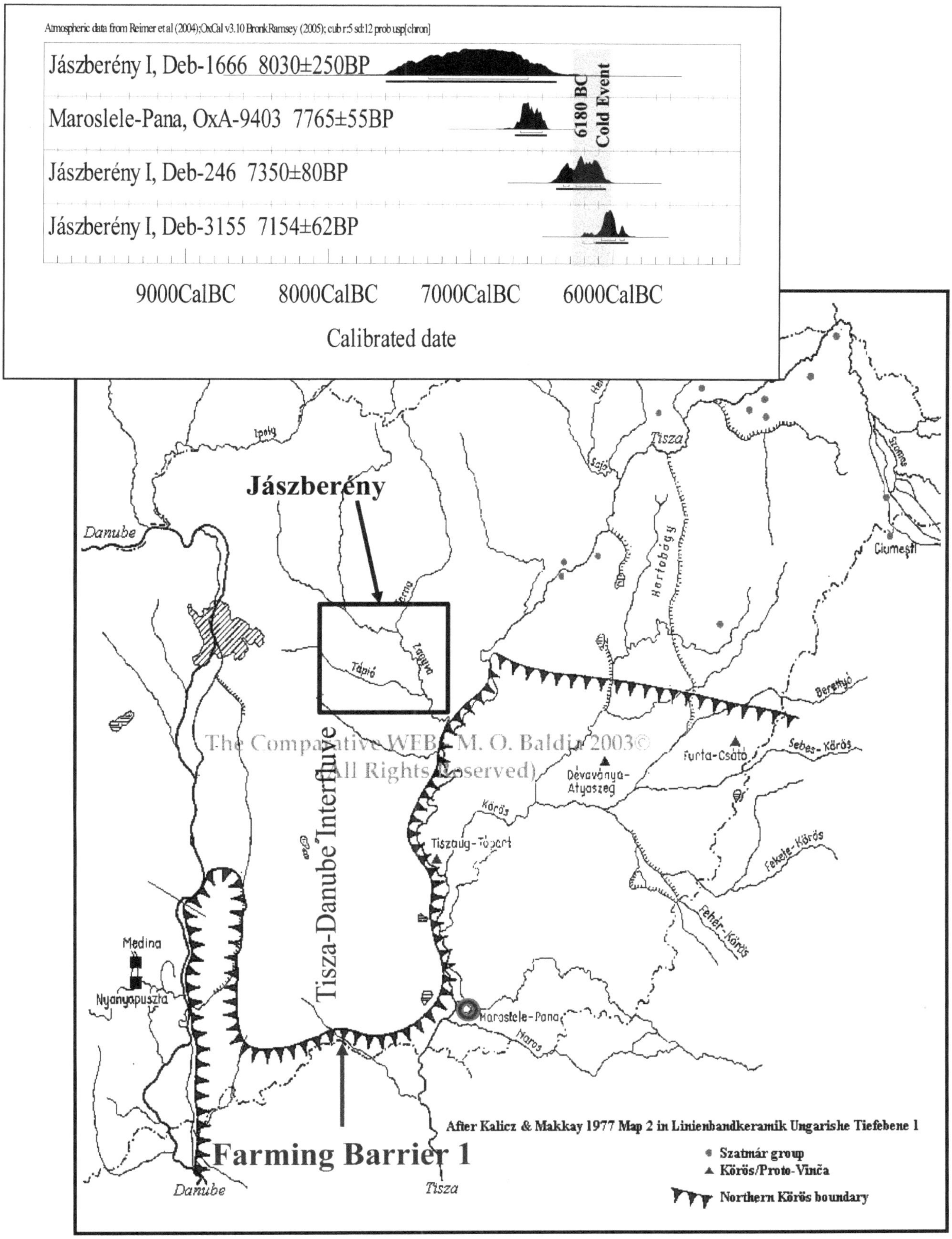

Figure 16.8. Location and ^{14}C dates of Mesolithic Jászberény sites and Maroslele-Pana (small circle) in relation to the traditional Körös-Starčevo boundary (map after Kalicz and Makkay 1977 Map 2)

calculated by Hertelendi *et al.* (1995) for Körös. This would imply that a Neolithic monochrome pottery horizon occurs close to Farming Barrier 1 during the 6180 BC cold event (Fig. 16.8). On the other hand, Budja

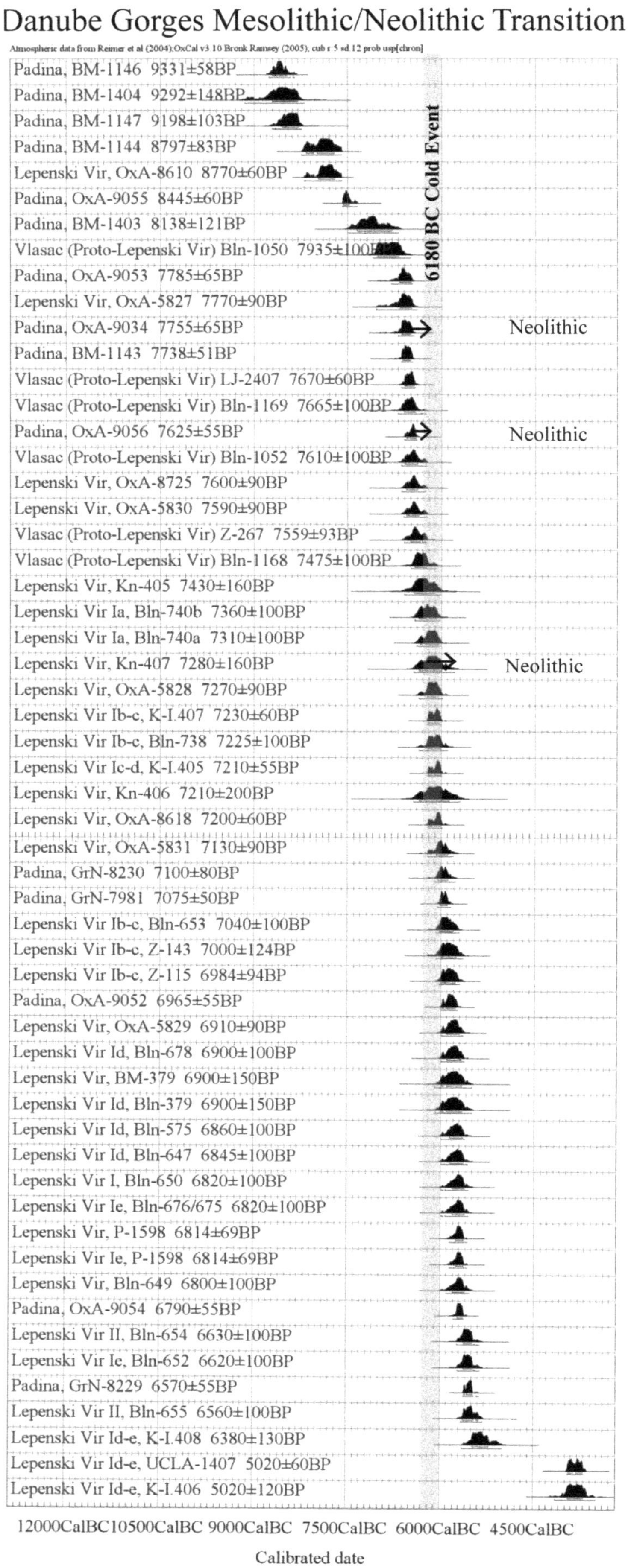

Figure 16.9. Radiocarbon dates of the Mesolithic-Neolithic transition near the Danube gorges (N = 56). Dates after Breunig 1987, Whittle *et al*. 2002. Neolithic: Earliest possible Neolithic dates. Arrow: Estimated reservoir correction, otherwise no correction

(2004b) maintains that the monochrome distribution is evidence of the indigenous Mesolithic population, which may or may not have engaged in farming. Whittle *et al.* (2002) have still another interpretation. They assert that monochrome and polychrome pottery are concomitant in some sites and maintain that the earliest Neolithic evidence occurs in the Danube gorges sites and at Blagotin-Poljna (Fig. 16.1).

Whittle *et al.* (2002) stress that Starčevo-like pottery occurs with the trapezoidal houses of Lepenski Vir. The long debated provenience of the ceramic pot with spiral motif is said to belong to House 54. It has an old radiocarbon assay (Kn-407 7280±160 bp), which yields 6360 (8.5%) 6290 BC and 6270 BC (59.7%) 6000 BC at 1σ. At nearby Padina two ^{14}C dates (OxA-9034: 7755±65 bp and OxA-9056: 7625±55 bp) from dog bones in the vicinity of House 12 and House 9 date to ca. 6500 BC (Fig. 16.9). This would suggest that the Neolithic technology was available at the threshold of the Carpathian Basin even before the 6180 cold event. However, the possibility of a still ill-defined reservoir effect of ca. 200-500 years would move the date closer to 6000 BC (Whittle *et al.* 2002:92).

Blagotin-Poljna is the other Starčevo site, which dates prior to 6000 BC (Whittle *et al.* 2002:87-89). It is located at the edge of the Carpathian Basin, to the south of the Danube-Tisza confluence. Its ^{14}C date (OxA-8608, 7480±55) calibrates to 6420 (42.9%) 6340 BC, 6310 (25.3%) 6260 BC at 1σ and 6440 (95.4%) 6230 BC at 2σ, placing the site before the cold event (Fig. 16.10). Within the basin Whittle *et al.* 2002 list the following sites: Topole-Bač, which is just north of the Danube and about 80 km west of the Danube-Tisza confluence, dates to 6000 BC.[5] Perlez-Batka, near the Danube-Tisza confluence in the Banat, Donja Branjevina in the Bačka area of Vojvodina, and Pitvaros, just north of the Maros River in southernmost Hungary, have an upper limit of ca. 6000 BC. This suggests to the authors that the Carpathian Basin begins to be settled by farmers from 6000 BC onwards.

The application of various methods to calculate the reservoir effect places the Mesolithic-Neolithic transition in the Danube gorges region between ca. 6300 to 5950 BC (Borić and Miracle 2004:341). Borić and Dimitrijević (2007) also provide reservoir corrected ^{14}C AMS dates to put the beginning of Lepenski Vir I/II, including the construction and occupation of trapezoidal houses, around 6300/6200 BC. The phase ends around 6000/5900 BC, when Phase II and evidence of domesticates starts. Even without considering the reservoir effect the ^{14}C dates indicate that the Starčevo-Körös-Criş complex probably appears during the 6180 BC climate event (Fig. 16.9 and 16.10). This is supported by Stadler's calculations (Table 16.1). Although, the number of precision ^{14}C dates for specific phases is still rather limited, the table indicates that the culture lasts from 6300/6200 to 5500/5400 BC.

[5] Topole-Bač dates stem from two flexed back-to-back skeletons of the same burial, but their dates differ widely.

Climate Change and Human Adjustments

Todorova (1989, 1998, Weninger *et al.* 2005) has suggested for some time that climate change drives the expansion of farming from the Near East into Southeastern Europe. This expansion is said to come to a halt in the Carpathian Basin. In Hungary the halt is initially attributed to a barrier of autochthonous foragers defined purely by the site distribution of the Körös culture as known in the 1970s (Fig. 16.8).[6] Eventually, the lack of Late Mesolithic sites begged for a different explanation. Hence, Farming Barrier 1 is later attributed to inhospitable soils for the Near Eastern domesticates, hindrance by the forest environment of Central Europe, the boundary between the Mediterranean and Continental climate zones, or a combination of these factors.

An early effort to support the existence of a climate barrier by Kosse (1970) correlates modern environmental and climatic conditions with the Neolithic site distribution in Hungary. A more recent model attributes the barrier to climatic and vegetation zones, growing-season temperatures, distribution of soil types, as well as the location of Mesolithic and Early Neolithic human impact sites (Chapter 11, this volume). The model states that the *mosaic* environment and climatic conditions conspired to create a complex barrier, resulting in an ecological trap for Starčevo-Körös farmers. It relies among other things on pollen analysis that is correlated with climatic and environmental change in the stratigraphy of bogs. However, it was not until recently that the 6180 cold event was confirmed through palynology at two bogs in the Austrian Alps (Kofler *et al.* 2005). Results from this analysis are comparable in time and magnitude to the GRIP Δ^{18}O data, sea-surface cooling in the North Atlantic, the Scandinavian glacier expansion and other proxies (ibid. Fig. 4 and 5).

The GRIP Δ^{18}O data demonstrate that the 6180 cold event is a complex anomaly (Fig. 16.4) so that estimates for its duration vary. Thomas *et al.* (2007) date it to 6297-6136 BC. For Central Europe the stable oxygen isotope record of ostracode valves found in a sediment core from the Ammersee, a lake in South Germany, points to a 180 to 200-year duration (Grafenstein *et al.* 1998:74). Data from a lake near Bern, Switzerland, indicate a 300 to 400 year cooling period (Heiri *et al.* 2003).

GISP 2 volcanic SO_4 accumulation indicates very high SO_4 concentrations indicative of strong volcanic activity before the 6180 cold event (Fig. 16.7). During the cold anomaly there is relatively little SO_4 activity. The 6180 cold event probably ends with the relatively high SO_4 accumulation around 5993 BC.

[6] The initial conception of the barrier is now questioned, because Körös sites occur beyond this boundary and Körös dates overlap with AVK dates (e.g. Bánffy 2004, 2006, Chapman 2003:91-92, Whittle *et al.* 2002).

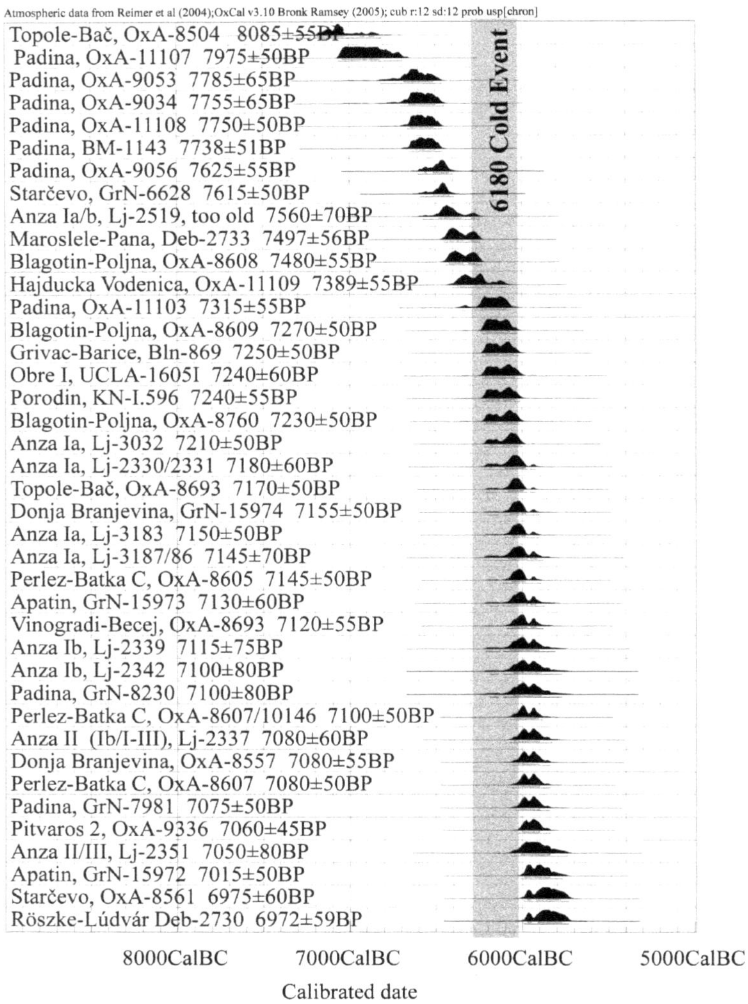

Figure 16.10. Early Starčevo-Körös-Criş ^{14}C dates ($\sigma < \pm 80$ year, N = 40). (Lepenski Vir dates see Fig. 16.9.)

Table 16.1. Starčevo phase sequencing using 41 ^{14}C dates at 1 Standard Deviation (Stadler personal communication 2009)

Starčevo	**^{14}C Samples**	**Beginning Time Interval (Highest Probability)**			**End Time Interval (Highest Probability)**		
Phase	N	Earliest	Latest	Mean	Earliest	Latest	Mean
Monochrome	3	6170	5990	6080	5765	5725	5750
White on Red	16	5765	5725	5750	5720	5685	5700
Linear	5	5720	5685	5700	5690	5650	5670
Spiraloid	13	5690	5650	5670	5500	5445	5470
Lengyel IV	4	5500	5445	5470	5450	5320	5390

In the Arctic, a cooling of up to ca. 3.3±1.1°C is estimated from the Greenland ice core data (Wiersma 2008). Other climate models indicate an air temperature drop of 3-7° C within ±50 years over the Atlantic. This is followed by a gradual ca. 200-year warming trend (cf. Weninger *et al.* 2005 with references). A cooling to 11.1-11.5°C in the July air temperature is projected for the Bern area of the Alpine piedmont (Heiri *et al.* 2003). Furthermore, Magny *et al.* (2003) propose wetter conditions for Europe's mid-latitudes during the 6180 cold event. Still, it is presently difficult to determine precisely how precipitation levels affected the Carpathian Basin hydrology and the introduction of farming (Budja 2007).

Dry conditions could imply that the historically rather wet Carpathian Basin drainage system (Budja 2007 Fig. 5) would make the region more conducive to planting the domesticated Near Eastern cereals. The assumption of dryer conditions may be bolstered by the construction of a water-well at Zadubravlje-Dužine, Slavonski Brod, Slovenia, which reaches a depth of over

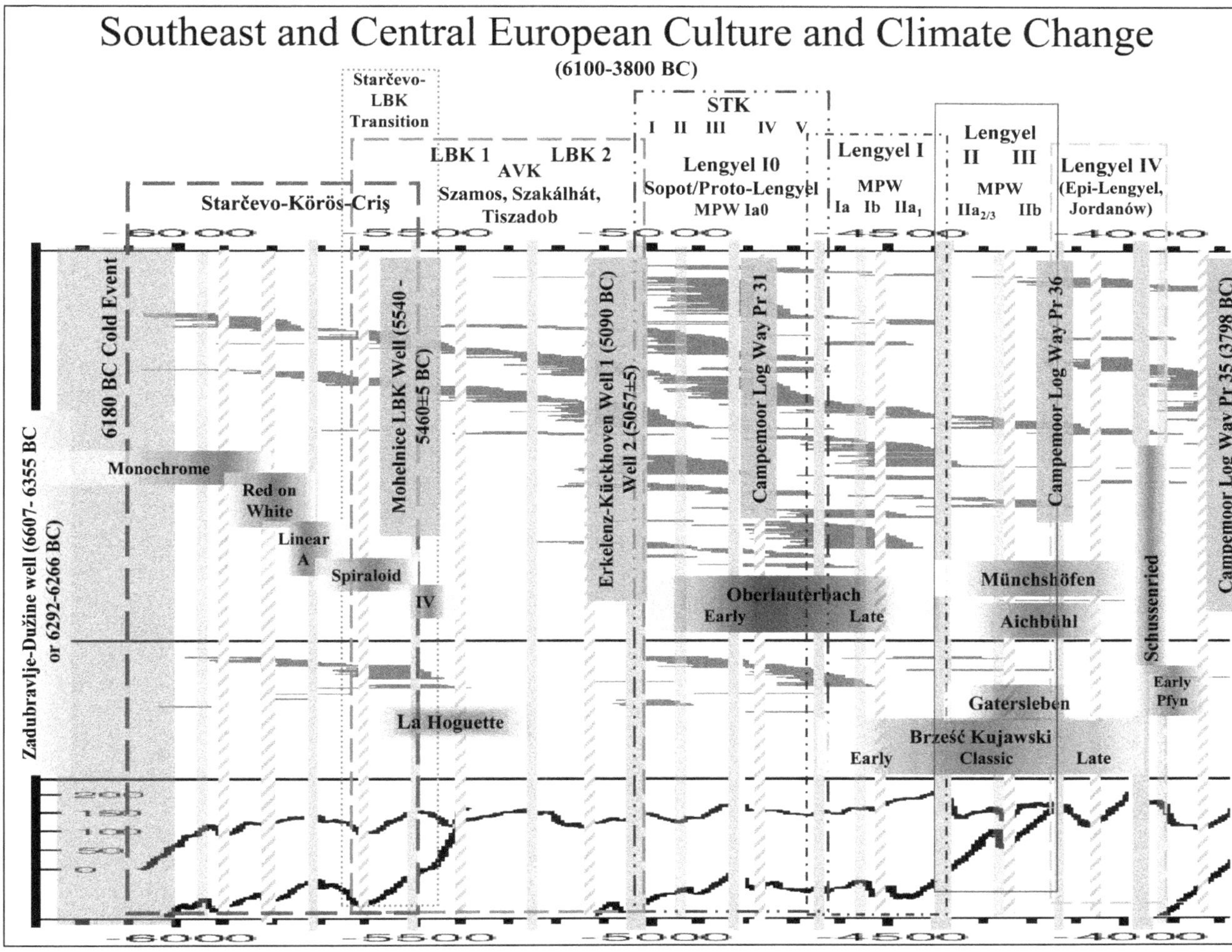

Figure 16.11. Southeast and central European culture and climate change (6100-3800 BC). Starčevo-Körös-Criş phases are rough interpolations of Table 1. Linienbandkeramik (LBK), Austro-Moravian Painted Ware (MPW). Alpine Piedmont tree-ring dated cultures or groups: Schussenried (3955 to at least 3871 BC, Hornstaad (4052 to at least 3890 BC), Early Pfyn (3870 to 3790 BC)

4.50 m, unless of course, the Starčevo village is built at a higher elevation to escape wet conditions, including river floods (Fig. 16.11).[7] Indeed, Budja (2007) proposes increased flooding, but points to likely geographical differences in precipitation. Regardless of the degree of precipitation, Borić *et al.* (2008) stress that Vlasac and similar Danube gorges sites were occupied during the 6180 cold event. Only more detailed site-specific data will sort out the specifics.

In spite of the fact that all pertinent proxies indicate strong climatic changes during the 6180 cold event, physical anthropology presently does not support the traditional hypothesis of a violent encroachment by allochthonous farmers on autochthonous Mesolithic foragers (Roksandic *et al.* 2006) during this period. Instead, foraging may have continued in areas with an environment rich in natural resources, such as the Danube gorges area. Therefore, it may be more prudent to distinguish between farming and foraging communities, rather than a monolithic allochthonous Neolithic versus an autochthonous Mesolithic ethnic group. This is accentuated by Borić *et al.* (2004), who challenge the assumption by Bonsall *et al.* (2004) that a largely fish-based subsistence in the Mesolithic is changed to a diet that is significantly altered by the introduction of farming. The former stress that even though Starčevo pottery is in use at Lepenski Vir, Padina and Vlasac, and there is an apparent increased reliance on terrestrial food sources, this is unrelated to agriculture and animal husbandry, even after 6000 BC (Borić *et al.* 2004, Budja 2004b:407). Thus, Budja (2007:196-197) concludes that the trapezoidal houses and the early pottery appears in hunter-gatherer contexts at Lepenski Vir and Padina before the cold event, while domesticated goat, pig and cattle are associated with later pits. In other words, despite apparent sedentism during Lepenski Vir I/II, it is not until after the start of Lepenski Vir III (6000/5900 BC) that domesticated animals and domed ovens appear.

[7] The precise elevation of Zadubravlje-Dužine vis-à-vis the prehistoric floodplain is not given. The well's date [Z-2924: 7610±140 bp; 68.2% probability (prob.) = 6607 (63.9%) 6355 BC, 6292 (4.3%) 6266 BC; 95.4% prob. = 6778 (94.2%) 6204 BC, 6144 (1.2%) 6101 BC] seems too early for the other dates from the Starčevo Linear A village (cf. Minichreiter 1998, Obelić 2002:618-619), which postdate the 6180 BC cold event.

If the start of Lepenski Vir III is placed at 6000/5900 BC, it roughly coincides with the GRIP temperature peak at ca. 5936 BC (Fig. 16.4), the fairly high GISP 2 volcanic SO_4 traces around 5993 BC (Fig. 16.7) and the Main River oak life expectancy peak around 5950 BC (Spurk *et al.* 2002 Fig. 4). However, if one accepts that the domed ovens date no earlier than 7891±38 BP (Borić and Dimitrijević's 2007), i.e. 5750 (68.2%) 5670 BC at 1σ, then Lepenski Vir III is concomitant with the peak in the oak curve just before 5700 BC, the third highest GRIP temperature peak around 5706 BC and the third highest GISP 2 volcanic SO_4 concentration at ca. 5676 BC. The German tree growth homogeneity index indicates a continental climate phase reaching a maximum around 5700 BC, followed by a maritime climate that peaks around 5660 BC (Schmidt and Gruhle 2003 Fig. 6).

Beyond the Danube gorges, i.e. within the Carpathian Basin, the period of infilling (Whittle *et al.* 2002:88) and cultural consolidation begins after the 6180 cold event. The process may involve the implementation of selected Neolithic features by various groups as they adjust their socio-cultural responses to local environments at particular times. This would result in the mosaic-like adjustment pattern observed by Tringham (2000:53-54). Unfortunately, such adjustments to climate oscillations can only be hypothetically matched with the still ill-dated phases of the Starčevo-Körös-Criş culture complex (Fig. 16.11) and the diverse regional cultural changes noted above. Nevertheless, the climatic oscillation from the hot peak to 5706 BC to the cold spike of 5607 BC may well foster the expansion of the farming territory, leading to occupations as far west as the Balaton and Vienna areas. Most importantly, this implies that Farming Barrier 1 was not entirely static.

The demise of the Starčevo-Körös-Criş complex between ca. 5600 and 5400 BC coincides with the notable climate anomalies of this time. At the same time new cultures or groups, including Vinča, the Transdanubian Bandkeramik (LBK) and the eastern Bandkeramik (AVK, Esztár, Szakálhát and Tiszadob) appear in the archaeological record. Of these, the LBK development is the most pertinent for the northwestward expansion of farming (Fig. 16.1).

FARMING BARRIERS 2A AND 2B

Originally the lack of Mesolithic sites triggered the notion that a Near Eastern Neolithic population migration or diffusion spread farming into a Central Europe that was devoid of people (e.g. Tringham 1971). The genesis of the Central European Bandkeramik is often seen as part of this process (e.g. Neustupný 2004), which is attributed to competing causes, such as population growth, soil depletion, etc. (e.g. Bogucki 2000).

In Hungary, the paradigm holds that the Near Eastern farming technology of the Starčevo-Körös-Criş culture complex is eventually modified by the continental climate. This causes the development of the Bandkeramik farmers who break the Farming Barrier 1, engendering an explosive folk movement that quickly spreads farming beyond the southern Carpathian Basin. Thus, the Hungarian Middle Alföld Linear Pottery (AVK) is seen as developing through contacts with Late Körös-Proto-Vinča culture. The distribution of AVK sites is traditionally seen as separated from the LBK by the ca. 100 km-wide Tisza-Danube interfluve. The LBK dominates the expansion of farming (Sherratt 1997:279-280).

More recent theories focus on a Mesolithic-Neolithic transition during which autochthonous foragers accept the Neolithic way of life through assimilation, adaptation, infiltration, absorption, etc. (e.g. Zvelebil 2004). Such theories imply a more complex process that go beyond simple migration or diffusion. They take into account the regional differentiation and interaction with autochthonous hunter-gatherers and advocate a multifaceted development (Gronenborn 2007). This interaction between the farmers and the autochthonous Mesolithic foragers takes place near Farming Barrier 1 in Western Hungary, adjacent eastern Austria (Bánffy 2000a, 2000b, 2001, 2002, 2004, 2006, Stadler 2005a), and perhaps also in Moravia and southwestern Slovakia (Nowak 2004:9, Pavúk 1994, 2004). Bánffy (2000a, 2000b, 2001, 2002, 2004, 2006) observes that the distribution of late Starčevo sites expand beyond their previous boundary into the area around Lake Balaton. The similarities between the pottery of the late Starčevo and the early LBK in Transdanubia and the shared use of red radiolarite as raw material for chipped stone tools is seen as evidence for the coalescence of farmers and autochthonous Late Mesolithic foragers, who control the prehistoric flint mine at nearby Bakony-Szentgál. Indeed, the radiolarite is distributed from south of Lake Balaton to the vicinity of Frankfurt on the Main River in Germany in the Earliest LBK (e.g. Gronenborn 2007 Fig. 2).

Bánffy allows forager input in the formation of the LBK, but Stadler supports the more traditional population expansion model (Stadler 2005, 2005b, Stadler and Kotova 2008). His arguments are derived from the archaeological excavations at Brunn-Wolfholtz (Brunn am Gebirge), near Vienna, Austria (see below), where the oldest occupation is represented by Late Starčevo-like pottery.

The competing views are profitable hypotheses (e.g. Bánffy 2004, 2006, Lukes 2004, Mateiciucová 2004, Nowak 2004, Oross and Bánffy 2009, and Pavúk 1994, 2004, Zvelebil 2004), but the precise timing and duration of the LBK genesis cannot be determined with great accuracy. The same is true of the pace of its expansion into central Europe. Nevertheless, the distribution of diagnostic artifacts suggests two successive expansion phases with concomitant boundaries. Farming Barrier 2a is established during LBK 1 (Fig. 16.1 and 16.2). During the LBK 2 the farming territory reaches its maximum and grinds to a sudden and long-lasting halt (e.g. Rowley-Conwy 1995:348). These boundaries are sometimes deemed to be an actual cultural barrier between farmers and Mesolithic foragers that were defended by LBK

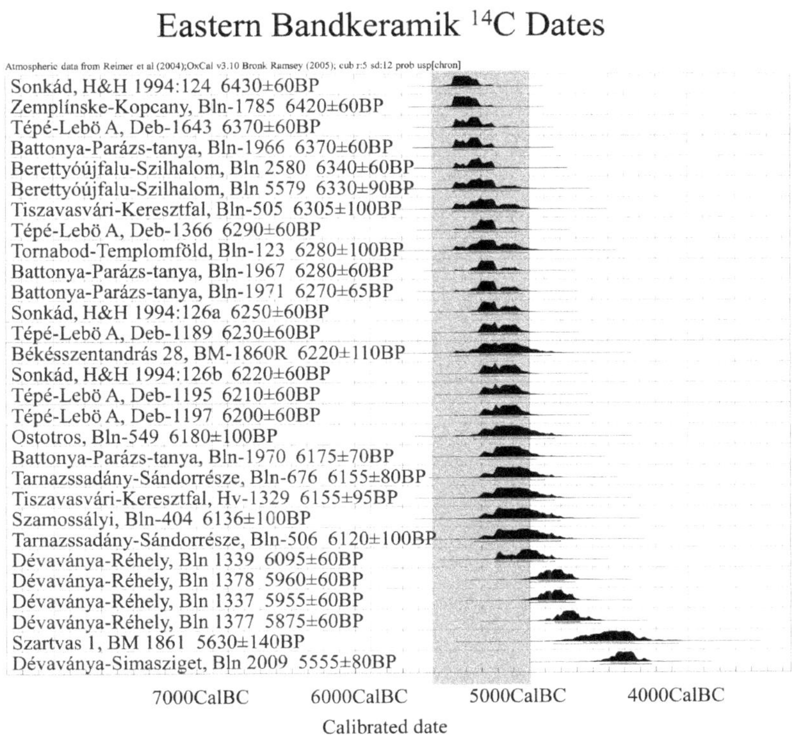

Figure 16.12. Eastern Bandkeramik ^{14}C dates, including AVK, Esztár, Tiszadob and Szakálhát (N = 23). Shaded area: Estimated Eastern Bandkeramik duration. Tiszadob: Tiszavasvári-Keresztfal (Bln-505) and Ostoros (Bln-549). The six most recent dates belong to the Szakálhát Group, but Szakálhát also has some of the oldest dates

earthworks (Golitko and Keeley 2007). However, others question whether there are militant relations between the two groups (e.g. Montfort 2008a). Irrespective of the view on violence, the LBK settlement pattern is concentrated in enclaves. Thus, there may not have been a solid cultural boundary, especially since individual sites, such as Brody Pomorskie and Kościelna Jania (Fig. 16.1) coexist with nearby Mesolithic sites (e.g. Hartz *et al.* 2007 Fig. 2, Nowak 2001, 2004, 2006, 2007).

Dating the Bandkeramik

Given the traditional theory of a farming expansion from the southeast to the northwest, one could expect the AVK to be older than the LBK. Indeed, Kozłowski *et al.* (2003) propose that the AVK develops to the north of the Körös culture as early as 5630 BC. This argument is based on only two ^{14}C dates from Moravany, Eastern Slovakia, which have an error range of 240 and 600 years. Conversely, the statistically determined range of ^{14}C dates for the Tisza Basin AVK is 5330-5000 BC, while a range of 5260-4880 BC is calculated for the Szakálhát, Esztár and Bükk Groups (Hertelendi *et al.* 1995 Fig. 1, Kalicz and Raczky 2000:28). My own analysis of the dates for the Eastern Bandkeramik in the *14CEurope* database, including twenty-nine ^{14}C dates from the AVK, Esztár, Szakálhát, Szamos and Tiszadob Groups, suggest a range of ca. 5500 to 5000 BC (Fig. 16.12). This implies a roughly simultaneous development of the eastern and western Bandkeramik.

Dating the western Bandkeramik development is also a work in progress. Stäuble (1995:229) proposes that the earliest acceptable LBK date is from the collagen assay of a burial in Schwanfeld, Germany. Unfortunately, the grave has no diagnostic artifacts, leaving the date open to debate. Calibration methods also leave room for disagreement. Thus, the ^{14}C sample was originally calibrated to 5521-5444 BC (Hd-14219). Strien and Gronenborn (2005 Fig. 1) recalculate the range to 5600-5480 BC at 1σ. The calibration used here yields 5540-5485 BC (Fig. 16.13).

Stäuble (1995:233, 235) divides the LBK into two phases and suggests a 100-year overlap between the Earliest LBK and the Middle LBK (5300-5200 BC). Thus, the beginning of the Earliest LBK is dated ca. 5500 BC and the Middle LBK ca. 5300 BC, while the Late LBK is dated ca. 5100-5000 BC (cf. Bentley *et al.* 2003:471-472). Many other chronologies deviate only mildly from this scheme, sometimes adding or subtracting phases (Gronenborn 2007 Fig. 1), Lenneis and Stadler 1995, Stöckli 2002, Strien and Gronenborn 2005 Fig. 1, Zvelebil 2004:194).

For Central Transdanubia Oross and Bánffy (2009) propose a Formative LBK Phase estimated to date to 5600/5500-5300 BC (Table 16.2). It is seen as contemporaneous with Late Starčevo in southern Transdanubia, as well as the Mesolithic in northern Transdanubia (ibid). Whittle at al. 2002:90 suggest a range of 5500 to 5400 BC for Szentgyörgyvölgy-Pityerdomb. My own analysis of ten charcoal-derived ^{14}C dates from this site (Bánffy 2000c) reveals that the majority of the dates range between ca. 5480 to 5310 BC. However, the oldest ^{14}C charcoal sample yields a calibrated range of 5570 to 5510 BC.[8]

[8] Szentgyörgyvölgy-Pityerdomb 20, VERA-218: 6611±41 bp; 68.2% prob. = 5620 (22.7%) 5580 BC, 5570 (45.5%) 5510 BC; 95.4% prob. = 5620 (95.4%) 5480 BC.

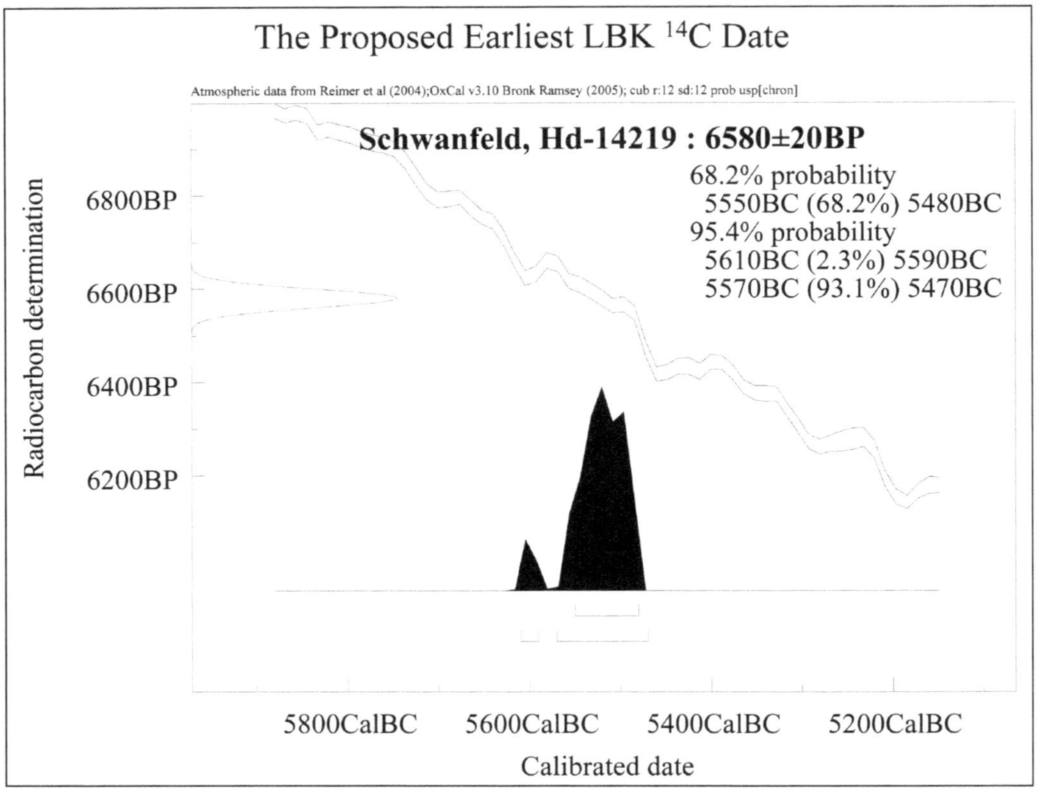

Figure 16.13. The LBK Schwanfeld Hd-14219 collagen based ^{14}C calibration

Table 16.2. Transdanubian phases and estimated absolute chronology (Oross & Bánffy 2009 Table 1)

<table>
<tr><th colspan="2">Transdanubian LBK</th><th>Northern</th><th>Central</th><th>Southern</th><th>Calibrated BC</th></tr>
<tr><td rowspan="3">Early</td><td>Earliest</td><td>Mesolithic and LBK</td><td>Formative LBK</td><td>Starčevo</td><td>5600/5500-5350</td></tr>
<tr><td rowspan="2">Earlier</td><td colspan="3">Bicske-Bíňa</td><td rowspan="2">5450-5300/5250</td></tr>
<tr><td colspan="3">Milanovce</td></tr>
<tr><td colspan="2" rowspan="2">Late</td><td>Notenkopf</td><td colspan="2">Keszthely</td><td rowspan="2">5300/5250-5000/4900</td></tr>
<tr><td>Želiezovce</td><td>Želiezovce & Keszthely</td><td>Keszthely</td></tr>
</table>

Table 16.3. Duration of the LBK settlement areas at Brunn-Wolfholz am Gebirge, Austria (after Stadler 2005a)

Area	Number of Samples	1 σ Range BC
IIa	12	5540-5210
IIb	14	5480-5280
III	24	5450-5200
IV	5	5390-5300
I	4	5310-5060
V	1	5305-5255
Total	**60**	**5480-5060**

For nearby Lower Austria Stadler (2005) calculates the absolute duration of the LBK at 5540-5060 BC, using 60 AMS dates from the LBK occupation of Brunn-Wolfholz (Table 16.3). Typologically, Area IIa is seen as the oldest settlement phase. It is associated with Starčevo-like pottery. Based on twelve AMS dates, the duration of this phase has a 1σ group calibration of 5540 to 5210 BC. However, this seems a little late.

My review of the earliest applicable charcoal-derived date from Brunn-Wolfholz II calibrates to 5734-5625 BC.[9] The next oldest date yields a range of 5638 to 5521 BC (cf. Lenneis and Stadler 1995)[10] and Stadler and Kotova (2008:1) note that Starčevo occupation of Site IIa could start between 5700-5600 BC. Furthermore, my analysis of ^{14}C dates from all pertinent regions of the LBK (Fig. 16.14) suggests that the LBK transition may happen as early as 6050/5600 and no later than 5500/5400 BC. Regrettably, even the tree-ring date of 5540±5 BC from the wood-lined LBK 1 water-well at

[9] Brunn-Wolfholz II, ETH-11148: 6785±75 bp; 68.2% prob. = 5740 (68.2%) 5620 BC; 95.4% prob. = 5840 (95.4%) 5550 BC.

[10] Brunn-Wolfholz II, ETH-11141: 6660±75 bp; 68.2% prob. = 5640 (68.2%) 5520 BC; at 95.4% prob. = 5710 (95.4%) 5480 BC.

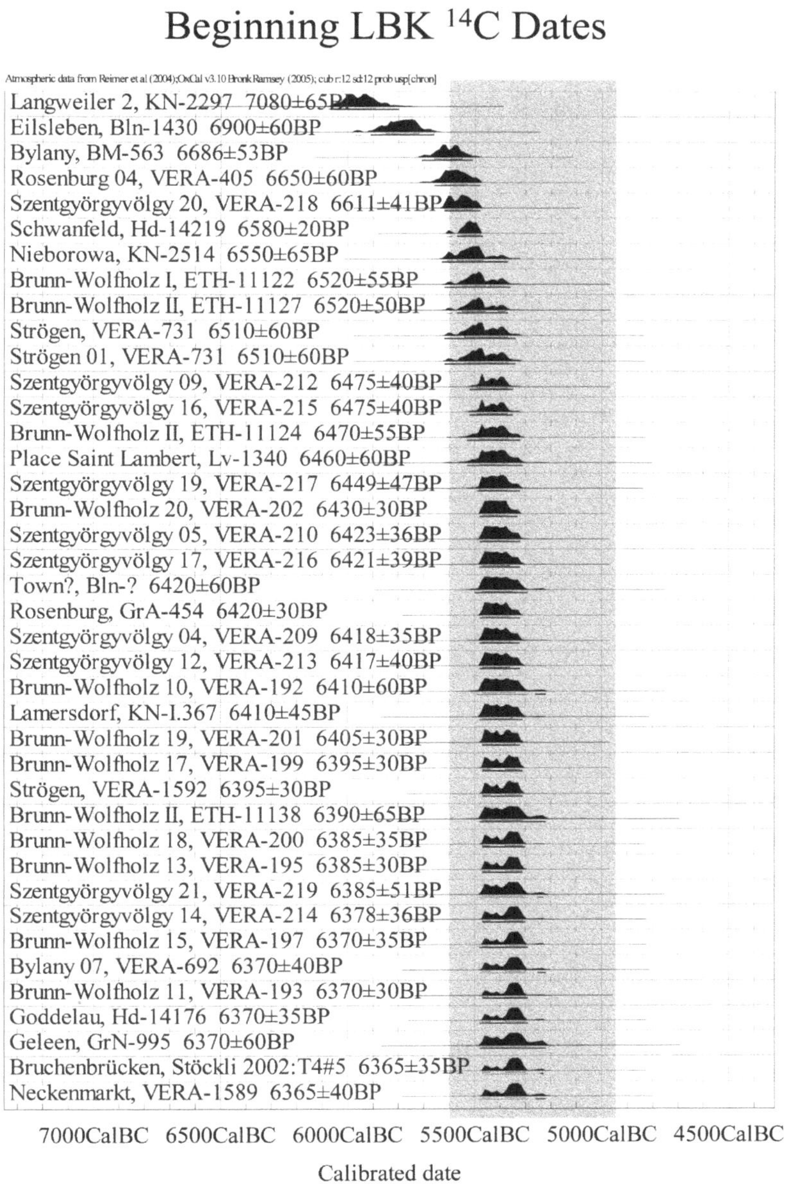

Figure 16.14. Beginning LBK ^{14}C dates ($\sigma < \pm 70$, N = 40). Shaded area: Estimated duration of the entire LBK

Mohelnice, Moravia, does not alleviate uncertainty (Fig. 16.11), since it may date as late as 5460±5 BC, if one compensates for missing later tree-rings (Schmidt and Gruhle 2003:289).

The transition from LBK 1 to LBK 2 transpires between ca. 5300 and 5200 BC, but exactly when Farming Barrier 2b is fully establishment is far from clear. This is exemplified in the Rheinland, adjacent to Dutch Limburg, parts of Belgium and France. In the Paris Basin the LBK is characterized by the Rubané récent Phase, which is thought to appear ca. 5300 BC in the Alsace and sometime later in Champagne, while most of the Paris Basin dates cluster around 5000 BC (Allard 2007:211). Furthermore, Allard (ibid. Fig. 4) dates Rubané récent final from ca. 5000 to 4800 BC and the Villeneuve-Saint-Germain Phase from 4950 to 4650 BC. In Belgium two farming groups are recognized: LBK 2 proper and the Groupe de Blicquy (Crombé and Vanmontfort 2007 Fig. 1). The sum calibration of 73 Belgian LBK dates at 1σ produces a range of ca. 5400 to 5000 BC (Crombé and Vanmontfort (2007 Fig. 2). Applying the same method to fourteen ^{14}C samples of the Groupe de Blicquy yields ca. 5250-4850 BC (ibid.). However, the distribution curve of the samples peaks between 5000 and 4900 BC. This and other evidence makes it likely that the LBK must have appeared in the Scheldt River Basin no earlier than ca. 5300/5200 BC. I calculate an end of the Groupe de Blicquy around 5000/4900 BC.[11]

The breakdown of the LBK in Germany is usually dated around 5000 BC, based on the tree-ring dates from the LBK water well near Erkelenz-Kückhoven, ca. 45 km northwest of Cologne, Germany (see below). A

[11] The *14CEurope* database has nineteen ^{14}C dates for the Blicquy Group, of which fourteen have a σ of less than ±100 years. The latest usable date is from Blicquy-Chaussée (Hv-8409: 6020±55 bp). It calibrates to 4990 (68.2%) 4840 BC at 68.2% prob.

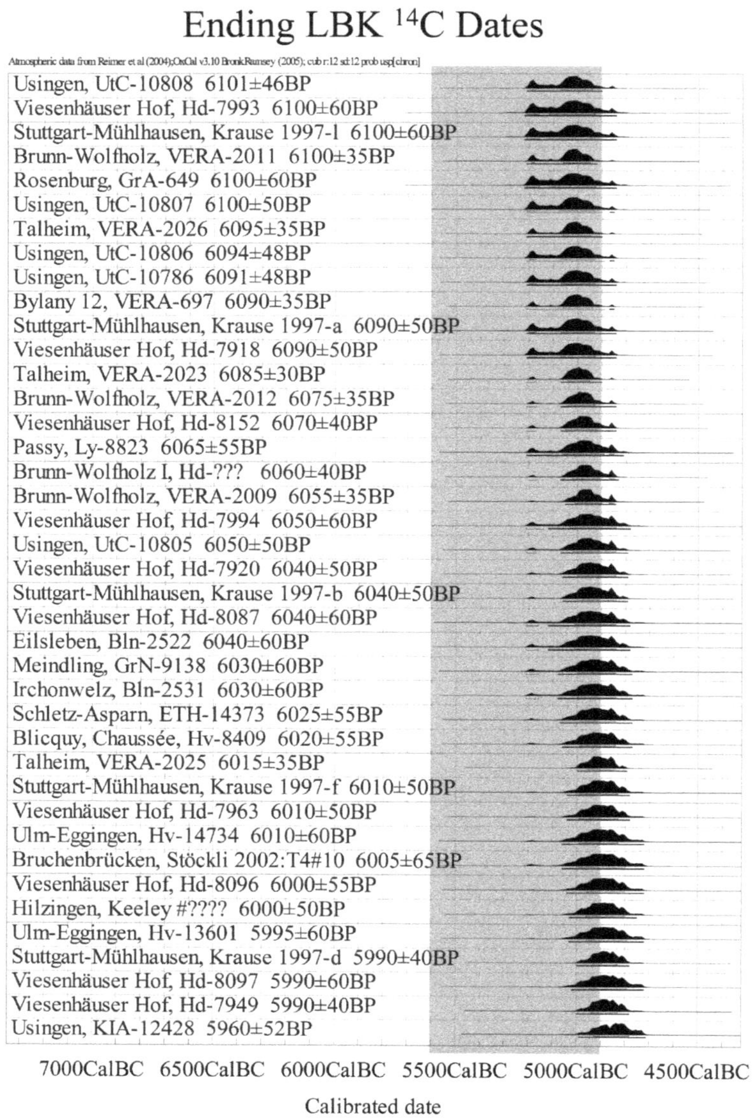

Figure 16.15. Ending LBK ^{14}C dates ($\sigma < \pm70$, N = 40). Shaded area: Estimated duration of the entire LBK

somewhat later dating is preferred in the Danube region (Želiezovce Phase), the Šarka Phase of the Upper Elbe River region of the Czech Republic and the Bükk culture in Northeast Hungary. Some researchers even suggest an end closer to 4800 BC (e.g. Stöckli 2002). My own analysis of the ^{14}C dates suggests an end between ca. 5000 and 4900 BC (Fig. 16.15). Given the uncertainties in absolute and indirect dating discussed above, I will use a two-part division, dating LBK 1 between 5600/5500 and 5300/5200 BC and LBK 2 between 5300/5200 and 5000/4900 BC.

Dating the Interaction of Foragers and Farmers

Evidence for the Late Mesolithic is relatively limited and largely lacks an adequate number of ^{14}C dates. This is particularly problematic in the crucial area of Transdanubia. In adjacent Lower Austria the Mesolithic and LBK occupations of Strögen and Mold are described by Lenneis *et al.* (2009). The lithic raw material indicates that the Mesolithic populations exploit locally available material and the LBK occupants imported Bakony-Szentgál radiolarite (ibid.), which is thought to imply little interaction between foragers and farmers.

In the neighboring lowlands of the Czech Republic the LBK sites are confined to the plain below the highlands that harbor Mesolithic rockshelters (Kacerik 2007 Fig. 3),[12] but their *stone industry may be an indirect indication of interactions* with the Mesolithic population (Kacerik (2007:7). Unfortunately, the lowland Mesolithic sites are deflated and cannot provide stratigraphic evidence regarding the relationship between the foragers and farmers. Furthermore, the stratigraphy of caves in the hill country of northern Bohemia is interpreted as a chronological hiatus between the Mesolithic and the LBK (Svoboda *et al.* 1998, Svoboda *et al.* 2002). Indeed, of the four Late Mesolithic ^{14}C dates from the sandstone rockshelters of Bezdéz, Dobký mlýn, and Pod zubem in

[12] The LBK 1 sites are at the edges of ecological zones on river floodplains or alluvial benches, permitting exploitation of a wide range of natural resources.

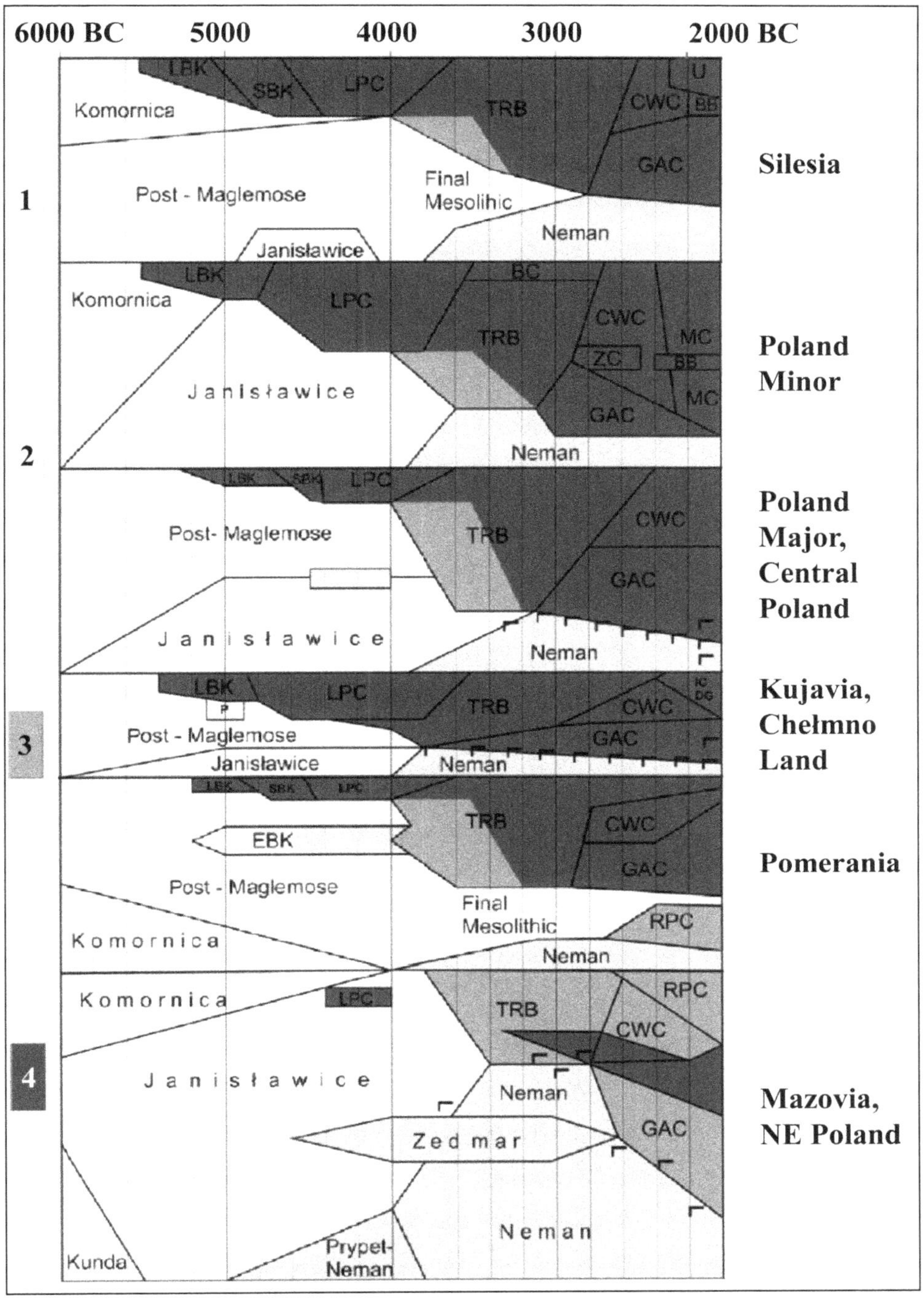

Figure 16.16. Polish Mesolithic and Neolithic chronology (after Nowak 2007 Fig. 7). 1: Late/Final Mesolithic foraging, 2: Pottery-use and foraging (Para-Neolithic), 3: Farming with significant foraging, 4: Farming. Cultures: Bell Beaker (BB), Baden (BC), Corded Ware (CWC), Globular Amphora (GAC), Dobre Group (DG), Ertebølle (EBK), Iwno (JC), Linear Band Pottery (LBK), Linin-type pottery (L), Lengyel-Polgar (LPC), Mierzanowice (MC), Podgaj 32-type sites (P), Rzucewoj-Pamariu (RPC), Stichbandkeramik (SBK), Funnel Beaker (TRB), Únětice (U), Złota (ZC)

Bohemia (Svoboda 2008 Table 9.2, Svoboda *et al.* 2002 Table 2), only the date from Pod zubem (GrN-23333) yields a calibration of 5560 to 5480 BC,[13] suggesting possible temporal if not geographical coexistence with reasonable statistical confidence.

North of the Czech and Slovak Republics, on the North European Plain, dating and interpretation of the evidence is strongly debated (cf. Zvelebil *et al.* 1998). Nowak (2001, 2004, 2006, 2007) develops a chronology of the farming expansion in Poland, simplifies the diverse Mesolithic nomenclature, and indicates that there is no pottery production beyond Farming Barrier 2b until after the demise of the LBK. (Fig. 16.16).

[13] Pod zubem, GrN-23333: 6580±50 bp; 68.2% prob. = 5610 BC (4.1%) 5590 BC, 5560 BC (64.1%) 5480 BC; 95.4% prob. = 5620 BC (95.4%) 5470 BC.

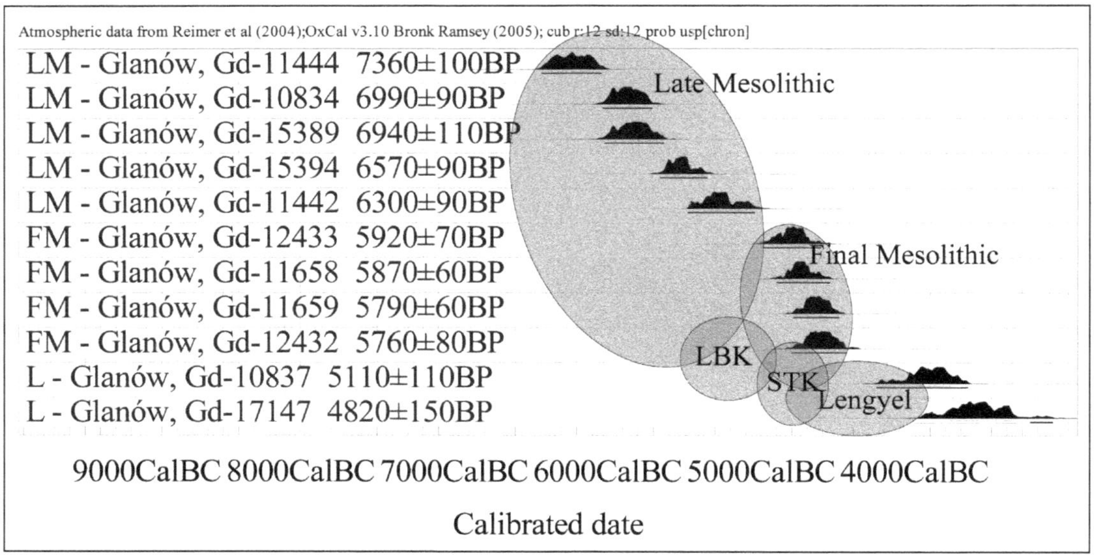

Figure 16.17. Radiocarbon dates from Glanów 3, Poland (Pazdur *et al.* 2004), and the generalized duration of Late Mesolithic and early Neolithic cultures (ovals). LBK: Linienbandkeramik, STK: Stichbandkeramik

Nowak (2007 Fig. 1) maps 91 Late Mesolithic sites. Of these, Glanów 3 is the southernmost. It is located close to an LBK 1 enclave and within the expanded LBK 2 settlement area. A series of ^{14}C dates by Pazdur *et al.* (2004) shows a long Mesolithic occupation and chronological coexistence with the LBK, as demonstrated by ^{14}C date Gd-11442, if not Gd-115394 (Fig. 16.17). Lithics and coarse pottery of several other Polish sites are thought to show farmer-forager interaction. However, Schild (1998) warns of a lack of stratigraphy for sites on sandy soils in which deflation results in the mixing of Mesolithic, Neolithic and even Bronze Age artifacts. Nowak (2007:91) acknowledges the probability that: *a large proportion or even the majority of the ... radiocarbon dates come from mixed context.* It is necessary to add that these dates have huge error ranges (cf. ibid. Fig. 2a and 2b). This includes the two earliest dates from Site 37[14] and Site 46[15] listed by Nowak (ibid. Fig. 2a). They may date either to the LBK or the Stroke Ornamented Pottery/Stichbandkeramik culture (STK) (Fig. 16.16). A similar problem exists at Dąbki 9, located north of Farming Barrier 2b on the Baltic coast (Fig. 16.1). It is considered to be the easternmost representative of the Ertebølle or part of the nearby Neman or the Chojnice-Pienki culture. In addition, it is reported to contain LBK-like pottery as well as domesticated animal bones (e.g. Jankowska 1998, Hallgren 2004a, Novak 2006, Shennan and Steele 2000). However, the site has only rudimentary stratigraphy according to Schild (1998:71). The earliest date for the cultural layer with LBK-like pottery ranges from 4840 to 4700 BC[16] (e.g. Hallgren 2004a, Novak 2007, Shennan and Steele 2000). This falls into the STK rather than the LBK timeframe. Furthermore, pottery making in northeast Poland and adjacent regions stems from a non-LBK tradition and there is no solid evidence that either cultivated plants or domesticated animals are kept by these foragers during the duration of the LBK.

The Mesolithic of Germany is addressed by (Conrad and Kind 1998) and Street *et al.* (2002:427-432). The latter summarize the evidence for the Mesolithic-Neolithic transition and discuss the possibility of a continuation of the Mesolithic chipped stone tradition in the LBK (Street *et al.* 2002 Fig. 27). Certain raw materials for the lithic artifacts of the Earliest LBK are found in Mesolithic territories and the artifact distribution is used to trace interaction with the farmers (e.g. Gronenborn 2003b, 2007). Proximity of sites is also used to hypothesize coexistence, if not outright interaction, as for example in the highlands of the Werra and Weser River, (Grote 1998). Furthermore, coexistence, possible Mesolithic acculturation, and even intermarriage are proposed (Bentley *et al.* 2003, 2005, Price *et al.* 2001, Price *et al.* 2002). However, such propositions are open to question (Bickle and Hofmann 2007).

Closer evaluation of the southern German Mesolithic ^{14}C dates compiled by Drafehn *et al.* (2003 Fig. 2 and 3) indicates three sparsely dated Mesolithic sites between ca. 5550 and 5000 BC. All three are close to the upper Danube River valley (Drafehn *et al.* 2003 Fig. 1). Felsdach Lautereck dates to LBK 1, but is located outside its southwestern distribution. At Hanauhof Nord II three of the site's five ^{14}C dates cover the entire duration of the LBK. The site is some distance from the LBK 1 territory, but within the vicinity of the later LBK's southwestern distribution. The third site (Abri am Galgenberg) is barely within the southern distribution area of both the LBK 1 and 2 territories. However, only one of the site's

[14] Łęczyn, Gd-14019: 6090±180 bp; 68.2% prob. = 5230 (68.2%) 4790 BC; 95.4% prob. = 5500 (95.4%) 4550 BC.

[15] Nowodworce 1, Lod-220: 6000±150 bp; 68.2% prob. = 5210 (3.2%) 5170 BC, 5070 (65.0%) 4710 BC; 95.4% prob. = 5300 (95.4%) 4500 BC.

[16] Dąbki 9, Gd-1278: 5890±60 bp; 68.2% prob. = 4840 (68.2%) 4700 BC; 95.4% prob. = 4930 (95.4%) 4600 BC.

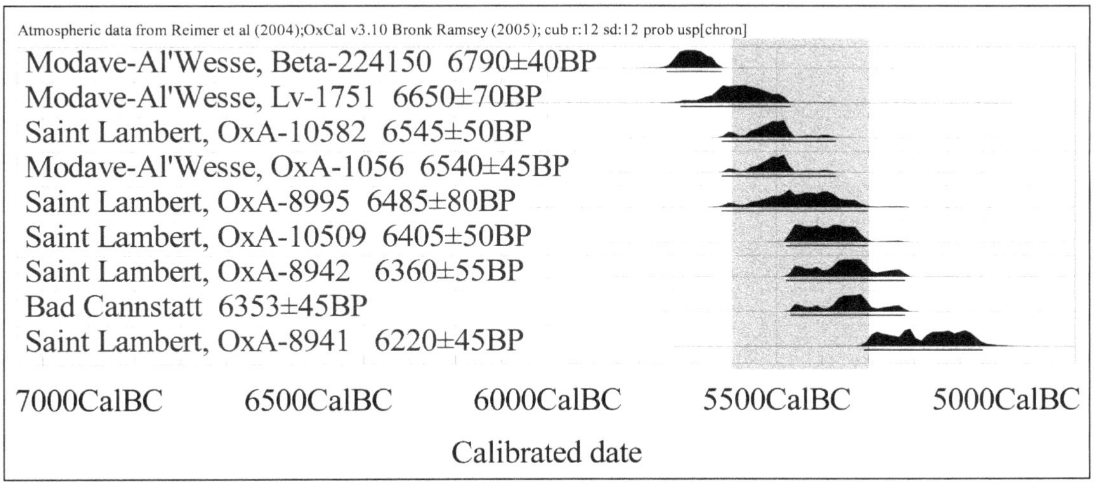

Figure 16.18. Calibrated ^{14}C dates of the La Hoguette sites from Belgium and Germany (N = 9). Shaded area: Likely duration

five ^{14}C dates is roughly concomitant with the LBK, and its error range of ±110 years minimizes the analytical value.

Most additional Mesolithic sites, ranging from Southern Germany via northern Switzerland to Western France, are undated. A few are on the west and northwest side of the LBK and some are even within the LBK settlement region. Most of the sites are attributed to La Hoguette hunter-sheep/goat herders (Street *et al.* 2002 Fig. 27) and sometimes contain conical to round-bottomed pottery tempered with crushed shells and bone. Occasionally sherds exhibit garlands of comb-like impressions. This pottery is thought to relate to the southern French Cardial Ware, but opinions about the exact relationship and age of the pottery vary (e.g. Lanting and van der Plicht 2000:14, Louwe-Kooijmans 2007 Fig. 2, Winiger 1998:7). The few available ^{14}C dates from France do not provide clarity. This is apparently even the case with the stratified, multi-phase site of Bavans, Franche-Comté (e.g. Lanting and van der Plicht 2000, Shennan and Steele 2000). The La Hoguette sites straddle the southwestern LBK 1 boundary (Knipper *et al* 2005 Fig. 1, Street *et al.* 2002 Fig. 27), where the interaction between the Terminal Mesolithic and the LBK has received much scrutiny as summarized by Bickle and Hofmann (2007).

Among the La Hoguette sites within the western LBK 1 boundary is Bad Cannstatt-Wilhelma near Stuttgart, Germany (e.g. Jochim 2000 Fig. 7.5, Price 2000a, Street *et al.* 2002, Winiger 1998). It is identified as a seasonal camp without LBK artifacts. The sole ^{14}C date of organic material from a potsherd of the lowest layer (Gronenborn 2003a) provides a calibration of 5380-5290 BC (Fig. 16.18), which makes it coeval with the later dates for LBK 1.

In the Rhein Delta, including the lower Maas (Meuse) and Scheldt Rivers, La Hoguette is discussed by several authors (Crombé and Vanmontfort 2007, Louwe-Kooijmans 2007, 2009, Lanting and van der Plicht 2000, Roever 2004, Vanmontfort 2008a, 2008b, Vanmonfort *et al.* 2010, Verhart 2008). Two Belgian inland sites (Modave Al'Wesse and Saint Lambert near Liège) offer a series of dates (Miller *et al.* 2009), which indicate that La Hoguette most likely begins between ca. 5600 and 5500 BC (Fig. 16.18). Closer to the coast, the site of Melsele in Flanders is seen as a possible evidence of interaction,[17] but it is a multi-phase site, leaving the relationship between the foragers and the LBK in Belgium a mystery according to Verhart (2008).[18] Nevertheless, the later dates from Saint Lambert fall in line with Bad Cannstatt-Wilhelma, implying that La Hoguette ends as LBK 1 transitions to LBK 2, i.e. when Farming Barrier 2a is shattered.

Louwe-Kooijmans (2007) proposes that La Hoguette develops in South and Central Netherland before 5500 BC, as does the subsequent Limburg Group, which he starts around 5300 BC. Limburg sites are most heavily concentrated within or near the later LBK's western Farming Boundary 2b (Street *et al.* 2002 Fig. 27).

The ^{14}C dates for foragers in Almere-Hoge Vaart, Jardinga, Niew Schoonebeek and Polderweg (Verhart 2008 Table 6.1) are partly coeval with the LBK. Contact is primarily deduced from the stylistic analysis of stone tools and their distributions, as well as the sources of their lithic raw material (e.g. Vanmontfort 2008a, 2008b). It is argued that the locations of LBK settlements are partly determined by environmental preferences and partly by avoidance of the hunter-gatherer territories. A hunter-gatherer territorial shift away from Neolithic villages is advocated, although, that Mesolithic-Neolithic interaction is not thought to be adversarial.

[17] The Groupe de Mesele is seen as Ertebølle-related. It may date to 5000 BC, but similar pottery is also reported from Rosheim, Alsace, in Eastern France (Gronenborn 2003a:85).

[18] Confusion in dating between Verhart (2008) and my article arises from Verhart's use of dates labeled as *cal BC*. The dates appear to be calibrated bp dates, making them 1950 years older than the dates used here.

Table 16.4. Chronology of the Late Mesolithic in Scandinavia, Netherland and Northern/Northwest Germany

Southern Scandinavia (Price, Gebauer *et al.* 2001)	**North Germany** (Hartz and Lübke 2006)	**Low Countries-Northwest Germany** (Roever 2004)
Ertebølle		**Swifterbant**
Trylleskove (5400-4800 BC)	Jäckelberg (5450-5100 BC)	Swifterbant 1 (5100/5000-4600 BC)
	Rosenfelde (5100-4750 BC)	
Stationsvej (4800-4300 BC)	Jarbock (4750-4450 BC)	Swifterbant 2 (4600-3900/3800 BC)
Alekistebro (4300-3900 BC)	Timmendorf (4450-4100 BC)	
		Swifterbant 3 (3900/3800-3400 BC)

Shaded area: The first evidence of autochthonous Mesolithic pottery production

For the German Highlands, Drafehn *et al.* (2003) list the site of Bettenroder Berg IX, located in the vicinity of an LBK enclave forming part of Farming Barrier 2a. One date yields a calibration of 5650 to 5470 BC, which overlaps with the beginning of LBK 1.[19] In Lower Saxony, closer to the North Sea, the site of Wehldorf 6 produced at least one charcoal based date whose calibrated range partly overlaps with the LBK 1, but the date has a huge error range. An AMS date from nearby Wehldorf 7 (Drafehn *et al.* 2003) yields a more useful calibrated range of 5550 to 5380 BC.[20] This means that the site exists as Farming Barrier 2a is being established a short distance away.

In northeast Germany, an AMS date from the Late Mesolithic burial at Steinhagen, Güstrow County, is roughly concomitant with LBK 1 (Terberger and Piek 1998 Table 1).[21] The site is to the north of both Farming Barrier 2a and 2b, but to the south of the Baltic coast. Near the present Baltic Sea, on the Island of Rügen, two dates (UZ-4093 and UtC-6938) from Drigge also date to the time of the LBK 2 (Terberger and Piek 1998 Table 1).[22] The sites are close enough to the LBK boundary that contact is possible, but proximity may not be a significant criterion for interaction, because a typical potsherd decorated in late LBK 2 style of the Rheinland is reported from the Mesolithic layer at Parow near Rügen (e.g. Hartz *et al.* 2007, Kaute *et al.* 2005). The distance is striking, even if, long-distance communication via boat may mitigate it, as the ca. 8 m and ca. 9 m long canoes dating around 5000 BC at the nearby site of Stralsund (*Mischwasserspeicher*) imply (Kaute *et al.* 2005). In spite of such contacts, the knowledge of pottery did not automatically lead to Mesolithic ceramic production since the earliest Mesolithic pottery in the region appears sometime between 4750 and 4450 BC, i.e. long after the LBK is replaced by the STK interaction sphere (see below). Presently, the only exception comes from Schlamersdorf-Travenbrück LA5 (Fig. 16.1). Here, pottery found in the Late Mesolithic (Early Ertebølle) context provides three ^{14}C dates from food crusts, yielding a calibrated range of ca. 5480 to 5050 BC (e.g. Hedges *et al.* 1995, Hallgren 2004a Fig. 7).[23] However, food crusts may suffer from fresh water reservoir effect (Hartz and Lübke 2006 Footnote 2, Hartz, Lübke and Terberger 2007:573, 575). Therefore, the pottery may actually be younger. This is supported by the fact that other Early Ertebølle sites of the southwestern Baltic coast, dating 5450 to 4750 BC, have not yielded pottery (Table 16.4).

It may be concluded that the location of the most northerly LBK enclaves together with a growing body of Mesolithic ^{14}C dated sites suggest that interaction was possible during the entire existence of the LBK. The meager evidence stems mostly from Mesolithic sites. However, the broken perforated piece of Baltic amber from the late LBK 2 context at Erkelenz-Kückhoven (Weiner 1995:187) and the LBK 2 potsherd from Mesolithic Drigge suggestive of two-way interaction. Yet, the quality and quantity of evidence is still too limited and variable to advocate for much more than casual communication, especially over longer distances.

Regarding pottery making beyond Farming Barrier 2b, the earliest Mesolithic pottery is found in Swifterbant 1 context (Table 16.4). In Germany and Denmark, the emergence of Ertebølle pottery occurs only after the demise of the LBK. Although, some pottery in the Polish Mesolithic (Para-Neolithic) has Ertebølle-like traits, it is largely linked to developments to the east. This also the case for the early pottery-making Mesolithic Comb Ware cultures of Finland and western Russia, which chronologically coincide with LBK 2 (Hallgren 2004a). Thus, the acquisition of pottery making seems to be derived from related cultures largely located within boundaries of the former Soviet Union. Some of these cultures have early dates for pottery and occasionally

[19] Bettenroder Berg IX KN-4149: 6610±110 bp; 68.2% prob. = 5650 (68.2%) 5470 BC; 95.4% prob. 5730 (95.4%) 5350 BC.

[20] Wehldorf 7/7 KIA-8937: 6524±49 bp; 68.2% prob. = 5550 (62.9%) 5460 BC, 5410 (5.3%) 5380 BC; 95.4% prob. 5610 (3.0%) 5590 BC, 5570 (92.4%) 5370 BC.

[21] Steinhagen, OxA-292: 6550±90 bp; 68.2% prob. = 5620 (63.9%) 5460 BC, 5410 (4.3%) 5380 BC; 95.4% prob. 5640 (95.4%) 5330 BC.

[22] Drigge, UZ-4093: 6250±80 bp; 68.2% prob. = 5320 (41.4%) 5200 BC, 5170 (26.8%) 5070 BC; 95.4% prob. = 5380 (95.4%) 4990 BC. Drigge, UtC-6938: 6070±60 bp; 68.2% prob. = 5060 (63.3%) 4890 BC 4870 (4.9%) 4850 BC; 95.4% prob. = 5210 (95.4%) 4830 BC.

[23] Schlamersdorf-Travenbrück LA 5, OxA-4802: 6385±60 bp; 68.2% prob. = 5470 (68.2%) 5310 BC; 95.4% prob. = 5480 (89.8%) 5290 BC, 5270 (5.6%) 5220 BC. OxA-4803: 6320±65 bp; 68.2% prob. = 5370 (68.2%) 5210 BC; 95.4% prob. = 5480 (91.3%) 5200 BC; 5170 (4.1%) 5080 BC. OxA-4801: 6155±60; 68.2% prob. 5210 (68.2%) 5040 BC; 95.4% prob. = 5300 (3.4%) 5250 BC; 5230 (92.0%) 4930 BC.

even domesticates (e.g. Dolukhanov *et al.* 2004, Kotova 2009), but they cannot be directly linked to the expansion of farming into Central or Northern Europe.

Climate Change and Human Adjustments

Around 5606 BC the coldest temperature since the 6180 BC cold event is logged in Greenland (Fig. 16.4). In Northwest Germany the oak mean-age curve drops to a low around 5600 BC and rises to a high around 5500 BC (Fig.16.6). The Main River oak mean tree lifespan curve drops below the mean by ca. 5650 BC (Spurk *et al.* 2002 Fig. 4). This is the lowest level since ca. 6700 BC. Around 5610 BC the curve climbs slightly above the mean and drops again at ca. 5550 BC. It stabilizes somewhat between 5500 and 5400 BC at the 5610 BC level. The German tree growth homogeneity index shows a continental climate trough around 5530 BC, a maritime peak shortly after 5500 BC, and another continental trough around 5480 BC (Schmidt and Gruhle 2003 Fig. 6 and 7).

Human adjustments seem to mirror the climatic ups and downs. Sometime between 5700 and 5600 BC Starčevo appears in the Balaton area of Transdanubia. In this area, palynology indicates a fairly dry Late Mesolithic, including a decrease in lake level and an increase in cereal pollen (Bánffy 2004). This is followed around 5600 BC by a drastic increase in hazel (*Corylus*) coupled with a decrease in cereal pollen. At 5500/5400 BC an intense rise in water level is observed. Similar changes are recorded in western Central Europe. In the state of Hessen, Germany, the pollen spectra of the sheltered and fertile intra-montane loess basin of the Wetterau near Frankfurt show an increase of *Corylus* from 5700 BC onwards (Kalis *et al.* 2003). This is attributed to woodland clearing by a Late Mesolithic La Hoguette population. Further changes at ca. 5500 BC are seen as forest clearing and farming by LBK farmers. Pollen spectra from the varves of the Meerfelder Maar, a volcanic lake in the western German Eifel highlands, indicate a disturbance shown by considerable change in arboreal pollen. Late succession species decline strongly around 5500 BC as *Corylus* and *Fraxinus* peak. Afterwards, a secondary forest succession with *Alnus*, *Betula* and *Quercus* develops, only to give way to a lime-dominated forest within a century, thus, returning to the state that existed prior to the disturbance. These environmental changes are also interpreted as being the result of woodland pasturage by LBK farmers from the nearby Mosel River Valley (Kalis *et al.* 2003).

Farther west, the oldest date (Beta-224150) from the La Hoguette site of Modave-Al'Wesse yields a calibrated 1σ range of 5720 to 5650 BC (Fig. 16.18). If one prefers this date as the start of the La Hoguette culture, it is possible to link the change from pure foraging to one that is supplemented by goat/sheep herding with the climate anomaly of ca. 5700 BC. However, the hypothesis still applies if La Hoguette appears in conjunction with the climate anomaly of ca. 5600 BC or perhaps even the one around 5500 BC. This underlines the lack of sufficient archaeological data for the La Hoguette formation. The end of La Hoguette is somewhat easier to define. The late dates at Saint Lambert and Bad Cannstatt roughly correlate with the second highest ice core volcanic SO_4 accumulations (ca. 5277 BC) and the roughly concomitant trough in the Main River oak tree lifespan (Spurk *et al.* 2002 Fig. 4). Around the same time there is a drop in the Northwest German bog oak curve and a sudden end of the Netherland curve (Fig. 16.11). This can be connected with the demise of La Hoguette, the start of the Limburg Group, and the transition from LBK 1 to LBK 2. Still, Mesolithic populations, especially those around the northern and eastern part of Farming 2b, continue without accepting most if any parts of the Neolithic package (Fig. 16.16, Table 16.4).

Returning to the Starčevo-LBK transition in the Balaton and Vienna areas, it cannot be fully ruled out that the LBK develops just before the 5600 BC climate anomaly mentioned above. However, if one emphasizes the later start-dates of Brunn-Wolfholz (Table 16.3), Starčevo-like pottery may appear there in response to the 5600 BC climate anomaly. In this case, the genesis of the LBK may occur in conjunction with the climatic anomalies between about 5550 and 5460 BC (Fig. 16.4, 16.6 and 16.7, Schmidt and Gruhle 2003 Fig. 7). Certainly, by 5500/5400 BC most LBK 1 sites around Lake Balaton are found in the marshy waterside area of the lake (Bánffy 2004). By ca. 5400 BC the Northwest German bog oak lifespan decline bottoms out and Netherland curve ends (Fig. 16.6). This correlates well with an extreme dry phase that bottoms out at ca. 5380 BC (Schmidt and Gruhle 2003 Fig. 7). The bottoming out represents the deepest trough between 6000 BC and 2000 AD (ibid. Fig. 7). Thus, the LBK 1 beginning and expansion seems to coincide with whet-dry oscillations between roughly 5550 and 5350 BC. Farming Barrier 2a must become established during this time.

Farming Barrier 2a is breached with the start of LBK 2, i.e. between 5300 and 5200 BC. During this time new climatic oscillations present themselves. The highest GRIP temperature spike of the entire LBK period occurs at 5332 BC. It is only slightly milder than the pre-LBK temperature peak of 5706 BC. This is followed by a temperature drop that bottoms out around 5303 BC, marking the coldest point since 5607 BC. Around the same time the SO_4 trend line shows considerable volcanic activity, including second highest accumulation mentioned above. The German bog oak lifespan curve registers a slight trough around 5289 BC and a peak around 5260 BC. By ca. 5220 BC a GDO event in the German bog pines is detected (Eckstein *et al.* 2009 Table 3).

The climate and human adjustment correlation just described for the LBK in general terms is also reflected at the archaeological site level. At Brunn-Wolfholz all six settlement areas of the site are established prior to 5300 BC (Table 16.3), but one of the sites is abandoned between ca. 5300 and 5200 BC. To be precise, Area IV is abandoned around 5300 BC and Area IIb around 5280 BC. The occupation of Area IIa ends by ca. 5210 BC. Area III ends by ca. 5200 BC. During the same period a denser settlement system with larger villages, modified house architecture and new subsistence strategies develops in nearby Transdanubia (Oross and Bánffy 2009). The final abandonment of Brunn is coincidental with climatic upheavals indicated by wooden planks from the water well of distant Erkelenz-Kückhoven, northwest of Cologne (Helle and Schleser 1998). Analysis of the plank reveals several climate oscillations of which the strongest cooling and precipitation increase occurs between 5220 and 5210 BC.

The end of the LBK also coincides with a complex series of rapid climatic oscillations. From ca. 5140 BC until around 5080 BC the GRIP running mean displays a steep trough. Extreme cold spikes occur at 5120 and 5086 BC. Another, slightly lesser spike is recorded around 5025 BC. Analysis of the Erkelenz-Kückhoven well's wooden planks yields an estimated average temperature drop of 4° C (Helle and Schleser 1998). This roughly coincides with the standardized tree-ring series of the German bog oaks, which indicates a long-lasting climate event dating between ca. 5040 and 4985 BC (Leuschner *et al.* 2002 Fig. 3a). A similarly long climate anomaly is indicated by the Main River oaks (Spurk *et al.* 2002 Fig. 4).

Human adjustments to the unsettled climate are also observed in the settlement pattern. Although, the main LBK settlement distribution continues to be centered in the basins, usually near creeks, new sites materialize in unusual locations. In some regions relatively remote settlements emerge at high elevations. Among four such sites in the Wetterau Basin is the short-lived 2.4 ha ditched LBK enclosure, which is constructed above 400 m asl near Usingen in the Taunus Highlands, ca. 25 km north of Frankfurt (Laufer 2002). Likewise, at Erkelenz-Kückhoven an LBK enclosure and village is built on a loess plateau, bypassing a nearby stream location (Weiner 1995).

The elevated locations of sites with wells are usually seen as a response to climate oscillations and resulting fluctuations in the water table (Smidt and Gruhle 2003). The wells at Erkelenz-Kückhoven and other LBK sites are discussed by Elburg (2011), in several related articles in Koschik (1998), and by Weiner (1995). Following Weiner (1995), the wood for the first well is cut in the winter of 5090 or 5089 BC. In the spring of 5089 BC a pit is dug to reach the ground water at a depth of 15 m. Eventually, the groundwater falls below the bottom of this well, leading to its abandonment. Afterwards, remains of a burnt roof somehow land in the dried out well and domestic debris is thrown in. In 5065 BC a second wooden well is constructed on top of the debris layer, inside the first one. By 5050 BC, or shortly thereafter, the groundwater falls again and a third wooden well is built inside the second one, reaching below the debris layer of the first well.

The complex development of the Erkelenz-Kückhoven well roughly coincides with the abandonment of the last known settlement at Brunn-Wolfholz, Austria (Table 16.3). In Poland the abandonment of LBK frontier sites is also linked to the climatic oscillations of this period (Grygiel and Bogucki 2008). Finally, these oscillations coincide with dramatic internal social upheaval and nutritional stress, which apparently culminates in mass-murder and destruction at the LBK earthworks of Asparn-Scheltz, Austria, Talheim, Vaihingen and perhaps Herxheim, all in Germany, and Meneville in the Paris Basin (Christensen 2004:136, Wild *et al.* 2004, Windl 1996:27-29, Windl 1999). Therefore, it may be concluded that the demise of the LBK is associated with the strong climatic oscillations around 5000 BC.

PERSISTENCE OF FARMING BARRIER 2B

In spite of the cost of violent social upheaval towards the end of the LBK, the cultural systems of both the farmers and the foragers manage to adjust to the climatic oscillations. The continuity is perhaps best expressed in the realm of ideology as reflected in the burial customs. The dominant LBK ritual of flexed inhumation persists within Barrier 2b while the most common forager's practice of extended supine inhumation largely continues north of that boundary (Häusler 1996).

With mostly minor dislocation of Barrier 2b, the ca. 500 year old LBK traditions find expression in new cultures, which occur in more or less the same enclaves used during LBK 2 times (Nowak 2000 Fig. 4, ibid. 2006 Fig. 1 and 3, Raetzel-Fabian 1986 Map 1-6, ibid. 2000 Fig. 72 and 111). Nonetheless, a ca. 100-year abandonment is proposed for the Polish lowlands and the upper Vistula region (Grygiel and Bogucki 2008:1997-2009). Subsequently, small houses are reported to precede the development of large trapezoidal ones in this area. Elsewhere, the LBK longhouse architecture seems to continue, but with modification (cf. Stöckli 2002).

The western unpainted pottery styles (Fig. 16.19) evolve out of the LBK 2 tradition (Stöckli 2002 Fig. 100). For instance, the early STK is thought to overlap chronologically and geographically with pottery that is still attributable to LBK 2. Later STK-derived Rössen pottery becomes elaborately decorated until the end of the Bischheim Group (Fig. 16.20). The situation differs in the southeast, where the pottery styles begin to follow the Carpathian Basin painted tradition. Even though, this development begins already in the Esztár-Szamos Group of the later Hungarian Bandkeramik (e.g. Bánffy 2006:127), it only becomes commonplace after the Szakálhát, Esztár, and Tiszadob Groups are replaced by the Tisza culture (Fig. 16.19). The Lengyel culture appears to the west of the Tisza culture (Fig. 16.20). It initially also employs painted pottery. Lengyel consists of various related pottery styles often also seen as individual cultures. Together they make up the Danube-centered Lengyel interaction sphere (Baldia, Frink and Boulanger 2008).

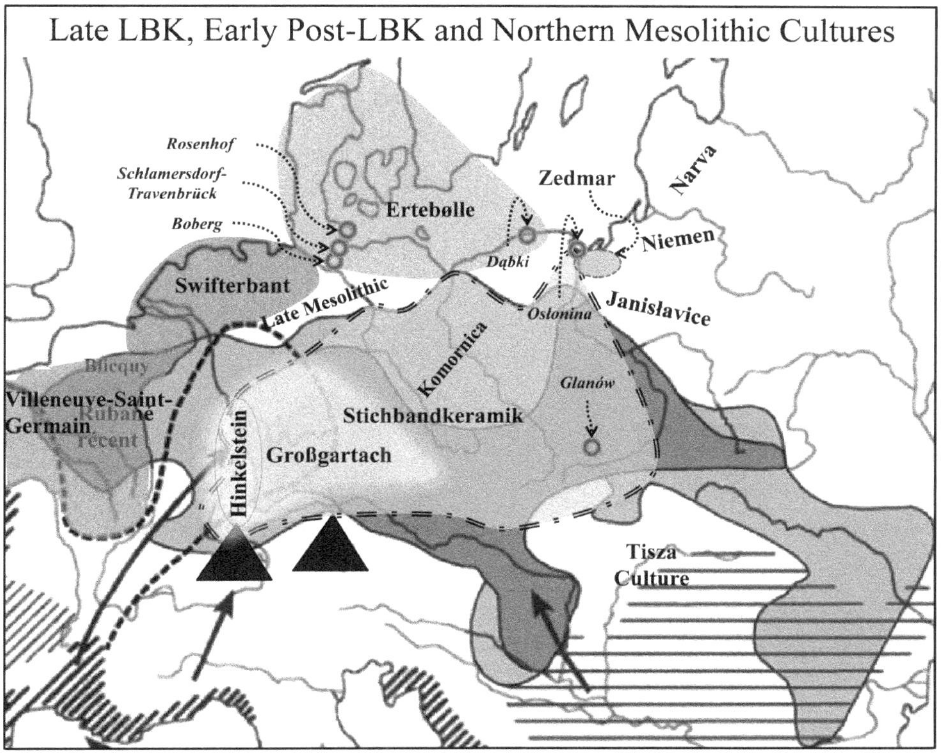

Figure 16.19. Late LBK, early Post-LBK and selected northern Mesolithic cultures. LBK-derived farming cultures: Grossgartach (Großgartach), Hinkelstein, Stichbandkeramik (STK), Villeneuve-Saint-Germain. Mesolithic territory (shaded): Ertebølle, Swifterbant, Zedmar. Other Mesolithic groups: German Late Mesolithic, Janisławice, Komornica, Narva, Niemen. Circles: Selected sites mentioned in the text

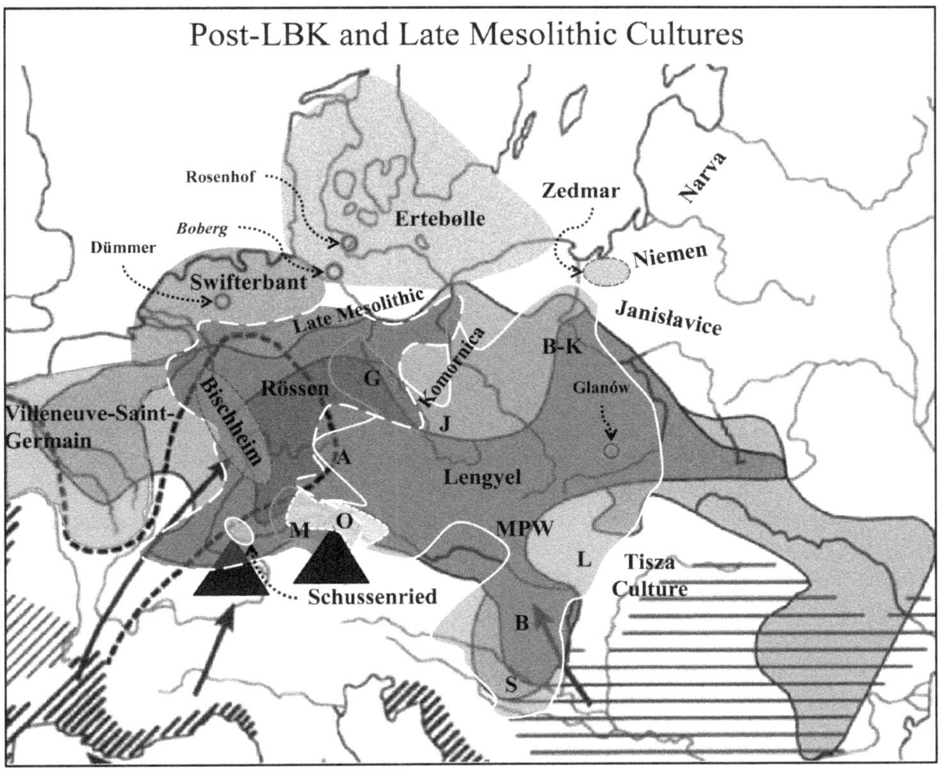

Figure 16.20. Stichbandkeramik-derived, Lengyel and Late Mesolithic cultures. Lengyel-related groups: Aichbühl (A), Balaton (B), Brześć-Kujawski (B-K), Gatersleben (G), Jordanów (J), Ludanice (L), Münchshöfen (M), Austrian/ Moravian Painted Ware (MPW), Oberlauterbach (O), Sopot (S). Stichbandkeramik-derived cultures: Bischheim, Planig-Friedberg/Rössen. Mesolithic territories (shaded): Ertebølle, Swifterbant, Zedmar. Other Mesolithic groups: German Late Mesolithic, Janisławice, Komornica, Narva, Niemen. Circles: Selected sites mentioned in the text

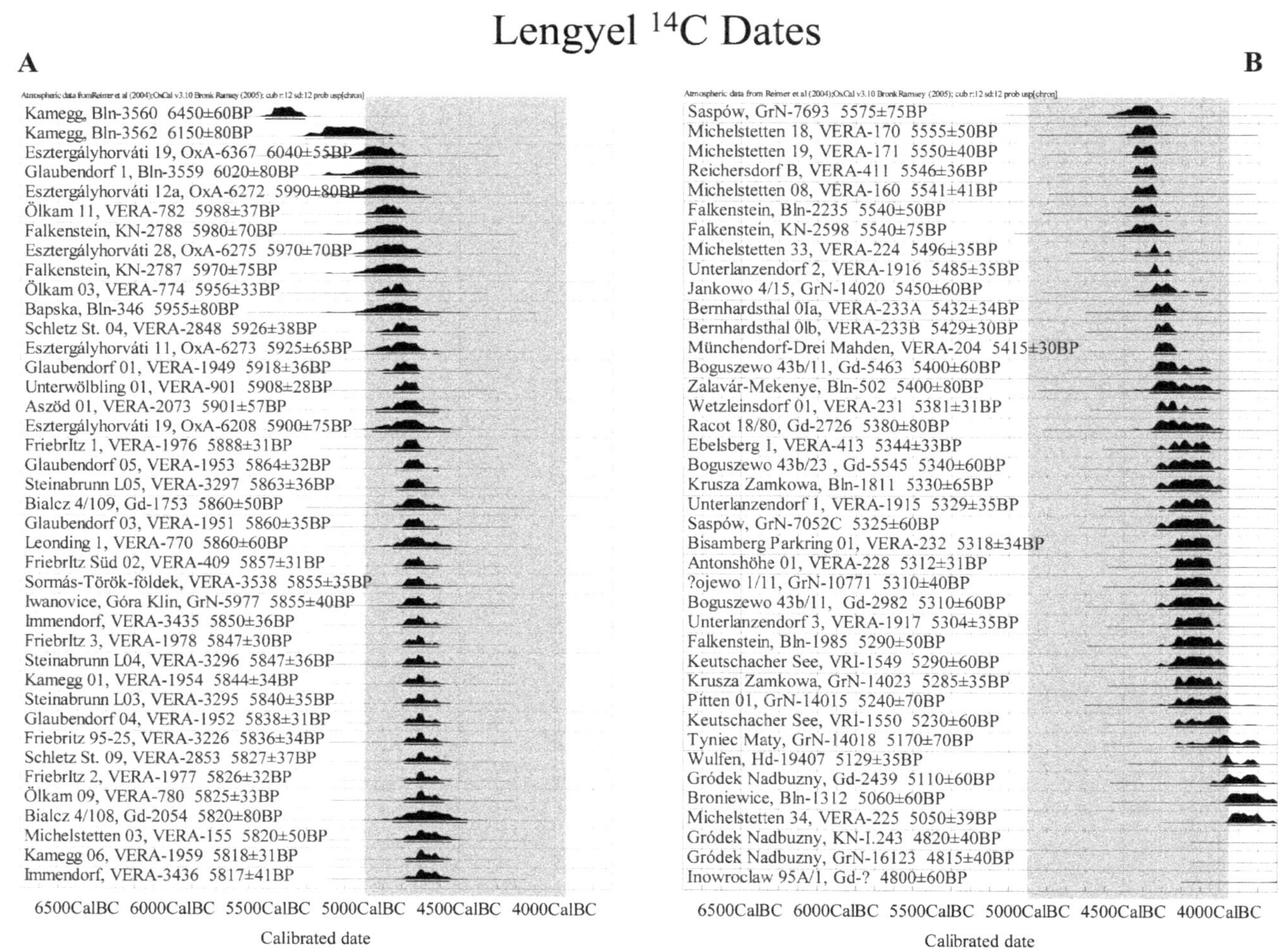

Figure 16.21. Lengyel ^{14}C dates ($\sigma < \pm 90$ years). A: Earliest dates (N=40), B: Latest dates (N=40), Shaded area: Lengyel duration

Dating the Post-Bandkeramik Farming Cultures

The Lengyel Interaction Sphere

Lengyel pottery has been seen as a development from the VIča via the Sopot cultures (Lazarovici 1999, Neugebauer-Maresch 1998, Podborský 1993). Others view the Sopot culture as the southernmost Lengyel Group. This dilemma is underlined by the fact that Sopot was originally called the Slavonian-Syrmian, Babska-Lengyel or Sopot-Lengyel culture. Following the European culture historical paradigm, Koštuřík (1973:5, Map 1, 1-3) and others seek the Lengyel's *tribal homeland* in Transdanubia and Southwest Slovakia. This region is roughly the same as the one proposed for the formation of the LBK. Even so, no direct link is seen between LBK and Proto-Lengyel pottery. The latter is thought to follow the polychrome painted tradition, which Čižmář *et al.* (2004) judge to be similar to the Tisza culture's pottery.

Many researchers envision a long duration for Lengyel (Breunig 1987, Obelić *et al.* 2002, 2004, Zvelebil and Lillie 2000 Fig. 3.8). However, my analysis of 250 ^{14}C dates from all parts of the interaction sphere indicates that Lengyel is first observable around 5000/4900 BC (Fig. 16.21) and ends about 4000/3900 BC.

Lengyel has four phases, although, a fifth phase or various sub-phases may also be employed (Čižmář *et al.* 2004, Podborský 1993). The phases were initially defined through separate regional pottery typologies (Čižmář *et al.* 2004, Podborský 1993:111) and their intra- and inter-cultural correlations vary (Grygiel and Bogucki 2008, Čižmář *et al.* 2004, Pavúk 2000, Lüning 1998, Neugebauer-Maresch 1995, M. Zápotocký 1998). For example, the Formative or Proto-Lengyel Phase may consist of the Sopot-Bicske, Bíňa and Lužianky Groups (Neugebauer-Maresch 1995) or Sopot-Bicske, Bíňa-Bicske, Lužianky, and Sé Groups (Čižmář *et al.* 2004). However, Oross and Bánffy (2009) assign the Bíňa-Bicske and a Milanovce Group to the Earlier LBK and date them to 5450-5300/5250 BC (Table 16.3). This is roughly the same range as the Sopot-Lengyel IB phase (Table 16.5), which would imply that Lengyel begins during Sopot-Lengyel Phase II-A or even II-B. A northward expansion into Austria, Moravia, East Slovakia and Poland Minor is usually stipulated after the development of Proto-Lengyel (Čižmář *et al.* 2004).

Table 16.6. Lengyel/Moravian Painted Ware Chronology (after Čižmář *et al.* 2004 and Stadler *et al.* 2006 Table 5)

Lengyel/Austrian-Moravian Painted Ware	**Samples**	**Beginning** (Highest Probability)			**Ending** (Highest Probability)		
Phase	***N***	***Earliest***	***Latest***	***Weighted Mean***	***Earliest***	***Latest***	***Weighted Mean***
Proto/Formative (Ia0)	11	4845	4755	4800	4715	4660	4688
Lengyel I/MPW Ia	10	4715	4660	4688	4650	4580	4615
Lengyel I/MPW Ib/IIa_1	10	4650	4580	4615	4550	4495	4523
Lengyel II/MPW $IIa_{2/3}$	23	4550	4495	4523	4405	4345	4375
Lengyel III/MPW IIb	8	4405	4345	4375	4180	4050	4115
Lengyel IV/Epi Lengyel	1	4180	4050	4115	4130	3920	4025
Total	**63**						

Table 16.5. Sopot-Lengyel chronology (after Obelić *et al.* 2004)

Sopot Phase	**Estimated Date Range**
I-B	5480-5070 BC
II-A	5030-4770 BC
II-B	4800-4250 BC
III	4340-3790 BC

In Austria and Moravia Lengyel is known as Austrian/Moravian Painted Ware (MPW). Čižmář *et al.* (2004) correlate MPW with the interregional Lengyel Phases I-IV, but do not mention any ^{14}C dates.[24] On the other hand, Stadler and Ruttkay (2006) illustrate their Lengyel pottery topology along with accompanying ^{14}C dates for Austria. Stadler *et al.* (2006) use the sequencing calibration method for 63 AMS dates of mostly short-lived materials to delineate their MPW chronology (Table 16.6). Accordingly, Lengyel starts no earlier than 4845 BC. An Austrian lake dwelling site attributed to the Lengyel IV Kanzianiberg-Lasinja Group, has two tree-ring felling dates (3947 and 3871 BC), which suggest an end sometime after 3870 BC (Samonig 2005).

North of Austria, the Moravian Lengyel is thought to gradually replace the STK (Čižmář *et al.* (2004 Table 1). Traditionally, this is seen as an expansion by *the people with Lengyel pottery*, which replace the *people with STK pottery* (Podborský 1993). This change is poorly dated. Excavations, such as the one of Olomouc-Slavonin I, merely indicate that pottery that is typologically defined as STK Phase IV occurs together with Lengyel Ia pottery in Pit 878 and K688 (cf. Kazdová *et al.* 1999). The Lengyel expansion into Bohemia is thought to be slightly later, occurring during MPW Ib-c (Čižmář *et al.* 2004). According to Table 16.6, this would be sometime between 4650/4580 and 4550/4495 BC).

For Poland Grygiel and Bogucki (2008) propose a coexistence of Stichbandkeramik-related and Lengyel sites at the beginning of the Lengyel. Nowak (2007) dates the beginning of Lengyel in Poland Minor to 4800/4700 and for the Brześć Kujawski Group around 4850/4800 (Fig. 16.16). However, Grygiel and Bogucki (2008:2032), who summarize the ^{14}C dates for the Brześć Kujawski Group, propose a later start.[25] Their three Lengyel phases are based on pottery typology, stratigraphic patterns, and ^{14}C dates (Grygiel and Bogucki 1986:136). Phase I starts at 4700/4670 BC, Phase II at ca. 4500 BC, and Phase III lasts from ca. 4300 to 4100/4000 BC (Grygiel and Bogucki 2008:1918). The *14CEurope* database contains some rather late dates, such as the two from Glanów 3 (Fig. 16.17). These dates merely indicate that Lengyel ends around 3950 BC or maximally 3850 BC (Fig. 16.21).

The Gatersleben Group of Central Germany (Koštuřík 1973 Map 1, No. 8) has two ^{14}C dates. They are insufficient to determine its duration, but they predict a range of ca. 4350-4000 BC.[26] This compares reasonably well with MPW IIb ^{14}C dates (Table 16.6). Therefore, one can coarsely date Gatersleben to ca. 4400-4150 BC. This opinion is roughly in line with that of Geschwinde *et al.* (2009 Fig. 143). Gatersleben is succeeded by the Jordanów Group (Czech: Jordanóv, German: Jordansmühl) in the March (Morava), the Upper Elbe and the Upper Odra (Oder) drainages (Koštuřík 1973 Map 1, No. 4, Müller 2001 Fig. 128-129). The number of dates is limited and most have large error ranges. Still, an existence of 4200/4180 to 3950/3900 seems plausible, given nine dates attributable directly to Jordanów and a total of twenty-one dates for other Lengyel IV groups (Baldia, Frink and Boulanger 2008 Fig. 14).

The appearance of Lengyel-related pottery styles in the upper Danube drainage of Southern Germany starts with Lengyel-like flat-bottomed pottery in the Münchshöfen Group (ca. 4450/4350-4000/3900 BC, N = 17) and the roughly contemporaneous Aichbühl Group (e.g. Hafner

[24] Čižmář *et al.* (2004) see Lengyel IV (Epi-Lengyel) as the beginning of the Czech and Slovak Eneolithic (Copper Age) and consider it a break from Lengyel proper, in contradistinction to the traditional perspective (Podborský 1993).

[25] The sum of 75 ^{14}C dates from all sites of the Brześć Kujawski Group in Kujavia is 4550 (68.2%) 3950 BC at 1δ. For the sites of Brześć Kujawski and Osłonki the sum of 53 dates at 1σ is 4800 (32.7%) 4450 BC and 4400 (35.5%) 4350 BC.

[26] Kmehlen, Bln-231: 5360±160 bp; 68.2% prob. = 4350 (64.2%) 4030 BC, 4020 BC (4.0%) 3990 BC.Wahlitz, GrN-433: 5300±200 bp; 68.2% prob. = 4350 (68.2%) 3800 BC.

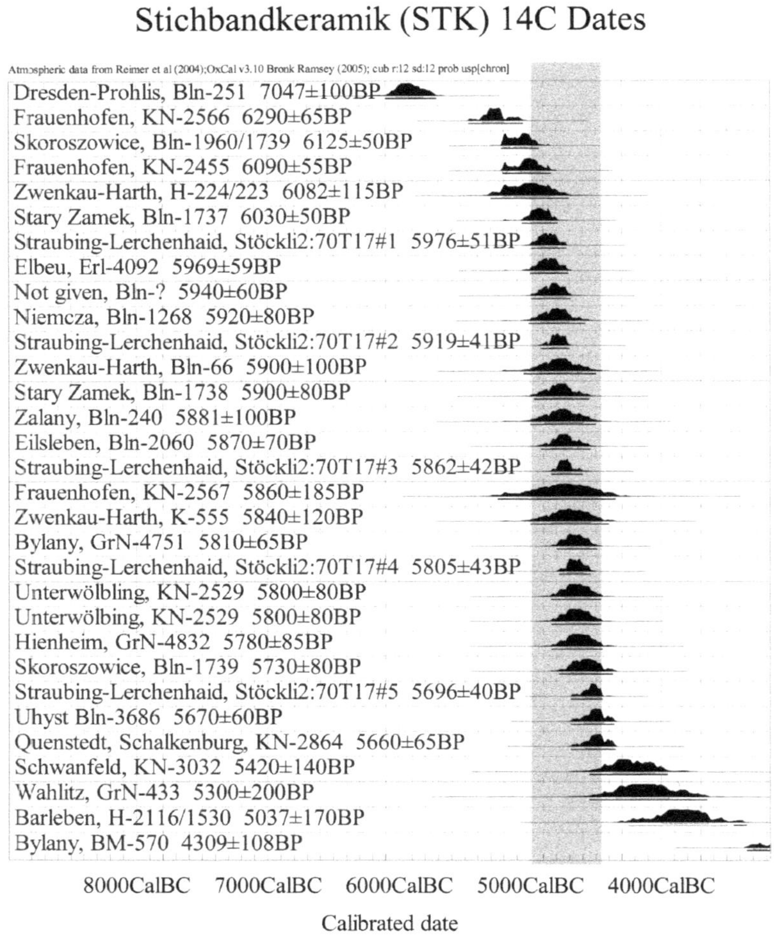

Figure 16.22. Stichbandkeramik (STK) calibrated ^{14}C dates (N=31)

and Suter 2001:292, Fig. 2 and 8). The authors assert that similar characteristics develop farther west in the Schussenried Group (ca. 4000 BC), the Early Pfyn culture on the Bodensee (ca. 3900 BC), and even in later groups of the upper Rhein Valley in Central Switzerland (ca. 3800 BC and later).

The Stichbandkeramik Interaction Sphere

The villages of the STK proper arise primarily in the same locations as the LBK. However, in parts of Poland Grygiel and Bogucki (2008) see a hiatus between the end of the LBK (ca. 4900) and the colonization of the same region by the STK (ca. 4800 BC). The STK interaction sphere consists of several cultures or groups characterized by related pottery styles with incised motifs. They carry various names in different regions and at different times. The pottery derives directly from LBK 2, although, typical stroke ornamented pottery, painted red after firing, also occurs in Early Tisza culture context, while diagnostic STK pottery occurs together with red painted Lengyel I/MPW Ia pottery in Moravia, Bohemia, Southern Germany, Hungary, Slovakia and Poland (Korek 1987, Raczky *et al.* 2002:841-843). In several of these cases the combinations occur in Lengyel *rondels* (geometric ditched earthworks) and even in some of the rather similar STK earthworks.

Overall the STK exists from 5000/4900 to 4600/4500 BC (Fig. 16.22). Typological sub-phases and their dating vary regionally (Grygiel and Bogucki 2008:2034). In Central Germany Phase I ends around 4800 BC and Phase II around 4700. A fivefold division is employed in the Czech Republic and adjacent areas. STK I-III ends at ca. 4800 BC, STK IV around 4700 BC and STK V ends about 4600 BC (Grygiel and Bogucki 2008:2034). In addition to the STK proper, related groups with incised pottery comprise the Samborzec-Opatów Group (4800-4600 BC) and the ill-dated Malice Group in Poland (e.g. Grygiel and Bogucki 2008, Hensel and Wiślański 1979, Kopacz 1974, Nowak 2006). In Germany they are the Hinkelstein, Großgartach, Planig-Friedberg/Rössen, and Oberlauterbach groups or cultures (Biermann 1997, 2000, 2003).[27]

Located at the western edge of the STK are the Grossgartach (Großgartach) and Hinkelstein Groups or cultures (Fig.16.19). Hinkelstein is traditionally seen as a successor to LBK 2. It overlaps geographically and chronologically with the more broadly distributed Großgartach pottery. The interpretation of the Hinkelstein

[27] Seventeen Oberlauterbach ^{14}C dates range from 4940 (68.2%) 4830 BC (Künzing-Unterberg, HD-11315-11106: 5990±40 bp) to 4530 (68.2%) 4450 BC (Künzing-Unterberg, HD-11321-11273: 5650±35 bp).

^{14}C dates and the precise chronological relationship with LBK 2 and Großgartach are open for discussion (e.g. Müller 2002, Stöckli 2002). I estimate Großgartach to range from 5000/4900 BC to 4600/4500 BC, based on 32 ^{14}C assays. Lenneis and Stadler (2003) date it to 4910-4520 BC at 1σ, using 46 ^{14}C assays and the sum method. However, Großgartach transitions to the Planig-Friedberg Phase of the Rössen culture earlier, since five Planig-Friedberg ^{14}C dates yield a duration of ca. 4800-4600 BC. Twenty-nine ^{14}C dates imply that Rössen proper dates to ca. 4600-4400 BC. Thereafter, Rössen's late eastern variant overlaps with the Lengyel interaction sphere in Central Germany. In Northwest Germany and adjacent regions the Rössen-derived Bischheim Group develops around 4400 BC and lasts until perhaps 4200 BC. However, the 34 ^{14}C dates in the *14CEurope* database do not provide a very precise picture.

The subsequent Michelsberg culture is frequently believed to develop from western roots, including the Bischheim Group, the French Chasséen culture etc. (Dubouloz 1998, Geschwinde *et al* (2008), Jeunesse 1998, Louwe-Kooijmans 2007, Scollar 1959, 1961, Vanmontfort 2001, Vanmontfort *et al.* 1977, Vermeersch and Burnez-Lanotte 1998). Michelsberg is usually divided into five phases (e.g. Lüning 1968, Raetzel-Fabian 2006). A west to east expansion during its earliest phases is often proposed. However, Michelsberg I was originally believed to start at 4400 BC, but a later start, roughly coincident with Phase II, seems presently more likely (Crombé and Vanmontfort 2007, Geschwinde *et al.* 2009, Lanting and van der Plicht 2000, Lüning 1998, Raetzel-Fabian 2000, Vermeersch and Burnez-Lanotte 1998, Vanmontfort 2001, Vanmontfort *et al.* 1997, Zápotocký 2000). Nevertheless, a transitional Bischheim/ Michels-berg-like phase starting around 4300 and ending before 4100 BC remains useful until a fuller understanding of the chronological and geographic development of Blicquy, Chasséen Septentrional, Noyen, Late Rössen/Bischheim and Michelsberg II is available. Michelsberg II itself starts at 4200/4100 BC and ends at 4000/3900 BC, based on my analysis of ^{14}C dates with an error of less than ±100 years. This is roughly in line with the projections of Geschwinde *et al.* (2009:187-188). Michelsberg III (ca. 4000-3850), IV (3850-3700 BC) and V (3700-3500 BC) fall into the period marked by the establishment of Farming Barrier 3 (Raetzel-Fabian 2000, Geschwinde *et al.* 2009) and will be discussed under that rubric.

Dating the Interaction of Foragers and Farmers

Foraging sites are found within and beyond Farming Barrier 2b between 5000 and 4000 BC (e.g. Drafehn *et al.* 2003, Nowak 2006, Street *et al.* 2001, Vanmontfort 2008a, 2008b, Zvelebil *et al.* 1998). Dating varies by region (Fig. 16.16, Table 16.4) and understanding the relationship between the farmers and foragers is problematic. For instance, Novak (2007) acknowledges research problems, but supports the notion of interaction between Mesolithic and farming communities throughout this period in Poland (Fig. 16.16). Among the pertinent sites are Glanów 3 and Dąbki. Glanów 3 is within the farming territory of the STK and Lengyel cultures. Its four Final Mesolithic ^{14}C dates overlap with these two cultures (Fig. 16.17), but there are no STK artifacts at the site and the Lengyel dates are rather late. Beyond Farming Barrier 2b, *individual sherds* of the late STK are reported from Dąbki on the Baltic coast (Fig. 16.19) (e.g. Grygiel and Bogucki 2008: 2026-2027).

Near the Oder River at Criewen, Germany, Mesolithic Grave 2 and 1, date to 4800/4720 BC and 4680/4530 BC respectively.[28] The grave goods are judged to show a relationship with nearby Lengyel farming enclaves (Grygiel and Bogucki 2008:2027). However, Grave 2 contains most of the skeleton of a 40 year old male in what appears to be a supine position (Street *et al.* 2002 Fig. 25). This form of burial is common among Mesolithic people, implying that contact did not have a substantive influence on the mortuary rituals of the foragers. Still, some form of interaction between farmers and foragers may be implied at the above mentioned Stralsund (*Mischwasserspeicher*) site on the Baltic coast, where a T-shaped antler ax is dated to 4950/4840 BC (Kaute *et al.* 2005).[29] Such axes are often thought to be evidence of interaction during the LBK/STK transition (cf. Grygiel and Bogucki 2008: 2025-2026).

Amphibolite shoe-last axes are always cited as evidence of occasional contact if not outright exchange between Ertebølle foragers and the farming territory to the south. They are commonly seen as dating from the LBK to ca. 4300 BC (e.g. Pedersen *et al.* 1997), but others propose that only one third of them may date before 4300 BC (e.g. Skak-Nielsen 2007). Still, at Rosenhof (Fig. 16.19) a large drilled amphibolite adze with parts of the wooden shaft is directly dated ca. 4900/4780 BC (Hartz *et al.* 2007:579), suggesting an STK timeframe. Unfortunately, different archaeologists point to different amphibolite raw material sources within Barrier 2b (e.g. Pedersen *et al.* 1997:204-205, Hartz *et al.* 2012, Schwabedissen 1994 Footnote 5), leaving the direction of the interaction uncertain.

The few heavy copper implements in Mesolithic northern Europe are thought to be imports from southeast European sources during the 5th millennium (Klassen 1997, 2000, 2004, Klassen and Pernicka 1998, Klassen, L. and S. Stürup 2001). Indeed, most of the copper artifacts from the late fifth and early fourth millennium seem to match ores from Serbia (Höppner *et al.* (2005:307). However, there is a lack of datable context for the Ertebølle artifacts according to Bartelheim (2003), who adds that Alpine copper sources should also be considered. Possible copper mining and smelting is attributed to the Lengyel-related Münchshöfen Group at Brixlegg-Mariahilfbergl, Upper Austria (Höppner *et al.*

[28] Criewen Grave 2, KIA-4347: 5890±40 bp; 68.2% prob. = 4800 (68.2%) 4710 BC; 95.4% prob. = 4890 (95.4%) 4680 BC; Grave 1, KIA-4346: 5740±40 bp; 68.2% prob. = 4680 (68.2%) 4530 BC; 95.4% prob. = 4710 (95.4%) 4480 BC.

[29] Stralsund, KIA-20436: 6010±35 bp; 68.2% prob. = 4950 (68.2%) 4840 BC; 95.4% prob. = 5000 (95.4%) 4790 BC.

2005). The two pertinent ^{14}C dates from the site fall into the Lengyel II or III timeframe.[30] Regardless of the precise dates and sources for the Ertebølle copper artifacts, they indicate that some level of cross-border interaction took place and that the artifacts most likely passed through the Lengyel interaction sphere.

Until recently, the impetus for Ertebølle pottery production, cattle keeping and domesticated cereal consumption was also thought to be attributable to cross-border interaction starting perhaps by ca. 4700 BC (e.g. Fischer 2002 Fig. 22.1, Kalis *et al.* 2003). However, domesticated sheep/goat and cattle, as well as cereal cultivation are currently seen as taking place no earlier than 4100/4000 BC (Behre 2007, 2009, Edwards *et al.* 2007, Hartz and Lübke 2006 Hartz *et al.* 2007, Fischer 2002, Scheu *et al.* 2007). This is at variance with the development in the Swifterbant culture (Ginkel *et al.* 1999:19-25), where the earliest pottery appears at 5300/5000 BC (Fig. 16.19 and 16.20).

In the Swifterbant culture domesticated animals occur by ca. 4700 BC and cultivated cereals by 4300/4000 BC (Cappers and Raemaekers 2008, Ginkel *et al.* 1999:23-27, Vanmontfort 2008b). This is interpreted as evidence of contact with the farming population to the south. In spite of these early dates, the effect of forager-farmer interaction appears to diminish with increased distance from the farming barrier (Vanmontfort 2008b:90) and farming activity seems to have been limited. East of the Rhein-Maas delta, the archaeological evidence from the Swifterbant-related site of Hüde I in the Dümmer Lake area is difficult to assess. The ^{14}C dates from this site center around 4700 BC and 4450 BC (Geschwinde *et al.* 2009, Lanting and van der Plicht 2000). The construction of the nearby corduroy road (Log Way Pr 31) at the Campemoor bog (see below) is close to the first cluster of dates. It coincides with the transition from the STK to Planig-Friedberg Phase of the Rössen culture to the south. The second cluster of Hüde I dates corresponds to the transition from the Rössen culture proper to the Bischheim Group. Some artifacts from the site do indicate contact with farmers to the south of Barrier 2b at these times (Geschwinde *et al.* 2009).

Climatic Change and Socio-Cultural Adjustment

To simplify the discussion, six subjective periods may be isolated. They can be arbitrary dated to start around 5000, 4800, 4650, 4400, 4250, and 4100 BC.

The earliest period coincides with GISP 2 volcanic SO_4 concentrations, which are relatively high around 4988 BC and even higher at 4803 BC (Fig. 16.7). GRIP temperature fluctuations occur at ca. 4909 BC (warm) and 4807 BC (cold) (Fig. 16.4). The 4909 BC peak is the warmest since the LBK's 5332 BC peak. The 4807 BC cold trough is the coldest since the trough of ca. 5120 to 5086 BC. The German tree growth homogeneity index indicates that continental conditions bottom out around 5000 BC and maritime-like wet conditions peak about 4940 BC (Schmidt and Gruhle 2003 Fig. 7). The peak is the highest recorded between 5600 and 4600 BC. It is followed by a series of oscillations, including strong moist conditions around 4880 and 4810 BC. The Main River oak mean lifespan reaches a rather sudden and high peak at ca. 4850 BC. These changes are concomitant with the archaeologically fuzzy period of transition from the LBK to Hinkelstein, STK, Lengyel, and Oberlauterbach, as well as the start of the Ertebølle's Rosenfelde Phase (Fig. 16.23).

Starting ca. 4800 BC new cultural developments begin. Hinkelstein is believed to end and the Planig-Friedberg Phase of the Rössen culture appears. The Sopot/Proto-Lengyel Phase (Lengyel I_0) leaves a better dated imprint in the archaeological record and the earliest Lengyel I (MPW Ia) pottery eventually develops (Table 16.6). The first Lengyel-related *rondels* may also be in evidence during this period (Barna and Pásztor 2010). The later STK becomes recognizable from Central Germany to the upper Oder region and the Polish Lowlands (Grygiel and Bogucki 2008:2034). In fact, Grygiel and Bogucki (2008:1998-1999) argue that the STK expansion of this phase breaks *all environmental boundaries* as it *colonizes* places such as Osłonina, located beyond Farming Barrier 2b (Fig. 16.19). Cultural adjustments in the Mesolithic territory lead to the development of the Ertebølle culture's Jarbock and Stationsvej Phases, which begin to exhibit the first pottery (Table 16.4).

These cultural developments correlate with fluctuations recorded by several climate proxies. The 4803 BC GISP 2 SO_4 accumulations represent the 7th highest peak between 6800 and 1500 BC. The Main River oak mean lifespan drops below its mean after 4800 BC (Spurk *et al.* 2002 Fig. 4). Similarly, a peak in the Northwest German and Dutch bog oak record is recognizable at ca. 4830 BC, followed by a trough bottoming out ca. 4770 BC. The latter coincides with a GDO phase (Eckstein *et al.* 2009:141) and the construction of the oldest known bog or corduroy road (Log Way Pr 31) in Lower Saxony (Table 16.7) (Metzler 1997, 2003, pers. comm. 2009). The track's construction represents the human response to soggy conditions in the Dümmer Lake region (Fig. 16.20). Ensuing continental (dry) conditions bottom out around 4740 BC with the deepest trough since 5380 BC (Schmidt and Gruhle 2003 Fig. 6 and 7).

Additional cultural shifts occur between ca. 4650 and 4500 BC (Fig. 16.11, 16.23, Table 16.4). They include the end of the STK and the Planig-Friedberg Phase of Rössen, as well as the start of Lengyel I (MPW Ib/IIa_1) and Swifterbant 2. However, the Ertebølle's transition from the Jarbock to the Timmendorf Phase in North Germany is placed around 4450 BC (Table 16.4), which suggests that it could also start in to the following period.

[30] Brixlegg-Mariahilfbergl 6/F2, GrN-22167: 5570±50 bp; 68.2% prob. = 4450 BC (68.2%) 4350 BC. Mariahilfbergl 5, GrN-21364: 5480±60 bp; 68.2% prob. = 4440 (3.4%) 4420 BC, 4370 (64.8%) 4250 BC GrN-213641 in Höppner *et al.* (2005:299) is a misprint. GrN-laboratory codes presently have only five digits (pers. com. Smith-Deenen, Centrum voor Isotopen Onderzoek, Groningen).

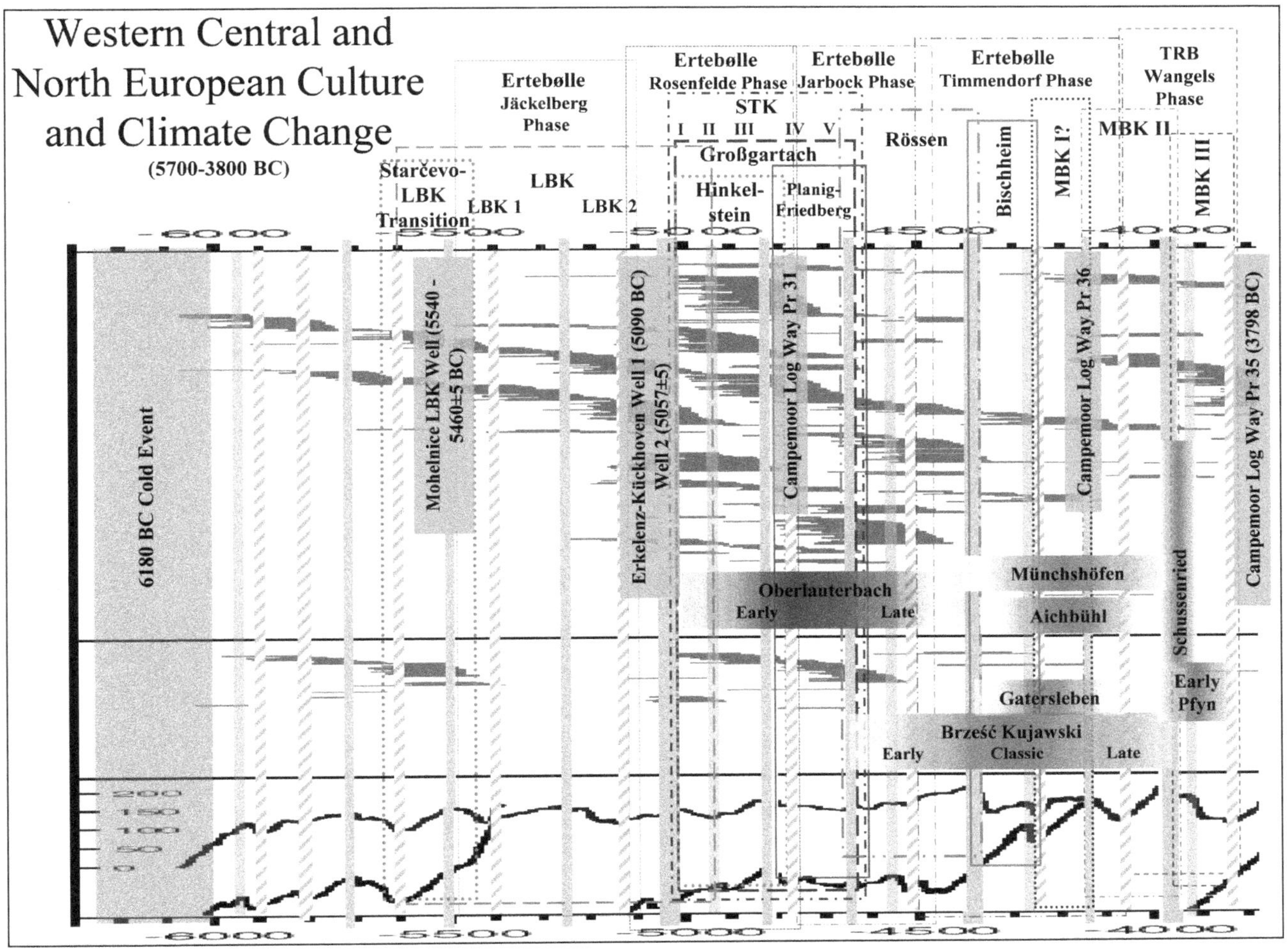

Figure 16.23. Western Central European culture and climate change (5700-3800 BC). Linienbandkeramik (LBK), Michelsberg (MBK), Stichbandkeramik (STK). Alpine Piedmont tree-ring dated cultures or groups: Schussenried (3955 to at least 3871 BC), Early Pfyn (3870 to 3790 BC)

Table 16.7. Dated log ways of the Campemoor, Dümmer Lake area, Germany (after Metzler personal communication)

Log Way	Date	Dating Method
Pr 31	4770 BC	^{14}C
Pr 36	4150 BC	^{14}C from peat cutting
Pr 35	3798 BC	Tree-ring
Pr 34	3701 BC	Tree-ring
Pr 32	2900 BC	^{14}C

During this period the Main River oak lifespan shows below mean oscillations (Spurk *et al.* 2002 Fig. 4). Around 4570 BC a GDO phase is recognizable in the bog pines of Northwest Germany (Eckstein *et al.* 2009:141). The Northwest German and Dutch oaks exhibit troughs following the peaks of ca. 4650 and 4550 BC (Fig. 16.6). In Greenland the GISP 2 SO_4 record indicates high concentrations around 4667, 4627, 4616 and 4596 BC. The last is the 11th highest concentration between 6800 and 1500 BC. The GRIP temperature trend line peaks around 4695 BC, which is only slightly lower than the one of 4909 BC. The subsequent temperature decline bottoms out around 4658 BC, but after a slight climb above the mean about 4519 BC, the trend line continues to stay largely below it. The strongest temperature decline occurs ca. 4505 BC, when it reaches the lowest point since the 6180 BC cold event.

The next period of adjustment starts around 4400 BC. In Eastern Hungary the traditional tell settlements are abandoned by about this time (Bánffy 2002, 2006). Farther west, the period is marked by the transitions from Oberlauterbach to Rössen proper, and from Lengyel I (MPW Ib/IIa_1) to Lengyel II (MPW $\text{IIa}_{2/3}$). The development of Münchshöfen, Aichbühl, Bischheim, Late Rössen, and perhaps Gatersleben and the proposed Swiss Proto/Early Pfyn (Hafner and Suter 2003) may also start around this time (Fig. 16.11, 16.23). In Poland the Lengyel's Classic Phase of the Brześć Kujawski Group is thought to transition to the Late Phase around 4300 BC, although this event may perhaps be placed into the next period. The transition is marked by the burning of numerous trapezoidal Lengyel longhouses (Grygiel and Bogucki 2008:1942), possibly signaling social turmoil. Around the same time the Danish Ertebølle culture transitions from the Stationsvej to the Alekistebro Phase (Table 16.4).

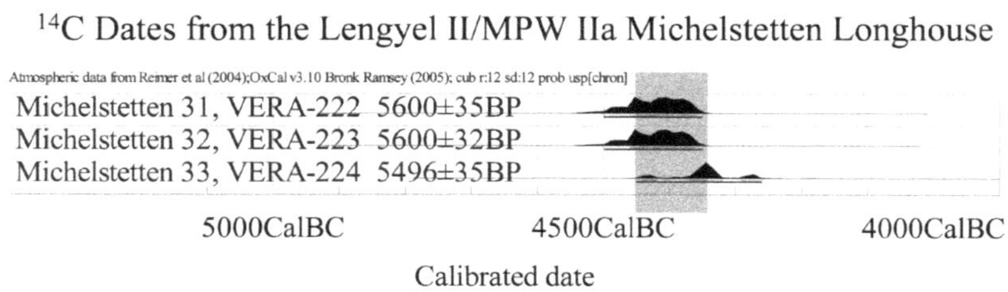

Figure 16.24. Radiocarbon dates from Lengyel II/MPW IIa Longhouse 1, Michelstetten, Austria (Stadler *et al.* 2006 Table 1). VERA-222 and VERA-223 are from the same material

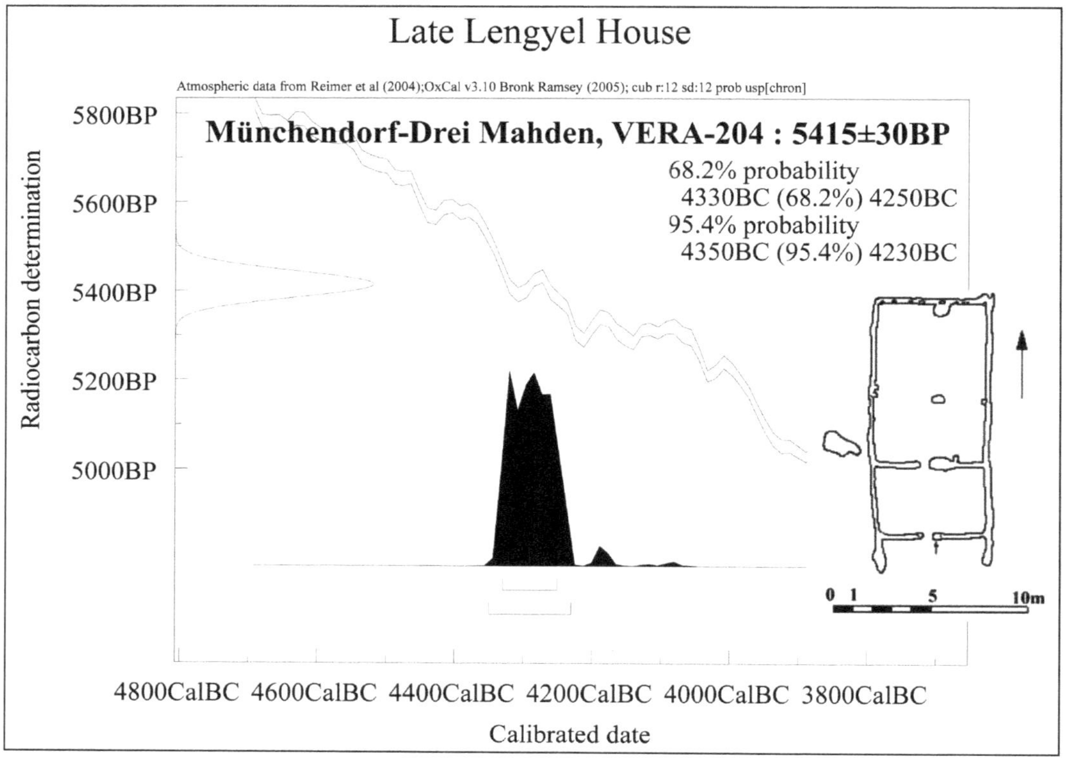

Figure 16.25. Münchendorf-Drei Mahden, Lower Austria, Lengyel III/IV house plan and ^{14}C date (plan after Carneiro and Stadler 2002)

The climatic upheavals of this phase includes the interruption of the GRIP temperature cold trend by a very high warm spike at ca. 4416 BC. It surpasses the 4695 BC spike as the strongest since 4909 BC. Thereafter, the sub-mean trend continues, hitting the lowest point at ca. 4345 BC. Meanwhile, the GISP 2 SO_4 accumulations indicate a renewed phase of volcanic activity. The 14th highest is measured around 4447 BC and the 12th highest around 4411 BC. The German tree growth homogeneity index begins a sudden drop just before 4400 BC, pointing to the beginning of wide-spread dry conditions, which last until ca. 4300 BC (Schmidt and Gruhle 2003 Fig. 6). The Main River mean oak lifespan peaks shortly after 4400 BC (Spurk *et al.* 2002 Fig. 4). It is the second highest peak between 6000 and 4000 BC and the sixth highest between 8000 and 1500 BC. The standardized tree-ring series of the Northwest German and Netherland bog oaks indicates a climate anomaly dated between ca. 4420 to 4405 BC (Leuschner *et al.* 2002 Fig. 3). The Northwest German bog pine data indicate a GDO event around 4390 BC (Eckstein *et al.* 2009).

The socio-cultural adjustments accelerate between ca. 4300 and 4250 BC. They may encompass not only the shift from Lengyel II to III (MPW IIb), but possibly also the above mentioned changes in the Brześć Kujawski Group and the Danish Ertebølle. In the west the ill dated transition from the Rössen-related Bischheim to the Michelsberg culture, with its affinity to the French Chasséen Septentrional, may also date to this period (Fig. 16.20 and 16.23). The previously open spaces between the *restricted Neolithic enclaves in Belgium are filled in* and Michelsberg begins to expand towards the north in the direction of the Swifterbant culture as *Neolithic society is restructured (*Louwe Kooijmans 2007:227-228).

The adjustments are perhaps best illustrated in changes of house size and architecture in the late phases of the Lengyel interaction sphere. At the Lengyel II (MPW IIa) *rondel* of Michelstetten, Austria, a traditional longhouse is still in use between ca. 4460 and 4360 BC (Fig. 16.24), but by 4330/4250 BC a small two room house with porch is used at Münchendorf-Drei Mahden, south of Vienna

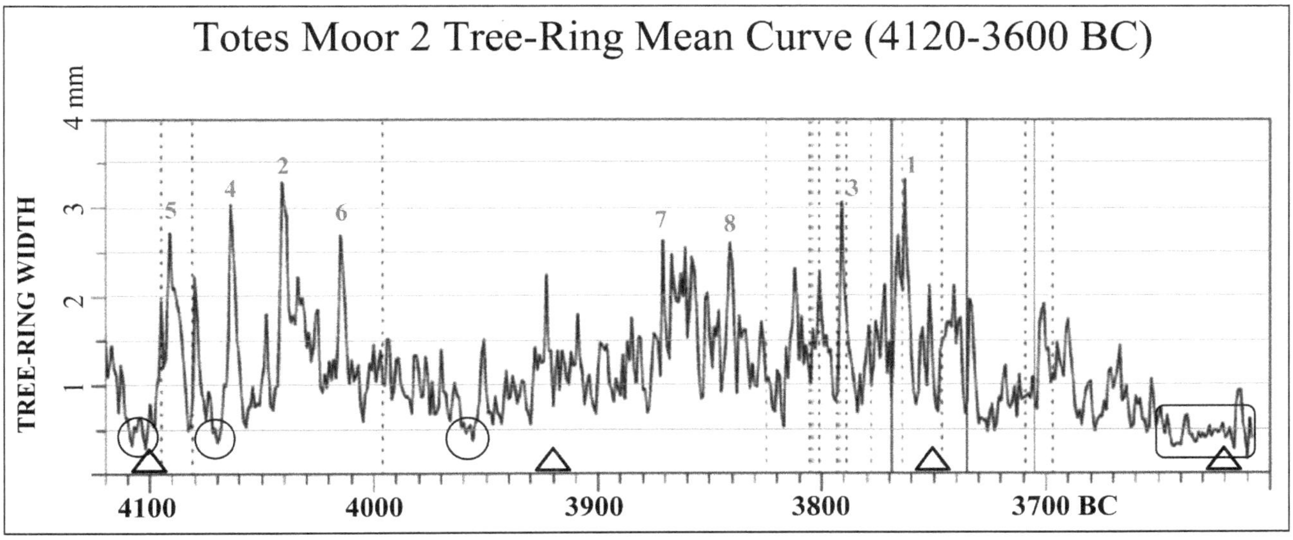

Figure 16.26. Totes Moor 2, Germany, pine tree-ring mean curve (after Eckstein *et al.* 2009 Fig. 8). Fire scared trees: Broken line = one tree, solid line = two trees. Order of greatest mean width: 1-8. Circle: Width below 0.5 mm. Rounded rectangle: Extended period of minimal ring growth. Triangle: Germination and/or dying-off phase (ibid. Table 3)

(Fig. 16.25). The latter measures only 14.87 x 6.70-6.15 m.[31] A similar house at Schleinbach, Lower Austria, is indirectly dated to the subsequent Lengyel IV Phase (Schwammenhöfer 1983:169-202).[32] The settlement has two other nearby two-room houses.[33] In adjacent Moravia the Lengyel IV Jordanów Group's house dimensions average only ca. 8.20 x 6.20 m (Šmíd 1997:123, 128).

In Germany the type-site of the Lengyel-related Aichbühl Group (Fig. 16.20) has over twenty small wooden houses. A single larger house is interpreted as a communal structure. Dating of the Aichbühl houses of the old excavation is limited (Billamboz 1998, Schlichtherle 1990, Strobel 1998), but a tree-ring date from an oak timber of House 15 implies construction or renovation shortly after 4260 BC (Billamboz 1998 Table 1). The Aichbühl Group Layer C at Felsdach Lautereck provides a ^{14}C date of 4335 (68.2%) 4255 BC, which coincides with the date of the Lengyel house from Münchendorf-Drei Mahden.[34] These dates are essentially the same as the earliest tree-ring and ^{14}C dates from the small lake dwelling houses of Egolzwil 3 and Zürich-Kleiner Hafner Layer 5A+B in Central Switzerland (Hafner and Suter 2003:34-35).[35]

[31] The partly enclosed porch-like area measures 1.57 m, the first room or antechamber is 3.95 m long and the back room is 8.52 m in length.

[32] The house itself could be traced for ca. 9 m. Its porch is a little less than 6 m wide and a little over two meters long, while the first room is ca. 4.00 m long.

[33] One measures ca. 13 x 7.5 m and the other ca. 17.00 x 7.00 m.

[34] Felsdach Lautereck, GrN-4666: 5430±40 bp; 68.2% prob. = 4335 (68.2%) 4255 BC; 95.4% prob. = 4360 (93.0%) 4220 BC, 4200 (2.4%) 4170 BC.

[35] The tree-ring dates from Egolzwil LU 3 are 4282-4275 BC and from Zürich Kleiner Hafner Layer 5A–B are 4384-4280 BC (Jacomet and Brombacher 2005:72).

The disappearance of LBK longhouse tradition[36] points to widespread socio-cultural adjustments. They are concomitant with climatic conditions illustrated by the oscillations in the Netherland and the German bog oak curves (Fig. 16.23). The peak around 4300 BC or slightly later is followed by a sudden trough that lasts until ca. 4250 BC. By 4190 BC a wide spread GDO phase occurs in Lower Saxony (Eckstein *et al.* 2009 Table 3). A short-lived GRIP Δ^{18}O warm phase is initiated by the temperature spike of ca. 4253 BC. Only the spikes of 5706 and 5332 BC are higher prior to this event. The short but strong warm phase collapses quickly into a cold phase those bottoms out around 4196 BC.

The final period of adjustment starts between 4150 and 4100 BC. In the Swifterbant culture of the Dümmer region climatic disturbances require the construction of Campemoor Log Way Pr 36 (Table 16.7) and Michelsberg Phase II becomes established. Lengyel Phase III transition to IV (Fig. 16.11). The likely switch from the Gatersleben to the Jordanów Group in Central Germany may occur around 4100 BC. In this area the Gatersleben site distribution partly overlaps geographically with that of the Jordanów Group (Müller 2001 Fig. 128-129), but Gatersleben has a greater northward extent than the Jordanów Group. The implied territorial shrinkage in the waning stages of the Lengyel interaction sphere could reflect adjustment to the increasing climatic oscillations. In Poland the Brześć Kujawski Group Late Phase ends by 4100/4000 BC. Similar end dates are found in other Lengyel regions, although some are as late as 3950/3850 BC.

Regarding climatic change of this period, there is a strong GRIP Δ^{18}O warm spike at ca. 4141 BC. It is followed by

[36] In Poland longhouses are traditionally thought to last until the end of the Lengyel culture.

Figure 16.27. The Northern European farming territory (base map from Microsoft World Atlas 1996). Interaction spheres/cultures: Baden, Chasséen Septentrional, Funnel Beaker (TRB), Globular Amphora (GAC), Michelsberg, Tripolie, Vlaardingen (V), Wartberg (W), Zedmar (Z). GAC regional groups: Central (GAC C), East (GAC E), West (GAC W). TRB regional groups: Altmark (A), Łupawa (Ł), Middle Elbe-Saale (MES), North 1 (TN 1), North 2 (TN 2), East (TE), South (TS), Southeast (TSE), West (TW), Oder/West Pomeranian (O). Circles: Selected sites mentioned in the text

a temperature collapse that bottoms out at ca. 3990 BC. GISP 2 SO_4 accumulations indicate strong volcanic activity between roughly 4100 to 3900 BC (Fig. 16.7). The Northwest German oak mean age graph bottoms out shortly after 4100 BC (Fig. 16.23) and Eckstein *et al.* (2009) observe another GDO phase. The rapid oscillations between ca. 4100 and 4000 BC are detailed by the tree-rings of Totes Moor Site 2 (Fig. 16.26).

To sum up, between ca. 5000 and 4000 BC cultural modifications, including changes in pottery styles and house architecture, cover wide areas of Europe. The human adjustments parallel climatic fluctuations, which turn into a major anomaly around 4000 BC. At this point, the Lengyel interaction sphere proper and with it the Danubian tradition, dating back to the earliest LBK, wither away. Simultaneously, the cross-barrier interactions with the Mesolithic territory to the north intensify, leading to a complete breach of Farming Barrier 2b.

THE ESTABLISHMENT OF FARMING BARRIER 3

The ca. 1500 year old Central European Farming Barrier 2b is fully breached by the Funnel Beaker interaction sphere (TRB). Farming Barrier 3 is established between ca. 4100 and 3900 BC (Fig. 16.27). Its extent is essentially defined by the northern limit of the TRB (Fig. 16.27). However, this boundary can only be roughly approximated since some of the most northerly TRB-like artifacts may be exchange-goods or items associated with TRB excursions into Mesolithic territory.

Dating the Funnel Beaker Interaction Sphere

In Austria and the Czech Republic the Baalberge or South Group is traditionally seen as the earliest manifestation of the TRB (Table 16.8).[37] TRB I exhibits affinities with Lengyel IV, including the Schussenried Group, as well as the Michelsberg culture (Čižmář 2004, Geschwinde *et al.* 2009, Lüning 1998, Zápotocký 1991, 1998, 2000). Sometimes this leads to different opinions when assigning the specific cultural affinity of archaeological evidence belonging to the transitional phase. This limits the utility of some of the early ^{14}C dates (Fig. 16.28).

[37] The Mondsee culture of Austria (3800/3700-3200/3100 BC) is currently seen as part of the TRB interaction sphere (Ruttkey *et al.* 2004), but detailed consideration of its development is beyond the scope of this chapter.

Table 16.8. Austrian-Moravian Lengyel IV and TRB South Group chronology and major climatic oscillations

Austrian-Moravian Lengyel IV and TRB Chronology

(after Baldia, Frink and Boulanger 2008, Čižmář et al. 2004, Ruttkay and Pucher 2006, Šmíd 2004)

Estimated cal. B.C.	Moravian Cultural Phase	Moravian Type-site	Moravian Mound Phase	Lower Austria Culture Phase	Austrian Type-site
2850/2750 3200/3100	Baden/Jevišovice Globular Amphora Culture	Jevišovice			
3200/3100 3350/3300	TRB IIB/Boleráz Ic-IIa	Jevišovice C1	Ohrozim (A-C)	Baalberge B2	Baierdorf?
3600/3400	TRB IIA/Boleráz Ia-Ib	Jevišovice C2	Drahanovice	Baalberge B1	
3600/3400	TRB IB_2 TRB IB_1	Rmíz Velatice	Slatinky 1 ?	Baalberge A2	Purbach Retz
3850/3750 4050/3950	TRB IA_2 TRB IA_1	Kobyli Božice/Kostelec na Hané		Baalberge A1	? Olgersdorf/Puch-Scheibenfeld
3950/3900 4200/4180	Lengyel IVC Lengyel IVB Lengyel IVA Jordanów Group	Křenovice Dolany Troubelice			

(Shaded) Major climatic oscillations

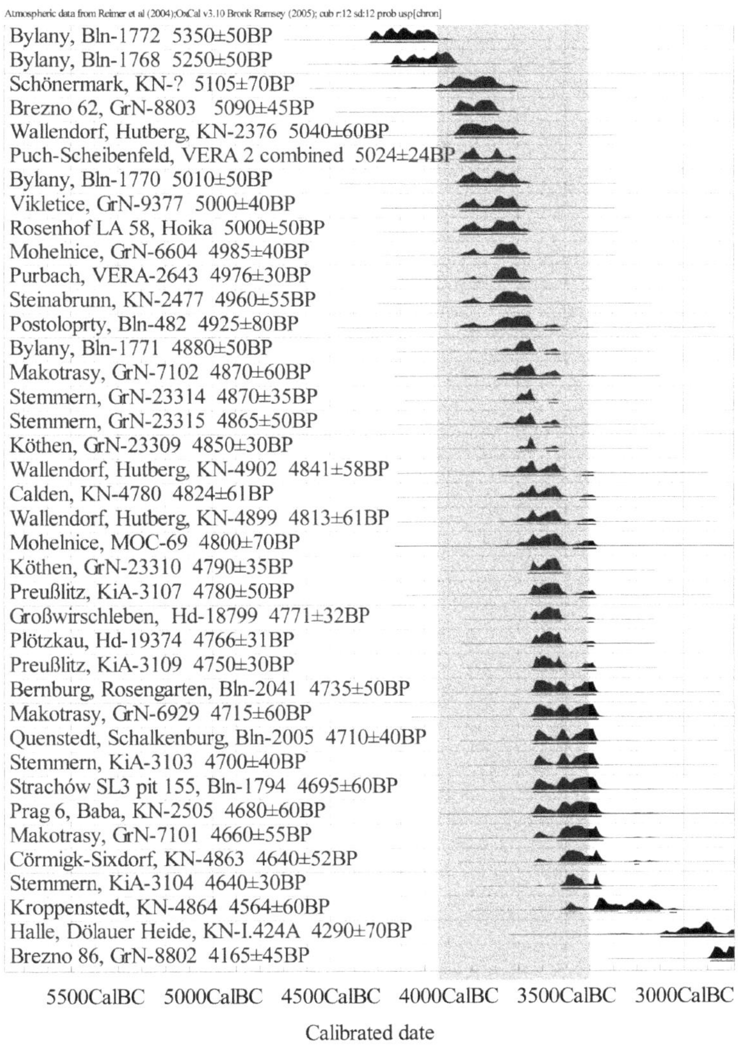

Figure 16.28. Baalberge-related ^{14}C dates ($\sigma < \pm 90$, N=40). Shaded area: Maximum Baalberge Group duration (ca. 4050/3950-3500/3350 BC)

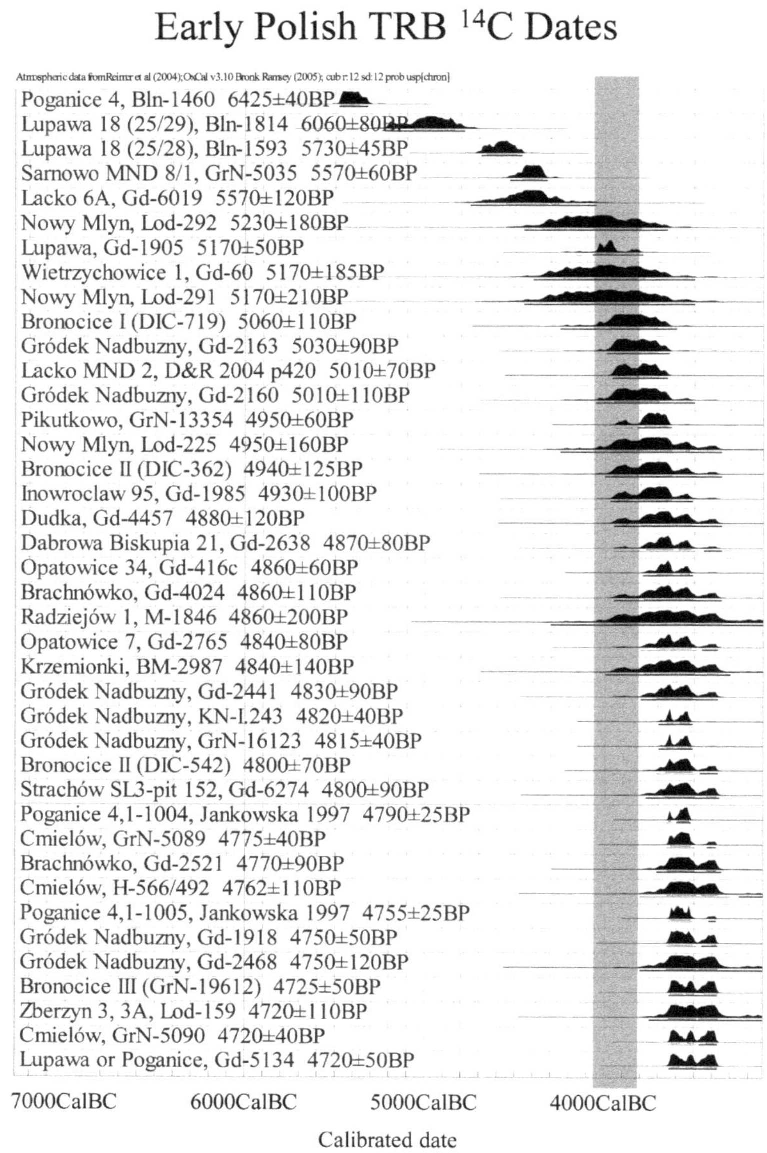

Figure 16.29. Early Polish TRB ^{14}C dates (N = 40). Shaded area: Probable TRB start (4050/3850 BC)

Nonetheless, the end of Lengyel IV roughly corresponds with the estimated beginning of the TRB IA around 4050/3950 BC, as indicated by two dates from Bylany, Czech Republic (Fig. 16.28, Table 16.6 and 16.8). For Austria two AMS dates from Puch-Scheibenfeld yield a combined range of ca. 3940 to 3870 BC for TRB IA_1 (Ruttkay and Pucher 2006 Fig. 11).[38] For TRB IB an AMS date (VERA-2653) from Purbach, Austria yields a range of 3780 to 3700 BC and the phase is estimated to last from roughly from 3850/3750 to 3600/3400 BC (Baldia, Boulanger and Frink 2008).

TRB II A starts in the difficult to date period of roughly 3600/3400 BC. TRB II B lasts from ca. 3350/3300-3200/3100 BC. During these phases the pottery begins to look more like that of the Baden interaction sphere in the middle Danube region (Baldia, Boulanger and Frink 2008, Baldia, Frink and Boulanger 2008,[39] Furholt 2008, Wild *et al.* 2001). In parts of Austria, the Czech Republic and Slovakia the TRB ends with the appearance of the Jevišovice Group of the Baden interaction sphere, which starts at 3200/3100 BC (Baldia, Boulanger and Frink 2008 Table 1, Mayer 2008, Šuteková 2008). During this time the Globular Amphora culture (GAC) occurs to the north of the Jevišovice Group in Central and North Moravia. It shares some attributes with the TRB and the Baden interaction sphere. With the demise of Jevišovice around 2850 BC the Baden interaction sphere ends.

North of Moravia in adjacent Poland numerous regional chronological schemes exist (e.g. Jankowska and Wiślański 1991, Kowalewska-Marszałek 2008, Midgley 1992, Prsybył 2008, Rybicka 2008, Szmyt 2008, Włodarczak 2008, Zastawny 2008). Early Polish TRB

[38] Group calibration by Stadler results in 5024±24 bp: 68.2% prob. = 3940 (42.3%) 3870 BC, 3810 (25.9%) 3710 BC; 95.4% prob. = 3950 (95.4%) 3720 BC.

[39] Baldia, Frink, and Boulanger (2008 Fig. 1) shows a more generalized Baden sphere of influence, which has a wider range than the hypothetical boundaries of the Baden interaction sphere shown in Fig. 16.27.

dates are shown in Figure 16.29. For the TRB's Southeast Group the Bronocice (Br) chronology has been most influential (Kruk *et al.* 1996 Table 2). Br I is represented by a single ^{14}C date, yielding 3970 to 3710 BC.[40] Br II roughly dates from 3800/3700 to 3600/3400 BC. For Phase III the oldest date (GrN-19612) overlaps with the ^{14}C dates of BR II (Bakker *et al.* 1999 Table 1). Analysis of eight ^{14}C dates reveals a range of ca. 3600/3400 to 3300/2900 BC. The calibration of eleven Bronocice IV ^{14}C dates yields a range of ca. 3300/3100-2900 BC. The distribution of five ^{14}C dates for Br V should probably be truncated at 2850/2750 BC, although, their error ranges permit a much later end. The relationship of the Polish TRB with the Baden interaction sphere becomes more pronounced in the later phases and its more southerly regions may well become an integral part of that sphere.

Farther north in Poland, the TRB is traditionally seen to develop from Mesolithic populations that come into contact with the Lengyel's Brześć-Kujawski Group (Fig. 16.16). Various TRB sub-phases have been envisioned (Jankowska and Wiślański 1991). Of these Kośko's pottery chronology, based on a widely used Scandinavian scheme (Jankowska and Wiślański 1991), has left the strongest imprint. Kośko first thought that Lengyel I and II as well as TRB I coexisted around 4600-4400 BC (Kośko 1980 Fig. 5). Later Lengyel I was thought to end around 4400 BC and Lengyel II around 3900 BC (Niesiołowska 1994 Table 1). Simultaneously, Kośko's hypothetical TRB Ia was dropped while TRB Ib (Łącko Phase) and Ic (Sarnowo Phase) were combined into the Sarnowo Phase. This was based on a single ^{14}C date (GrN-5035: 5570±60 bp) from Sarnowo Mound 8 (Bakker 2002, Bakker, Vogel and Wiślański 1969). Currently the date calibrates to 4460 (68.2%) 4350 BC at 1σ. Later Czerniak, Domańska and Kośko (1991 Fig. 1) added the date (Gd-6019, 5570±120 bp) from Łącko Mound 2, which now calibrates to 4550 (64.0%) 4320 BC at 1σ (Fig. 16.29).

The very early Polish dates have always been problematic, in part because the ^{14}C samples from Sarnowo Mound 8 and Łącko Mound 2 come from underneath the mounds. The date from Sarnowo Mound 8 may actually date a Lengyel-related Brześć Kujawski Group occupation (Bakker 2002, Baldia 1995, 2010). The Łącko Mound 2 date is now reinterpreted as coming from a wooden post of an earlier TRB site preceding the mound (Domańska and Rzepecki 2004).[41] Nonetheless, it falls into the Lengyel timeframe. The trapezoidal mound itself is now dated to 3820 (41.1%) 3700 BC by a more recent ^{14}C date (Domańska and Rzepecki 2004) (Fig. 16.29).[42] Therefore, the extraordinarily early beginning of the Polish TRB and the derivation of earthen long-barrows from Lengyel longhouses are thrown into doubt. This all the more so, since Sarnowo pottery can be linked with TRB's Baalberge pottery of Central Germany and Oxi pottery of the TRB North Group 2 in Scandinavia (e.g. Koch 1998:45, 181-182). Both Oxi and Baalberge are dated ca. 3800 to 3500 BC (Müller *et al.* 2010 BC). Thus, the earliest mound construction occurs not in the putative Sarnowo phase, but in the Pikutkowo or early Wiórek Phase (Kośko's TRB II/IIIa) (Jankowska 1999:224) as indicated by the later Łącko assay (Fig. 16.29). Given the end of Lengyel IV, as noted above, one is forced to conclude that the transition to the TRB begins in Poland around 4050/3950 BC and the construction of earthen long-barrows starts at 3800/3700 BC (Baldia 2010:204-205).

The GAC overlaps geographically with much of the TRB (Fig. 16.30) and shares many traits with the later TRB and the Baden interaction sphere (e.g. Szmyt 2008). Unfortunately, contradictory interpretations are associated with the development of the GAC. Wiślański (1966) divides it into three regional groups (Fig. 16.27). The East Group reaches beyond the TRB distribution (Szmyt 1998), while the West Group ends beyond the west bank of the Elbe River in Germany (Dirks 2000 Fig. 64). Wiś-lański's Central GAC is centered in Kujavia, where two ^{14}C dates are seen to *dovetail* with the end of the Late Phase of the Brześć Kujawski Group (Grygiel and Bogucki 2008:2031).[43] There the GAC is also thought to have built megalithic chambers (Jankowska 1999). Nevertheless, the duration of the GAC (Fig. 16.16) as well as its construction of the chambers is contested (Bakker 1992, Baldia 2010, Baldia, Boulanger and Frink 2008, Jankowska 1999). The oldest date (Ki-6238: 4970±30 bp) from Kuczkowo Site 1/C2 is often used to justify the antiquity of the GAC, but it is a statistical outlier (Fig. 16.30). The cultural affinity of the dates next two dates (Bronocice DIC-718 and Zarębowo GrN-5044) is problematic. The subsequent two dates (Dęby 29 Gd-2148 and Bronocice III DIC-360) have large calibrated ranges and do not unequivocally favor a beginning prior to 3300/3100 BC. Dates after ca. 2850 BC fall into the reverse S tail of the distribution and should probably be ignored. Thus, Baldia, Boulanger and Frink (2008:271-273) conservatively date its duration to ca. 3300/3100-2850/2750 BC, but the narrower range of ca. 3100-2850 BC strikes me currently most appropriate, especially since Włodarczak (2008 Fig. 2) expresses a similar view.

As in Austria, Moravia and southeast Poland, communication between the TRB in Kujavia and the Baden interaction sphere becomes evident around 3600/3400 BC. As new regional groups develop, those closer to the central Danube have greater links with the core of the Baden interaction sphere (e.g. Jankowska and Wiślański 1991) and a few groups may actually be part of it. On the Baltic coast of Poland the Pomeranian TRB's Łupawa Group (Fig. 16.27) displays similarities with other regions

[40] Bronocice I, DIC-719: 5060±110 bp; 68.2% prob. = 3970 (68.2%) 3710 BC; 95.4% prob. = 4250 (95.4%) 3600 BC.

[41] Łącko 6A, Gd-6019: 5570±120 bp; 68.2% prob. = 4550 (64.0%) 4320 BC, 4300 (4.2%) 4260 BC; 95.4% prob. = 4750 (95.4%) 4050 BC. Lacko Mound 2 (Domańska and Rzepecki 2004:420): 5010±70 bp; 68.2% prob. = 3940 (27.1%) 3860 BC, 3820 (41.1%) 3700 BC; 95.4% prob. = 3960 (95.4%) 3650BC.

[42] Łącko Mound 2 (Domańska and Rzepecki 2004:420): 5010±70 bp; 68.2% prob. = 3940 (27.1%) 3860 BC, 3820 (41.1%) 3700 BC; 95.4% prob. = 3960 (95.4%) 3650 BC.

[43] The pertinent ^{14}C dates appear in a later volume Grygiel and Bogucki (2008:2031), which is presently not available to me.

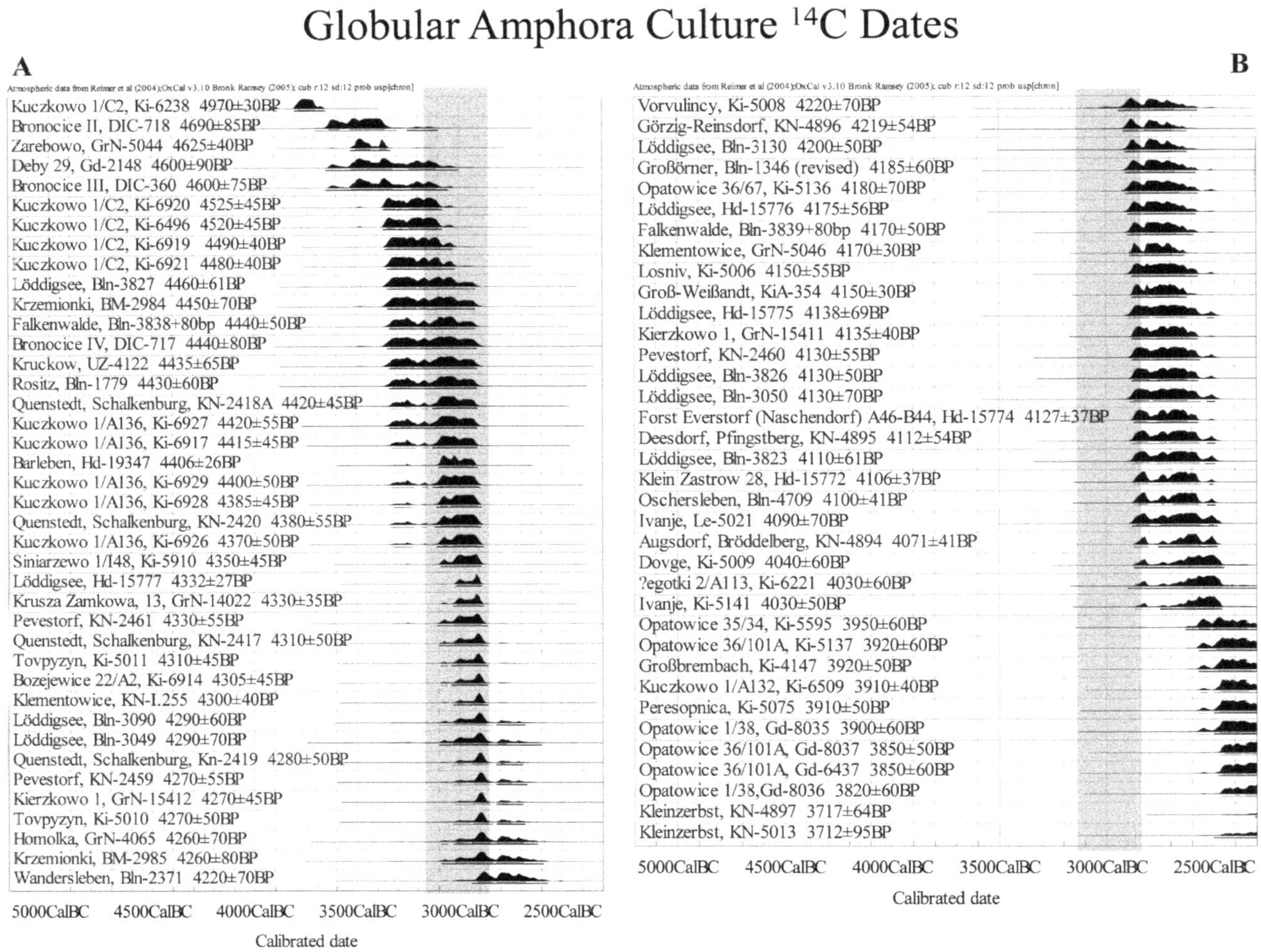

Figure 16.30. Globular Amphora culture (GAC) ^{14}C dates ($\sigma < \pm 100$, N = 76). Cultural association is not always exclusively attributable to the GAC. Shaded area: GAC duration, A: Beginning dates, B: Continuation

in Poland as well as the TRB in Germany and Scandinavia (Jankowska, 1980, 1994 Table 2, Jankowska and Wiślański 1991 Fig. 5). Granting that many Łupawa Group ^{14}C dates are problematic (Baldia, Boulanger and Frink 2008, Baldia, 1995, 2010), the transition from a local more or less Ertebølle-like forager tradition to the TRB may nonetheless be captured by sample Gd-1905: 5170±50 bp (Fig. 16.29). It calibrates to 4050 (68.2%) 3940 BC at 1σ. The TRB ends no later than ca. 2850/2750 BC in Pomerania and elsewhere in Poland according to 164 Polish ^{14}C dates in the *14CEurope* database, but a later end is traditionally preferred in some regions (Fig. 16.16).

West of Poland, the TRB development is documented in Central Germany by Müller (2001). Technically, this is the Middle Elbe-Saale River region (MES), which forms the center of the German Baalberge Group (Fig. 16.28). Geschwinde *et al.* (2009:185-206) view the Baalberge Group as a culture distinct from the TRB. In addition, they contrast it with the Michelsberg culture (Fig. 16.27), using a Michelsberg centered pottery typology. The authors date Baalberge to ca. 3700-3500 BC. However, in the south this timeframe only includes the later Baalberge phases, i.e. Baalberge A2 in Austria and TRB IB in Moravia (Table 16.8). Thus, Moravian Baalberge (TRB IA) equates with the very sparsely dated TRB-MES I (Lengyel IV), which dates to ca. 4100-3800 BC according to Müller (2001) and Müller *et al.* (2010). Central German Baalberge proper (TRB-MES II) starts ca. 3800 BC and ends between 3500 to 3400 BC (Müller 2001 Fig. 140). A later *Baalberge Grave Pottery Phase* (TRB-MES III) is dated ca. 3500-3300 BC (Müller *et al.* (2010 Fig. 1). This assessment hinges on the literal interpretation of the ^{14}C dates from the individual interments at Stemmern (KiA-3104) and Kroppenstedt (KN-4864). However, the calibrations do not necessarily date much later than 3400 BC (Fig. 16.28). In addition, Geschwinde *et al.* (2008) find this pottery type in both funerary and profane context and thus, negate its distinction as an independent later Baalberge phase. Therefore, I currently prefer to combine both phases into MES-II/III and end Baalberge along with Michelsberg in the chronologically opaque period of 3650/3600 to 3400/3350 BC.[44] Still, placing the end of Baalberge in

[44] MES-II/III may also include the Schöninger Group, because Raetzel-Fabian and Furholt (2006) suggest that the undated regional ceramic assemblage, named after the type-site of Schöningen, Germany (Fig. 16.27), shows Baden-Boleráz attributes. However, Grygiel and Bogucki 2008:2030 continue to prefer to associate it with the late Lengyel interaction sphere.

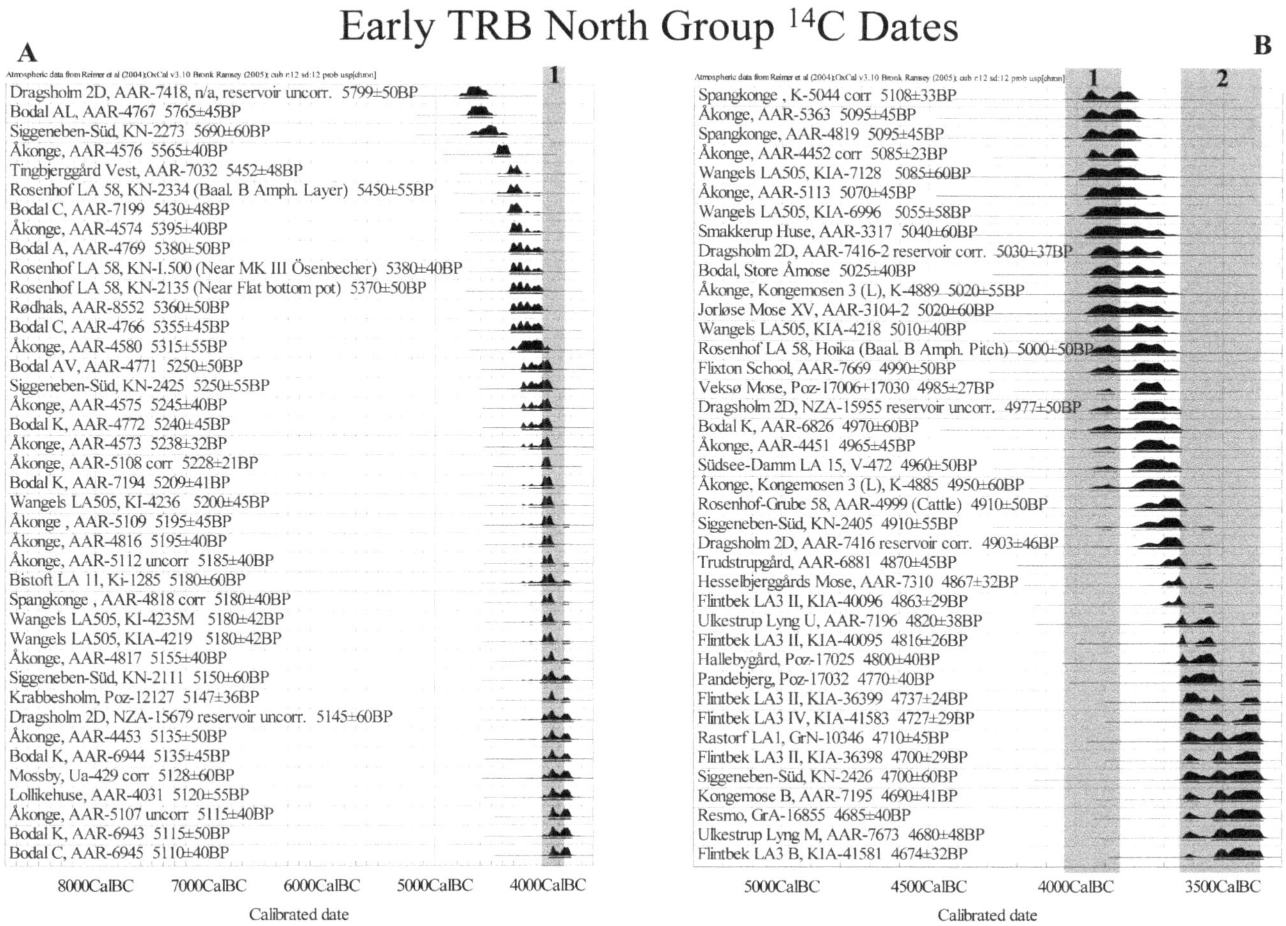

Figure 16.31. Early TRB North Group ^{14}C dates ($\sigma < \pm 65$, N = 80). Shaded areas: 1. TRB EN I start (4050-3850 BC), 2. Duration of EN II (3650/3600 to 3400/3350 BC). Some dates are attributable to the Ertebølle culture and some are uncorrected for reservoir effect

this period requires that Müller's TRB-MES IV (the Salzmünde Group) starts up to 100 years earlier. Therefore, it should exist from 3400/3350 to 3200/3100 BC. Müller's TRB-MES IV is followed by TRB-MES V (ca. 3100-2800 BC), which corresponds to my conservative estimate of 3200/3100-2850/2750 BC. This final phase is characterized by the TRB's Bernburg pottery, which has strong Baden attributes and coexists with the GAC in the region.

Next to the MES region (Fig. 16.27) is the Altmark district, for which Müller *et al.* (2010 Fig. 1) propose a seven fold chronology ranging from 3600 to 2800 BC. West of the Lower Elbe River the TRB West Group (ca. 3400-2850/2750 BC) develops in most of the Swifterbant culture's territory (Bakker 1992, Brindley 1986, Lanting and van der Plicht 2000). Immediately to the south, the northern Michelsberg culture seems to transition directly to the Wartberg culture between ca. 3500 and 3400 BC.[45] The latter is divided into an Early and a Late Phase according to Raetzel-Fabian (2000). Glossing over several outliers, I date Phase I to ca. 3500/3400-3100/3000 BC and Phase II to ca. 3100/3000-2850/2750 BC, using ten ^{14}C dates with a 1σ range of less than ±100 years respectively.

TRB West Group and the Wartberg culture pottery occasionally exhibit Baden and GAC features (Bakker 1979:125-126, 134-135, Dirks 2000, Geschwinde *et al.* 2008, Laux 1990). Interaction between the Wartberg culture and neighboring cultures (Fig. 16.27) also includes the Horgen culture and related groups (Geschwinde *et al.* 2008, Raetzel-Fabian 2000, Schlichtherle 2003).[46] The Horgen culture is centered on the Bodensee, where the tree-ring dated site of Arbon Bleiche 3 (3384-3370 BC) includes locally made Pfyn/Horgen and Baden-Boleráz pottery, as well as a small sample of allochthonous bone tempered pottery (Baldia, Frink and Boulanger 2008, Bonzon 2005, Capitani *et al.* 2002).

[45] The distribution of 39 Wartberg culture's ^{14}C dates of < ±100 years does not start with the characteristic reversed S-shape, which theoretically implies a continuous development out of the northern Michelsberg culture. The end of Wartberg around 2850/2750 BC is supported by an inverse S-shaped tail in the ^{14}C distribution.

[46] The distribution of 69 Horgen culture ^{14}C dates does not start with the characteristic reversed S-shape, which theoretically implies a continuous development from the Pfyn to the Horgen culture via a Pfyn/Horgen phase. However, the end of Horgen around 2850/2750 BC is supported by a reverse S-shaped tail in the ^{14}C distribution.

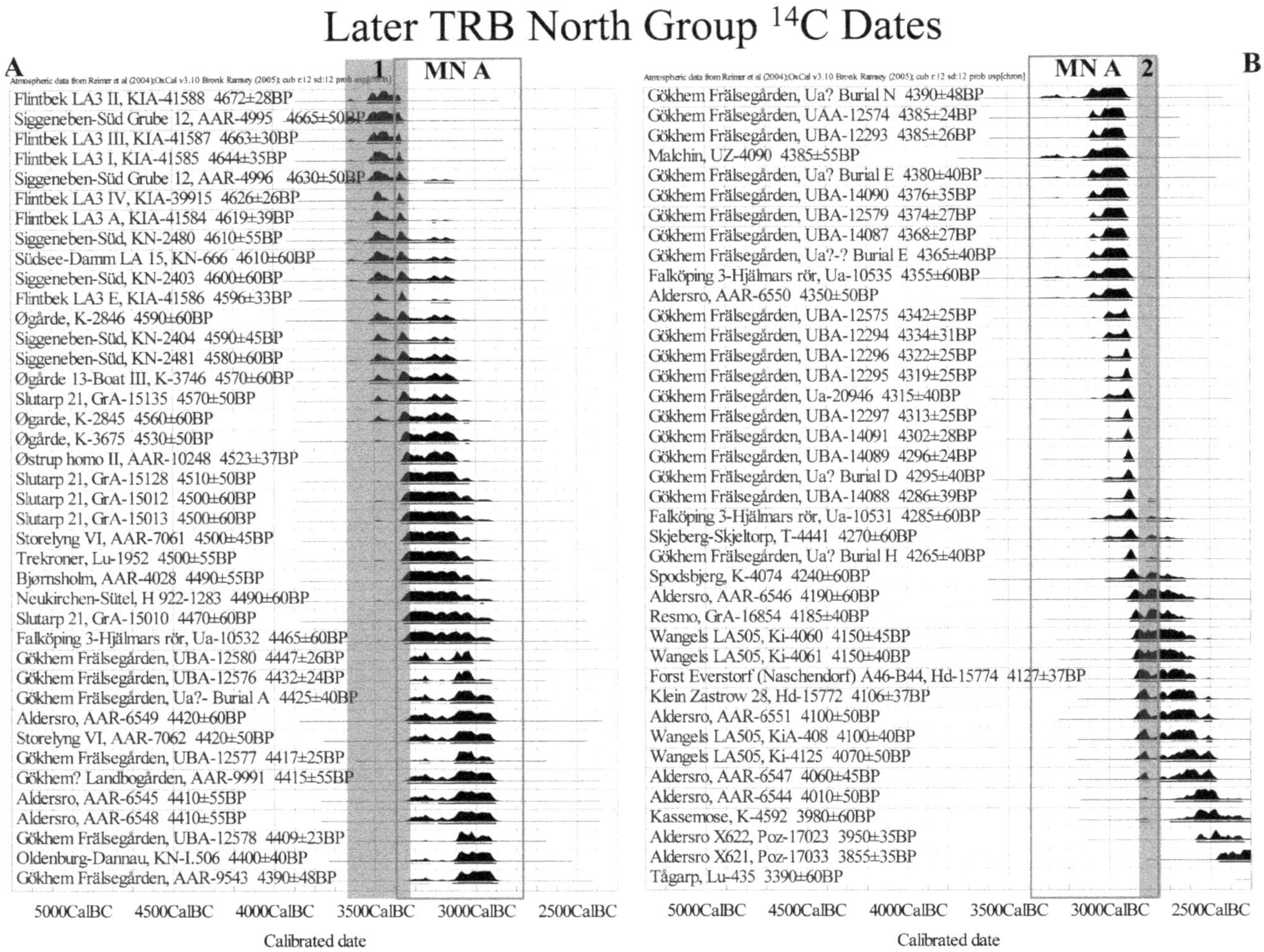

Figure 16.32. Later TRB North Group ^{14}C dates ($\sigma < \pm 65$, N = 80). A: Beginning dates, B: Continuation. Shaded area: 1. Duration of EN II (3650/3600-3400/3350 BC), 2. End of TRB MN A (2850-2750 BC). Unshaded box: Overall duration of MN A (3400/3300 to 2850/2750 BC). Some dates are uncorrected for reservoir effect

The TRB North Group develops in the Ertebølle territory (Fig. 16.27) and is minimally split into two geographic sub-groups (e.g. Midgley 1992). As in other cultural transitions, it is difficult to distinguish the late Ertebølle from earliest TRB artifacts (Fischer 2002:350-355). Legacy dates from the North Group 1 site of Rosenhof once placed its beginning at 4400/4300 BC. However, concentrated research programs, using precision dating, put the beginning around 4100 BC (Hartz and Lübke 2006). North Group 2 develops within roughly the same timeframe as Group 1, but is traditionally seen as starting slightly later (cf. Fischer 2002, Fischer and Heinemeier 2003, Fischer *et al.* 2007, Østmo 1991, 2007, Koch 1998, Knutsson and Knutsson 2003). My evaluation of the TRB ^{14}C dates from both groups in the *14CEurope* database suggest a start range between ca. 4050 and 3850 BC (Fig. 16.31, No. 1).

Overall, the chronology of the two North Groups is divided into the Early Neolithic (EN) and the Middle Neolithic (MN). The EN is subdivided into Phase I and II. In North Group 1 the EN I is separated into EN Ia (Wangels Phase), dating ca. 4100 to 3800 BC, and EN Ib (Siggeneben Phase), dating ca. 3800 to 3500 BC (Hartz and Lübke 2006). However, my analysis of the 40 latest ^{14}C samples with a σ of less than ±90 years in the *14CEurope* database suggests that the EN may end as early as 3650/3600 BC. In both groups EN II is usually dated to ca. 3500-3300 BC, but my calculation of 36 ^{14}C samples with a σ of less than ±90 years indicates that the calibrations fall mostly into the 3600-3400 BC range (Fig. 16.31, No. 2). Therefore, I tentatively propose a range of 3650/3600 to 3400/3350 BC for EN II.

The Nordic Middle Neolithic A (MN A) is traditionally dated from ca. 3300 to 2800 BC and divided into MN A Ia through MN A V using roughly 100-year increments (e.g. N. Andersen 1997, Müller *et al.* 2010, Nielsen 1993). However, the ^{14}C dates overlap considerably and do not easily support the typological divisions.[47] Therefore, I generally treat the MN A as a unit and date it from 3400/3300 to 2850/2750 BC (Fig. 16.32 and 16.33).

[47] The *14CEurope* database contains nineteen EN II/MN A transition and MN A dates with less than ±80 years at 1σ.

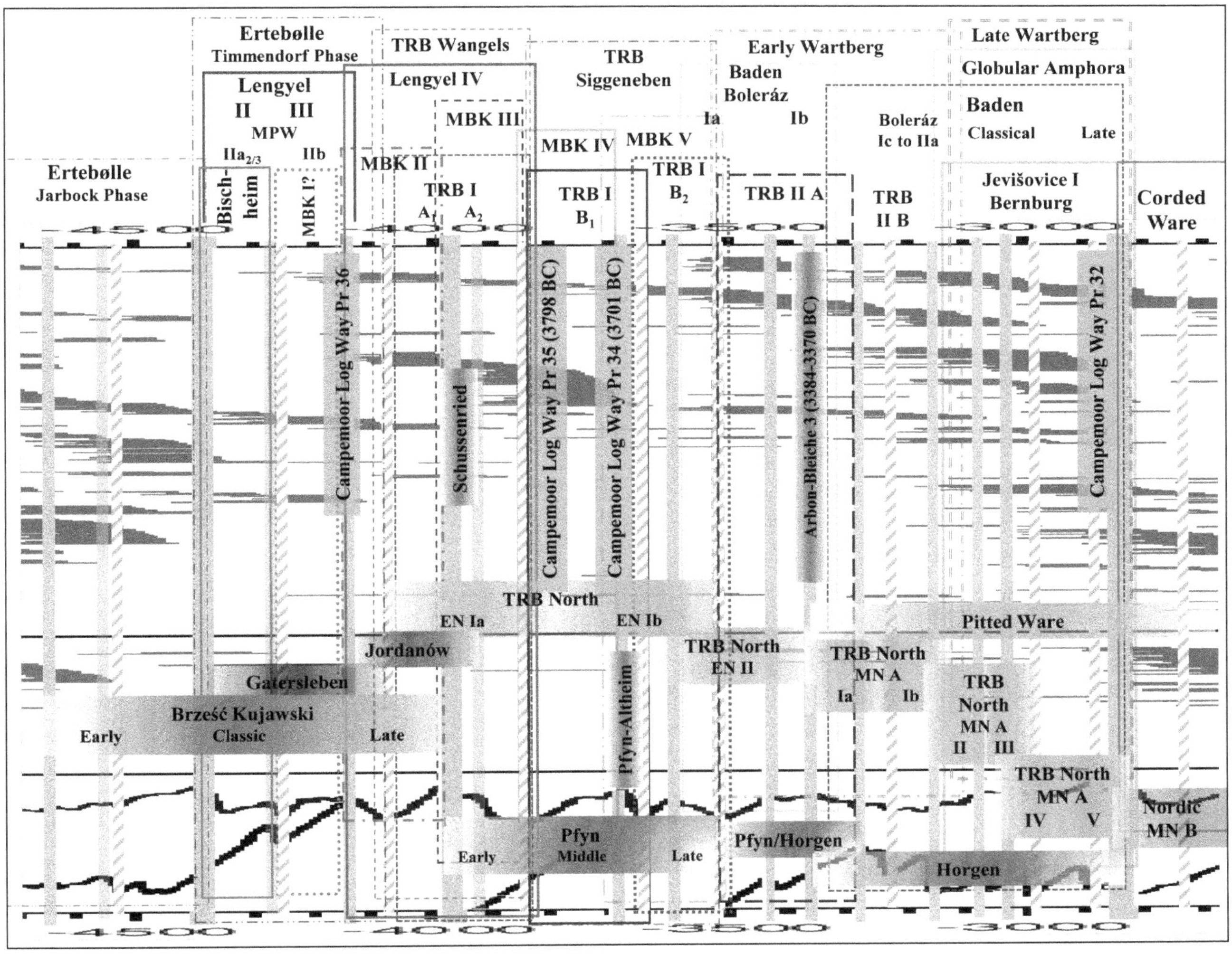

Figure 16.33. Northern Central and North European culture and climate change (4700-2600 BC). Michelsberg culture (MBK), Funnel Beaker culture (TRB): North Group Early Neolithic (EN Ia – Ib), Middle Neolithic (MN A Ia – MN A V), South Group (TRB I A_1 – II B). Alpine Piedmont tree-ring dated groups: Schussenried (3955 to at least 3871 BC), Hornstaad (4052 to at least 3890 BC), Pfyn, Early (3870 to 3790 BC), Middle-Late (3701-3517 BC), Altheim-Pfyn (3745-3650 BC), Arbon-Bleiche 3 (3384-3370 BC)

During the EN II and MN A Baden-Boleráz-like pottery attributes can be traced from Moravia to the Lower Elbe and into North Group 1 (Baldia, Boulanger and Frink 2008, Baldia, Frink and Boulanger 2008). Towards the end of this period, i.e. MN A IV/V, GAC-related pottery is recognized as far north as Denmark (Davidsen 1973, 1978).

Typical North Group 2 EN pottery is found throughout northern Denmark and the tip of southern-most Sweden (Skåne). North of Skåne the artifacts are not as easily pigeonholed, even though, the ^{14}C dates do not differ appreciably. In Southeast Sweden the site of Kv. Seglaren, Växjö, has several ^{14}C dates,[48] of which two (Ua-21169, Ua-21181) indicate that the Mesolithic-Neolithic transition begins around 4100/4000 BC (Åstrand 2005 Fig. 7). By 3600 BC settlement activities intensify. Sample Ua-21176 dates the transition to the local MN to 3400/3300 BC. Farther north in Södermanland, south of Stockholm, a serious of dates indicates the beginning of the TRB at 4000/3900 BC (Hallgren 2004b).

The location of the Swedish sites are mapped by Hallgren (2004b Fig. 11) and Knutsson and Knutsson (2003 Fig. 4a). East of Stockholm and Uppsala, EN sites are found on the present islands of Öland and Gotland in the East Baltic Sea. Inhabitants of these coastal sites subsist primarily on fishing and seal hunting, but domesticated animals are also found early on. The ^{14}C assay from a sheep in Stora Förvar cave on the current Island of Stora Karlsö near Gotland dates to 3960 (68.2%) 3780 BC (Rundkvist *et al.* 2004 Table 9e).[49]

[48] Kv. Seglaren, Växjö, Ua-21169: 5360±60 bp; 68.2% prob. = 4330 (47.0%) 4160 BC, 4130 (21.2%) 4070 BC. Ua-21181: 5220±45 bp; 68.2% prob. = 4150 (0.9%) 4130 BC, 4060 (67.3%) 3960 BC. Ua-21176: 4555±50 bp; 68.2% prob. = 3370 (22.0%) 3310 BC, 3240 (46.2%) 3100 BC.

[49] Stora Karlsö-Stora Förvar, Ua-4952: 5070±75 bp; 68.2% prob. = 3960 (68.2%) 3780 BC.

Inland from the coast near Uppsala and Stockholm early TRB sites are distributed in a southwesterly direction to Lake Vättern. The sites mark Farming Barrier 3 and exhibit remains of cattle, sheep, goat, wheat, barley, peas, beans and even grapes (Knutsson and Knutsson 2003:51). Father west, between Lake Vättern and Lake Vänern in the Falköping region is a small cluster of EN TRB sites. By the MN this area becomes the most densely settled region in Sweden, probably due to its ideal farming environment.

For west coast TRB sites Clark (1977) proposes a fishing subsistence. However, Bradley and Phillips (2004), who analyze the location of MN megalithic tombs in relation to the coastline around the time of the TRB, conclude that a fishing subsistence was likely on the island of Tjörn, but domesticated ungulates could have been pastured on Örust Island. Finally, Sjögren (2003) concludes that the ocean seems to have contributed little and terrestrial food, most likely from domesticates, dominates the diet.

Sjögren (2003) proposes three phases of development in Southwest Sweden: Formative (ca. 3800-3500/3400 BC), Climax (3500/3400-3100/3000 BC), and Degenerative (3100/3000-2800 BC). For neighboring South Norway, Østmo (2007:111-112) prefers an early (ca. 4000-3500 BC), a middle (3500-3300 BC) and a late phase (3300-2800 BC), with the proviso that the middle phase should probably be split in two.

In South Norway farming substance is practiced around Oslo Fiord as early as ca. 4000 BC (Østmo 1991, 2007). Sites to the west generally follow the coast line at least as far north as Bergen (Fig. 16.27). Stray finds may be found even farther north. The coastal sites are said to focus on marine resources and the pottery is not fully identical to the TRB. Two of these TRB-related sites are reported inland in Middle Norway. Their location on the east-west watershed may hint at an inland communication line between the west coast sites and the more typical TRB farmsteads around Oslo Bay. During the local Late MN, the TRB site distribution moves towards the coast, implying a farming hiatus (Østmo 2007:115).

The Pitted Ware culture appears at Oslo Bay and the adjacent West Coast of Sweden around 3200 BC (Bengtsson *et al.* 2010). It is thought to develop from the same EN substrate as the TRB (Knutsson and Knutsson 2003). The Pitted Ware culture is customarily seen as having an aquatic diet. Although, such a diet often predominates, this is not always the case according to Lidén *et al.* (2004). Even in the Gotland Island region of the distant East Baltic the archaeological record exhibits a combination of seal, porpoise, codfish, (wild?) pig, cattle and sheep, but their ratios vary in time (Rundkvist *et al.* 2004), perhaps suggesting adjustments to climatic oscillations.

As the TRB retreats towards the southwest, the Pitted Ware culture's site distribution reaches north beyond Farming Barrier 3 in eastern Sweden (Knutsson and Knutsson 2003 Fig. 10). From Lake Vättern to the south and west, both cultural entities largely overlap or exist in close proximity. Across from the Swedish west coast in the Danish coastal zone Pitted Ware-related sites appear by 3100/2870 BC.[50] The TRB North Group 2 ends by 2850/2750 BC, but the Pitted Ware culture seems to persist for several centuries longer, at least in parts of Sweden.

Dating the Mesolithic-Neolithic Transition

Much emphasis has been placed on the long period during which domesticates have been available to the foragers beyond Farming Barrier 2b. For Poland Nowak (2001, 2004, 2006, 2007) proposes centuries of co-existence with various kinds of interaction between Mesolithic foragers and Central European farmers. He sees this as the basis for regional differences in the degree and the kind of farming in different parts of Poland (Fig. 16.16).

In adjacent East Germany the Mesolithic-Neolithic transition is outlined by Terberger and Piek (1998) and Hartz *et al.* (2007 Fig. 3). At the previously mentioned Ertebølle sites of Parow and Stralsund (*Mischwasser-speicher*), near the island of Rügen, the final manifestation of the Mesolithic is dated to ca. 4000 BC (Fisher and Heinemeier 2003 Fig. 6, Kaute *et al.* 2005, Terberger and Piek 1998) and the transition to TRB North Group 1 seems to occur around this time in the region (Hartz *et al.* 2007 Fig. 4).

In the west, the Swifterbant culture uses pottery and domesticated animals rather early and exhibits cereal cultivation by 4300/4000 BC (Cappers and Raemaekers 2008). By 4000/3800 BC there is evidence of human impact at Brandwijk-Kerkhof, Netherland (Out 2008).

Near the eastern border of the Swifterbant culture is the Ertebølle dune site of Hamburg-Boberg (Fig. 16.19, 16.20 and 16.27). Past excavations yielded Mesolithic as well as stroke ornamented pottery, of which the latter was traditionally attributed to the STK and/or Rössen cultures (Schindler 1953, 1961, Schwabedissen 1997, 1994). This suggested early interaction with the farming cultures south of Barrier 2b. However, three ^{14}C assays from food crusts of pointed- and round-bottomed pottery date between ca. 4150 to 4050 BC (Fisher and Heinemeier 2003 Fig. 6, Lanting and van der Plicht 2000:59, Hedges *et al.* 1995). Furthermore, similar stroke ornamented pottery occurs at other Late Ertebølle sites of the region (Raetzel-Fabian and Furholt 2006:2). This contradicts the traditional interpretation. Finally, evidence associating the first domesticates with the Ertebølle culture has also been negated by new research in North Germany and South Scandinavia, which links the beginning of farming with TRB EN I (Behre 2007, 2009, Bollongino and Burger 2007, Edwards *et al.* 2007, Hartz, Lübke and Terberger 2007, Scheu *et al.* 2007). Thus, the use of domesticates appears to be later than in the Swifterbant

[50] Kainsbakke, K-4553: 4320±85 bp; 68.2% prob. = 3100 (68.2%) 2870 BC; 95.4% prob. = 3350 (95.4%) 2650 BC.

culture, in spite of a similarly long period of availability (Cappers and Raemaekers 2008:392).

The simultaneous appearance of the TRB and domesticates in the Ertebølle territory may be understood as an invasion by farmers from other regions (e.g. Fischer 2002, Klassen 2004). Yet, in spite of the rapid change in material culture, the Mesolithic-Neolithic transition is usually seen as an autochthonous development by Scandinavian and German researchers. This notion is buttressed by the continually stable marine biotope between 5000 and 3700 BC, as well as the persistence of a Mesolithic hunting, gathering and fishing economy, which is only supplemented by a minor reliance on domesticates during TRB EN I (S.H. Andersen 2008, Fischer 2002, Fischer and Heinemeier 2003, Fischer *et al.* 2007). In addition, Fischer (2002 Fig. 22.15) stresses that the decline of shellfish consumption occurs before the appearance of farming, as indicated by the initial drop in the number of ^{14}C dated Danish shell middens. A closer look at Fisher's diagram shows that shellfish consumption plateaus around 4400 BC and begins a gradual decline by ca. 4300 BC. The decline accelerates around 4100/4000 BC and drops drastically after ca. 3900/3800 BC. These dates correlate well with the environmental, climatic and socio-cultural changes before and during the Mesolithic-Neolithic transition.

To sum up, the desire to establish a single, datable event, leading to the Mesolithic/Neolithic transition, may be unwise or even unrealistic. Instead, the range of ca. 4100/4000 to ca. 3900/3800 BC is currently most appropriate for the events leading to the establishment of Farming Barrier 3 (Fig. 16.33).

Climate Change and Human Adjustments

Due largely to the aftereffects of deglaciation, the environmental alterations impact the regions around the Baltic and the North Sea most strongly. This leads to a complicated evolution of the landscape, characterized by differential regional uplift and subsidence, causing repeated marine transgressions and considerable environmental alterations (e.g. Andrén *et al.* 2000, Burman and Schmitz 2005, Christensen 1993, Gerdes *et al.* 2003, Fisher 2002, Hartz and Lübke 2006, Jakobsen *et al.* 2004, Kalis *et al.* 2003, Poska and Saarse 2002, Price *et al.* 2001). Vivid examples of these drastic changes are the differences between modern and Ertebølle-TRB sea levels. For instance, the Baltic Sea near Rügen Island is ca. 1.7 m below the current sea level, but some 100 km to the west (Wismar Bay) it is about 3.5 m below (Hartz *et al.* 2007:572). In South Sweden the ancient coast is now ca. 5 to 6 m above current sea level (Fahlander 2008:35). However, even Central Europe did not escape its share of environmental alterations. This includes floods and course changes of the Morava River (Boulanger, this volume). Turning to the climatic and concomitant socio-cultural changes between 4100/4000 and 2800/2700 BC (Fig. 16.33), I divide this period of more than a millennium somewhat arbitrarily into four phases of change.

Change ca. 4100-3800 BC

The GRIP temperature trend line falls below the mean between ca. 4100 BC and 3950 BC (Fig. 16.4). The strongest temperature drop occurs around 3980 BC. This is the worst cold spike since the 6180 BC cold event. Above normal temperatures exist from ca. 3950 to about 3826 BC. They are succeeded by another cold phase. It bottoms out with the cold spike of 3821 BC. This extreme spike is even worse than the one of ca. 3980 BC. Shortly thereafter, the temperature climbs above the mean, peaking around 3809 BC.

As mentioned previously, the SO_4 data indicate an uptick in volcanic activity starting just before 4100 BC. The highest peak since ca. 5275 BC occurs around 4035 BC (Fig. 16.7). This is the 6th highest peak between 6800 and 1500 BC. It is followed by more peaks, including the 18th highest (ca. 3977 BC) and the 19th highest (ca. 3905 BC). The high volcanic activity ends shortly after 3870 BC.

The climatic oscillations are echoed in the German oak tree record (Fig. 16.5). The Main River trees register a severe anomaly between ca. 4100 to 3900 BC (Spurk *et al.* 2002 Fig. 4). The detailed pine tree-ring mean curve of Totes Moor 2 exhibits a GDO phase around 4100 BC, hits the 2nd highest mean-treeing width at 4040 BC and the 6th highest around 4015 BC (Fig. 16.26). Starting ca. 4010 BC the width oscillates around 1 mm, drops below 0.5 mm around 3960 BC, reaches 1.5 mm at ca. 3950 BC and then stays largely below 1 mm until just before 3920 BC.[51] The GDO of ca. 3920 BC is followed by width oscillations that stay mostly between 0.7 and 1.7 mm until ca. 3870 BC, when they bounce back to nearly optimal growth conditions with the 7th highest growth peak. The 8th highest peak around 3840 BC. Good growth conditions continue with minor interruptions until ca. 3825 BC. Shortly afterwards tree-ring widths fall again to ca. 1 mm. Just before 3810 BC the tree-ring width begins to oscillate rapidly, reaching the 3rd highest peak at 3790 BC.

The Northwest German Totes Moor 2 trees show fire scarring around 4095, 4081 and 3997 BC. The highest number of detected scars occurs between ca. 3825 and 3790 BC. One is led to wonder if anthropogenic activity may have a hand in what may amount to several phases of fire activity. Certainly, the tree-ring dated Campemoor Log Way Pr 35 is built around 3798 BC in the Dümmer Lake area (Table 16.7), suggesting human adjustment to wet climatic conditions during the transition from this period of change to the next one.

Beyond the Dümmer region, at the western edge of the adjacent Michelsberg culture, the appearance of domesticated pigs of European ancestry at roughly 4000 BC starts the replacement of the Near East-derived

[51] The standardized tree-ring series of Northwest German bog oaks (Leuschner *et al.* 2002 Fig. 3a) indicates a roughly 30 year growth depression around 3900 BC.

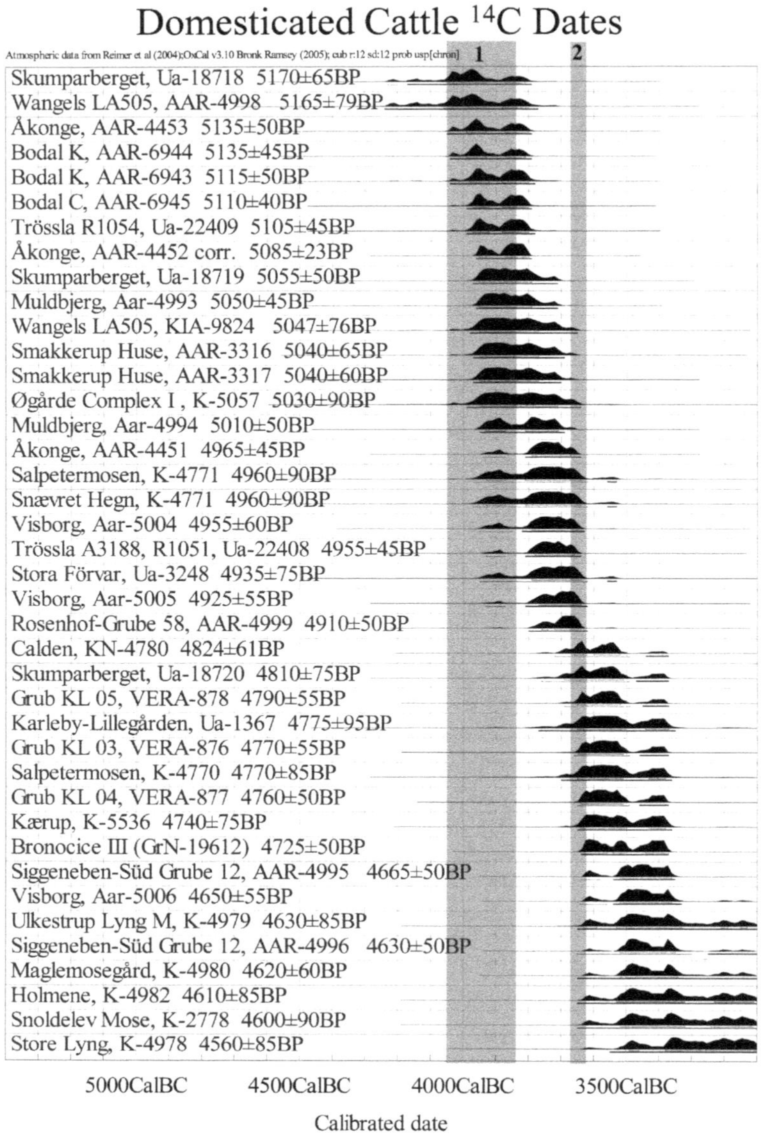

Figure 16.34. Domesticated Cattle ^{14}C dates (N = 40). Shaded areas: 1. Introduction of cattle into the TRB North Groups (ca. 4100/4000-3950/3850 BC), 2. Hypothetical cattle hiatus (ca. 3650 BC). All dates are from the TRB North Groups except Calden, Germany (TRB West Group/Wartberg culture), Grub, Austria (TRB South Group/Baden-Boleráz) and Bronocice, Poland (TRB Southeast Group/Baden-Boleráz)

stock throughout Europe (Larson *et al.* 2007:15277).[52] In southwest Germany's Michelsberg territory farming is heavily supplemented by hunting around this time. Steppan (1998) observes that this increase in hunting is an interregional adjustment to cold weather in marginal and non-marginal agricultural zones of the region.

In Northern Europe the climatic fluctuations coincide with human adjustments manifested in the transition from the Ertebølle's Timmendorf to the TRB's Wangels Phase (EN Ia) and the introduction of farming. The first evidence of cattle in the Nordic TRB dates between 4100/4000-3950/3850 BC (Fig. 16.34, 1). The smaller number of ^{14}C dates of sheep/goat and domesticated cereals show a similar introductory phase (Fig. 16.35). Simultaneously, the decline of ^{14}C dated Danish shell middens accelerates (Fischer 2002 Fig. 22.5).

Whether or not the possible dichotomy in climate regimes of Central and Northern Europe (e.g. Magny *et al.* 2003, Spurk *et al.* 2002:713-714) provides the impetus for the sudden interaction between Central European farmers and the northern foragers requires further multi-disciplinary research. Nevertheless, one can conclude that there is strong correlation with the highly uncertain climate and pronounced socio-cultural adjustments between ca. 4100 and 3800 BC in Central as well as Northern Europe. This supports the proposition that *the decedents*[53] *of the Linearbandkeramik farmers ... in Central Europe intensify mutually beneficial interaction with their neighbors – the northern Mesolithic hunter-gatherers* (Baldia 2004) during this period. The likelihood that the change in emphasis from foraging to farming in Northern Europe appears roughly simultaneously in the British Isles (cf. Whittle 2007) lends strength to this postulate and implies a wider European response to the strong climatic oscillations of this period.

[52] Tree-ring dates from Bercy 44 and Bercy 58 range respectively from at least 4064 to at least 3795 BC and between ca. 3860 and about 3800 BC (Bernard 1998 Fig. 76 and 78).

[53] *Descendants* refers to the inheritance of the Danubian farming tradition, but not necessarily to human biological (genetic) descent.

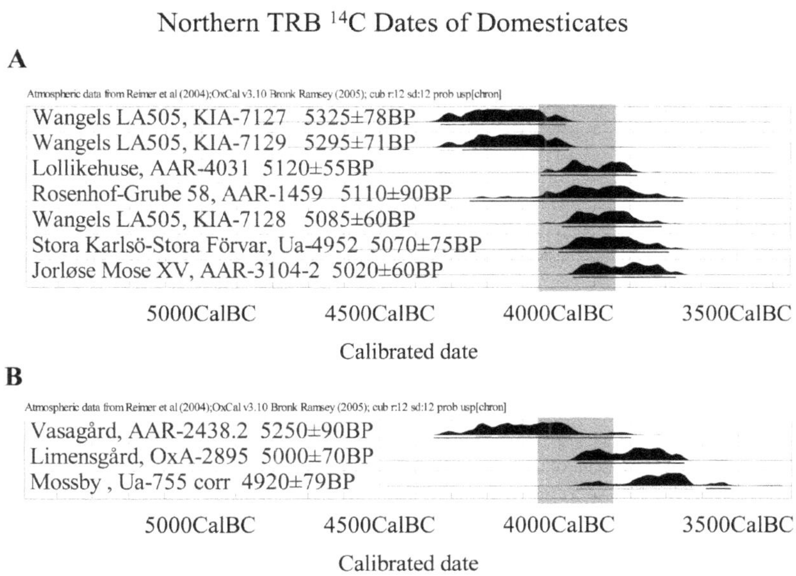

Figure 16.35. Northern TRB ^{14}C dates for the introduction of sheep/goat and domesticated cereals. Shaded areas: Estimated range (4100/4000-3950/3850 BC). A: sheep/goat (N = 7), B: cereals (N = 3)

Change ca. 3800-3650 BC

The 3809 BC GRIP $\Delta^{18}O$ warm spike initiates a phase of relatively high mean temperatures, which end with a strong cold spike ca. 3752 BC (Fig. 16.4). Afterwards, the temperature trend line stays close to the mean. Starting ca. 3700 BC the trend line begins to swing between normal and warm until the warm peak of ca. 3687 BC. This peak is immediately succeeded by a strong temperature drop, which hits bottom only about five years later. After ca. 3670 BC temperature fluctuations narrow until shortly after a mild warm spike around 3650 BC. In spite of the mostly warm climate, short-term fluctuations are relatively drastic. The volcanic SO_4 concentrations, which are relatively high between ca. 3793 and 3720 BC, peak at 3775 BC (Fig. 16.7). Between ca. 3720 and 3650 BC the concentrations remain comparatively low.

The German tree homogeneity index points to a wet or humid phase, which peaks around 3800 BC (Schmidt and Gruhle 2003 Fig. 6). The German and Netherland bog oak lifespan curve exhibits a low just before 3800 BC, a high around 3700 BC, and another low about 3650 BC (Fig. 16.23). at Totes Moor 2 numerous fire scarred trees appear between ca. 3810 and 3700 BC (Fig. 16.26). The moor's tree-ring mean width reaches the third largest width around 3790 BC. The largest width is attained around 3762 BC, but it shrinks almost immediately. By ca. 3750 BC a pine bog GDO event is detected (Eckstein *et al.* 2009 Table 3). Shortly thereafter, the tree-ring mean improves, indicating less soggy conditions. However, it drops again around 3685 BC and largely hovers between 0.5 and 1 mm until 3650 BC, suggesting another wet or humid climate phase.

The cultural changes between 3800 and 3650 BC are signified by most of Michelsberg IV (3850-3700 BC) and perhaps the beginning of Michelsberg V (3700-3500 BC) in the west, the appearance of Phase IB in the TRB South Group, the beginning of Baalberge Group in Central Germany, and the start of EN Ib in TRB North Group 1 (Fig. 16.33). Although, EN Ib is traditionally not recognized in North Group 2, there is a perceptible change in the declining trajectory of ^{14}C dated Danish shell middens during this phase (Fischer 2002 Fig. 22.5).

In the Dümmer Lake area Campemoor Log Way Pr 35 (ca. 3798 BC) and Pr 34 By (ca. 3701 BC) appear to be built by the Swifterbant culture in reaction to rising bogs. Such log way construction is indicative of inter-communal cooperation during climate anomalies that threaten inter-village communication. In adjacent areas inter-communal cooperation is better known through the archaeological remains of monumental enclosed sites (earthworks) and burial mounds.

The first monumental burial structures are the ill-defined earthen long-barrows of the TRB in the Czech Republic, Poland and Denmark (Baldia 2010:204-205, Fischer 2002 Table 22.3, 22.4). It may be worth contemplating if they are a response to the climate oscillations around 3800 BC. Certainly, the dating of Łącko Mound 2 (3820/3700 BC) could support this view. If so, propitiation of the ancestors may well have been the religious response to the fluctuating climate. From a functionalist perspective ancestor worship provides generational knowledge transfer (Midgley 2010). From a social perspective, the felling, transporting and working of huge trees, such as those found in the architecture of long-mounds at Lindebjerg and Rude, Denmark, suggest inter-communal cooperation. These and other construction chores indubitably foster rights and obligations between people, which in time may become enshrined in various religious rituals. Towards the end of the period of change under discussion and especially during the next one, TRB tombs begin to be built in alignments in parts of Poland, Central and North Germany, western Netherland, and southern Scandinavia. These arrangements suggest prehistoric lines of communication, but the supposition that these alignments follow early track ways is disputed (Baldia 1995, 1998, 2008). Nevertheless, the necessary inter-communal planning for tomb construction seems to have nurtured regional or even interregional cooperation, cohesion, and communication.

Geschwinde *et al.* (2008:225-249) argue that the monumental enclosed sites of the eastern Michelsberg culture are built near longstanding communication lines and see an apparent building boom around 3700 BC near the posited cultural boundary with the German TRB's Baalberge Group. Geschwinde *et al.* (2008:198) also note a cultural dichotomy in which the Michelsberg culture builds immense enclosed sites and the adjacent TRB concentrates on monumental burial architecture. To buttress this view, the authors note that the closest causewayed TRB enclosure at Uelzen-Walmsdorf, Germany, exhibits pottery, which could be attributable to the Michelsberg culture (Geschwinde *et al.* 2008:191-192). Nevertheless, the TRB South Group also sports enclosed sites. In Bohemia the oldest date from the Baalberge earthwork of Makotřasy (GrN-7102) calibrates to 3720 (57.5%) 3620 BC at 1σ (Fig. 16.28). Its semicircular ditch (Pleslová-Štiková 1985 Plan A-B) is reminicent of the single ditch enclosures in the Michelsberg culture (Geschwinde *et al.* 2008:209). In Moravia the stonewalled hilltop site of Rmíz u Laškov, which may appear as early as TRB IA, is in full use during Phase IB (Baldia, Frink and Boulanger 2008).

The construction of various enclosed sites in the Michelsberg culture and the TRB South Group coincides with rapidly oscillating climatic conditions, when shorter term temperatures fluctuate relatively widely. The climatic fluctuations must create social and economic uncertainty that is mitigated through regional interaction and exchange, reaching beyond possible cultural boundaries. This view may support the hypothesis by Geschwinde *et al.* (2008:225-249), which sees the Michelsberg earthworks as inter-communal cattle corrals. The outermost TRB enclosure, a palisade, at Rmíz u Laškov may have had a similar purpose (Šmíd pers. comm.).

As inter-communal cattle corrals the enclosed sites would have prevented cattle from escaping. As exchange centers they could have facilitated the distribution of animals and perhaps other goods needed to preserve communities. They may also have been useful in warding off raids in times of social upheaval. As religious and ritual centers N. Andersen (1997) ascribes a mortuary function to such enclosed sites and links them with the megalithic TRB tombs of the next period. From this standpoint, both enclosed sites and monumental tombs may indeed have been designed to appease ancestors or divinities deemed responsible for the climatic anomalies.

Overall, the burgeoning construction of the monumental architecture, the well documented log ways, and the purported track ways hint at the earliest effort to organize interaction between communities in a more efficient manner. This may be seen as the response to the generally warm climate with relatively strong and frequent short-term fluctuations between ca. 3800 and 3650 BC.

Change 3650-3350 BC

This is the period when the variability of the solar energy output produces multiple high amplitude wiggles in the ^{14}C curve (Fig. 16.3). Although, they are probably the major cause of climatic change during this time, there are also significant volcanic SO_4 spikes, one around 3648 BC and a much higher one around 3588 BC. The highest SO_4 concentration since 4035 BC occurs about 3518 BC (Fig. 16.7, 13). Subsequently, the concentrations remain relatively insignificant.

The GRIP temperature proxy trend line shows a warm period between ca. 3650 and 3580 BC, reaching its highest peak at 3593 BC (Fig. 16.4). After that, it drops below the mean, bottoming out around 3516 BC. An extreme warm spike occurs at ca. 3419 BC. This peak is higher than the 5706 BC peak noted in conjunction with the Starčevo-Körös-Criş culture. In fact, it is the third highest between 6500 and 1500 BC. Moreover, it is closely followed by several other spikes. The succeeding temperature decline bottoms out at ca. 3383 BC with the third coldest spike since the 6180 BC cold event. Shortly before 3350 BC the temperature trend line climbs briefly above the mean.

The Main River oak lifespan drops to the mean by ca. 3650 BC (Spurk *et al.* 2002 Fig. 4). At ca. 3450 BC it hits the highest lifespan between 8000 and 1000 BC. By ca. 3350 BC it drops to the lowest level since 5600 BC. This drop is as severe as the one at the beginning of the LBK. The Northwest German bog oak annual mean age graph, which rises to a peak around 3650 BC, drops sharply and bottoms out around 3600 BC (Fig. 16.6). It peaks again after ca. 3450 BC. The Totes Moor 2 pine tree-ring mean curve drops to 0.5 mm shortly after 3650 BC. The record ends around 3620/3610 BC (Fig. 16.26), which coincides with a GDO event (Eckstein *et al.* 2009 Table 3). The Northwest German oak tree data set, which supplements the pine data, shows another anomaly just before 3350 BC (Fig. 16.6). It may coincide with the slightly later GDO phase indicated by Eckstein *et al.* (2009 Fig. 6).

The time between roughly 3650 and 3350 BC witnesses substantial culture change (Fig. 16.33). Regrettably, the most significant and widespread changes appear to be arbitrarily placed at 3500 BC due to the uncertainty created by the strong wiggles in the ^{14}C curve. However, the later dating of the transitions from the Swifterbant culture to the TRB West Group suggests that this cultural change coincides with the huge temperature swings between 3419 and 3383 BC (Table 16.4).

Turning to the socio-economic changes of this period, it is important to note that the number of dated shell middens hits bottom around 3600/3500 BC in the TRB of

Denmark (Fischer 2002 Fig. 22.5). Similarly significant is the fact that the replacement of domesticated Near East-derived pigs with those of European ancestry is virtually completed throughout Europe at this time (Larson *et al.* 2007:15277).

For southern Germany and adjacent Switzerland Steppan (1998:148) proposes cold and wet summers for the 37th/36th centuries BC, which result in poor harvests and a loss of calories. This spawns a renewed upsurge in hunting in the southern part of Michelsberg culture and the nearby tree-ring dated Swiss lake dwellings. For the latter Hafner and Suter (2003 Fig. 6) illustrate the ratio of hunting versus domesticated animals from various villages of Lake Bielersee. Between 3649 and 3631 BC hunting begins to dominate. Between 3622 and 3607 BC it reaches the highest level, only to decrease somewhat between 3596 and 3560 BC. By 3532 BC hunting increases to the previous level and around 3400 BC the use of domesticated animals begins to dominate again.[54] Shortly thereafter, the lake dwelling record at the Bielersee is interrupted until about 3200 BC, which falls into the subsequent period of change.

In the Pfyn-Michelsberg culture site of Thayngen-Weier near the Bodensee in Switzerland a wooden byre is ^{14}C dated between ca. 3640 and 3580 BC (Nielsen *et al.* 2000). It is shown to house cattle and leaf fodder remains. Zimmermann (1999) notes byres are not so much the response to cold temperatures as to the thickness of the snow cover. Indeed, the strong cold spike around 3609 BC is quickly succeeded by the warm peak of ca. 3593 BC (Fig. 16.4). Therefore, I propose that byres could be the human response to a highly variable climate. At any rate, they are evidence of socio-economic adjustments, including using oxen as draft animals, feeding practices, milking, and protection against cattle raids (ibid.).

In terms of ideology and religion, 3650 BC marks the beginning of the most intensive and diverse period of ritual activity, which is best documented in the Nordic TRB (Koch 1998:155). It includes sacrifices of pottery and stone axes in bogs. The construction of burial mounds increases and the first megalithic chambers are built. The Flintbek LA 3 tomb of North Group 1 demonstrates the agglutination of mildly mounded graves and early megalithic chambers, which are eventually covered by an evolving long-mound. The ^{14}C dates for this process range from ca. 3650 to 3400 BC (Mischka 2010). Perhaps more importantly, wagon tracks occur under the extension of the long-mound. Mathematical modeling indirectly dates the tracks to 3460/3385 BC (Mischka 2010). Baldia (1998) observes that this long-mound and the wagon tracks align with a large number of other nearby tombs and hypothesizes that the alignments follow prehistoric communication lines, if not outright ways.

The evidence for wheels and wagons is not limited to Flintbek. Another set of wagon tracks occurs under the soil eroded from a later TRB West Group mound at Helvesiek, but it remains undated (Baldia 1995, 1998). The megalithic gallery-grave of Züschen (Wartberg culture) and other TRB and Baden-Boleráz sites, such as Budakálasz, Bronocice,[55] Ćmielów etc. depict wheels and wagons, sometimes pulled by oxen (e.g. Bakker *et al.* 1999, Baldia 1995, 1998, Bogucki 1993, Häusler 1992, Schlichtherle 2003, Vosteen 1999). The use of oxen as draft animals is also stipulated from plow marks found below burial chambers and burial mounds. Even though, the earliest plow marks are reported from the Sarnowo long-mound mentioned above, not only its ^{14}C date but also its marks are controversial (Milisauskas 2002:206). According to Schlichtherle (2003) the yoke from the tree-ring dated site of Arbon Bleiche 3 (3384-3370 BC) is the most solid evidence for the use of draft animals within the Baden-Boleráz interaction sphere. This together with multicultural ceramics at the site implies that people from other regions, or intimately familiar with them, made the foreign pottery locally. Local production of such heavy fragile items seems more cost-effective. Nevertheless, the proliferation of Boleráz style pottery outside of the central Danube drainage implies increased communication with the Baden interaction sphere. This intensification may be part of the continued inter-cultural effort to mitigate the effects of climatic oscillations through a far reaching exchange network.

Unfortunately, it is presently impossible to trace the network used for the most important exchange goods, which are necessary to sustain society during climate anomalies. Such goods should include domesticated cereals and animals as well as preserved fish, wild fruits and berries, mushrooms, pelts, woven textiles, etc. Only copper artifacts have left rough traces of such a network throughout the TRB (cf. Klassen 1997, 2000, 2004).

For the TRB copper melting or actual smelting evidence comes only from sites in the South and Southeast Group. These include Makotřasy (Bohemia), Laškov (Moravia), Gródek and Cmielów (both in southeast Poland). In Moravia copper artifacts are found in the hilltop sites of Hlinsko and Rmíz u Laškov (Baldia 2010, Baldia, Frink and Boulanger 2008). In the greater vicinity of Rmíz u Laškov they occur in TRB II A long-mounds with low sub-megalithic stone frames and cremation urns. Although, the artifacts are usually copper wire, a celt or flat-axe covered in a woven textile pseudomorph has recently be unearth from Džbán Long-mound 1 (Baldia, Boulanger and Frink 2008 Fig. 2, Frink this volume). In the Danish TRB a few copper artifacts occur in Konens-Høj type wooden burial chambers. The chamber type is dated between 3500 and 3430 BC at Flintbek.[56] The EN II

[54] A temporary increase of hunting could be suggested by a gap in ^{14}C dated cattle remains of the TRB North Groups between 3650 and 3600 BC (Fig. 16.34), but this may simply be the spurious result of insufficient sample size.

[55] The Bronocice III pot, depicting two-axel wagons, is dated by association with a bone sample GrN-19612 (Fig. 16.29). Calibration at 68.2% prob. = 3630 (27.3%) 3560 BC, 3540 (15.2%) 3500 BC, and 3440 (25.7%) 3370 BC.

[56] Flintbek LA3, Grave A, KIA-41584: 4619±39 bp, 68.2% prob. = 3500 (48.0%) 3430 BC, 3380 BC (20.2%) 3350 BC; 95.4% prob. = 3520 BC (92.6%) 3330 BC.

copper hoard of Bygholm, Denmark, contains several copper artifacts, including clets, spiral bracelets, and a dagger. The last is compared to a similar dagger from the Austrian Mondsee culture by Klassen and Stürup (2001 Fig. 9). Regardless of the precise copper source, the hoard points to an extensive exchange network between Northern and southern Central Europe.

The construction and use of walled hilltop sites increases between 3650 and 3350 BC in the TRB South Group (Baldia, Frink and Boulanger 2008) and the building of related earthworks begins in other TRB regions during this time.[57] This and the simultaneous increase of monumental burial architecture may be linked to this network. However, it is presently unclear to me if these developments simply promote exchange and integrate communities, or if individual communities actually begin to exert greater control over the exchange network in their territory. Nevertheless, the complex developments between 3650 and 3350 BC seem to be an outgrowth of the technological, social and ideological trajectory that has roots in the previous centuries, but they become amplified during a period of more drastic climatic disturbances.

Change 3350-2750 BC

Between ca. 3350 and 3200 BC the GRIP temperature stays mostly below the mean (Fig. 16.4). Then it climbs above the trend line for ca. 50 years. Around 3156 BC an abrupt cold spike initiates another cold phase. By ca. 3100 BC the temperature climbs again above the mean and a strong warm phase ensues, which lasts until the robust cold spike of 3021 BC. Around 2992 BC a major warm peak occurs. Afterwards, the trend line begins to oscillate below the mean with ever deeper cold troughs. The end of this period of change coincides with the coldest temperature since 3383 BC, which is reached around 2847 BC. Subsequently there is a temperature peak at 2791 BC followed by a deep cold trough that bottoms out at ca. 2751 BC.

GISP 2 SO_4 concentrations point to sustained volcanic activity between ca. 3300 BC and 3100 BC and the 11th highest SO_4 peak between 6800 and 1500 BC occurs around 3202 BC. The relatively isolated 20th highest peak is attained at about 2958 BC. From then on the SO_4 accumulations remain relatively insignificant until ca. 1700 BC.

By ca. 3350 BC the Main River oak tree lifespan drops to the lowest level since 5600 BC, but recovers to its second highest peak within about 50 years (Spurk *et al.* 2002 Fig. 4). Another less substantial peak is reached around 3100 BC (ibid.). Thereafter, the curve fluctuates above the mean until the climate anomaly around 2740 BC (Fig. 16.5 and 16.6). The Northwest German bog trees supplement the Main data with GDO phases at ca. 3000 BC and 2880 BC (Eckstein *et al.* 2009 Table 5). The dying off phase of 2880 is followed by a germination phase, peaking around 2823 BC (ibid. Fig. 9).

Between ca. 3400/3350 and 3300/3200 BC the material culture is expressed by elaborately decorated pottery as well as the construction of megalithic and related tombs. New megalithic construction wanes between 3200 and 3100 BC.[58] This is coincident with the peak of the volcanic SO_4 concentrations and the cold spike of ca. 3156 BC. Around the same time shell midden use exhibits a secondary peak in the costal zones of Denmark (Fischer 2002 Fig. 22.5). Farming Barrier 3 shrinks. The Pitted Ware culture becomes visible around Oslo Bay and the Swedish west coast. The GAC appears in parts of Germany and southern Denmark, and in Moravia the TRB is fully replaced by the Baden interaction sphere. The climatic fluctuations between ca. 2900 and 2750 BC are coincident with the end of the remaining TRB, Baden, GAC, Wartberg and Horgen manifestations (Fig 16.33). Campemoor Log Way Pr 32 bears witness to the climatic fluctuations datable to ca. 2880 and 2823 BC. Its construction and maintenance is described as a ca. 20 year battle with the rising moor, which is given up only after the track is completely covered (Leuschner *et al.* 2007).

Archaeologically, the use of communication lines as a tool for human adjustment becomes more visible between 3350-2750 BC. The construction of the Campemoor Pr 32 track illustrates the desire to keep communication open between villages within the region (ibid). It is a well-constructed, wide but winding way and, therefore, not thought to have been built for vehicle traffic. The 1.6 m wide way leading into the TRB-Boleráz hilltop village of Hlinsko, Moravia (Baldia 2010, Baldia, Frink and Boulanger 2008), indicates the ability to use stone-pavements, but it does not provide evidence of inter-site connectivity. The ca. 50 m long wagon tracks at Anloo, Netherland, which run roughly parallel to a small enclosed TRB earthwork (Ginkel *et al.* 1999:96), hint at longer communication lines associated with enclosed sites. A ^{14}C date from the site (Lanting and van der Plicht 2000) calibrates to and 3110 (62.3%) 2910 BC at 1σ. Coles and Coles (1989:162) describe the slightly later 150 m long TRB way of Tibirke, Denmark, and suggest that its ca. 3 m width is sufficient to accommodate wagons traffic. In the northeastern Alpine region the existence of ways leading beyond villages, sometimes traceable for several hundred meters, is chronicled by Pétrequin and Pétrequin (2005 Fig. 3). However, the most solid evidence for an inter-community way occurs at Seekirch-Stockwiesen in South Germany. Here, substantial houses are aligned along a wooden way with a special substructure, which is designed to support wagon traffic (Schlichtherle 2003). A wooden wheel from the substructure dates ca. 3000/2900 BC.[59] Schlichtherle maintains that the roadway connects several villages. It

[57] The range of 3650 and 3350 BC for the beginning of TRB enclosed sites is based on 76 ^{14}C dates. Earthworks with Michelsberg culture components noted by Geschwinde *et al.* (2008) are excludes.

[58] The construction of gallery-graves in the Wartberg culture seems to end somewhat later (Raetzel-Fabian 2000 Fig. 135).

[59] Seekirch-Stockwiesen Hd14711: 4340±45 bp; 68.2% prob. = 3020 (68.2%) 2900 BC; 95.4% prob. = 3090 (95.4%) 2880 BC.

has also been proposed that enclosed sites and megalithic tombs in much of the TRB are associated with lines of communication (e.g. Baldia 1995, 1998). Therefore, it is possible that the communication network becomes relatively structured between 3400/3350 and 2850/2750 BC.

The drastic climatic oscillations around ca. 2900 and 2750 BC are not only concomitant with the demise of numerous cultural traditions, but also with the birth of the huge Corded Ware interaction sphere. This ushers in a new social structure (Haak *et al.* 2008), which seems to be designed to further consolidate the integration between communities, while exerting greater control over the communication network. Likely tribal or chiefly boundaries may even replace the old farming barriers. Undoubtedly, these adjustments allow a quicker and more varied human response to the continuing, almost cyclical climatic anomalies of the following centuries (Fig. 16.6).

CONCLUSIONS

To understand humanity's ability to overcome barriers created by a changing world, Comparative Archaeology must bridge the artificial barriers to multidisciplinary research. The present cross-cultural diachronic analysis bridges the divide by analyzing archaeological and paleoclimatic data pertaining to the Mesolithic-Neolithic transition of Central and Northern Europe.

The initial cultural changes and the amount of time it takes to reach each farming barrier are difficult to pinpoint precisely, because climate anomalies influence the deposition and preservation of archaeological remains. The problem is compounded by the low critical mass of diagnostic artifacts created during transitional phases. Nevertheless, major cultural transitions appear to be largely stochastic and coincide with severe changes in the climate. On the other hand, good preservation conditions and diligent data analysis occasionally also show more or less continual socio-cultural responses. Such changes can occur within two to three generations (cf. Eckstein *et al.* 2009:193).

The major cultural transitions begin with the appearance of pottery-using Starčevo-Körös-Criş culture complex, which develops around the time of the 6180 BC cold event. Around 5600 to 5500 BC strong climate oscillations occur as Farming Barrier 1 is broken and LBK 1 establishes Barrier 2a. LBK 2 develops during the extreme volcanic activity of 5300/5200 BC and establishes Farming Barrier 2b. The culture ends between 5000 and 4900 BC, as a major climatic anomaly occurs. Farming Barrier 2b remains relatively static for centuries, although, recent discoveries suggest that the frontier fluctuates during the STK and early Lengyel periods in accordance with climatic oscillations.

The final breakup of Barrier 2b transpires a millennium after the end of the LBK when Nordic Mesolithic foragers and southern Copper Age farmers coalesce into the TRB interaction sphere to ensure long-term economic security. This coincides with climatic oscillations and increased hunting in the south between 4040 and 3820 BC and the simultaneous spread of farming to Barrier 3 in the north. Afterwards, animal husbandry increases and monumental architecture develops in parts of the TRB interaction sphere. However, a series of solar energy fluctuations occur between ca. 3650/3600 and 3400/3350 BC as interregional socio-cultural changes lead to a renewed emphasis on hunting in Central Europe and the intensification of farming in much of Northern Europe. Other human adjustments around this time include the innovations of cattle byres, ox-drawn ploughs, wheeled vehicles, and megalithic tombs. Between 3300 and 3100 BC the climate is influenced primarily by a prolonged period of volcanic activity. During this time Farming Barrier 3 seems to contract and the construction of megalithic tombs largely ends. Between ca. 2900 and 2750 BC strong climatic fluctuations can be linked with the demise of the TRB and the Baden interaction spheres as well as other cultures.

The establishment of the Danubian longhouse tradition and subsequent changes in size and construction style may correlate with climatic fluctuations. The abandonment of the tradition in favor of the cross-cultural use of smaller houses correlates with the climatic oscillations around 4350/4200 BC. This is usually seen as a socio-cultural adjustment, which shifts the social organization from extended to the nuclear family. House size reduction increases the flexibility of households by reducing the cost of relocation when adjusting to frequent changes in the climate. On the other hand, the resulting mobility has the potential to interfere with necessary group cohesion and the ability to follow the agricultural calendar. This in turn makes additional adjustments of the interaction network through more efficient lines of communication a social requirement.

When Farming Barrier 2b is broken, lines of communication begin to tie together North and Central Europe. The resulting communication network seems to facilitate not only the production and exchange of copper artifacts, but also various other essential goods and services, which are usually less easily traceable in the archaeological record. During the climatically unsettled period of ca. 3650 to 3350 BC, when individual households within the same village probably exploit different environments (Doppler *et al.* 2011), the lines of communication may start to become somewhat more formalized. Around 3000 BC, if not earlier, the development of inter-communal, vehicle-bearing log ways lined with substantial houses become visible in the archaeological record.

The climatic upheavals after 3000 BC are concomitant with considerable culture change, leading to the huge Corded Ware/Single Grave Interaction sphere. It exhibits a new social order as adjustments are made to almost cyclical climatic upheavals (Fig. 16.3-16.6). In spite of this change, tenuous archaeological evidence suggests that some of the ancient communication lines are kept in use and may even have been extended (e.g. Jager 1985). The continuing investments in the land, which includes arable fields, represented in the archaeological record by ancient plough marks and prehistoric bounded Celtic fields (cf. Jager and Ginkel 2005:119-125), results in man-made barriers. These seem to become more prominent than those related to climate. Simultaneously human impact on the environment becomes ever more pronounced. For instance, traces of copper mining and smelting begin to leave an increasing footprint as far away as Greenland after ca. 3000 BC (Boutron *et al.* 1998). Thus, the refinements in socioeconomic strategies, aimed at coping with the ever changing world, begin to create a feedback loop that amplifies environmental and climatic changes right down to the present.

Acknowledgements

My gratitude for discussion, advice, ^{14}C dates, and literature goes to Alf Metzler, B. Obelić, Dirk Raetzel-Fabian, Peter Stadler and H.E. Smith-Deenen. I thank the following for providing literature and references: Eszter Bánffy, Peter Bogucki, Sake Jager, Svend Hansen, Nicky Milner, P.J. Richerson, Ute Sass-Klaassen, Pál Sümegi, and Alasdair Whittle. Thanks also go to the Institute for the Study of Earth and Man at Southern Methodist University and the staff of The Ohio State University Library. Last but not least I thank Christel Baldia and Suzanne Dice-Goldberg for editing this paper. However, I am entirely responsible for its content, including data analysis, interpretations and errors.

References

AccuWeather.com 2002 – European Satellite Weather Map. www.accuweather.com (Accessed 2002).

ALLARD, P. 2007 – The Mesolithic-Neolithic Transition in the Paris Basin: A Review. *Proceedings of the British Academy* 144, 2007:211-223.

AMMERMAN, A.J., and L.L. CAVALLI-SFORZA 1971 – Measuring the Rate of Spread of early Farming in Europe. *Man* 6, 1971:674-688.

AMMERMAN, A.J., and L.L. CAVALLI-SFORZA 1973 – A Population Model for the Diffusion of Early Farming in Europe. C. Renfrew (ed.), *The Explanation of Culture Change*. Duckworth, London, 1973:343-357.

AMMERMAN, A.J., and L.L. CAVALLI-SFORZA 1984 – *The Neolithic Transition and the Genetics of Population in Europe*. Princeton University Press, Princeton, N.J., 1984.

ANDEL, T.H. van, and C.N. RUNNELS 1995 – The Earliest Farmers in Europe. *Antiquity* 69, 1995:481-500.

ANDERSEN, N.H. 1997 – *Sarup* Vol. 1. *The Sarup Enclosures*. Jutland Archaeological Society Publications XXXIII:1, Aarhus University Press, Århus, 1997.

ANDERSEN, S.H. 2008 – A report on Recent Excavations at the Shell Midden of Havnø in Denmark. *Mesolithic Miscellany* 19/1, 2008:3-6.

ANDRÉN, E., T. ANDRÉN and H. KUNZENDORF 2000 – Holocene History of the Baltic Sea as a Background for Assessing Records of Human Impact in the Sediments of the Gotland Basin. *The Holocene* 10/6, 2000:687-702.

ARBOGAST, R.-M., S. JACOMET, M. MAGNY and J. SCHIBLER 2006 – The Significance of Climate Fluctuations for Lake Level Changes and Shifts in Subsistence Economy During the Late Neolithic (4300-2400 B.C.) in Central Europe. *Vegetation History and Archaeobotany* 15/4, 2006:403-418.

ASCOUGH, P.L., G.T. COOK, and A.J. DUGMORE 2009 – North Atlantic Marine ^{14}C Reservoir Effects: Implications for Late-Holocene Chronological Studies *Quaternary Geochronology* 4, 2009:171-180.

ÅSTRAND, J. – Kvarteret Seglaren: Skärvor från ett svårfångat förflutet. *Urminne*, 2005:23-39.

BAILEY, G.N., and P. SPIKINS (eds.) 2008 – *Mesolithic Europe*. Cambridge University Press, New York, 2008.

BAKKER, J.A. 1979 – *The TRB West Group: Studies in the Chronology and Geography of the Makers of Hunebeds and Tiefstich Pottery*. Universiteit van Amsterdam, Subfaculteit der Pre- en Protohistorie. *Cingula* V, De Bussey Ellerman Harms, Amsterdam, 1979.

BAKKER, J.A. 1992 – *The Dutch Hunebedden: Megalithic tombs of the Funnel Beaker culture*. Archaeological Series 2, International Monographs in Prehistory, Ann Arbor, Michigan, 1992.

BAKKER, J.A. 2002 – Polish Neolithic Archaeology in 1966: A Personal Recollection. *Archaeologia Polana*, 40, 2002:163-170.

BAKKER, J.A., J. KRUK, A.E. LANTING and S. MILISAUSKAS 1999 – The Earliest Evidence of Wheeled Vehicles in Europe and the Near East. *Antiquity* 73, 1999:778-790.

BAKKER, J.A., J.C. VOGEL and T. WIŚLAnski 1969 – TRB and other C-14 Dates from Poland (c.4350-1350 BC and 800-900 AD). *Helenium*, IX/3, 1969:209-238.

BALDIA, M.O. 1995 – *A Spatial Analysis of Megalithic Tombs*. Vol. 1-2. Ph. D. Dissertation. Southern Methodist University, Dallas, 1995.

BALDIA, M.O. 1998 – Megalithic Tombs and Interregional Communication. Paper presented at the international symposium Megaliths and Social Geography, 13-17 May, 1994, Falköping, Sweden. *The Comparative Archaeology* WEB 1998/2007 (comp-archaeology.org/Falbytxt.htm).

BALDIA, M.O. 2004 – The Oldest Stone Rampart: Enclosures and Megalithic Tombs of the Funnel Beaker Culture (4100-2800 cal BC) in Central Europe. In I. Jadin, A. Hauzeur, N. Cauwe, M. Vander Linde *et al.* (eds.), *Section 9. Le Néolithique au Proche Orient et en Europe / The Neolithic in the*

Near East and Europe. Section 10: L'âge du cuivre au Proche Orient et en Europe / The Copper Age in the Near East and Europe. Actes du XIVème Congrès UISPP, Université de Liège, Belgique, 2-8 Septembre 2001. BAR S1303, BAR Publishing, Oxford, 2004:153-161.

BALDIA, M.O. 2010 – Monumental Questions: Prehistoric Megaliths, Mounds and Enclosures of Central and Northern Europe. In D. Calado, M.O. Baldia, M. Boulanger (eds.), 2010:195-212.

BALDIA, M.O., M.T. BOULANGER and D.S. FRINK 2008 – The Earthen Long-Barrow of Džbán, Moravia, Czech Republic and its Implications for the Interaction between the Nordic Funnel Beaker and the Southern Baden Culture. In M. Furholt *et al.* (eds.), 2008:263-287.

BALDIA, M.O., D.S. FRINK and M.T. BOULANGER 2008 – Problems in the Archaeological Legacy: The TRB/Lengyel-Baden Conundrum. In M. Furholt *et al.* (eds.), 2008:25-48.

BALDIA, M.O., J.P. STAECK and M. ŠMÍD 2001 – Hügelgräber der Trichterbecherkultur in Ludérov. *Pravěk*, Nova rada 11, 2001:43-59.

BÁNFFY, E. 2000a – Starčevo und/oder LBK? *Varia Neolithica I. Beiträge zur Ur- und Frügeschichte Mitteleuropas* 22, 2000:47-60.

BÁNFFY, E. 2000b – The Late Starčevo and the Earliest Linear Pottery Groups in Western Transdanubia. *Documenta Praehistorica* 2, 2000:173-185.

BÁNFFY, E. 2000c – Szentgyörgyvölgy, District Zala, Western Hungary. In P. Stadler, S. Draxler, H. Friesinger, W. Kutschera, *et al.* (eds.), *Status of the Austrian Science Fund Project P12253-PHY: Absolute Chronology for Early Civilisations in Austria and Central Europe using ^{14}C Dating with Accelerator Mass Spectrometry.* University of Vienna, Wien, 2000.

BÁNFFY, E. 2001 – Emberkezes edény a vonaldőizez kultura kialakulánsának idejéből. *Zalai Museum* 10, 2001.

BÁNFFY, E. 2002 – A Stuck Process: Urbanisation in the Carpathian Basin. In K. Jones-Bley, D.G. Zdanovich (eds.), *Complex Societies of Central Eurasia from the 3rd to the 1st Millennium BC. Regional Specifics in light of Global Models.* JIES Monograph 45. Washington D.C., 2002:492-503.

BÁNFFY, E. 2004 – Advances in the Research of the Neolithic Transition in the Carpathian Basin. In Lukes and Zvelebil (eds.), 2004:49-70.

BÁNFFY, E. 2006 – Eastern, Central and Western Hungary: Variations of Neolithisation Models. *Documenta Praehistorica* XXXIII, 2006:125-142.

BARNA, J.P and E. PÁSZTOR 2010 – Two Neolithic Enclosures at Sormás-Török-Földek (Southwest-Transdanubia, Hungary) and their Possible Geometrical and Astronomical Role: A Case Study. In D. Calado, M.O. Baldia, M. Boulanger (eds.), 2010:119-125.

BARNETT, W.K. 2000 – Cardial Pottery and the Agricultural Transition in Mediteranian Europe. In Price (ed.), 2000:93-116.

BARTELHEIM, M. 2003 – Book review: Lutz Klassen, "Frühes Kupfer im Norden. Untersuchungen zu Chronologie, Herkunft und Bedeutung der Kupferfunde der Nordgruppe der Trichterbecherkultur." *Praehistorische Zeitschrift* 78/2, 2003:231-235.

BEHRE, K.-E. 2007 – Evidence for Mesolithic Agriculture in and Around Central Europe? *Vegetation History and Archaeobotany* 16, 2007:203-219.

BEHRE, K.-E. 2009 – Wo sind die Nachweise für mesolithischen Ackerbau in Mitteleuropa? Zum Diskussionsbeitrag von W. Schön und B. Gehlen in Archäologische Informationen 29, 1 & 2, 2006. *Archäologische Informationen* 30/2, 2007:59-66.

BENGTSSON, L., B. WENNSTEDT EDVINGER and M. REUTERDAHL 2010 – Forntiden på Kattleberg: Belyst genom arkeologisk undersökning av två förhistoriska boplatser, Skepplanda 32 och 230, Västergötland, Ale kommun, Västra Götalands län. *Rapport från Arkeologicentrum* 20, Arkeologicentrum, Göteborg, 2010 (www.arkeologicentrum.se).

BENTLEY, R.A., R. KRAUSE and T.D. PRICE 2003 – Human Mobility at the Early Neolithic Settlement of Vaihingen, Germany; Evidence from Strontium Isotope Analysis. *Archaeometry* 45/3, 2003:471-486.

BENTLEY, R.A., and C. KNIPPER 2005 – Transhumance at the Early Neolithic Settlement at Vaihingen (Germany). *Antiquity* 79/306, 2005 (www.antiquity.ac.uk/ProjGall/bentley/index.html (Accessed 12/16/2005).

BERNARD, V. 1998 – *L'Homme, le Bois et la Forêt dans la France du Nord entre le Mésolithique et le Haut Moyen-Age.* BAR S733, Oxford, 1998.

BICKLE, P., and D. HOFMANN 2007 – Moving on: The Contribution of Isotope Studies to the Early Neolithic of Central Europe. *Antiquity* 81, 2007:1029-1041.

BIEL, J., H. SCHLICHTHERLE, M. STROBEL and A. ZEE (eds.) 1998 – *Die Michelsberger Kultur und ihre Randgebiete: Probleme der Entstehung, Chronologie und des Siedlungswesens.* Materialhefte zur Archäologie in Baden-Württemberg, 43, Theiss, Stuttgart, 1998.

BIERMANN, E. 1997 – *Großgaretach und Oberlauterbach: Interregionale Beziehungen im süddeutschen Mittleneolithikum.* Archäologische Berichte, 8. Deutsche Gesellschaft für Ur- u. Frühgeschichte. Habelt, Bonn, 1997.

BIERMANN, E. 2000 – *Einige Anmerkungen zur Chronologie des „Südostbayerischen Mittelneolithikums" (SOB)*, Köln, 2000 (www.rheinland-archäologie.com/Alt-%20und%20Mittelneolithikum%20in%20Mitteleuropa.pdf).

BIERMANN, E. 2003 – *Alt- und Mittelneolithikum in Mitteleuropa Untersuchungen zur Verbreitung verschiedener Artefakt- und Materialgruppen und zu Hinweisen auf regionale Tradierungen.* 2001/2003 (www.rheinland-archäologie.com/Alt-und Mittelneolithikum in Mitteleuropa.pdf).

BILLAMBOZ, A. 1998 – Die jungneolithischen Dendrodaten der Pfahlbausiedlungen Südwestdeutschlands als Zeitrahmen für die Einflüsse der Michelsberger Kultur in ihrem südlichen Randgebiet. In Biel *et al.* (eds.), 1998:159-169.

BINDER, D. 2000 – Mesolithic and Neolithic Interaction in Southern France and Northern Italy: New Data and Current Hypotheses. In Price (ed.), 2000:117-143.

BLAAUW, M., J.A. CHRISTEN, D. MAUQUOY, J. van der PLICHT and K.D. BENNETT 2007 – Testing the Timing of Radiocarbon-Dated Events Between Proxy Archives. *The Holocene* 17/2, 2007:283-288.

BODRI, L., and V. CERMAK 2003 – High Frequency Variability in Recent Climate and the North Atlantic Oscillation. *Theoretical and Applied Climatology* 1-2, 2003:33-40.

BOGAARD, A. 2002 – Questioning the Relevance of Shifting Cultivation to Neolithic Farming in the Loess Belt of Europe: Evidence from the Hambach Forest Experiment. *Vegetation History and Archaeobotany* 11, 2002:155-168.

BOGAARD, A. 2004 – The Nature of Early Farming in Central and South-East Europe. *Documenta Praehistorica* 31, 2004:49-58.

BOGAARD, A. 2005 – 'Garden Agriculture' and the Nature of Early Farming in Europe and the Near East. *World Archaeology* 37/2, 2005:177-196.

BOGUCKI, P. 1993 – Animal Traction and Household Economics in Neolithic Europe. *Antiquity*, 67:492-503.

BOGUCKI, P. 1998 – Holocene Climatic Variability and Early Agriculturein Temperate Europe: the case of Northern Europe. In Zvelebil *et al.* (eds.), 1998:77-85.

BOGUCKI, P. 2000 – How Agriculture came to North-Central Europe. In Price (ed.), 2000:197-218.

BOGUCKI, P., and R. GRYGIEL 1993 – The First Farmers of Central Europe: A Survey Article. *Journal of Field Archaeology* 20/4, 1993:399-426.

BOLLONGINO, R., and J. BURGER 2007 – Neolithic Cattle Domestication as seen from Ancient DNA, In A. Whittle and V. Cummings (eds.), 2007:165-187.

BONSALL, C., M. MACKLIN, R. PAYTON and A. BORONEANŢ 2002 – Climate, floods and river gods: environmental change and the Meso-Neolithic transition in southeast Europe. *Before Farming* 3-4/2, 2002:1-15.

BONSALL, C., G.T. COOK, R.E.M. HEDGES, T.F.G. HIGHAM, and C. PICKARD 2004 – Radiocarbon and Stable Isotope Evidence of Dietary Change from the Mesolithic to the Middle Ages in the Iron Gates: New Results from Lepenski Vir. *Radiocarbon* 46/1, 2004:293-300.

BONZON, J. Freudiger 2005 – *Archaeometrical study (petrography, mineralogy and chemistry) of Neolithic Ceramics from Arbon Bleiche 3, (Canton of Thurgau, Switzerland).* Thèse n°1471, Présentée à la Faculté des Sciences de l'Université de Fribourg (Suisse), Multiprint SA, Fribourg, Switzerland, 2005.

BORIĆ, D., and V. DIMITRIJEVIĆ 2007 – When did the 'Neolithic Package' Reach Lepenski Vir? Radiometric and Faunal Evidence. *Documenta Praehistorica* XXXIV, Neolithic Studies 14, 2007:52-71.

BORIĆ, D., C. FRENCH and V. DIMITRIJEVIĆ 2008 – Vlasac Revisited: Formation Processes, Stratigrapy and Dating. *Documenta Paehistorica* XXXV, 2008:262-287.

BORIĆ, D., G. GRUPE, J. PETERS and Ž. MIKIĆ 2004 – Is the Mesolithic–Neolithic Subsistence Dichotomy Real? New Stable Isotope Evidence from the Danube Gorges. *European Journal of Archaeology* 7/3, 2004:221-248.

BORIĆ, D., and P. MIRACLE 2004 – Mesolithic and Neolithic (Dis)Continuities in the Danube Gorges: New AMS Dates from Padina and Hajdučka Vodenica (Serbia). *Oxford Journal of Archaeology* 23/4, 2004:341-371.

BORONEANŢ, A., and A. DINU 2006 – The Romanian Mesolithic and the Transition to Farming. A Case Study: The Iron Gates. *Studii de Preistorie* 3, 2006:41-76.

BOS, J.A.A., and R. URZ 2003 – Late Glacial and Early Holocene Environment in the Middle Lahn River Valley (Hessen, Central-West Germany) and the Local Impact of Early Mesolithic People: Pollen and Macrofossil Evidence. *Vegetation History and Archaeobotany* 12, 2003:19-36.

BOUZEK, J. 1993 – Climatic Changes: New Archaeological Evidence from the Bohemian Karst and Other Areas. *Antiquity* 67, 1993:386-93.

BOUZEK, J. 2001 – Klimatische Entwicklung im Äneolithikum und in der Frühbronzezeit in Böhmen und Mähren. (Climatic Development during the Eneolithic and Early Bronze Age of Bohemia and Moravia) In Lippert *et al.*, 2001:21-26.

BOUZEK, J. 2005 – Klimatické změny ve středoevropském pravěku. [Climatic Changes and Central European Prehistory.] *Archeologické rozhledy* LVII, 2005:493-528.

BOUTRON, C.F., S. HONG, J.-P. CANDELONE and C.C. PATTERSON 1998 – History of Ancient Copper Smelting recorded in Central Greenland Ice. In C. Mordant, M. Pernot and V. Rychner (eds.), *L'Atelier du bronzier en Europe du XXe au VIIIe siècle avant notre ère.* Actes du colloque international Bronze '96, Neuchâtel et Dijon. II: Du minerai au métal, du métal à l'objet. Paris, CTHS, 1998:93-100.

BRADLEY, R., and T. PHILLIPS 2004 – The High-Water Mark: The Siting of Megalithic Tombs on the Swedish Island of Tjörn. *Oxford Journal of Archaeology* 23/2, 2004:123-133.

BREUNIG, P. 1987 – *^{14}C-Chronologie des vorderasiatischen, südost- und mitteleuropäischen Neolithikums.* Fundamenta Reihe A, Band 13. Böhlau Verlag, Köln, Wien, 1987.

BRINDLEY, A.L. 1986 – The Typochronology of TRB West Group Pottery. *Paleohistoria* 28, 1986:93-132.

BUDJA, M. 2001 – The Transition to Farming in Southeast Europe: Perspectives from Pottery. *Documenta Praehistorica* XXVIII, 2001:27-47.

BUDJA, M. 2004a – The Neolithic of the Balkans: Where in the Puzzle? In Lukes and Zvelebil (eds.), 2004:37-48.

BUDJA, M. 2004b – The Neolithisation of South-eastern Europe. In Knutsson (ed.), 2004:419-433.

BUDJA, M. 2006 – The Transition to Farming and the Ceramic Trajectories in Western Eurasia: From Ceramic Figurines to Vessels. *Documenta Praehistorica* XXXIII, 2006:183-201.

BUDJA, M. 2007 – The 8200 calBP 'Climate Event' and the Process of Neolithisation in South-Eastern Europe. *Documenta Praehistorica* XXXIV, 2007:191-201.

BURMAN, J., and B. SCHMITZ 2005 – Periwinkle (*Littorina Littorea*) Intrashell δ^{18}O and δ^{18}C Records from Mid-Holocene Limfjord Region, Denmark: A new High-Resolution Palaeoenvironmental Proxy Approach. *The Holocene* 15-4, 2005:567-575.

CALADO, D., M.O. BALDIA, M. BOULANGER (eds.) 2010 – *Monumental Questions: Prehistoric Megaliths, Mounds and Enclosures,* Section 68 (*Part II*), Actes du XV Congrès Mondial de l'Union Internationale des Sciences Préhistoriques et Protohistoriques 8, Lisbon 2006, BAR S2123. BAR Publishing, Oxford,

CAPITANI, A. de, S. DESCHLER-ERB, U. LEUZINGER, E. MARTI-GRÄDEL, and J. SCHIBLER 2002 – *Die jungsteinzeitliche Seeufersiedlung Arbon/Bleiche 3.* Archäologie im Thurgau 11, Departement für Erziehung und Kultur des Kantons Thurgau. Frauenfeld, Switzerland, 2002.

CAPPERS, R.T.J., and D.C.M. RAEMAEKERS 2008 – Cereal Cultivation at Swifterbant? Neolithic Wetland Farming on the North European Plain. *Current Anthropology* 49/3, 2008:385-401.

CARNEIRO, Â., and P. STADLER 2002 – Das neolithische Haus von Münchendorf-Drei Mahden in Niederösterreich. In P. Kaserer (ed.), *Chronik Münchendorf von den Anfängen bis 2002.* Münchendorf, Austria, 2002:7-12.

CHAMBERS, F.M. 1993 – Late-Quaternary Climatic Change and Human Impact: Commentary and Conclusion. In F.M. Chambers (ed.), *Climate Change and Human Impact on the Landscape.* Chapman and Hall, London and New York, 1993:248ff.

CHAPMAN, J. 2003 – From Franchthi to the Tiszazug: Two Early Neolithic Worlds. In Jerem, Erzsébet and Pál Raczky (eds.), *Morgenrot der Kulturen. Frühe Etappen der Menschheitsgeschichte in Mittel- und Südosteuropa.* Festschrift für Nándor Kalicz zum 75. Geburtstag. Archaeolingua Main Series 15, Budapest, 2003:89-108.

CHRISTENSEN, C. 1993 – Land and Sea. In S. Hvass and B. Storgaard (eds.), *Digging into the Past: 25 Years of Archaeology in Denmark.* Aarhus Universitetsforlag, Århus, 1993:20-23.

ČIŽMÁŘ, Z., J. PAVÚK, P. PROCHÁZKOVÁ and M. ŠMÍD 2004 – K problému definováni finálního stádia lengyelské kultury (New observations on problems in the relationship between the Stroked Pottery and Lengyel cultures). In B. Hänsel and E. Studeníková (eds.), *Zwischen Karpaten und Ägäis: Neolithikum und ältere Bronzezeit; Gedenkschrift für Viera Němejcová Pavúková.* Internationale Archäologie. Studia honoraria 21. Rahden, Westf. VML, 2004:207-232.

CLARK, G. 1977 – The Economic Context of Dolemens and Passage Graves in Sweden. In *Ancient Europe and the Mediterranean.* V. Markotic (ed.), *Ancient Europe and the Mediterranean.* Aris & Phillips, Warminster, England, 1977:35-49.

COLES, B. and J. COLES 1989 – *People of the Wetlands: Bogs, Bodies and Lake-Dwellers.* Ancient Peoples and Places, Thames and Hudson, London and New York.

CONARD, N.J., and C.-J. KIND (eds.) 1998 – *Aktuelle Froschung zum Mesolithikum: Current Mesolithic Research.* Mo Vince Verl., Tübingen, 1998.

CRAIG, O.E. 2004 – Organic Analysis of "Food Crusts" from Sites in the Schelde Valley, Belgium: a Preliminary Evaluation. *Notae Praehistoricae* 24, 2008:209-217.

CROMBÉ, P., and B. VANMONTFORT 2007 – The Neolithisation of the Scheldt Basin in Western Belgium. In A. Whittle and V. Cummings (eds.), 2007:263-285.

CZERNIAK, L., L. DOMAŃSKA, A. KOŚKO and D. PRINKE 1991 – The Funnel Beaker Culture in Kujavia. In Jankowska (ed.), 1991:68-75.

DAVIDSEN, K. 1973 – Valbykeramik und Kugelamphorenkultur: Zur Verbreitung zweier mittelneolithischer Keramikgruppen in Schleswig-Holstein und Nachbargebieten. *Offa* 29, 1972:133-137.

DAVIDSEN, K. 1978 – *The Final TRB Culture in Denmark: A Settlement Study.* Arkæologiske Studier V. Akademisk Forlag, København, 1978.

DAVISON, K., P.M. DOLUKHANOV, G.R. SARSON and A. SHUKUROV 2006 – The role of waterways in the spread of the Neolithic. *Journal of Archaeological Science* 33, 2006:641-652.

DAVISON, K., P.M. DOLUKHANOV, G.R. SARSON, A. SHUKUROV and G.I. ZAITSEVA 2007 – A Pan-European Model of the Neolithic. *Documenta Praehistorica* XXXIV, 2007:139-154.

DINU, A., S. ANDREI and D. MIRITOIU 2007 – The Mesolithic at the Danube's Iron Gates: New Radiocarbon Dates and Old Stratigraphies. *Documenta Praehistorica* XXXIV, 2007:31-52.

DIRKS, U. 2000 – *Die Bernburger Kultur in Niedersachsen.* Beiträge zur Archäologie in Niedersachsen 1. VLM, Rahden, Westf., Germany, 2000.

DOLUKHANOV, P., and A. SHUKUROV 2004 – Modelling the Neolithic Dispersal in Northern Eurasia. *Documenta Praehistorica* XXXII, 2004:35-47.

DOLUKHANOV, P., A. SHUKUROV, D. GRONENBORN, D. SOKOLOFF, *et al.* 2005 – The Chronology of Neolithic Dispersal in Central and Eastern Europe. *Journal of Archaeological Science* 32, 2005:1441-1458.

DOMAŃSKA, L., and S. RZEPECKI 2004 – Łącko, Site 6, Pakość Commune, Poland. In Knutsson (ed.), 2004:2004:419-433.

DOPPLER, T., B. POLLMANN, S. PICHLER, S. JACOMET, J. SCHIBLER and B. RÖDER 2011 – Bauern, Fischerinnen und Jäger: Unterschiedliche Ressourcen- und Landschaftsnutzung in der neolithischen Siedlung Arbon Bleiche 3 (Thurgau, Schweiz)? In J. Studer, M. David-Elbiali and M. Besse, *Paysage... Landschaft... Paesaggi:. L'impact des activités humaines sur l'environnement du Paléolithique à la période romaine.* Cahiers d'Archéologie Romande 120, 2011:143-158, Lausanne.

DRAFEHN, A., A.-L. FISCHER, T. FRANK, B. GEHLEN, *et al.* 2003 – Zum Stand der Mittelsteinzeit-Forschung in Deutschland: Erste Ergebnisse einer umfassenden Untersuchung. *Archäologische Informationen* 26/1, 2003:33-62.

DRESLEROVÁ, D. 2005 – Klima v pravěku: mýtus a skutečnost. Několik poznámek k článku Jana Bouzka [Climate in Prehistory: Myth and Reality.] *Archeologické rozhledy* LVII, 2005:534-548.

DUBOULOZ, J. 1998 – Réflexions sur le Michelsberg Ancien en Bassin Parisien. In Biel *et al.* (eds.), 1998:9-20.

ELBURG, R. 2011 – Weihwasser oder Brauchwasser? Einige Gedanken zur Funktion bandkeramischer Brunnen. *Archäologische Informationen* 34/1, 2011:25–37.

ECKSTEIN, J., H.H. LEUSCHNER, A. BAUEROCHSE and U. SASS-KLAASSEN 2009 – Subfossil Bog-Pine Horizons Document Climate and Ecosystem Changes During the Mid-Holocene. *Dendrochronologia* 27, 2009:129-146.

EDWARDS, C.J., R. BOLLONGINO, A. SCHEU, A. CHAMBERLAIN, A. TRESSET, *et al.* 2007 – Mitochondrial DNA Analysis Shows a Near Eastern Neolithic Origin for Domestic Cattle and no Indication of Domestication of European Aurochs. *Proceedings of the Royal Society B* 274, 2007:1377–1385. doi:10.1098/rspb.2007.0020, Published online 3 April 2007.

FAGAN, B.M. 2004 – *The Long Summer: How Climate Changed Civilization.* Basic Books, New York, 2004.

FAHLANDER, F. 2008 – A Piece of the Mesolithic: Horizontal Stratigraphy and Bodily Manipulations at Skateholm. In F. Fahlander and T. Oestigaard (eds.), *The Materiality of Death: Bodies, Burials, Beliefs.* BAR, London, 2008:29-45.

FISCHER, A. 2002 – Food or Feasting? An Evaluation of Explanations of the Neolithisation of Denmark and Southern Sweden. In A. Fischer and K. Kristiansen (eds.), *The Neolithisation of Denmark: 150 Years of Debate*, J.R. Collins Publications, Sheffield, England, 2002:343-393.

FISCHER, A., and J. HEINEMEIER 2003 – Freshwater Reservoir Effect in ^{14}C Dates of Food Residue on Pottery. *Radiocarbon* 45/3, 2003:449-466.

FISCHER, A., J. OLSEN, M. RICHARDS, J. HEINEMEIER, A.E. SVEINBÖRNSDÓTTIR and P. BENNIKE 2007 – Coaste-Inland Mobility and Diet in the Danish Mesolithic and Neolithic: Evidence from Stable Isotope Values of Humans and Dogs. *Journal of Archaeological Science* 34, 2007:2125-2150.

FOKKENS, H. 2008 – The Temporality of Culture Change. In H. Fokkens, B.J. Coles, A.L. Van Gijn, J.P. Kleijne *et al.* (eds.), *Between Foraging and Farming. An Extended Broad Spectrum of Papers Presented to Leendert Louwe Kooijmans.* The Faculty of Archaeology, Leiden University, Leiden, 2008:15-24.

FRIEDRICH, M., S. REMMELE, B. KROMER, J. HOFMANN *et al.* 2004 – The 12,460-Year Hohenheim Oak and Pine Tree-Ring Chronology from Central Europe: A Unique Annual Record for Radiocarbon Calibration and Paleoenvironment Reconstruction. *Radiocarbon* 46/3, 2004:1111-1122.

FURHOLT, M. 2008 – Culture History Beyond Cultures: The Case of the Baden Complex. In Furholt *et al.*, 2008:13-24.

FUHOLT, M., J. MÜLLER, D. RAETZEL-FABIAN, C. RINNE, and H.-P. WOTZKA 2002 – *Radon-Radiocarbon online Datenbank mitteleuropäischer ^{14}C Daten für das Neolithikum und die Bronzezeit.* www.jungsteinSITE.de, Article of August 6, 2002.

FURHOLT, M., M. SZMYT and A. ZASTAWNY (eds.) 2008 – *The Baden Complex and the Outside World.* Proceedings of the 12th Annual Meeting of the EAA in Cracow, 19-24th September 2006. Studien zur Archäologie in Ostmitteleuropa (SAOISPEŚ) 4, Habelt, Bonn, 2008.

GERDES, G., B.E.M. PETZELBERGER, B.M. SCHOLZ-BÖTCHER and H. STREIF 2003 – The Record of Climatic Change in the Geological Archives of Shallow Marine, Coastal, and Adjacent Lowland Areas of North Germany. *Quaternary Science Reviews* 22, 2003:104-124.

GESCHWINDE, M., D. RAETZEL-FABIAN, E. GEHRT, S. GREFEN-PETERS and W. WIMMER 2009 – *EWBSL: Eine Fallstudie zu den jungneolithischen Erdwerken am Nordrand der Mittelgebirge.* Beiträge zur Archäologie Niedesachsens 14, Archäologische Komission für

Niedersachsen, e. V., Leidorf, Rahden, Westf., Germany, 2009.

GINKEL, E. van, S. JAGER and W. van der SANDEN 1999 – *Hunebedden: Monumenten van een Steentijdcultuur*. Uitgeverij Unipers Abcoude, Rijksdienst voor het Oudheidkundig Bodemonderzoek, Utrecht, 1999.

GKIASTA, M., T. RUSSEL, S. SHENNAN and J. STEEL 2003 – The Neolithic Transition in Europe: The Radiocarbon Record Revisited. *Antiquity* 77/295, 2003:45-62.

GOLITKO, M., and L. KEELEY 2007 – Beating Ploughshares back into Swords: Warfare in the Linearbandkeramik. *Antiquity* 81, 2007:332-342.

GRAFENSTEIN, U. von, H. ERLENKEUSER, J. MÜLLER, J. JOUZEL and S. JOHNSEN 1998 – The Cold Event 8000 Years Ago Documented in Oxygen Isotope Records of Precipitation in Europe and Greenland. *Climate Dynamics* 14, 1998:73-81.

GRONENBORN, D. 2003a – Migration, Acculturation and Culture Change in Western Temperate Eurasia, 6500-5000 cal BC. *Documenta Praehistorica* XXX, 2003:79-91.

GRONENBORN, D. 2003b – Lithic Raw Material Distribution Networks and the Neolithization of Central Europe. In Laurence Burnez-Lanotte (ed.), *Production and Management of Lithic Materials in the European Linearbandkeramik.* [*Gestion des matériaux lithiques dans et Rubané européen.*] Actes du XIVème Congrés UISPP, Université de Liège, Belgique, 2-8 septembre 2001. BAR International Series 1200. Hadrian Books, Oxford, 2003:45-50.

GRONENBORN, D. 2007 – Beyond the Models: 'Neolithisation' in Central Europe. In A. Whittle and V. Cummings (eds.), 2007:73-98.

GRONENBORN, D. 2009 – Climate Fluctuations and Trajectories to Complexity in the Neolithic: Towards a Theory. *Documenta Praehistorica* 36, 2009:97-110.

GRONENBORN, D. (ed.) 2005 – *Klimaschwankungen und Kulturwandel in neolithischen Gesellschaften Mitteleuropas*. Römisch-Germaniaches Zentralmuseum, Mainz, 2005.

GROSS-KLEE, E. 1998 – Michelsberg: Heterogenität und kulturelle Einbindung in Raum und Zeit. In Biel *et al.* (eds), 1998:249-259.

GROTE, K. 1998 – Laubach 7: Eine spätmesolithische Station im unteren Werra Tal (Süd Niedersachsen). In N.J. Conard and C.-J. Kind (ed.), *Aktuelle Froschung zum Mesolithikum. Current Mesolithic Research.* Mo Vince, Tübingen, 1998:203-221.

GRYGIEL, R., and P. BOGUCKI 1986 – Early Neolithic Sites at Brzesc Kujawski, Poland: Preliminary Report on the 1980-1984 Excavations. *Journal of Field Archaeology* 13/2, 1986:121-137.

GRYGIEL, R., and P. BOGUCKI 2008 – *Neolit i Początki Epoki Brązu w rejonie Brześcia Kujawskiego i Osłonek: Środkowy Neolit grupa brzesko-kujawska kultury lendzielskiejśrodkowy neolit grupa brzesko-kujawska kultury lendzielskiej.* [*The Neolithic and Early Bronze Age in the Brześć Kujawski and Osłonki Region: Middle Neolithic; The Brześć Kujawski Group of the Lengyel Culture*] II/3. Fundacja Badań Archeologicznych Imienia Profesora Konrada Jażdżewskiego, Muzeum Archeologiczne i Etnograficzne w Łodzi, Łódź, 2008.

HAAK, W., G. BRANDT, H.N. de JONG, *et al.* 2008 – Ancient DNA, Strontium isotopes, and osteological analyses shed light on social and kinship organizationof the Later Stone Age. PNAS 105/47, 2008:18226-18231.

HAFNER, A., and P. SUTER 2001 – Das Neolithikum im 4. Jahrtausend v. Chr.: Tradition, Einflüsse und Entwicklung. *Archäeologische Informationenen* 24/2, 2001:291-309.

HAFNER, A., and P. SUTER 2003 – *Das Neolithikum in der Schweiz*. www.jungsteinSITE.de Article of November 27, 2003.

HALLGREN, F. 2004a – The Introduction of Ceramic Technology Around the Baltic Sea in the 6th Millennium. In Knutsson (ed.), 2004:123-142.

HALLGREN, F. 2004b – Trössla: Efterundersökning av en grustagen Trattbägarboplats i Östra Södermanland. *SAU Rapport* 7, Uppsala Universitet, Uppsala, 2004.

HART, J.P., and W.A. LOVIS 2007 – The Freshwater Reservoir and Radiocarbon Dates on Cooking Residues: Old Apparent Ages or a Single Outlier? Comments on Fischer and Heinemeier 2003. *Radiocarbon* 49/3, 2007:1403-1410.

HARTZ, S., A.J. KALIS, L. KLASSEN and J. MEURERS-BALKE 2012 – Neue Ausgrabungen zur Ertebøllekultur in Ostholstein und der Fund von vier stratifizierten durchlochten donauländischen Äxten. *Archäologische Berichte* 22, 2012:25-62.

HARTZ, S., and H. LÜBKE 2006 – New Evidence for a Chronostratigraphic Division of the Ertebølle Culture and the Earliest Funnel Beaker Culture on the Southern Mecklenburgian Bay. In C.J. Kind, (ed.), *After the Ice Age: Settlements, Subsistence and Social Development in the Mesolithic of Central Europe.* Materialhefte zur Archäologie in Baden-Württemberg 78, Theiss, Stuttgart, 2006:61-77.

HARTZ, S., H. LÜBKE and T. TERBERGER 2007 – From Fish and Seal to Sheep and Cattle: New Research into the Process of Neolithisation in Northern Germany. In A. Whittle and V. Cummings (eds.), 2007:567-594.

HÄUSLER, A. 1992 – Der Ursprung des Wagens in der Diskussion der Gegenwart. *Archäologische Mitteilungen aus Nordwestdeutschland,* 15:179-190, Oldenburg.

HÄUSLER, A. 1996 – Totenorientierung und geographischer Raum. In S. Ostriz and R. Einicke (eds.), *Terra & Praehistoria: Festschrift für Klaus-Dierter Jäger*. Beiträge zur Ur- und Frühgeschichte Mitteleuropas 9. Beier & Beran. Wilkau-Hasslay, 1996:61-92.

HEDGES, R.E.M., R.A. HOUSLEY, C. BRONK RAMSEY and G.J. van KLINKEN 1995 –

Radiocarbon Dates from the Oxford AMS System: Archaeometry Datelist 19. *Archaeometry* 37, 1995:195-214.

HENSEL, W, and T. WIŚLAŃSKI (eds.) 1979 – Prahistoria Ziem Polskich, Tomb 2: Neolit. Polska Akademia Nauk Instytut Historii Kultury Materialnej, Wrocław, Warszawa, Kraków, Gdańsk, 1979.

HEIRI, O., A.F. LOTTER, S. HAUSMANN and F. KIENAST 2003 – A Chironomid-based Holocene Summer Air Temperature Reconstruction from the Swiss Alps. *The Holocene* 13/4, 2003:477-484.

HELLE, G., and G.H. SCHLESER 1998 – Die Eichenbohlen des neolithischen Brunnens von Erkelenz-Kückhoven als Klimaarchive. In H. Koschik (ed.), 1998:271-277.

HERTELENDI, E., N. KALICZ, P. RACZKY, F. HORVÁTH *et al.* 1995 – Re-evaluation of the Neolithic in Eastern Hungary Based on Calibrated Radiocarbon Dates. In G.T. Cook, D.D. Harkness, B.F. Miller and E.M. Scott (eds.), *Proceedings of the 15th International ^{14}C Conference.* Radiocarbon 37/2, 1995:239-244.

HOIKA, J. und J. MEURERS-BALKE (eds.) 1994 – *Beiträge zur frühneolithischen Trichterbecherkultur im westlichen Ostseegebiet,* 1. Internationales Trichterbechersymposium in Schleswig vom 4. bis 7. März 1985. Untersuchungen und Materialien zur Steinzeit in Schleswig-Holstein aus dem Archäologischen Landesmuseum der Christian-Albrechts-Universität, Band 1. Wachholz, Neumünster, 1994.

HÖPPNER, B., M. BARTELHEIM, M. HUIJSMANS, R. KRAUSS *et al.* 2005 – Prehistoric Copper Production in the Inn Valley (Austria), and the Earliest Copper in Central Europe. *Archaeometry* 47/2, 2005:293-315.

HORVÁTH, T. 2009 – The Intercultural Connections of the Baden "Culture" *MOMOS* VI, 2009:101-149 (www.archeo.mta.hu/hun/munkatars/horvathtunde/MOMOS2009.pdf).

HORVÁTH, F., and E. HERTELENDI 1994 – Contributions to ^{14}C Based Absolute Chronology of the Early and Middle Neolithic Tisza Region. *Jósa András Múzeum Évkönye* XXXVI, 1994:111-133.

HORVÁTHOVÁ, E. 2008 – About the Development of the Baden Culture in the Region of the Northern Tisza River in Slovakia. In M. Furholt *et al.* (eds.), 2008:111-130.

IVERSEN, J. 1941 – *Landnam i Danmarks Stenalder.* Danmarks geologiske undersøgelse, Række 66, II, København, 1941.

IVERSEN, J. 1949 – *The influence of Prehistoric Man on Vegetation.* Danmarks geologiske undersøgelse, Række 3, IV, København, 1949.

JACOMET, S., and C. BROMBACHER 2005 – Reconstructing Intra-Site Patterns in Neolithic Lakeshore Settlements: The State of Archaeobotanical Research and Future Prospects. In P. Della Casa and M. Trachsel (eds.), *WES'04: Wetland Economies and Societies.* Proceedings of the International Conference in Zürich, 10-13 March, 2004. Collectio Archæologica 3, 69–94. Chronos, Zurich, 2005:69-94.

JAGER, S.W. 1985 – A prehistoric Route and Ancient Cart-Tracks in the Gemeente Anloo (Province Derenthe). *Paleohistoria,* 27:185-202.

JAGER, S.W. and E. van GINKEL 2005 – *Archeologie van de Stellenwerven: Rendierjager, boern en ontginners in het stroomgebied von Tjonger en Linde.* Stichting Stellingwarver Schrieversronte. Koninklijke Van Gorcum BV, Assen, Netherland.

JAKOBSEN, O., J. MEURERS-BALKE, G. HOFFMANN-WIECK and J. THIEDE 2004 – Postglazialer Meeresspiegelanstieg in der südwestlichen Ostsee: Geoarchäologische Ergebnisse aus der Niederung des Oldenburger Grabens (Ostholstein). [Postglacial sea level rise in the south-western Baltic – Geoarchaeological results from the lowland of the Oldenburger Graben (Ostholstein).] *Geographie der Meere und Küsten, Coastline Reports* 1, 2004:9-21.

JANKOWSKA, D. 1980 – *Kultura pucharów lejkowatych na Pomorzu Środkowym: Grupa Lupawska.* Wydawnictwo Naukowe Universitetu im. Adama Mickiewicza w Poznańiu, Poznań, Poland, 1980.

JANKOWSKA, D. 1998 – Environmental Conditions during the Neolithization of Pomerania. In Zvelebil *et al.* (eds.), 1998:121-128.

JANKOWSKA, D. 1999 – Megalithik and kujawische Graber. (Megalithism and Kujavian Tombs). In K.W. Beinhauer, G. Cooney, C.E. Guksch and S. Kus (eds.), *Studien zur Megalithik: Forschungsstand und ethnoarchäologische Perspektiven.* Beiträge zur Ur- und Frühgeschichte Mitteleuropas 21. Beier und Beran. Weissbach, Germany, 1999:215-226.

JANKOWSKA, D. (ed.) 1990 – *Die Trichterbecherkultur: Neue Forschungen und Hypothesen* I. Poznań, Instyitut Prahistorii Universitetu im. Adam Mickiewicza w Poznańiu Zakład Archeologii Wielkopolski IHKM PAN w Poznańiu, Poznań, Poland, 1990.

JANKOWSKA, D. (ed.) 1991 – *Die Trichter-becherkultur: Neue Forschungen und Hypothesen* II. Poznań, Instyitut Prahistorii Universitetu im. Adam Mickiewicza w Poznańiu Zakład Archeologii Wielkopolski IHKM PAN w Poznańiu, Poznań, Poland, 1991.

JANKOWSKA, D., and T.S. WIŚLAŃSKI 1991 – Trichterbecherkultur im polnischen Tiefland: Die wichtigsten Forschungsprobleme. In Jankowska (ed.), 1991:53-65.

JANKOWSKA, D., and T.S. WIŚLAŃSKI 1994 – Frühneolithische Elemente in der mittelpommerischen Lupa-Gruppe der Trichterbecherkultur. In Hoika and Meurers-Balke 1994:137-154.

JEUNESSE, C. 1998 – Pour une Origine Occidentale de la Culture de Michelsberg? In Biel *et al.* (eds.), 1998:29-45.

JOCHIM, M. 2000 – The origins of agriculture in south-central Europe. In Price, 2000:183-196.

JOHNSEN, S.J., D. DAHL-JENSEN, N. GUNDESTRUP, J.P. STEFFENSENET *et al.* 2001 – Oxygen Isotope and Palaeotemperature Records from six Greenland Ice-Core Stations: Camp Century, Dye-3, GRIP, GISP2, Renland and NorthGRIP. *Journal of Quaternary Science* 16/4, 2001:299-307.

KACERIK, A. 2007 – Linear Pottery Settlement Structure in Northwest Bohemia, Czech Republic: A Study of Socio-Economic Dynamics in Micro-Regions. *The European Archaeologist* 26, 2006/2007:6-7.

KALICZ, N., and J. MAKKAY 1977 – *Die Linienbandkeramik in der Großen ungarischen Tiefebene.* Akadémiai Kiadó, Budapest, 1977.

KALIS, A.J., J. MERKT and J. WUNDERLICH 2003 – Environmental Changes During the Holocene Climatic Optimum in Central Europe: Human Impact and Natural Causes. *Quaternary Science Reviews* 22, 2003:33-79.

KAUTE, P., G. SCHINDLER and H. LÜBKE 2005 – Der endmesolithisch/frühneolithische Fundplatz Stralsund-Mischwasserspeicher: Zeugnisse früher Bootsbautechnologie an der Ostseeküste Mecklenburg-Vorpommerns. In H. Lübke, F. Lüth und T. Terberger (eds.), *Neue Forschungen zur Steinzeit im südlichen Ostseegebiet.* 46. Jahrestagung der Hugo Obermaier-Gesellschaft zur Erforschung des Eiszeitalters und der Steinzeit e.V., Hansestadt Greifswald 13.-17. April 2004. Bodendenkmalpflege in Mecklenburg-Vorpommern, Jahrbuch 52, Lübstorf, 2004:221-241.

KAZDOVÁ, E., J. PEŠKA, and I. MATEICIUCOVÁ 1999 – *Olomouc-Slavonin I: Sídliště Kultury s Vypíchanou Keramilou.* Archaeologiae Regionalis Fontes 2. Vlastivědné museum v Olomouci, Archeofické centrum. Olomouc, Czech Republic, 1999.

KLASSEN, L. 1997 – Die Kupferfunde der Nordgruppe der Trichterbecherkultur. *Archaeologische Informationen* 20/1, 1997:189-93.

KLASSEN, L. 2000 – *Frühes Kupfer im Norden: Untersuchungen zu Chronologie, Herkunft und Bedeutung der Kupferfunde der Nordgruppe der Trichterbecherkultur*, Aarhus University Press, Aarhus, 2000.

KLASSEN, L. 2004 – *Jade und Kuppfer: Untersuchungen zum Neolithisierungsprozess im westlichen Ostseeraum unter besonderer Berücksichtigung der Kulturenentwicklung Europas 5500-3500 BC.*, Aarhus University Press, Aarhus, 2004.

KLASSEN, L., and E. PERNICKA 1998 – Eine kreuzschneidige Axthacke aus Südskandinavien: Ein Beispiel für die Anwendungsmöglichkeiten der Stuttgarter Analysendatenbank. *Archäologisches Korresbondenzblatt* 28/1, 1998:35-45.

KNIPPER, C., S. HARRIS, L. FISHER, R. SCHREG *et al.* 2005 – *The Neolithic Settlement Landscape of the Southeastern Swabian Alb (Germany).* www.jungsteinSITE.de, Article of May 25, 2005.

KLASSEN, L., and S. STÜRUP 2001 – Decoding the Riesebusch-copper: lead-isotope analysis applied to early neolithic copper finds from south Scandinavia. *Praehistorische Zeitschrift* 76/1, 2001:55-73.

KLOPFSTEIN, S., M CURRAT, and L. EXCOFFIER 2006 – The Fate of Mutations Surfing on the Wave of a Range Expansion Molecular Biology and Evolution. *Molecular Biology and Evolution* 23/3, 2006:482-490.

KNUTSSON, H. 2004 – *Coast to Coast: Arrival, Results and Reflections.* Proceedings of the final Coast to Coast Conference, 1-5 October 2002 in Falköping, Sweden, Coast to Coast Books, Uppsala, 2004.

KNUTSSON, H., and K. KNUTSSON 2003 – Stone Age Transitlons: Neolithization in Central Scandinavia. *Documenta Praehistorica* XXX, 2003:48-78.

KOFLER, W., V. KRAPF, W. OBERHUBER and S. BORTENSCHLAGER 2005 – Vegetation responses to the 8200 cal. BP cold event and to long-term climatic changes in the Eastern Alps: possible influence of solar activity and North Atlantic freshwater pulses. *The Holocene* 5/6, 2005:779-788.

KOPACZ, J. 1974 – Views Concerning Typological Classification and the Relative Chronology of Stroke Ornamented Band Pottery in Poland. *Archaeoiogia Polona* XV, 1974:29-50.

KOREK, J. 1987 – Szegvár-Tűzköves: A Settlement of the Tisza Culture. In P. Raczky, (ed.), *The Late Neolithic of the Tisza Region: A Survey of Recent Excavations and their Findings: Hódmezovásárhely-Gorzsa, Szegvár Tuzköves, Öcsöd-Kováshalom, Vészto-Mágor, Berettyóújfalu-Herpály.* Budapest, Szolnok, 1987:47-51.

KOSCHIK H. (ed.) 1998 – *Brunnen der Jungsteinzeit.* Rheinisches Amt für Bodendenkmalpflege. Rheinland-Verlag, Habelt, Bonn, 1998.

KOŚKO, A. 1980 – The position of Funnel Beaker Culture in the Lowland Model of Neolithization. In J.K. Kozlowski and J. Machnik (eds.), *Prob lèmes de la néolithisation dans certaines régions de l'Europe*, Polska Akademia Nauk, Ossolineum, Wrocław, Poland, 1980:123-138.

KOSSE, K. 1979 – *Settlement and Ecology of the Early and Middle Neolithic Körös and Linear Pottery Cultures in Hungary.* BAR, Oxford, 1979.

KOŠTUŘÍK, P. 1973 – *Die Lengyel-Kultur in Mähren: Die jüngere mährische bemalte Keramik.* Studie Archeologického Ústavu Českolovenske Akademie věd v Brně. Academia Praha, Praha, 1973.

KOTOVA, N. 2009 – The Neolithization of Northern Black Sea Area in the Context of Climate Changes. *Documenta Praehistorica* XXXVI (2009).

KOWALEWSKA-MARSZAŁEK, H. 2008 – The most Distant Outskirts: The Baden Elements in the Zlota Culture (Little Poland). In M. Furholt *et al.* (eds.), 2008:233-246.

KOZŁOWSKI, J.K, M. NOWAK and M. VIZDAL 2003 – A Settlement of the Early Eastern Linear Pottery Culture at Moravany (Eastern Slovakia) within the Context of the Neolithization of the Upper Tisza Basin. In E. Jerem and P. Raczky (eds.), *Morgenrot der Kulturen: Frühe Etappen der Menschheitsgeschichte in Mittel- und Südosteuropa. Festschrift für Nándor Kalicz*. Archaeolingua, Budapest, 2003:127-143.

KRUK, J., S. MILISAUSKAS, W. ALEXANDROWICZ and Z. ŚNIESZKO 1996 – *Osadnictwo i zmiany środowiska naturalego wyżyn lessowych: Studium archeologiczne i paleogeograficzne nad neolitem w dorzeczu Nidzicy*. Instytut Archeologii i Etnologii Polskiej Akademii Nauk. Kraków, 1996.

LANTING, J.N., and J. van der PLICHT 2000 – De ^{14}C-Chronologie van de nederlandse Pre- en Protohistorie III: Neolithicum. *Palaeohistoria* 41/42, 1999/2000:1-110.

LARSSON, L. 2007 – Mistrust Traditions, Consider Innovations? The Mesolithic-Neolithic Transition in Southern Scandinavia. *Proceedings of the British Academy* 144, 2007:595-616.

LARSONA, G., U. ALBARELLA, K. DOBNEY, P. ROWLEY-CONWY *et al.* 2007 – Ancient DNA, Pig Domestication, and the Spread of the Neolithic into Europe. *Procedings of the National Academy of Sciences* 104/39, *PNAS*, 2007 (www.pnas.org_cgi_doi_10.1073_pnas.0703411104, 2007:15276–15281).

LAUFER, E. 2002 – *Ein spätbandkeramisches Erdwerk bei Usingen im Taunus*. www.jungsteinSITE.de, Article of February 12, 2002.

LAUX, F. 1990 – Die Trichterbechergruppen zwischen Elbe und Ems (Niedersachsen). In Jankowska, D. (ed.), *Die Trichterbecherkultur: Neue Forschungen und Hypothesen* I. Poznań, Instyitut Prahistorii Universitetu im. Adam Mickiewicza w Poznańiu Zakład Archeologii Wielkopolski IHKM PAN w Poznańiu, Poznań, Poland, 1990:181-195.

LAZAROVICI, G. 1999 – Vinča and Lengyel, *Lengyel '99 – 2nd Workshop Meeting of IGCP-442* (11-13 October, 1999), Veszprém, Hungary, 1999, www.ace.hu/igcp442/ws2/lazar.html (Accessed June 8, 2002).

LEEUW, S. van der, and C.L. REDMAN. 2002 – Placing Archaeology at the Center of Socio-Natural Studies. *American Antiquity*.6/4, 2002:579-605.

LENNEIS, E., and P. STADLER 1995 – Zur Absolutchronologie der Linearbandkeramik auf Grund von ^{14}C-Daten. (Also published in Archäologie Österreichs 6/2, 1995:4-13, www.nhm-wien.ac.at/NHM/Prehist/Stadler/LVAS/QAM/14C/C14-Datierung_der_LBK.html (Accessed July 17, 2003).

LENNEIS, E., O. BRINKKEMPER, O. CICHOCKI, C. FRANK *et al.* 2009 – *Rosenburg im Kamptal, Niederösterreich: Ein Sonderplatz der älteren Linearbandkeramik*. Hablet, Bonn.

LEUSCHNER, H.H., A. BAUEROCHSE and A. METZLER 2007 – Environmental Change, Bog History and Human Impact Around 2900 B.C. in NW Germany: Preliminary Results from a Dendroecological Study of a Sub-Fossil Pine Woodland at Campemoor, Dümmer Basin. *Vegetation History and Archaeobotany* 16, 2007:183-195.

LEUSCHNER, H.H., U. SASS-KLAASSEN, E. JANSMA, M. BAILLIE and M. SPURK 2002 – Subfossil European Bog Oaks: Population Dynamics and Long-Term Growth Depressions as Indicators of Changes in the Holocene Hydro-Regime and Climate. *The Holocene* 12/6, November 1, 2002:695-706.

LIDÉN, K., G. ERIKSSON, B. NORDQVIST, A. GÖTHERSTRÖM and E. BENDIXEN 2004 – "The Wet and the Wild Followed by the Dry and the Tame" – Or did They Occur at the Same Time? Diet in Mesolithic-Neolithic Southern Sweden. *Antiquity* 78/299, 2004:23-33.

LIPPERT, A., M. SCHULTZ, S. SHENNAN and M. TESCHLER-NICOLA (eds.) 2001 – *Mensch und Umwelt während des Neolithikums und der Frühbronzezeit in Mitteleuropa. Ergebnisse interdisziplinärer Zusammenarbeit zwischen Archäologie, Klimatologie, Biologie und Medizin.* Internationaler Workshop vom 9.-12. November 1995 im Archäologiezentrum Wien. (People and their Environment during the Neolithic and the Early Bronze Age in Central Europe. Results of interdisciplinary cooperation between archaeology, climatology, biology and medicine. International workshop at Vienna Archaeological Centre, November 9-12, 1995.) Internationale Archäologie: Arbeitsgemeinschaft, Symposium, Tagung, Kongress 2, Leidorf, Rahden, Westf., Germany, 2001.

LITT, T. 2003 – Environmental Response to Climate and Human Impact in Central Europe During the last 15,000 years: A German Contribution to PAGES-PEPIII. *Quaternary Science Reviews* 22, 2003:1-4.

LOUWE-KOOIJMANS, L.P. 2007 – The Gradual Transition to Farming in the Lower Rhine Basin. In A. Whittle and V. Cummings (eds.), 2007:287-309.

LOUWE-KOOIJMANS, L.P. 2009 – The Agency Factor in the Process of Neolithisation: A Dutch Case Study. *Journal of Archaeology in the Low Countries* 1/1, May 2009 (dpc.uba.uva.nl/jalc/01/nr01/a03).

LÜNING, J. 1968 – Die Michelsberger Kultur: Ihre Funde in zeitlicher und räumlicher Gliederung. *Bericht der Römisch-Germanischen Kommission*, 48, 1968:1-350.

LÜNING, J. 1998 – Betrachtungen über die Michelsberg Kultur. In Biel *et al.* (eds.), 1998:277-290.

LUKES, A. 2004 – Social Perspectives on the Constitution of the Linear Pottery Culture (LBK). In Lukes and Zvelebil (eds.), 2004:17-33.

LUKES, A., and M. ZVELEBIL (eds.) 2004 – *LBK Dialogues: Studies in the Formation of the Linear Pottery Culture*. BAR S1304, BAR Publishing, Oxford, 2004.

MAGNY, M., C. BÉGEOT, J. GUIOT and O. PEYRON 2003 – Contrasting Patterns of Hydrological Changes in Europe in Response to Holocene Climate Cooling Phases. *Quaternary Science Reviews* 22/15-17, 2003:1589-1596.

MATEICIUCOVÁ, I. 2004 – Mesolithic Traditions and the Origin of the Linear Pottery Culture (LBK). In Lukes and Zvelebil (eds.), 2004:9-107.

MAROLLE, C. 1998 – Le site de Michelsberg des « Hautes Chanvrières » avec bâtiments et enceinte à Mairy, Ardennes, France. In Biel *et al.* (eds.), 1998:21-28.

MAYER, C. 2008 – Mappings of the Late Neolithic Cultures in the Austrian Danube Region. In M. Furholt *et al.* (eds.), 2008:167-176.

METZLER, A. 1997 – *Aktuelles zur Moorarchäologie in Niedersachsen.* Urgeschichts-Abteilung des Niedersächsischen, Niedersächsisches Landesmuseum Hannover, 1997, www.urgeschichte.de/artikel/moorarch.htm (Accessed 17 March 2002).

METZLER, A. 2003 – Early Neolithic Peatland Sites Around Lake Dümmer. [Frühneolithische Fundstellen in Mooren der Dümmerniederung.] In A. Bauerochse and H. Haßmann (eds.), Peatlands: Archaeological Sites – Archives of Nature – Nature Conservation – Wise Use. Proceedings of the Peatland Conference 2002 in Hannover, Germany. Moorlandschaften: Archäologische Fundstellen – Archive der Natur – Tagungsband der Moorlandschafts Naturschutz – Nachhaltige Nutzung. Konferenz im Jahre 2002 in Hannover, Deutschland. VLM, Rhaden, Westf., 2003:62-67.

MICHCZYNSKI, A. 2004 – Influence of ^{14}C Concentration Changes in the Past on Statistical Inference of Time Intervals. *Radiocarbon* 46/2, 2004:997-1004.

MIDGLEY, M.S. 1985 – *The Origin and Function of the Earthen Long Barrows of Northern Europe.* BAR S259, Oxford, 1985.

MIDGLEY, M.S. 1992 – *TRB Culture: The First Farmers of the North European Plain.* Edinburgh University Press, Edinburgh, 1992.

MIDGLEY, M.S. 2002 – Early neolithic farming communities in northern Europe: Reconsideration of TRB culture. *Archeologicke rozhledy* 54/1, 2002:208-22.

MIDGLEY, M.S. 2010 – *Who was who in the Neolithic?* www.jungsteinSITE.de, Aricle of November 9, 2010:1-10.

MILISAUSKAS, S., and J. KRUK 2002 – Continuity, Diversity, Innovations, and Greater Complexity, 5500/5000-3500/3000 BC. In S. Milisauskas, (ed.), *European Prehistory: A Survey.* Kluwer Academic / Planum Publishers, New York, Boston, Dordrecht, London, Moscow, 2002:193-246.

MILLER, R., M. OTTE and J. STEWART 2009 – Le Mesolithique recent du *Trau A/'Wesse* (Modave, Provode Liege): Decouverte de tessons non rubanes ou «Bereitkeramiek». *Notae Praehistoricae* 29, 2009:5-14.

MINICHREITER, K. 1998 – The Oldest Neolithic Water-Well in Croatia from the Early Starčevo Settlement near Slavonski Brod. In H. Koschik (ed.), 1998:25-31.

MINICHREITER, K. 2001 – The Architecture of Early and Middle Neolithic Settlements of the Starcevo Culture in Northern Croatia. *Documenta Praehistorica* XXVIII, 2001:199-214 (arheologija.ff.uni-lj.si/documenta/pdf28/28minichreiter.pdf).

MISCHKA, D. 2010 – *Flintbek LA 3: Biography of a Monument.* www.jungsteinSITE.de, Aricle of December 20, 2010:1-36.

MÜLLER, J. 2001 – *Soziochronologische Studien zum Jung- und Spätneolithikum im Mittelelbe-Saale-Gebiet (4100-2700 v. Chr.): Eine sozialhistorische Interpretation prähistorischer Quellen.* Vorgeschichtliche Forschungen 21. VLM, Rahden, Westf., Germany, 2001.

MÜLLER, J. 2002 – *Zur Belegungsabfolge des Gräberfeldes von Trebur: Argumente der typologieunabhängigen Datierungen.* www.jungsteinSITE.de, Artikel of October 1, 2002.

MÜLLER, J., J.-P. BROZIO, D. DEMNICK, H. DIBBERN *et al.* 2010 – *Periodisierung der Trichterbecher-Gesellschaften: Ein Arbeitsentwurf.* www.jungsteinSITE.de, Aricle of October 26, 2010:1-6.

NASH, S.E. 2002 – Archaeological Tree-Ring Dating at the Millenium. *Journal of Archaeological Research* 10/3, 2002:243-275.

NEUGEBAUER-MARESCH, C. 1995 – Mittelneolithikum: Die Bemaltkeramik. In E. Lenneis, C. Neugebauer-Maresch and E. Ruttkay, *Jungsteinzeit im Osten Österreichs.* Forschungsberichte zur Ur- u. Frühgeschichte 17. Niederösterreichisches Pressehaus u. Verlagsgesellschaft, St. Pölten and Wien, 1995:57-107.

NEUSTUPNÝ, E. 2004 – Remarks on the Origin of the Linear Pottery Culture. In Lukes and Zvelebil (eds.), 2004:3-5.

NIELSEN, P.O. 1993 – Yngere stenalder. In S. Hvass and B. Storgaard (eds.), *Digging into the Past: 25 Years of Archaeology in Denmark.* Aarhus Universitetsforlag, Århus, 1993:84-87.

NIELSEN, E.H. 2009 – The Mesolithic Background for the Neolithisation Process. *Documenta Praehistorica* XXXVI, 2009:151-158.

NIELSON, B.O, V. MAHLER and P. RASMUSSEN 2000 – An Arthropod Assemblage and the Ecological Conditions in a Byre at the Neolithic – Settlement of Weier, Switzerland. *Journal of Archaeological Science* 27, 2000:209-218.

NIESIOŁOWSKA, E. 1994 – Einige Probleme der frühen Trichterbecherkultur in Polen: Die Sarnowo-Stufe und die Pikutkowo-Phase. In Hoika and Meurers-Balke 1994:325-336.

NOWAK, M. 2001 – The second phase of Neolithization in east-central Europe. *Antiquity* 75, 2001:582-592.

NOWAK, M. 2004 – Is 'Pot Prehistory' Real Prehistory. In Lukes and Zvelebil (eds.), 2004:7-15.

NOWAK, M. 2006 – Transformations in East-Central Europe from 6000 to 3000 BC: Local vs. Foreign Patterns. *Documenta Praehistorica* XXXIII, 2006:143-158.

NOWAK, M. 2007 – Middle and Late Holocene Hunter-Gatherers in East Central Europe: Changing Paradigms of the 'non-Neolithic' way of life. *Documenta Praehistorica* XXXIV, Neolithic Studies 14, 2007:89-103.

OBELIĆ, B., I.K. BRONIĆ and N. HORVANTINČIĆ 2002 – Rudjer Bošković Institute Radiocarbon Measurements XV. *Radiocarbon* 44/2, 2002:601-630.

OBELIĆ, B., M.K. ŠKRIVANKO, B. MARIJAN, and I.K. BRONIĆ 2004 – Radiocarbon Dating of Sopot Culture Sites (Late Neolithic) in Eastern Croatia. Radiocarbon 46/1:245-258.

ØSTMO, E. 1991 – The rise and fall of the TRB in Southeast Norway. In Die Trichterbecherkultur: Neue Forschungen und Hypothesen, I, Jankowska, Dabrochna (ed.), Poznan, Poland, 1991:19-26.

ØSTMO, E. 2007 – The Northern Periphery of the TRB: Graves and Ritual Deposits in Norway, *Acta Archaeologica* 78/2, 2007:111-142.

OROSS, K., and E. BÁNFFY 2009 – Three Successive Waves of Neolithisation: LBK Development in Transdanubia. *Documenta Praehistorica* XXXVI, 2009:175-189.

OTTE, M., and P. NOIRET 2001 – La Mésolithique du Bassin Pannonien et la formation du Rubané. *L'Anthropologie*, 105, 2001:409-419.

OUT, W.A. 2008 – Neolithisation at the Site Brandwijk-Kerkhof, the Netherlands: Natural Vegetation, Human Impact and Plant Food Subsistence. *Vegetation Hististory and Archaeobotany* 17, 2008:25-39.

PAETZOLD, D. 1992 – Bemerkungen zum Siedlungsverhalten neolithischer bis latènzeitlicher Bevölkerungen zwischen Regensburg und Deggendorf. *Bayrische Vorgeschichtsblätter* 57, 1992:77-102. Beck'sche Verlagshandlung, München.

PAVÚK, J. 1994 – Zur relativen Chronologie der älteren Linearkeramik. *Jósa András Múzeum Évkönyve* 1994:135-149.

PAVÚK, J. 2000 – Epilengyel-Lengyel IV als Kulturhistorische Einheit. (Epilengyel/Lengyel IV ako kultúrnohistorický celok.) *Slovenská Archeológia* XLVIII-1, 2000:1-28.

PAVÚK, J. 2004 – Early Linear Pottery Culture in Slovakia and the Neolithisation of Central Europe. In Lukes and Zvelebil (eds.), 2004:71-82).

PAVÚK, J. 2004 – Early Linear Pottery Culture in Slovakia and the Neolithisation of Central Europe. In Lukes and Zvelebil (eds.), 1994:71-82).

PAZDUR, A., M. FOGTMAN, A. MICHCZYŃSKI, J. PAWLYTA, and M. ZAJĄC 2004 – ^{14}C Chronology of Mesolithic Sites from Poland and the Background of Environmental Changes. *Radiocarbon* 46/2, 2004:809-826.

PEDERSEN, L, A. FISCHER and B. AABY (eds.) 1997 – *The Danish Storebælt since the Ice Age: Man, Sea and Forest.* A/S Storebælt Fixed Link, Kalundborg Regional Museum, National Forest and Nature Agency, National Museum of Denmark, København, 1997.

PETREQUIN, P. and A.-M. PETREQUIN 2005 – Premiers Attelages de Boeufs à Chalain (Jura, France) au XXXIe Siècle av. J.-C. *Revue de Paléobiologie, Genèv,* 10, 2005:197-207.

PERLÉS, C. 2001 – *The Early Neolithic in Greece: The First Farming Communities in Europe.* Cambridge University Press. Cambridge, 2001.

PLESLOVÁ-ŠTIKOVÁ, E. 1985 – *Makotřasy: A TRB site in Bohemia.* Fontes Archaeologici Pragenenses 17, Museum Nationale Pragae. Praha 1985.

PODBORSKÝ, V. 1993 – *Praveké Dejiny Moravy.* Vlastiveda Moravská Zeme a Lid, Nová řada 3. Muzejní a vlastivedna spolecnost, Brno, 1993.

POSKA, A., and L. SAARSE 2002 – Vegetation Development and Introduction of Agriculture to Saaremaa Island, Estonia: The Human Response to Shore Displacement. *The Holocene* 12/5, 2002:555-568.

PRICE, T.D. 2000a – The Introduction of Farming in Northern Europe. In Price (ed.), 2000:301-318.

PRICE, T.D. 2000b – The Introduction to Farming in Northern Europe. In Price (ed.), 2000:260-300.

PRICE, T.D. 2000c – Lessons in the Transition to Agriculture. In Price (ed.), 2000:301-318.

PRICE, T.D. (ed.) 2000 – *Europe's First Farmers.* Cambridge University Press, Cambridge, 2000:260-300.

PRICE, T.D., R.A BENTLEY, D. GRONENBORN, J. LÜNING and J. WAHL 2001 – Human migration in the Linearbandkeramik of central Europe, *Antiquity* 75, 2001:593-603.

PRICE, T.D., J.H. BURTON and R.A. BENTLEY 2002, The Characterisation of Biologically-Available Strontium Isotope Ratios for Investigation of Prehistoric Migration, *Archaeometry* 44, 2002:117-135.

PRICE, T.D., B. GEBAUER, S. ULFELDT HEDE, C. SEDLACEK LARSEN, N. NOE-NYGAARD, S. MASON, J. NIELSEN, and D. PERRY 2001 – Sakkerup Huse: A Mesolithic Settlement in NW Zealand, Denmark. *Journal of Field Archaeology* 28/1-2, 2001:45-67.

PRZYBYŁ, A. 2008 – Badenization of the Late Neolithic Funnel Beaker Culture Societies between Oder and Vistula Basins in the Light of ^{14}C-Datings. In M. Furholt *et al.* (eds.), 2008:189-204.

RACZKY, P., W. MEIER-ARENDT, A. ANDER, Z. HAJDÚ *et al.* 2002 – Polgár-Csőszhalom (1989-2000): Summary of the Hungarian-German Excavations on a Neolithic Settlement in Eastern Hungary. In R. Asian, S. Blum, G. Kastl, F. Schweizer and D. Thumm (eds.), *Mauerschau: Festschrift für Manfred Korfmann.* Greiner, Remshalden-Grunbach, Germany, 2002:833-860.

RADOVANOVIĆ, I. 2006 – Further Notes on Mesolithic-Neolithic Contacts in the Iron Gates Region and the Central Balkans. *Documenta Praehistorica* XXXIII, 2006:107-124.

RAETZEL-FABIAN, D. 1986 – *Phasenkartierung des mitteleuropäischen Neolithikums: Chronologie und Chorologie.* B.A.R. S316, 1986.

RAETZEL-FABIAN, D. 2000 – *Calden: Erdwerk und Bestattungsplätze des Jungneolithikums; Architektur – Ritual – Chronologie.* Habelt, Bonn, 2000.

RAETZEL-FABIAN, D. 2006 – Bookreview: Birgit Höhn, Michelsberger Kultur in der Wetterau. Universitätsforschungen zur prähistorischen, Archäologie 87, Habelt, Bonn, 2002. *Germania,* 2006:177-182.

RAETZEL-FABIAN, D., and M. FURHOLT 2006 – *Frühbadener Elemente im Neolithikum Mitteldeutschlands: Die Schöninger Gruppe.* www.jungsteinSITE.de, Article of October 10, 2006.

REIMER, P.J., M.G.L. BAILLIE, E. BARD *et al.* 2004 – IntCal04 Terrestrial Radiocarbon Age Calibration, 0–26 Cal Kyr BP. *Radiocarbon* 46/3, 2004:1029-1058.

REGENYE, J. 2005 – A Vizsgálandó Közép-Dunántúli Újkőkori Kerámia és Környezete. *Archeometriai Műhely* 2, 2005:16-23.

RICHERSON, P.J., and R. BOYD 2001 – Institutional Evolution in the Holocene: The Rise of Complex Societie, In W.G. Runciman (ed.), *The Origin of Human Social Institutions,* 2001:197-204, Proceedings of the British Academy 110.

RICHERSON, P.J., R. BOYD and R.L. BETTINGER 2001 – Was Agriculture Impossible during the Pleistocene but Mandatory during the Holocene? A Climate Change Hypothesis. *American Antiquity* 66/3, 2001:387-411.

ROEVER, J.P. de 2004 – *Swifterbant-Aardewerk: Een analyse van de Neolithische Nderzetting bij Swifterbant, 5e millennium voor Christus.* Groningen Archaeological Studies 2, 2004. Rijksuniversiteit Groningen. Barkhuis, Elde, Netherland.

ROKSANDIC, M., M. DJURIĆ, Z. RAKOČEVIĆ and K. SEGUIN 2006 – Interpersonal Violence at Lepenski Vir Mesolithic/Neolithic Complex of the Iron Gates Gorge (Serbia-Romania). *American Journal of Physical Anthropology* 129, 2006:339-348.

ROWLEY-CONWY, P. 1995 – Making First Farmers Younger: The West European Evidence. *Current Anthropology* 36-2, 1995:346-353.

RUNDKVIST, M., C. LINDQVIST and K. THORSBERG 2004 – Barshalder 3: Rojrhage in Grötlingbo – A multi-component Neolithic shore site on Gotland. *Stockholm Archaeological Reports* 41, Department of Archaeology, University of Stockholm, Stockholm, 2004.

RUOFF, U., and S. JACOMET 2001 – Die Datierung des Rades von Zürich-Akad und die stratigraphische Beziehung zu den Rädern von Zürich-Pressehaus *Hemmenhofener Skripte* 3, 2001:35-37.

RUTTKAY, E., O. CICHOCKI, E. PERNICKA and E. PUCHER 2004 – Prehistoric Villages on the Austrian Lakes. In F. Menotti (ed.), *Living on the Lake in Prehistoric Europe: 150 Years of Lake-Dwelling Research.* Routledge. London, 2004:50-68.

RUTTKAY, E., and E. PUCHER 2006 – Eine Siedlungsgrube mit jungneolithischer inkrustierter Keramik aus Puch-Scheibenfeld, SG und VB Hollabrunn, Niederösterreich: Neue Beiträge zur Furchenstichkeramik und zum Scheibenhenkel. *Annalen des Naturhistorischen Museums in Wien* 107A, 2006:267-304.

RYBICKA, M. 2008 – Settlement, Chronology and Economy of the Funnel Beaker-Baden Society in Kujavia and the Gostynin Lake District. In M. Furholt *et al.* (eds.), 2008:205-216.

SAMONIG, B. 2005 – *Studien zur Pfahlbauforschung in Österreich. Materialien II: Die Pfahlbaustation des Keutschacher Sees.* Mitteilungen der Prähistorischen Kommission 51, Österreichische Akademie der Wissenschaften, Wien, 2005.

SCHEU, A., S. HARTZ, U. SCHMÖLCKE, A. TRESSET, J. BURGER and R. BOLLONGINO 2007 – Ancient DNA Provides no Evidence for Independent Domestication of Cattle in Mesolithic Rosenhof, Northern Germany. *Journal of Archaeological Science* XX, 2007:1-8.

SCHIBLER, J.S., S. JACOMET, H. HÜSTER-PLOGMANN and C. BROMBACHER 1997 – Synthese – Synthesis. In J.S. Schibler, S. Jacomet, H. Hüster-Plogmann C. Brombacher, *et al.* (eds.), *Ökonomie und Ökologie neolithischer und bronzezeitlicher Ufersiedlungen am Zürichsee.* Monographien der Kantonsarchäologie Zürich 20, Zürich und Egg, 1997:329-361.

SCHIBLER, J.S., and C. BROMBACHER 1995 – Wirtschaft. In W.E. Stöckli, U. Niffeler and E. Gross-Klee (eds.), *Die Schweiz vom Palaeolithikum bis zum frühen Mittelalter, Band II: Neolithikum* (SPM II). Verlag Schweizerische Gesellschaft für Ur- und Frühgeschichte, Basel, 1995:85-118.

SCHIBLER, J.S., and S. JACOMET 2005 – Fair-weather Archaeology? A possible Relationship between Climate and the Quality of Archaeological Sources. In Gronenborn (ed.), *Klimaveränderung und Kulturwandel in neolithischen Gesellschaften Mitteleuropas, 6700-2200 v. Chr.* RGZM-Tagungen 1, Mainz, 2005:27-39.

SCHILD, R. 1998 – The Perils of Dating Open-air Sandy Sites of the North European Plain. In Zvelebil *et al.* (eds), 1998:87-94.

SCHINDLER, R. 1953 – Die Entdeckung zweier jungsteinzeitlicher Wohnplätze unter dem Marschenschlick im Vorgelände der Boberger Dünen und ihre Bedeutung für die Steinzeitforschung Nordwestdeutschlands. *Hammaburg* 4, 1953:1-17.

SCHINDLER, R. 1961 – Rössener Elemente im Frühneolithikum von Boberg. *Hammaburg* 7/13, 1961:9-29.

SCHLICHTHERLE, H. 1990 – Siedlungen und Funde jungsteinzeitlicher Kulturgruppen zwischen Bodensee und Federsee. In M. Höneisen (ed.). *Die Ersten Bauern 2: Einführung, Balkan, angrenzende Regionen der Schweiz*. Schweizerisches Landesmuseum, Zürich, 1990:134-156.

SCHLICHTHERLE, H. 2003 – Die jungsteinzeitlichen Radfunde vom Federsee und ihre kulturgeschichtliche Bedeutung. *Hammenhofener Skripte* 3, 2003:9-34.

SCHMIDT, B and W. GRUHLE 2003 – Niederschlalagsschwankungen in Westeuropa während der letzten 8000 Jahre: Versuch einer Rekonstruktion. *Archäologisches Korrespondenzblatt* 33, 2003:281-300.

SCHUBERT, H. 1999 – *Die bemalte Keramik des Frühneolithikums in Südosteuropa, Italien und Westanatolien*. Internationale Archäologie 47. VLM, Rahden, Westf., Germany, 1999.

SCHWABEDISSEN, H. 1979 – Der Beginn des Neolithikums im nordewestlichen Deutschland. In H. Schirnig (ed.), *Großsteingräber in Niedersachsen*. Lax, Hildesheim, 1979:203-222.

SCHWABEDISSEN, H. 1994 – Die Ellerbek-Kultur in Schleswig-Holstein und das Vordringen des Neolithikums über die Elbe nach Norden. In Hoika and Meurers-Balke, 1994:361-401.

SCHWAMMENHÖFER, H. 1983 – Eine Gehöftgruppe der Epilengyelzeit in der Ziegelei Schleinbach, NO. Fundberichte aus Österreich 22, 1983:169-202.

SKAK-NIELSEN, N.V. 2007 – The Function of Perforated Shoe – Last Axes. *Acta Archaeologica* 78/2, 2007:79-86.

SHENNAN, S., and J. STEELE 2000 – Spatial and Chronological Patterns in the Neolithisation of Europe. ads.ahds.ac.uk/catalogue/resources.html?c14_meso (Accessed December 16, 2005).

SCOLLAR, I. 1959 – Regional Groups in the Michelsberg Culture: A Study in the Middle Neolithic of Central Europe. *Proceedings of the Prehistoric Society*, XXV, 1959:52-134.

SCOLLAR, I. 1961 – The Late Neolithic in Belgium, Western Germany and Alsace. *L'Europe à la fin de l'âge de la pierre. Acte du Symposium consacré aux problemes du Néolithique européen*. Prágue-Liblice-Brno 5-12 octobre 1959, L'Académie tchécoslovaque des Sciences, Praha, 1961:519-548.

SHERRATT, A. 1997 – *Economy and society in prehistoric Europe: changing perspectives*. Princeton University Press, Princeton, N.J., 1997.

SJÖGREN, K.-G. 2003 – *"Mångfalldige uhrminnes grafvar..." Megalitgravar och samhälle i Västsverige*. GOTARC B/ 27, Coast to Coast Books 9, Göteborg universitet, Intitutionen för arkeologi, Göteborg, 2003.

SJÖGREN, K.-G. 2010 – Anonymous Ancestors? The Tilley/Shanks Hypothesis Revisited. In D. Calado, M. Baldia and M. Boulanger, *Monumental Questions: Prehistoric Megaliths, Mounds and Enclosures*. Proceedings of Session C68/C21, Actes du XVème Congrès UISPP, Lisbon, 2006. BAR Publishing, Oxford, 2010:111-118.

ŠMÍD, M. 1997 – Sídlište kultury s nálevkovitými poháry u Laškova, okr. Prostejov. *Pravěk* N. ř. 6, 1996:97-138.

ŠMÍD, M. 2004 – *Středomoravská mohylová porhřebiště KNP*. Archeologickě Památky Střední Moravi 7, Olomouc, 2004.

SMITH, B. 2001 – Low Level Food Production. *Journal of Archaeological Research* 9/1, 2001:1-43.

SOUTHON, J. 2004 – A Radiocarbon Perspective on Greenland ice-core Chronologies: can we use ice cores for ^{14}C Calibration? *Radiocarbon* 46/3, 2004:1239-1259.

SPURK, M., LEUSCHNER, H. HUBERT; M. BAILLIE, K. BRIFFA and M. FRIEDRICH 2002 – Depositional frequency of German subfossil oaks: climatically and non-climatically induced fluctuations in the Holocene. *The Holocene* 12/6, November 1, 2002:695-706.

STADLER, P. 1995 – Ein Beitrag zur Absolutchronologie des Neolithikums in Österreich auf Grund der ^{14}C-Daten. In Lenneis, E., C. Neugebauer-Maresch, E. Ruttkay, *Jungsteinzeit im Osten Österreichs*. Forschungsberichte zur Ur- u. Frühgeschichte 17. Niederösterreichisches Pressehaus u. Verlagsgesellschaft, St. Pölten and Wien, 1995:210-224.

STADLER, P. 2005a – Settlement of the Early Linear Ceramics Culture at Brunn am Gebirge, Wolfholz Site. *Documenta Praehistorica* XXXII, 2005:269-278.

STADLER, P. 2005b – Die Frühneolithische Siedlung von Brunn am Gebirge, Flur Wolfholz: Aktuelle Forschungsergebnisse. Verein für NÖ Landeskunde (in press), (homepage.univie.ac.at/Peter.Stadler/Stadler/StadlerP_2004c.pdf).

STADLER P. and N. KOTOVA 2008 – *Early Neolithic Contacts between Croatia and the Vienna Basin from 5700-5200 BC*. (homepage.univie.ac.at/Peter.Stadler/Stadler/StadlerP_2009i.pdf).

STADLER, P., S. DRAXLER, H. FRIESINGER, W. KUTSCHERA *et al.* 2000 – *Status of the Austrian Science Fund Project P-12253-PHY: Absolute Chronology for Early Civilisations in Austria and Central Europe using ^{14}C Dating with Accelerator Mass Spectrometry*. Projektbericht an den FWF als Manuskript (winserion.org/Stadler/2000d.pdf).

STADLER, P., and E. RUTTKAY 2006 – ^{14}C (AMS)-datierte Fundkomplexe der MOG in Österreich: Eine typologische Anordnung des Materials nach Phasen. Suplement to Stadler *et al.*, 2006 (homepage.univie. ac.at/Peter.Stadler/Stadler/StadlerP_2006b.pdf).

STADLER, P., E. RUTTKAY, M. DONEUS, H. FRIESINGER *et al.* 2006 – Absolutchronologie der Mährisch-Ostösterreichischen Gruppe (MOG) der bemalten Keramik aufgrund von neuen ^{14}C-Datierungen. *Archäologie Österreichs* 16/17 Sonderausgabe, 2005/2006:53-67.

STÄUBLE, H. 1995 – Radiocarbon Dates of the Earliest Neolithic in Central Europe. In G.T. Cook, D.D. Harkness, B.F. Miller and E.M. Scott (eds.), *Proceedings of the 15th International UC Conference.* Radiocarbon 37/2, 1995:227-237.

STEIG, E.J. 1999 – Mid-Holocene Climate Change. *Science* 286/5444, 1999:1485-1487.

STEPPAN, K. 1998 – Archäologisch Untersuchungen der Säugetiere aus den Graben der Michelsberger ✎Erdwerke' in Bruchsal, Landkreis Karlsruhe. In Biel *et al.* (eds.), 1998:143-158.

STREET, M., M. BAALES, E. CZIESLA, S. HARTZ *et al.* 2002 – Final Paleolithic and Mesolithic Research in Reunified Germany. *Journal of World Prehistory* 15/4, 2001:355-453.

STRIEN, H.-C., and D. GRONENBORN 2005 – Klima- und Kulturwandel während des Mitteleuropäischen Altneolithikums. In Gronenborn (ed.), 2005:131-149.

STROBEL, M. 1998 – Diskontinuität oder Forschungslücke? Die Gruppen Aichbühl und Schussenried in Oberschwaben. In Biel *et al.* (eds.), 1998:201-212.

STÖCKLI, W.E. 2002 – *Absolute und relative Chronologie des Früh- und Mittelneolithikums in Westdeutschland (Rheinland und Rhein-Main-Gebiet).* Baseler Hefte zur Archäologe 1. Archäologie Verlag, Basel.

STUIVER, M., P.J. REIMER, E. BARD, J.W. BECK *et al.* 1998. INTCAL98: Radiocarbon age calibration 24,000 – 0 cal BP. *Radiocarbon* 40:1041-1083.

SÜMEGI, P. 2004 – The Results of Paleoenvironmental reconstructions and Comparative Geoarchaeological Analysis for the Examined Area. In P. Sümegi and S. Gulyas (eds.), *The Geohistory of Bátorliget Marshland: An Example for the Reconstruction of Late Quaternary Environmental Changes and Past Human Impact from the Northestern Part of the Carpathian Basin.* Archaeolingua, Budapest, 2004:302-348.

ŠUTEKOVÁ, J. 2008 – The Jevišovice Culture in Slovakia. In M. Furholt *et al.* (eds.), 2008:131-138.

SVOBODA, J.A. 2008 – The Mesolithic of the Middle Danube and the Upper Elbe Rivers. In Bailey and Spikins (eds.), 2008:221-237.

SVOBODA, J.A., V. CÍLEK and L. JAROŠOVA 1998 – Zum Mesolithikum in den Sandsteingebieten Nordböhmens. *Archäologisches Korrespondenzblatt*, 25/3, 1998:357-372.

SVOBODA, J.A., J. van der PLICHT and V. KUZELKA 2002 – Upper Paleolithic and Mesolithic human fossils from Moravia and Bohemia (Czech Republic): some new ^{14}C Dates. *Antiquity* 76/294, 2002:957-962.

SZMYT, M. 1998 – Die Kugelamphorenkultur und die Gemeinschaften der Steppenwald- und Steppenzone Osteuropas. Der Forschungsstand und die Forschungsperspektiven im Grundriß. In B. Hänsel and J. Machnik (eds.), *Das Karpatenbecken und die osteuropäische Steppe. Nomadenbewegungen und Kulturaustausch in den vorchristlichen Metallzeiten (4000-500 v.Chr.).* Prähistorische Archäologie in Südosteuropa 12 – Südosteuropa-Schriften 20. VLM, Rahden, Westf., Germany, 1998:221-231.

SZMYT, M. 2008 – Baden Patterns in the Milieu of the Globular Amphorae: Transformation, Incorporation and Long Continuity: A case study from the Kujavian Region, Polish Lowland. In M. Furholt *et al.* (eds.), 2008:217-231.

TELFORD, R.J., E. HEEGAARD and H.J.B. BIRKS 2004 – The intercept is a poor estimate of a calibrated radiocarbon age. *The Holocene* 14/2, 2004:296-298.

TERBERGER, T., and J. PIEK 1998 – Zur absoluten Chronologie der Steinzeit in Mecklenburg-Vorpommern. *Bodendenkmalpfege in Mecklienburg-Vorpommern* 45, 1997:7-39.

THOMAS, J. 1987 – Relations of Production and Social Change in the Neolithic of North-West Europe. *Man,* New Series 22/3, 1987:405-430.

THOMAS, E.R., E.W. WOLFF, R. MULVANEY, J.P. STEFFENSEN, S.J. JOHNSEN, C. ARROWSMITH, J.W.C. WHITE, B. VAUGHN and T. POPP 2007 – The 8.2 ka Event from Greenland Ice Cores. *Quaternary Science Reviews* 26/1-2, 2007:70-81.

TINNER, W., E.H. NIELSEN, and A.F. LOTTER 2007 – Mesolithic Agriculture in Switzerland? A Critical Review of the Evidence. *Quaternary Science Reviews* 26, 2007:1416-1431.

TODOROVA, H. 1989 – Ein Korrelationsversuch zwischen Klimaänderungen und prähistorischen Angaben. In E. Pleslová-Štiková and M. Buchwaldek (eds.), *Das Äneolithikum und die früheste Bronzezeit (C^{14} 3000-2000 b.c.) in Mitteleuropa: Kulturelle und chronologische Beziehungen.* Acta des XIV. Internationalen Symposiums, Prag-Liblice 20-24. 10. 1986. Praehistorica XV. Acta Instituti Praehistorici Universitatis Carolinae Pragensis. Univerzita Karlova, Praha, 1989:25-28.

TODOROVA, H. 1998 – Probleme der Umwelt der prähistorischen Kulturen zwischen 7000 und 100 v.Chr. In B. Hänsel, and J. Machnik (eds.), *Das Karpatenbecken und die osteuropäische Steppe. Nomadenbewegungen und Kulturaustausch in den vorchristlichen Metallzeiten (4000-500 v.Chr.).* Prähistorische Archäologie in Südosteuropa 12, Südosteuropa-Schriften 20. VLM, Rahden, Westf., Germany, 1998:65-70.

TRINGHAM, R. 1971 – *Hunters, Fishers and Farmers of Eastern Europe: 6000-3000 B.C.*, Hutchinson, London, 1971.

TRINGHAM, R. 2000 – Southeastern Europe in the Transition to Agriculture in Europe: Bridge, Buffer and Mosaic. In Price (ed.), 2000:19-56.

Van BUREN, M. 2001 – The Archaeology of El Niño Events and Other "Natural Disasters." *Journal of Archaeological Method and Theory* 8/2, 2001:129-149.

VANMONTFORT, B. 2001 – The Group of Spiere as a New Stylistic Entity in the Middle Neolithic Scheldt Basin. *Notae Praehistoricae* 21, 2001:139-143.

VANMONTFORT, B. 2008a – Forager-Farmer Connections in an 'Unoccupied' Land: First Contact on the Western Edge of LBK Territory. *Journal of Anthropological Archaeology* 27, 2008:149-160.

VANMONTFORT, B. 2008b – A Southern View on North-South Interaction During the Mesolithic-Neolithic Transition in the Lower Rhine Area. In H. Fokkens, B.J. Coles, A.L. van Gijn, J.P. Kleijne *et al.* (eds.), *Between Foraging and Farming: An Extended Broad Spectrum of Papers Presented to Leendert Louwe Kooijmans*. Faculty of Archaeology, Leiden University, Leiden, 2008:85-97.

VANMONTFORT, B., C. CASSEYAS and P.M. VERMEERSCH 1997 – Neolithic Ceramics from Spiere "De Hel" and Their Contribution to the Understanding of the Earliest Michelsberg Culture. *Notae Praehistoricae* 17, 1997:123-134.

VANMONFORT, B., L. LOUWE KOOIJMANS, L. AMKREUTZ and L. VERHART 2010 – Pots, Farmers and Foragers: Pottery Traditions and Social Interaction in the Earliest Neolithic of the Lower Rhine Area. Leiden University Press. Leiden, 2010.

VERHART, L. 2008 – New Developments in the Study of the Mesolithic in the Low Countries. In Bailey and Spikins (eds.), 2008:182-202.

VERMEERSCH, P. M., and L. BURNEZ-LANOTTE 1998 – La Culture de Michelsberg en Belgique: État de la Question. In Biel *et al.* (eds.), 1998:47-54.

VOSTEEN, M. 1999 – *Urgeschichtliche Wagen in Mitteleuropa: Eine archäologische und religionswissenschaftliche Untersuchung neolithischer bis hallstattzeitlicher Befunde*. VLM, Rahden, Westf., 1999.

WATKINS, T. 2006 – Neolithisation in Southwest Asia: The Path to Modernity. *Documenta Praehistorica* XXXIII, 2006:71-88.

WEINER, J. 1995 – Eine Zimmermannstechnische Glanzleistung: Der 7000 Jahre alte Eichenholzbrunnen aus Erkelenz-Kückhoven. In H. Hellenkemper *et al.* (eds.), *Ein Land Macht Geschichte. Archäologie in Nordrhein-Westfalen*. Römisch-Germanisches Museum, Köln, 1995:179-187.

WENINGER, B., E. ALRAM-STERN, E. BAUER, L. CLARE *et al.* 2005 – Die Neolithisierung von Sodosteuropa als Folge des abrupten Klimawandels um 8200 cal BP. In Gronenborn (ed.), 2005:75-117.

WHITTLE, A. 1996 – *Europe in the Neolithic: The Creation of New Worlds*. Cambridge University Press, Cambridge, 1996.

WHITTLE, A. 2007 – The Temporality of Transformation: Dating the Early Development of the Southern British Neolithic. A. Whittle and V. Cummings (eds.), 2007:377-398.

WHITTLE, A., L. BARTOSIEWICZ, D. BORIĆ, P. PETTITT and M. RICHARDS 2002 – In the beginning: New Radiocarbon date for the Early Neolithic in northern Serbia and South-East Hungary. *Antaeus: Communicationesex Instituto Archaeologica Academiae Scientiarum* 25, 2002:63-117.

WHITTLE, A and V. CUMMINGS (eds.) 2007 – *Going Over: The Mesolithic-Neolithic Transition in North-West Europe*. Proceedings of the British Academy 144, Oxford University Press, Oxford, 2007.

WIERSMA, A. 2008 – *Character and Causes of the 8.2 ka Climate Event: Comparing Coupled Climate Model Results and Palaeoclimate Reconstructions*. Ph.D. Dissertation. Vrije Universiteit Amsterdam, Department of Paleoclimatology and Geomorphology, 2008 (dare.ubvu.vu.nl/bitstream/1871/12690/1/full-text.pdf).

WILD, E.M., P. STADLER, M. BONDÁR, S. DRAXLER *et al.* 2001 – New Chronological Frame for the Young Neolithic Baden Culture in Central Europe (4th Millennium BC). *Radiocarbon* 43/2B, 2001:1057-1064.

WILD, E.M, P. STADLER, A. HÄUßER, W. KUTSCHERA *et al.* 2004 – Neolithic Massacres: Local Skirmishes or General Warfare in Europe? *Radiocarbon* 46/1 2004:377-385.

WINDL, H.J. 1996 – Archäologie einer Katastrophe und deren Vorgeschichte. In H. Windl (ed.), *Rätsel um Gewalt und Tod vor 7.000 Jahren: Eine Spurensicheurng*. Ausstellung im Museum für Urgeschichte Asparn a. d. Zaya. Katalog des NÖ Landesmuseums, N.F. 393. Asparn a. d. Zaya, 1996:7-45.

WINDL, H.J. 1999 – Makaberes Ende einer Kultur? *Archäologie in Deutschland* 1/99, 1999:54-57.

WINIGER, J. 1998 – *Ethnoarchaeologische Studien zum Neolithikum Südwesteuropas*. BAR S701, BAR Publishing, Oxford, 1998.

WIŚLAŃSKI, T.S. 1966 – *Kultura amfor kulistych w. Polsce północno-zachodniej*. Zakład Narodowy Imienia Ossolińskich Wydawnictwo Polskiej Akademii Nauk, Wroclaw, Warszawa, Krakow, 1966.

WŁODARCZAK, P. 2008 – Corded Ware and Baden Cultures: Outline of Chronological and Genetic Relations Based on Finds from Western Little Poland. In M. Furholt *et al.* (eds.), 2008:247-261.

ZÁPOTOCKÝ, M. 1991 – Michelsberg und die relative Chronologie des Frühen und Älteren Äneolithikums in Böhmen. In Jankowska (ed.), 1991:127-131.

ZÁPOTOCKÝ, M. 1998 – Zum Forschungsstand über die relative Chronologie des frühen Äneolithikums in Böhmen. In Biel *et al.* (eds.), 1998:303-302.

ZÁPOTOCKÝ, M. 2000 – Keramika s Brázděným Vpichem a synchronizace Čech, Morav a Středního Poduynají ve Starším Eneolitu. *Archeologické Rozhledy* 52, 2000:595-622.

ZASTAWNY A. 2008 – The Baden and the Funnel Beaker-Baden Settlement in Lesser Poland. In M. Furholt *et al.* (eds.), 2008:177-188.

ZEDER, M.A. 2008 – Domestication and Early Agriculture in the Mediterranean Basin: Origins, Diffusion, and Impact. *PNAS* 105/33, 2008:11597-11604 (www.pnas.org_cgi_doi_10.1073_pnas.0801317105).

ZILHÃO, J. 2000 – From the Mesolithic to the Neolithic in the Iberian Peninsula. In Price (ed.), 2000:19-56.

ZVELEBIL, M. 2004 – The Many Origins of the LBK. In Lukes and Zvelebil (eds.), 2004:183-205.

ZVELEBIL, M. 2005 – Homo Habitus: Agency, Structure and the Transformation of Tradition in Constitution of the TRB foraging-farming communities in the North European Plain (ca. 4500-2000 BC). *Documenta Praehistorica* 32, 2005:87-101.

ZVELEBIL, M., R. DENNELL and L. DOMAŃSKA (eds.) 1998 – *Harvesting the Sea, Farming the Forest: The Emergence of Neolithic Societies in the Baltic Region.* Sheffield Archaeological Monographs 10. Sheffield Academic Press, Ltd. Sheffield, England, 1998.

ZVELEBIL, M., and M. LILLIE 2000 – Transition to Agriculture in Eastern Europe. In Price (ed.), 2000:57-92.

Chapter 17
CULTURAL GEOGRAPHY IN THE CONTEXT OF CLIMATIC AND ENVIRONMENTAL CHANGE IN THE LATE NEOLITHIC AND ENEOLITHIC OF THE MORAVA VALLEY

Matthew T. BOULANGER
Archaeometry Laboratory, University of Missouri Research Reactor, Columbia, MO, USA

Abstract: *Geographic, climatic, and archaeological data provide contextual information to complement theories on the emergence of monumental architecture and the development of prehistoric social complexity in Central Europe. Environmental, climatic and archaeological data are analyzed at a variety of scales. Spatial relationships between monumental and non-monumental sites, which were created during the transitions from the Late Lengyel, the Funnel Beaker (TRB) and to the Early Baden interaction spheres in Central Moravia, Czech Republic, are examined. It is concluded that Neolithic and Eneolithic (Copper Age) cultures behaved in an organic fashion, evolving with and responding to, a dynamic environment.*

INTRODUCTION

Comparative archaeology requires environmental, climatic and archaeological datasets to generate a broader contextual understanding of the archaeological record. This chapter focuses on changes in site types and site locations during the dynamic climate and paleoenvironment of the Morava River Valley in Central Moravia, Czech Republic (Fig. 17.1) between ca. 5000 and 3000 BC, i.e. during the Neolithic and the Copper Age (Eneolithic).

Data describing site locations and types are gathered from sources, which include survey maps, aerial photographs, publications, and coordinates from hand-held Global Positioning Satellite Systems (GPS).[1] Environmental data incorporating topography, soil types, and river systems are combined with archaeological site data and analyzed with Geographic Information Software (GIS). Additional data are drawn from artifact typologies, radiocarbon dates, and climate proxies derived from ice core and tree-ring data.

CULTURAL CONTEXT OF THE MORAVA VALLEY

During the 5th and 4th millennium BC, the Morava valley is occupied primarily by the Lengyel, the Funnel Beaker (TRB) and the Boleráz Phase of the Baden interaction spheres (Baldia 2010, Baldia, Boulanger and Frink 2008, Baldia, Frink and Boulanger 2008). Figure 17.2 represents a simplified chronology. Detailed chronological analysis and references are presented in Baldia (this volume). Essentially the interregional Lengyel phases (I-IV) are locally known as Moravian Painted Ware (MPW) Phases 0, I, IIa, IIb and IIc or Epi-Lengyel (e.g. Podborský 1993:111, 1999, Čižmář *et al.* 2004).[2] The TRB is divided into Phase I and II, as well as several possible subphases (Baldia, this volume). TRB II shows signs of increasingly closer interaction with the Boleráz Phase of the Baden interaction sphere (ibid.).

Based on the appearance of the first copper artifacts in Moravia, MPW IIa marks the end of the Moravian Neolithic and the beginning of the Eneolithic or Copper Age (Podborský 1993). These artifacts are likely imported from adjacent regions, because there is little to no evidence for local production of copper implements during the Moravian MPW (Podborský 1993, Buchvaldek *et al.* 1988). On the other hand, Čižmář *et al.* (2004) propose that the Czech and Slovak Eneolithic does not start until Epi-Lengyel (Lengyel IV), which they consider a break from Lengyel proper. Nevertheless, we continue to use the first appearance of copper as the transition to the Eneolithic (Fig. 17.2).

During the Neolithic, the Lengyel interaction sphere of Moravia is represented by monumental architecture consisting of Danubian longhouses and geometric ditched and causewayed earthworks (*rondels*), while the Copper age TRB interaction sphere introduces earthworks with non-geometric stone walls and ditches, as well as long-barrows (Baldia 2010). The emergence of monumental architecture is considered to be a diagnostic trait of the Neolithic in Central Europe. Podborský (1993:287-288) provides a concise English-language summary of various ideas about the function of these earthworks as expressed over the past 100 years. Without engaging the long-standing debate over the function of monumental sites, we join Oliva (2004) in his observation that monumental architecture reflects cooperative efforts arising from ever-increasing cultural complexity, regardless of its function or intent.

[1] We thank Miroslav Šmíd, Head of the Prostějov branch of the Moravian Cultural Resource Management Office, for liberally sharing his knowledge on the location of archaeological sites in his district.

[2] Lengyel Phase 0 (Formative Lengyel) may occur outside of Moravia according to Baldia (this volume).

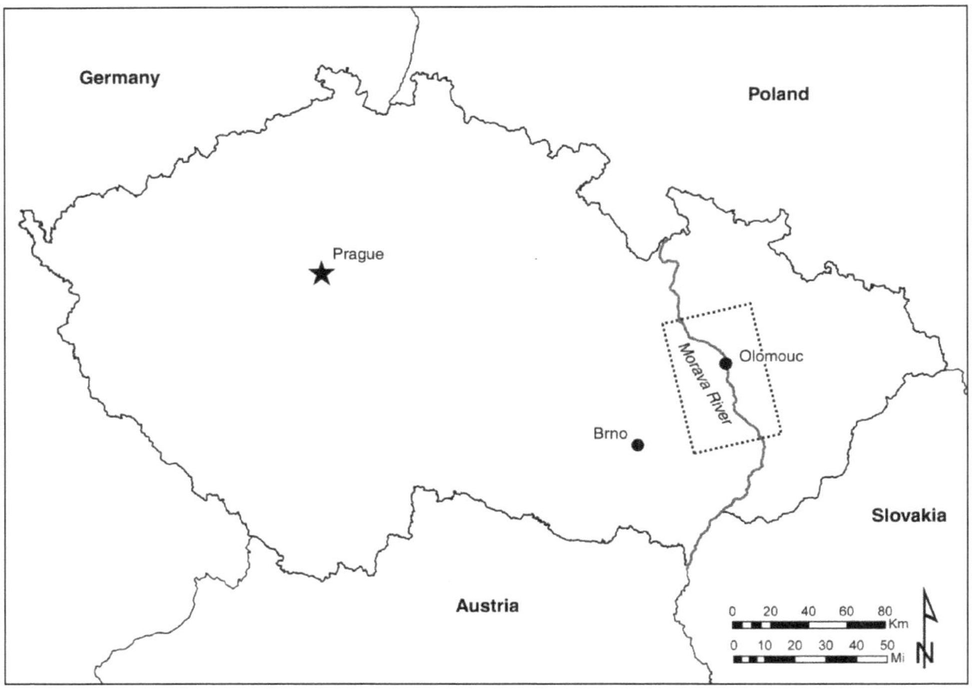

Figure 17.1. Map of the Czech Republic showing major cities, the Morava River, and the study area

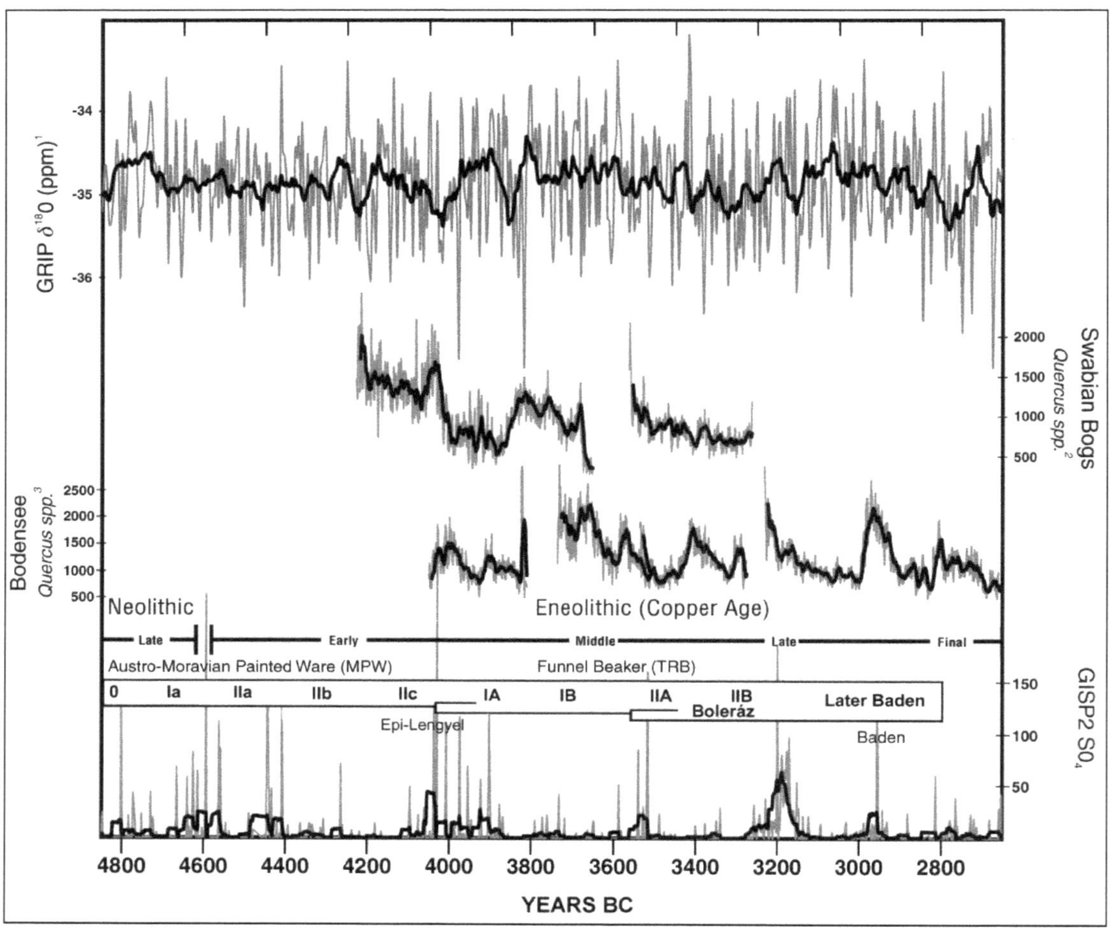

Figure 17.2. Lengyel, Funnel Beaker and Baden chronology in the upper Morava valley and climate proxies. Light grey lines represent raw data, while dark black lines indicate a smoothed 10-point average. 1: $\delta^{18}O$ record from the GRIP ice core (Cross 2003, Dansgaard *et al.* 1989, 1993, Grootes *et al.* 1993, Johnsen *et al.* 1997); 2: Upper Swabian Bogs Quercus spp chronology (Billamboz n.d. b); 3: Lake Bodensee Quercus spp. chronology (Billamboz n.d. a); 4: SO_4 (volcanic sulfate) record from the GISP2 ice core (Cross 2003, Zielinski *et al.* 1994, 1997)

PROCEDURES

The sites in the spatial analysis are classified into four categories based on the presence or absence of monumental architecture. Sites containing remnants of habitation are divided into two groups: *Wall and Ditch Enclosures* (earthworks that possess monumental architecture) and *Settlements* (those that do not). Mound sites containing human burials or cremations are classified as *Mound Groups* (clusters of burials with monumental architecture) and *Burials* (simple graves without evidence of elaborate architecture). Although relatively simple, this classification system is consistent with how the majority of sites are reported within the region, therefore allowing the maximum number of site locations to be drawn from reports, articles, and manuscripts.

Sites are plotted over soil units defined using the standard FAO/UNESCO soil classification scheme (Němeček *et al.* 2001). Pedological descriptions of these soils are examined, and each soil type is assigned a general physiographic description based upon topography, soil taxonomy (Buol *et al.* 1980, Driessen and Deckers 2001), and past geomorphic studies (Frink and Hathaway 1999, Havlíček 1988, 1994, Růžičková and Zeman 1994).

Fluvial soils, including Fluvisols, Histosols and Gleysols, represent *Valley Floor* soils. Loess-derived soils such as Chernozems may have undergone a boggy period during their initial formation, and are classified as *Steppe* soils. Some Phaeozems and Greyzems have pedogenic characteristics of both *Valley Floor* and *Steppe* soils, and therefore are considered to represent intergraded soils found between the two physiographic zones. Well-drained Luvisols at the transition from valley to highland environments represent *Valley Wall* soils. Weathered soils with limited pedogenic development, such as Cambisols, represent *Highland* soils (Fig. 17.3).

Climate data are examined for correlations with the Neolithic and Copper Age in Central Europe. The data include oxygen-isotope ratios ($\delta^{18}O$ or $\Delta^{18}O$), volcanic-sulfate (SO_4) concentrations, and ice-accumulation rates from the Greenland summit ice core projects (Cross 2003, Zielinski *et al.* 1994) as well as tree-ring records from Lake Constance (Bodensee) and nearby Upper Swabian (Schwaben) Bogs in Germany (Billamboz n.d. a, b).

RESULTS

The climate in Central Europe is primarily governed by the North Atlantic Oscillation. Referred to as the NAO, this is a punctuated oscillation of atmospheric pressure over Iceland and the Azores. A high NAO indicates high pressure over the Azores and low pressure over Greenland, and results in warm-westerly winds across Central Europe. Conversely a low NAO, with high pressure over Iceland and low pressure over the Azores, permits Arctic winds to push across Europe creating bitterly cold winters. Oscillations of the NAO account for much of the winter temperature and summer precipitation variability in Central Europe (cf. Fagan 2000). Oxygen-isotope ratios ($\delta^{18}O$), measured from the Greenland ice cores, provide one method of identifying fluctuations of temperature and precipitation (Mayewski and White 2002). Examination of the $\delta^{18}O$ curve is therefore one way that climatologists can reconstruct the NAO, linking the ice-core data of Greenland with the climate record of Central Europe (Luterbacher *et al.* 2002, Schulz and Paul 2002). A comparison of the $\delta^{18}O$ record to several European climate records suggests that between approximately 4650 and 2050 BC, negative $\delta^{18}O$ values (colder temperatures in Greenland) correlate with generally warmer summer temperatures and increased rainfall in Central Europe (Schulz and Paul 2002).

During the MPW habitation of the upper Morava Valley, changes in $\delta^{18}O$ are relatively moderate and characterized by regularly occurring small peaks (Fig. 17.2). Increased SO_4 is evident from 4650-4550 BC, and again around 4450 BC, and corresponds well with downward trends in the $\delta^{18}O$ curve, suggesting that these SO_4-producing volcanic events had a measurable effect on global climate patterns.

A change in the $\delta^{18}O$ curve is evident at approximately 4300 BC. The relatively steady isotope curve assumes a new pattern, characterized by extreme high and low peaks. Between 4300-3850 BC, three downward spikes exceeding two standard deviations are evident in the $\delta^{18}O$ curve. These three spikes each last between 50 and 100 years, suggesting sudden and pronounced changes in the climate. The period around 4050 BC correlates with increased rainfall in Central Europe (Schettler *et al.* 1999, Schulz and Paul 2002) and glacial advance in Scandinavia (Dahl and Nesje 1996). This period also correlates with increased volcanic sulfate levels between 4050-3850 BC (Zielinski *et al.* 1994).

The Bodensee and Upper Swabian Bogs' oak chronologies produced by Billamboz (n.d. a, b) provide additional lines of evidence to infer the effects of mid-Holocene climatic variability in Central Europe. The Bodensee chronology indicates increased oak growth at approximately 4250 BC, and pronounced downward spikes in the Greenland $\delta^{18}O$ curve are also evident at this time. A hypothesis of negative $\delta^{18}O$ values correlating with generally warmer summers and increased rainfall in Northern and Central Europe as indicated by increased oak growth, fits well with Schulz and Paul's (2002) conclusions.

The $\delta^{18}O$ curve changes phase at approximately 3750 BC, returning to a pattern characterized by above-average values with few extreme peaks. Approximately 200 years later (3550 BC), the $\delta^{18}O$ curve exhibits a roughly 400-year span of below-average values characterized by sudden and extreme increases and decreases. This period of below-average isotope levels ends at approximately 3200 BC with a sudden increase in isotope concentrations followed promptly by an immediate (and extreme) decrease in values. The erratic spike at 3200 BC

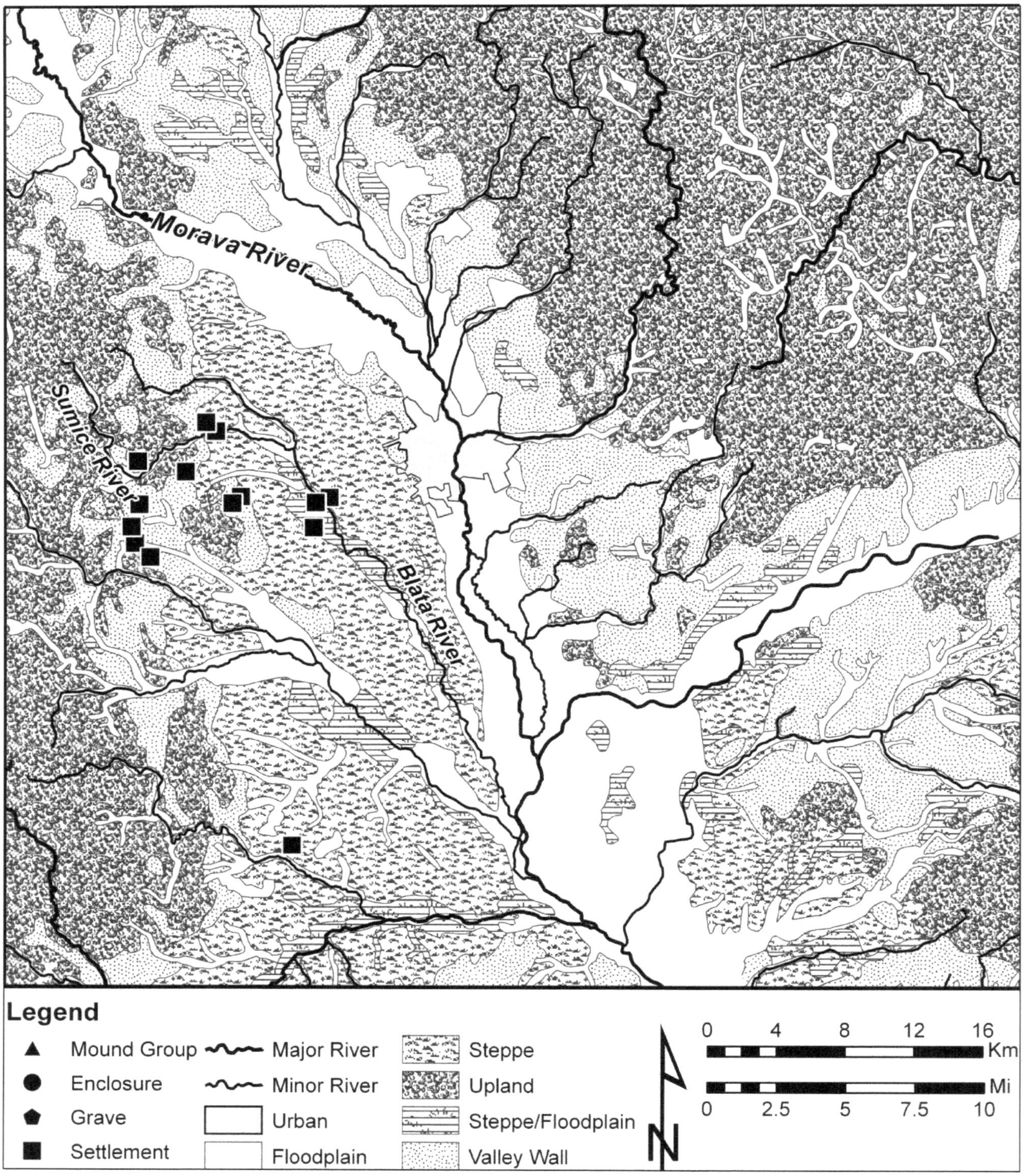

Figure 17.3: Locations of Austro-Moravian Painted Ware (MPW) era sites within the central-northern Morava Valley near Olomouc, Czech Republic

coincides with an increase in SO_4 associated with pronounced warming over Greenland (Zielinski *et al.* 1994).

Environment

Geomorphological studies suggest that the Morava River has actively evolved towards its current course. During the early and middle Holocene, the Morava and its tributaries are characterized as actively shifting with significant erosion and deposition (Havlíček 1988). Within the research area, the Morava did not assume its current course until at least the middle to late Holocene (ca. 2300 BC) based on dates from Břeclav-Poštorná and Veselí nad Moravou published in Havlíček (1994), recalibrated using OxCal v.3.10 and Reimer *et al.* (2004).

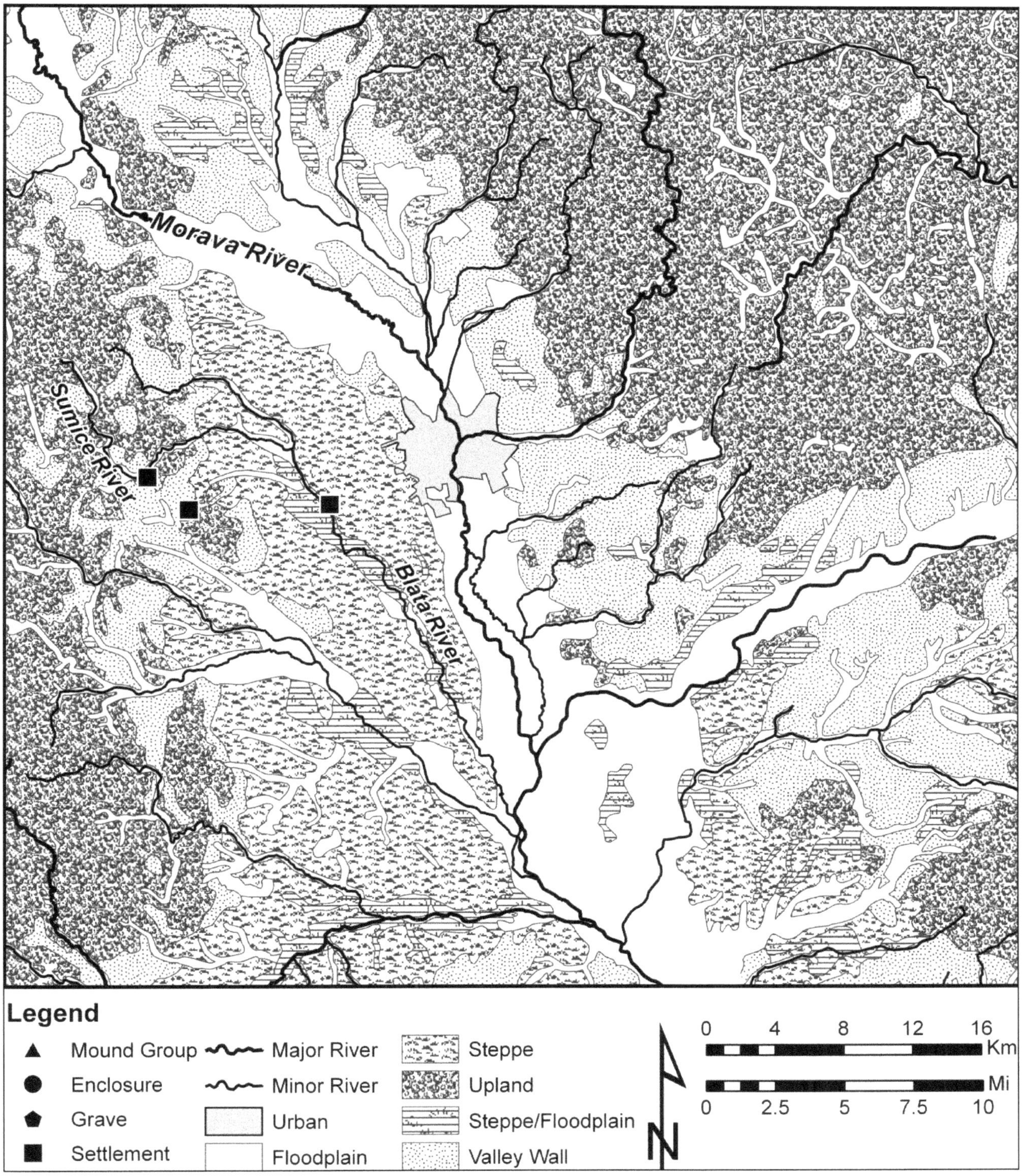

Figure 17.4. Locations of Epi-Lengyel period sites within the central-northern Morava Valley near Olomouc, Czech Republic

Morphological attributes of soils within the Morava valley further suggest that the valley has undergone significant changes since the early Holocene. Currently, the Morava is flanked by a wide floodplain of valley floor soils, reflecting middle to late Holocene fluvial action. The western side of the Morava valley is relatively flat and gently slopes toward the Bohemian Massif highlands. Steppe soils cover much of this side of the valley, extending to the valley's edge. This relatively flat steppe environment appears to have been part of the Morava's floodplain at some time during the early to middle Holocene. In contrast, the eastern side of the Morava valley exhibits a sudden-upward slope and contains predominantly well-developed and geologically stable soils. This side of the valley is characterized by an abrupt rise in elevation and contact between fluvial valley floor

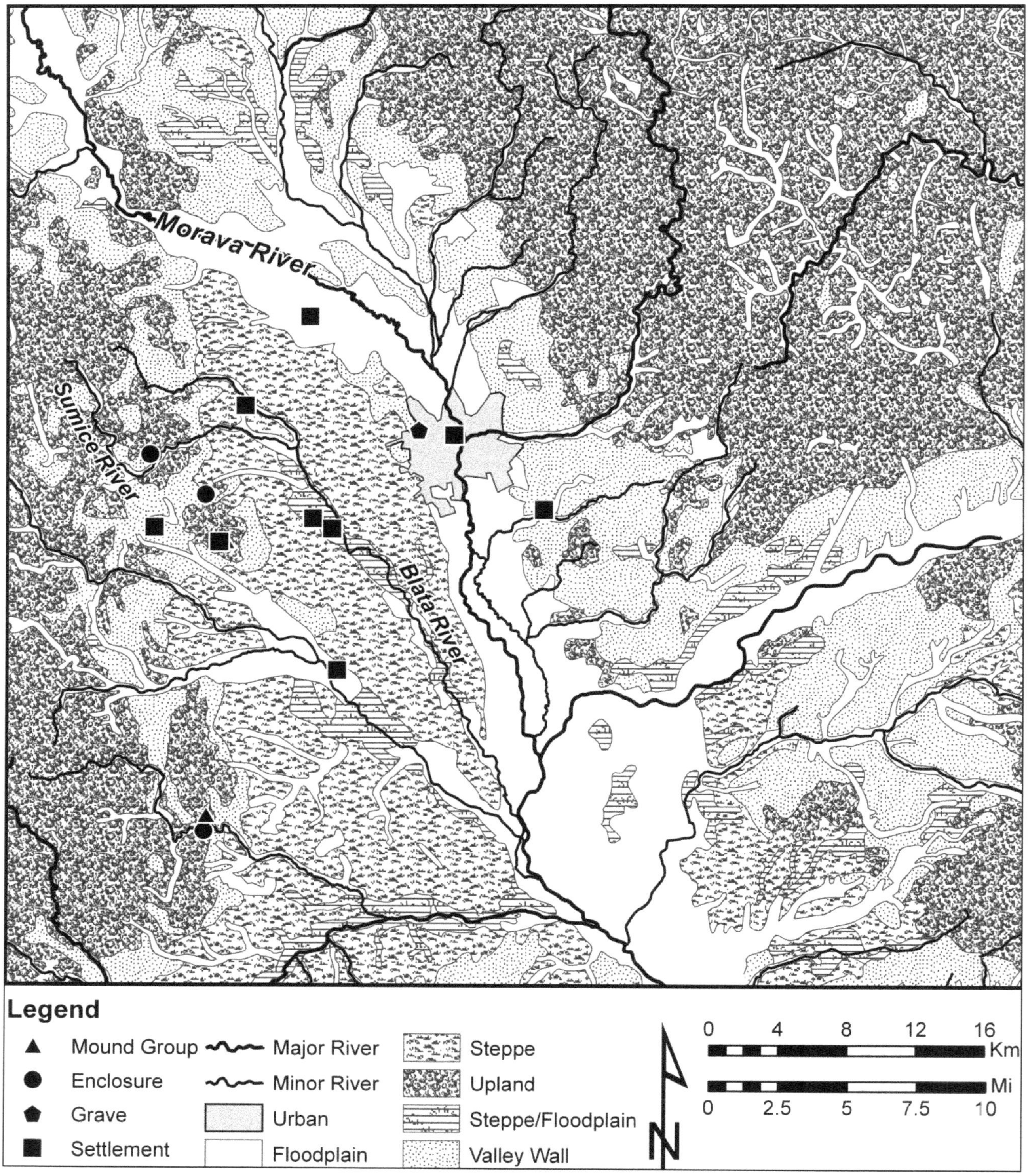

Figure 17.5. Locations of Funnel-Beaker Phase I (TRB I) sites within the central-northern Morava Valley near Olomouc, Czech Republic

soils and valley wall soils, suggesting that the Morava has meandered to its current position from the western, rather than the eastern, side of the valley.

Culture

The Lengyel-related MPW is well established within the upper Morava Valley by at least 4750 BC. MPW settlements are located predominantly on the edges of the river valley in the valley wall and steppe zones (Fig. 17.3). Settlements are organized into small clusters consisting of up to four sites spaced 1-1.5 km apart. Settlement sites located in the steppe are adjacent to the Blata and Šumice rivers, and sites in the valley wall are adjacent to the headwaters of intermittent drainages. Numerous MPW-period geometric enclosures (*rondels*)

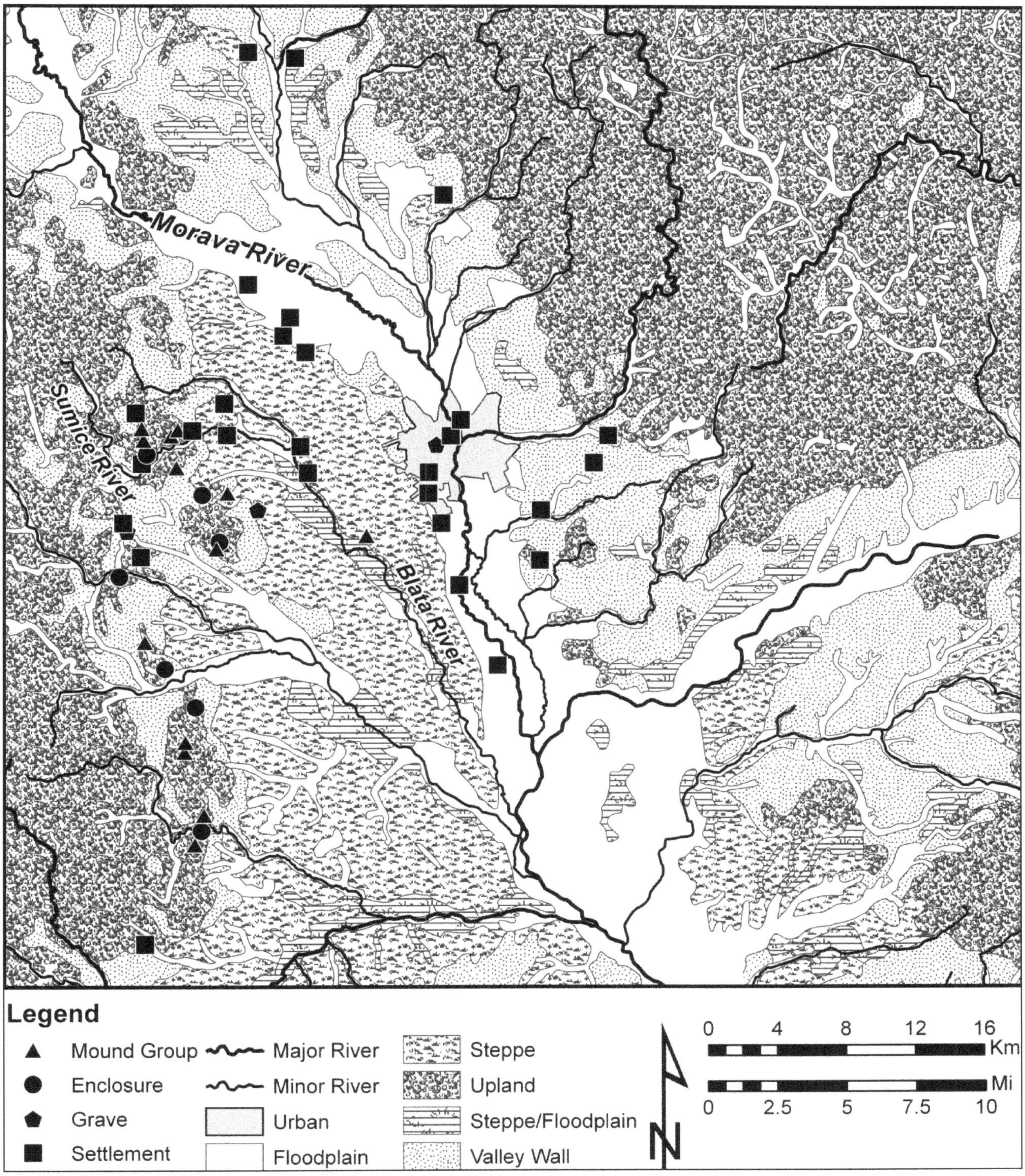

Figure 17.6. Locations of Funnel-Beaker Phase II (TRB II) sites within the northern-central Morava Valley near Olomouc, Czech Republic

have been identified in southern Moravia (Podborský 1999, Whittle 1988), but none have been identified within the study area.

There are only three Epi-Lengyel sites within the study area and all are classified as settlements (Fig. 17.4). The poor representation may be the result of reduced visibility in the archaeological record, although survey bias and poor definition of the cultural complex may also be contributing factors. The decline of highly decorated ceramics and identified MPW-period sites coincides with the dramatic changes in the $\delta^{18}O$ curve and oak growth chronologies discussed above. This suggests that climate changes sufficient to affect oak growth in Germany may have also affected cultural organization in Moravia.

Ceramics attributed to the TRB begin to appear in the Morava Valley ca. 4000 BC. TRB I settlements are situated in the valley wall, steppe, and valley floor zones (Fig. 17.5). TRB I settlement sites are also located further into the valley than those of previous cultures. As with MPW-era settlements, some of the TRB I settlements are located adjacent to the Blata and Šumice rivers. A few TRB I settlements have also been identified near the modern course of the Morava River, suggesting that the river may have assumed portions of its present course by this time.

Distances between TRB I settlements are approximately 5 km on average, which is notably greater than distances between Lengyel culture sites. Former locations of MPW settlement clusters contain single and larger TRB I settlements. TRB I enclosures are located at strategic locations in the highlands overlooking secondary river valleys. Virtually all mound groups are also located in the highlands, suggesting a functional and ideological partitioning of the landscape.

TRB II sites (Fig. 17.6), like those of the TRB I, are located in the valley wall, steppe, and valley floor zones. TRB II sites extend lower into the valley system, and form distinct chains of sites on terraces overlooking the Blata, Šumice, and Morava rivers. This suggests that these rivers were continuing to assume their modern courses. TRB II sites are 1-3 km from their nearest neighbors. The number of monumental sites increases dramatically during the TRB II. Wall and ditch enclosures and mound groups are located almost exclusively in the highlands. Enclosures, like those of the TRB I, are strategically located on prominent hills in the highlands overlooking moderate-sized drainages in the steppe. Ten of the 12 TRB II mound groups in the study area are located within 2 km of a wall and ditch enclosure, and all are located within 6 km of such an enclosure.

DISCUSSION

Distributions of Neolithic and Eneolithic sites change through time in Central Moravia, suggesting among other things a cultural response to environmental change. Furthermore, periods of cultural change within the Morava Valley coincide with climatic fluctuations evident in both the Greenland ice-core record and Central European oak-growth records. Archaeological investigations conducted by the Czech-American Research Project (CARPro) provide one method of examining how the proposed culture-climate-landscape interrelationships can be applied to our interpretations of the archaeological record.

CARPro's investigations focus on the Eneolithic wall-and-ditch enclosed sites of Rmíz by Laškov and Hrad u Bílovice, the burial mounds of Ludéřov and Džbán, both near Náměšť na Hané, as well as the surrounding landscape (Baldia 2010, Baldia *et al.* 2001, Baldia, Frink and Boulanger 2008, Baldia, Boulanger and Frink 2008).

Rmíz is considered the oldest and largest hilltop enclosure in a continuous chain of similar sites extending along the Morava River Valley highlands from Olomouc to Brno. Four walls with corresponding ditches have been identified surrounding the 17.5 hectare (43.2 acre) site. Of these walls, one consists of two dry-wall stone structures. This appears to be the oldest stone wall construction, rivaled only by the stone-faced retaining wall at Hlinsko, some 40 km east of Rmíz. The stone enclosure of Hrad u. Bílovice, located ca. 7.3 km (4.5 Miles) south south-west of Rmíz is associated with TRB II artifacts.

Few MPW-period sites are known within CARPro's research area. The closest MPW site is located immediately below Rmíz (Šmíd 1994:203, Fig. 17.2), this is the only known site within 3 km. Scattered finds of Epi-Lengyel ceramics have been reported from Rmíz, but the first clearly definable occupation of the site occurs at approximately 4000 BC, and deemed to be attributable to a very early part of TRB I.

The initial TRB I habitation at Rmíz occurs during a prolonged period of climatic fluctuation identified from the $\delta^{18}O$ and SO_4 records as well as the German oak chronologies. The 400-year span from 4250-3850 BC has been identified as generally colder and wetter in Central Europe (Schulz and Paul 2002) and falls into the Epi-Lengyel/TRB transition. Based on the limited test excavations conducted by CARPro, the first episode of wall construction at Rmíz occurred during the TRB I (Frink, this volume, 1999, 2011, Frink and Dorn 2002, Šmíd 1994). Sub-megalithic tombs appear in the research area during the later part of TRB I. These rare TRB I burial mounds and stone cists are known to occur only near the enclosed TRB II site of Slatinki, some 7 km (4 Miles) to the south east (Šmíd 1993).

CARPro excavations suggest a brief abandonment near the end of TRB I for reasons not yet understood (Frink and Dorn 2002). However, the volcanic SO_4, the oscillations in the Bodensee oak curve, and the gap in the Swabian oak curve hint at climatic changes between ca. 3650 and 3450 BC (Fig. 17.2, No. 2, 3, 4 respectively). These and other events coincide with transition from TRB I to TRB II (Baldia, this volume), suggesting a correlation with culture change.

Rmíz was reoccupied during the TRB II period and the earthen ramparts at the site were refurbished and expanded (Šmíd 1994, M. Baldia 2004, Frink and Dorn 2002). This reoccupation corresponds with a prolonged period of colder-than-average temperatures in Central Europe indicated by the previously mentioned decreased $\delta^{18}O$ concentrations and below-average winter temperatures. Unlike the noted absence of nearby tombs during TRB I, at least five TRB II mound groups (each group containing over 30 individual mounds) have been identified less than 2 km from Rmíz.

Although Rmíz is located in the highlands, evidence from the site and nearby tombs implies cultural ties with settlements in the valley floor. Linen textiles and einkorn-wheat kernels recovered from a cremation burial at the Křemela mound groups, only 0.5 km north of Rmíz (C. Baldia 2004) attest to the importance of agriculture in life and death. Crop-growing likely took place at settlements on the valley floor, but production of textiles and storage of grain are evident at Rmíz in the form of loom weights, spindle whorls, and large thick-walled amphorae. Additional evidence of food processing and local trade is manifested through ground stone querns. These querns are macroscopically identical to those found at the wall-and-ditch enclosure of Hrad u Bílovice, where quarrying and quern production has been identified (Baldia, Frink and Boulanger 2008, Frink 2002, Staeck *et al.* 2004).

In addition to local trade and/or the sharing of resources, Rmíz also appears to have taken part in a regional exchange network that reached beyond the Morava Valley. Evidence of copper smelting is identified at the TRB II settlement of Laškov located immediately below Rmíz (Šmíd 1996). Copper artifacts have been recovered from Rmíz and from several of the mound groups (Baldia, Boulanger and Frink 2008). Additionally, tools made from Polish and German flint have been recovered from TRB II levels at Rmíz.

Evidence for local and regional trade found at Rmíz suggests that during the second phase of the TRB the site was part of an extensive exchange and communication networks, which connected settlements within the valley and spanned Central Europe. The strategic locations of these enclosures (in the highlands) were likely advantageous for the control of exchange and communication (M. Baldia 2004). Given the monumental architecture of Rmíz and nearby long-barrows, it seems likely that the site served an important cultural role during TRB II in Moravia.

During the TRB, the Central European climate was generally colder and wetter than it had been during the preceding Lengyel period. As noted above, the Morava River did not assume its modern course until well after the Eneolithic period (Havlíček 1988, 1994), and it was actively shifting courses throughout much of the early to middle Holocene. Arable land, specifically the gently sloping steppe region and the valley floor, was likely used for agriculture and browse for herd animals. The uplands may have been forested, supplying wood for waddle and daub house construction, palisades etc. as well as fire for cooking, leaf fodder for cattle, and forage for pigs, not to mention the fuel for known pottery and probable copper smelting kilns.

The TRB occupation of Rmíz correlates well with the aforementioned climatic fluctuations, specifically to periods of colder winters and increased rainfall. The appearance of wall and ditch enclosed sites such as Rmíz may in part be a cultural adjustment to fluctuating climate and environment. However, climatic and environmental changes are not the sole causes of the widespread cultural changes that occurred during the Neolithic and Eneolithic of the region. Archaeological data from Rmíz indicates that the site was a large and regionally significant cultural center, not solely a refuge from cold winters or wet floodplains. The still modest amount of archaeological information available from Rmíz (and similar sites) limits conclusions about those cultural factors that influenced people's decisions. However, by incorporating the dynamic nature of past environments and climates with the dynamic archaeological record, archaeologists are able to better contextualize the cultural phenomena of the Central European Neolithic and Eneolithic.

References

BALDIA, C. 2004 – The Oldest Woven Textile of the Funnel Beaker Culture (4000-2900 cal BC) in North and Central Europe. In I. Jadin, A. Hauzeur, N. Cauwe, M. Van der Linde, T. Onhan and M. Lebeau (ed.) *Acts of the XIVth UISPP Congress, University of Liege, Belgium, 2-8 September 2001, Section 9: Le Neolithique au Proche Orient en Europe/The Neolithic in the Near East and Europe; Section 10: L'age du Cuivre au Proche Orient et en Europe/The Copper Age in the Near East and Europe; Sessions Generales et posters/General Sessions and Posters,* British Archaeological Reports International Series 1303, BAR Publishing, Oxford, 2004:67-70.

BALDIA, M.O. 2002 – The Central and North European Neolithic/Copper Age Chronology, With Emphasis on the Funnel Beaker Culture (TRB). Revised February 4, 2009, The Comparative Archaeology Web (www.comp-archaeology.org).

BALDIA, M.O. 2004 – The Oldest Stone Rampart: Enclosures and Megalithic Tombs of the Funnel Beaker Culture (4100-2800 cal BC) in North and Central Europe. In I. Jadin, A. Hauzeur, N. Cauwe, M. Van der Linde, T. Onhan and M. Lebeau (ed.), *Acts of the XIVth UISPP Congress, University of Liege, Belgium, 2-8 September 2001, Section 9: Le neolithique au Proche Orient en Europe/The Neolithic in the Near East and Europe; Section 10: L'age du cuivre au Proche Orient et en Europe/The Copper Age in the Near East and Europe; Sessions generales et posters/General Sessions and Posters,* British Archaeological Reports International Series 1303, BAR Publishing, Oxford, 2004:153-161.

BALDIA, M.O. 2010 – Monumental Questions: Prehistoric Megaliths, Mounds and Enclosures of Central and Northern Europe. In D. Calado, M.O. Baldia, M. Boulanger (eds.), *Monumental Questions: Prehistoric Megaliths, Mounds and Enclosures,* Section 68 (*Part II*), Actes du XV Congrès Mondial de l'Union Internationale des Sciences Préhistoriques et Protohistoriques 8, Lisbon 2006, BAR S2123. BAR Publishing, Oxford, 2010:195-212.

BALDIA, M.O., C. Baldia and D. Frink 1998 – *Czech-American Rmíz Research Program: 1998 Activity Report.* (http://www.comp-archaeology.org/RMIZ1998REPORT.HTM) Version 2.4 (January 14, 2007 Update).

BALDIA, M.O., M.T. BOULANGER and D.S. FRINK 2008 – The Earthen Long-Barrow of Džbán, Moravia, Czech Republic and its Implications for the Interaction Between the Nordic Funnel Beaker and the Southern Baden Culture. In M. Furholt, M. Szmyt and A. Zastawny (eds.) in coop. with E. Schalk, *The Baden Complex and the Outside World.* Proceedings of the 12th Annual Meeting of the EAA in Cracow, 19th-24th September 2006. Studien zur Archäologie in Ostmitteleuropa, Habelt, Bonn. 2008:263-287.

BALDIA, M.O., D.S. FRINK and M.T. BOULANGER 2008 – Problems in the Archaeological Legacy: The TRB/Lengyel-Baden Conundrum. In M. Furholt, M. Szmyt and A. Zastawny (eds.) in coop. with E. Schalk, *The Baden Complex and the Outside World.* Proceedings of the 12th Annual Meeting of the EAA in Cracow, 19th-24th September 2006. Studien zur Archäologie in Ostmitteleuropa, Habelt, Bonn. 2008:25-48.

BALDIA, M.O., J.P. STAECK and M. ŠMÍD 2001 – Hügelgräber der Trichterbecherkultur in Luděřov. *Pravěk*, Nová řada 11, 2001:43-59.

BILLAMBOZ, A. n.d. a – *Bodensee 1-6 Archaeological Tree Ring Data. Unpublished Archaeological Tree Ring Data.* International Tree-Ring Data Bank, IGBP PAGES/World Data Center for Paleoclimatology, NOAA/NGDC Paleoclimatology Program, Boulder, Colorado, USA (ftp://ftp.ncdc.noaa.gov/pub/data/paleo/treering/chronologies/europe/) Files: germ021.crn, germ022.crn, germ023.crn, germ024.crn, germ025.crn, germ026.crn, germ032.crn.

BILLAMBOZ, A. n.d. b – *Upper Suevia Bog 1-5 Archaeological Tree Ring Data.* Unpublished Archaeological Tree Ring Data. International Tree-Ring Data Bank, IGBP PAGES/World Data Center for Paleoclimatology, NOAA/NGDC Paleoclimatology Program, Boulder, Colorado, USA (ftp://ftp.ncdc.noaa.gov/pub/data/paleo/treering/chronologies/europe/) Files: germ027.crn, germ028.crn, germ029.crn, germ030.crn, germ031.crn.

BUOL, S.W., F.D. HOLE and R.J. McCRACKEN 1980 – *Soil Genesis and Classification.* The Iowa State University Press, Ames, Iowa, 1980.

BUCHVALDEK, M., B. NOVOTNY and E. PLESLOVA-STIKOVA 1988 – The Copper Age in Czechoslovakia. *Rassegna di Archeologia* 7, 1988:105-142.

ČIŽMÁŘ, Z., J. PAVÚK, P. PROCHÁZKOVÁ and M. ŠMÍD 2004 – K problému definováni finálního stádia lengyelské kultury (New observations on problems in the relationship between the Stroked Pottery and Lengyel cultures). In B. Hänsel and E. Studeníková (eds.), *Zwischen Karpaten und Ägäis: Neolithikum und ältere Bronzezeit; Gedenkschrift für Viera Němejcová Pavúková.* Internationale Archäologie. Studia honoraria 21. Rahden, Westf. VML, 2004:207-232.

CROSS, M. (ed.) 2003 – *Greenland Summit Ice Cores,* CD-ROM, updated January 2003. National Snow and Ice Data Center, World Data Center for Paleoclimatology at NOAA-NGDC and the Institute of Arctic and Alpine Research, 2003.

DAHL, S.O., and A. NESJE 1996 – A new Approach to Calculating Holocene Winter Precipitation by Combining Glacier Equilibrium-Line Altitudes and Pine-Tree Limits: A Case Study from Hardangerjokulen, Central Southern Norway. *The Holocene* 6, 1996:381-398.

DANSGAARD, W., S.J. JOHNSEN, H.B. CLAUSEN, D. DAHL-JENSEN, N.S. GUNDESTRUP, C.U. HAMMER, C.S. HVIDBERG, J.P. STEFFENSEN, A.E. SVEINBJÖRNSDOTTIR, J. JOUZEL and G.C. BOND 1993 – Evidence for General Instability of Past Climate from a 250 Kyr Ice-Core Record. *Nature* 264, 1993:218-220.

DANSGAARD, W., J.W.C. WHITE and S.J. JOHNSEN 1989 – The Abrupt Termination of the Younger Dryas Climate Event. *Nature* 339, 1989:532-533.

De CAPITANI, A., S. DESCHLER-ERB, U. LEUZINGER, E. MARTI-GRÄDEL and J. SCHIBLER 2002 – *Die Jungsteinzeitliche Seeufersiedlung Arbon/Bleiche 3: Funde.* Archäologie im Thurgau 11, Departement für Erziehung und Kultur des Kantons Thurgau, Frauenfeld, 2002.

FAGAN, B. 2000 – The Little Ice Age: How Climate Made History. Basic Books, New York, 2000.

DRIESSEN, P., and J. DECKERS (ed.) 2001 – *Food and Agricultural Organization of the United Nations (FAO).* Lecture Notes on the Major Soils of the World. FAO, Rome, 2001.

FRINK, D. 1999 – The OCR Carbon Dating Procedure and its Application at the Rmíz Site, Czech Republic. Paper presented at the symposium entitled "*Prehistoric Technology and its Social Implications: New Theories and Methods*" at the 64th Annual Meeting of the Society of American Archaeology, Illinois, Chicago, 1999.

FRINK, D. 2002 – *OCR Analyses at Hrad Bílovice.* OCR Carbon Dating, Inc. Submitted to the Czech-American Research Program. 2002.

FRINK, D. 2011 – Explorations into a Dynamic Process-Oriented Soil Science. Waltham, MA, Elsevier Press, 2011.

FRINK, D., and R. DORN 2002 – Beyond Taphonomy: Pedogenic Transformations of the Archaeological Record in Monumental Earthworks. *Journal of the Arizona-Nevada Academy of Science* 34/1, 2002:24-44.

FRINK, D., and A. HATHAWAY 1999 – *Pedomorphic understanding* of *river systems' behavior through space* and *time*: *key* to *managing cultural resources.* Paper presented at the 64th Annual Meeting of the Society for American Archaeology, Chicago, 1999.

GROOTES, P.M., M. STUIVER, J.W.C. WHITE, S.J. JOHNSEN and J. JOUZEL 1993 – Comparison of Oxygen Isotope Records from the GISP2 and GRIP Greenland Ice Cores. *Nature* 366, 1993:552-554.

HAVLÍČEK, P. 1988 – Evolution of the Morava River in Late Pleistocene and Holocene Time. In G. Lang and C. Schlüchter (ed.), *Lake, Mire and River Environments During the last 15000 Years*. Balkema, Rotterdam, 1988:133-136.

HAVLÍČEK, P. 1994 – State of Research of the Moravian Rivers in Holocene Time. In E. Růžičková and A. Zeman (ed.), *Holocene Flood Plain of the Labe River*. Geological Institute of Academy of Sciences of the Czech Republic, Prague, 1994:98-99.

JOHNSEN, S.J, H.B. CLAUSEN, W. DANSGAARD, N.S. GUNDESTRUP *et al.* 1997 – The $\delta^{18}O$ Record Along the Greenland Ice Core Project Deep Ice Core and the Problem of Possible Eemian Climatic Instability. *Journal of Geophysical Research* 102, 1997:26397-26410.

LUTERBACHER, J., E. XOPLAKI, D. DIETRICH, P.D. JONES *et al.* 2002 – Extending North Atlantic Oscillation Reconstructions back to 1500. *Atmospheric Science Letters* 2, 2002:114-124.

MAYEWSKI, P., and F. WHITE 2002 – *The Ice Chronicles: The Quest to Understand Global Climate Change*. University Press of New England, Hanover, New Hampshire, 2002.

NĚMEČEK, J., and the Department of Soil Science and Geology 2001 – Digitální Mapa Půd 1:200.000. ČZU, Praha, 2001.

OLIVA, M. 2004 – Flint Mining, Rondels, Hillforts ... Symbolic Works or Too Much Free Time?. *Archeologické rozhledy* LVI, 2004:499-531, 2004.

PODBORSKÝ, V. (ed.) 1993 – *Pravěké Dějiny Moravy* (Vlastivěda Moravská; Země a Lid, Nová Řada 3. Muzejní a vlastivědna společnost, Brno.

PODBORSKÝ, V. (ed.) 1999 – *Pravěka Sociokultovní Architektura na Moravé. [Primeval Socio-Ritual Architecture in Moravia.]* Ústav archeologie a muzeologie. Filozofická fakulta, Masaryk University, Brno, 1999.

REIMER, P.J., M.G.L. BAILLIE, E. BARD, A. BAYLISS *et al.* 2004 – IntCal04 Terrestrial Radiocarbon Age Calibration, 0-26 Cal Kyr BP. *Radiocarbon* 46/3, 2004:1029-1058.

RŮŽIČKOVÁ, E., and A. ZEMAN 1994 – Holocene Fluvial Sediments of the Labe River. In E. Růžičková and A. Zeman (eds.), *Holocene Flood Plain of the Labe River*. Geological Institute of Academy of Sciences of the Czech Republic, Prague, 1994.

SCHETTLER G., B. REIN and J. NEGENDANK 1999 – Geochemical Evidence for Holocene Paleodischarge Variations in Lacustrine Records from the Westeifel Volcanic Field, Germany: Schnalkenmehrener Maar and Meerfelder Maar. *The Holocene* 9, 1999:381-400, 1999.

SCHULZ, M., and A. PAUL 2002 – Holocene Climate Variability on Centennial-To-Millennial Time Scales: 1. Climate Records from the North-Atlantic Realm. In G. Wefer, W. Berger, K.E. Behre and E. Jansen (eds.), *Climate Development and History of the North Atlantic Realm*. Springer, Berlin and Heidelberg, 2002:41-54.

ŠMÍD, M. 1991 – Nové Nálezy Kultury s Moravskou Malovaou Keramikou na Prostejovsku. *Archeologické rozhledy* XLIII, 1991:185-205.

ŠMÍD, M. 1993 – Nástin Periodizace Kultury s Nálevkovitými Poháry na Moravé. *Pravěk*, Nová řada 2, 1992:131-157.

ŠMÍD, M. 1994 – Ein Burgwall mit steinerner Stirnmauer aus der älteren Stufe der Trichterbecherkultur auf dem Burgwall Rmíz bei Laškov im Kataster der Gemeinde Náměšt na Hané, Kreis Olomouc, Land Mähren. *Jahresschrift für mitteldeutsche Vorgeschichte* 76, 1994:201-230.

ŠMÍD, M. 1996 – Sídliště Kultury s Nálevkovitými Poháry u Laškova, okr. Přostějov. *Pravěk*, Nova řada 6, 1996:97-138.

ŠMÍD, M. 1999 – Das dritte Hügelgräberfeld der Trichterbecherkultur im Kataster von Náměšť na Hané, Bez. Olomouc. *Pravěk*, Nova řada 8, 1998:115-140.

ŠMÍD, M. 2000 – Drobné Zjišťovací Výzkumy na Eneolitických Mohylových Pohřebištých Střední Moravy. [Eine Suchgrabung auf äneolithischen Hügelgräberfeldern in Mittelmähren.] *Památky archeologické*, Supplementum 13, 2000:389-404.

ŠMÍD, M. 2001 – Nálezy Kultury s Nálevkovitými poháry z Kostelec na Hané a Typologie Keramiky Straší Fáze I, Stupne KNP na Moravě. [Entwurf der Keramiktypologie für die Mährische Trichterbecher-kultur.] *Pravěk* Supplementum 8, 2001:276-298.

ŠMÍD, M., M.O. BALDIA and E. PERNICKA 1999 – Neolithic/Eneolithic Copper technology, settlement and trade in Central Moravia,' Paper presented at the symposium entitled "*Prehistoric Technology and its Social Implications: New Theories and Methods*" at the 64th Annual Meeting of the Society of American Archaeology, Chicago, Illinois, USA, 1999.

STAECK, J.P. 1999 – Defining the Playing Field: Implications of GIS Analyses to the Emergence of Prestige Technology. Paper presented at the symposium entitled *Prehistoric Technology and its Social Implications: New Theories and Methods* at the 64th Annual Meeting of the Society of American Archaeology, Chicago, Illinois, USA, 1999.

STAECK, J.P., T. KOPECKY and M. ŠMÍD 2004 – TRB Walled Settlements and Economy: The View from Hrad Bilovice. Paper presented in the general session *European Studies* at the 69th Annual Meeting of the Society of American Archaeologists, Montreal, Quebec, Canada, 2004.

WHITTLE, A. 1988 – Contexts, Activities, Events: Aspects of Neolithic and Copper Age Enclosures in Central and Western Europe. In C. Burgess, P. Topping, C. Mordant and M. Maddison (ed.), Enclosures and Defenses in the Neolithic of Western Europe, BAR S403(i). BAR Publishing, Oxford, 1988:1-19.

ZIELINSKI, G.A., P.A. MAYEWSKI, L.D. MEEKER, S.I. WHITLOW *et al.* 1994 – Record of Volcanism Since 7000 BC from the GISP2 Greenland Ice Core and Implications for the Volcano-Climate System. *Science* 264, 1994:948-952.

Chapter 18
TAPHONOMIC PROCESSES AFFECTING MONUMENTAL EARTHEN ARCHITECTURE AS A PROXY FOR CLIMATIC CHANGE

Douglas S. FRINK
Dept. of Physical and Earth Sciences, Worcester State College, Worcester, MA, USA

Abstract: *Pedogenic processes primarily issue from the interface between the soil and the atmosphere, and through time evolve downwards through the soil profile. The OCR Carbon Dating (OCR) procedure takes into account a number of variables reflecting the pedogenic evolution of the soil, and models pedogenic processes through a dynamic, non-linear formula. As such, the OCR procedure is well suited to describing the physiology of the soil body. With close interval sampling along a vertical soil column, the archaeological and temporal context of artifacts and perturbation features may be determined. The variables used in the OCR Formula describe various related and temporally separate production processes in the soil network that participate in the transformation of the soil body through time. Together, these variables and processes describe the history and evolution of the soil. Using data from the excavations at the Džbán Mound Cluster, Moravia, Czech Republic, correlations between defined perturbation events and climatic change, based on proxy measures of tree ring growth and residual atmospheric ^{14}C, are seen to correlate with cultural activities as well. The consilience of these independent and dynamic studies, paleoclimatic, pedogenic, and archaeological, provides unique opportunities for hypothesis building and testing.*

INTRODUCTION

Monumental architecture, whether mounds, enclosure walls, standing stones, or buildings, is meant to be seen. It has other functions within a culture – religious, ceremonial, defensive, habitational – but their size, design, position on the landscape, and organization with other structures serve as a visual element. They convey messages to the observer about the culture and people who built them; they also describe status, wealth, lineage, alliances, vocations, and other relative positions within a broad regional culture. As such, these structures are messages expressed in a symbolic language of style, shape, size, motifs, material, and context.

Time has the effect of obscuring these messages, introducing noise resulting from taphonomic changes to the structure. When the structures are composed of soil sediments and exposed to the natural environment, such taphonomic changes are processes of normal pedogenesis (Johnson 1993, Johnson and Watson-Stegner 1987, 1990, Johnson and Watson-Stegner *et al.* 1987, Keller *et al.* 1990). Soil is a living entity, a dynamic system evolving along trajectories constrained by the five factors of soil formation (Jenny 1941, Phillips 1998).

Soils are complex and dynamic systems that evolve from initial conditions towards an unpredictable mature end point (Phillips 1995, 1997, 1999, 2000). The end point is not predictable because, although closed organizationally, soils are open to energy and matter exchange with their environment, which is itself complex and unpredictable. Human activity and climatic change are constituents of the soil's environment.

Soil science in general, and as it has been specifically applied to archaeology (Cornwall 1958, Limbrey 1975, Holliday 1992, Vogel 2002), emphasizes a descriptive rather than interpretive approach (Retallack 2001). Soil classification describes multivariate, but linear, progressions of raw sediments to mature end products. Haploidalization (the mixing of soil horizons) is considered, but primarily as a minor interruption of the entropic process. Yet, it is precisely these haploidal, or perturbation, events that are of vital interest to the archaeologist (Wood and Johnson 1978). Human activity, the archaeological site itself, is a perturbation event. While such events rarely alter the organization of a soil and its trajectory to reaching a metastable end state, soils do adjust structurally to perturbation events.

The anatomic, or descriptive, approach normal to pedological analysis fails to capture these events. What is needed is a physiological approach that explores the processes fundamental to all soils, and their taphonomic effects on cultural material. For example, the observed processes of lessivage (downward movement of fine particles forming horizons) and pedoperturbation (mixing of horizons) are fundamental to all soils, but they also have specific and measurable effects on the provenience of artifacts and ecofacts of particular interest to the archaeologist. Anatomically, they define the degree of horizonation (the forming of horizons). However, they may also be used to describe the coeval physiological processes of soil. Lessivage occurs, not in opposition to perturbation, but because of it (Johnson and Watson-Stegner *et al.* 1987). Fine particles of clay and silt are differentially sorted out of the A-horizon and into the lower portions of the B-horizon. Coarse particles, including cultural artifacts, are translocated to levels just above the evolving argillic (clay) horizon. Organic matter, sesquioxides, carbonates, and other soil constituents are similarly sorted as a natural result of soil development and growth (Frink 2011).

Examination of these kinds of physiological alterations allows for the reconstruction of a fuller history of the soil body that constitutes the context of cultural material. This physiological history incorporates cultural, natural, and environmental events that effect interpretations made of the archaeological record.

The OCR Carbon Dating procedure, which takes into account a number of variables reflecting the pedogenic evolution of the soil, is well suited to describing the physiology of the soil body (Frink 1992, 1994, 1995, 2001, 2007). The biological recycling of organic carbon is fundamental to nearly all biological systems on this planet. While some forms of organic carbon, such as fresh organic matter, are quickly recycled, other more resistant forms, such as humus and charcoal, are recycled at a much slower rate. The effect of the biochemical degradation of charcoal and soil humic material can be measured by the ratio of the total carbon to the readily oxidizable carbon in the sample. This ratio is called the Oxidizable Carbon Ratio (OCR). The rate of biochemical degradation of the relatively stable forms of organic matter varies within the specific physical and environmental contexts of the sample. To determine an age for the carbon sample, a systems formula was designed to account for the biological influences of oxygen, moisture, temperature, and the media's (soil) reactivity. These variables are measured by soil texture, depth below the soil surface, the site specific mean annual temperature and rainfall, and the soil pH. Residual influences on this system are included through a statistically derived constant.

With close interval sampling along a vertical soil column, the archeological and temporal context of artifacts (and associated features) may be determined within the context of the living soil system. The variables used in the OCR Formula describe various related production processes in the soil network that participate in the transformation of the soil body. A comparison between samples in the soil column reveals certain individual processes and their related participation in the other processes. Variations in soil textures between samples show the evolving stone lines (coarse particles) and argillic horizons (fine particles). Soil reactivity (pH) and total organic carbon describing consumption, digestion, and waste elimination processes, and are attributable to specific temporally contextualized events of aggradations, or perturbations, such as mound building, and the soil's recovery following such events. Together, these variables describe the history and evolution of the soil, before, during and after the formation of the archaeological site.

CASE STUDY

Excavations at the Džbán Mound Cluster at Náměšť na Hané, Olocmouc County, Central Moravia, Czech Republic, were conducted to determine physical changes that had altered the mound's appearance through time – pedogenesis, erosion, deposition, intrusions, as well as renovations (Baldia, Boulanger and Frink 2008, Baldia, Frink and Boulanger 2008).[1] As the construction material of burial mounds consist primarily of soil, and secondarily of stone, our approach relied heavily on understanding both pedogenic and taphonomic processes. Excavations proceeded by pedogenic and depositional levels within meter units, defining present provenience of cultural material. OCR column samples were obtained to evaluate the taphonomic processes that resulted in the present provenience.

Three individual full column samples, extending from the surface to sub-mound buried subsoil, along with several partial columns, were obtained from various excavation units within Mound 1 (Fig. 18.1). Additionally, two full column samples were obtained from off-mound locations.

The Off-Mound control soil consist of a 10 cm thick A/E surface horizon overlying a 14 cm thick E-horizon. A Bs-horizon extends from 24 to 37 cm, below which is a Bt-horizon from 37 cm to the base of excavation (50 cm). The soils overlay bedrock within 1 to 2 meters of the surface, are clay loams and stony clay loams, and are well drained and acidic throughout.

The OCR data suggests a more complex history of these soils than is evident from the profile description (Table 18.1). Samples from the upper portion of the profile (A and E horizons) evidence relatively higher pH values than those from lower portions. The descending pH values in these upper samples suggest the likelihood that soil amendments have been added in the recent past. A discontinuity, suggestive of a perturbation event sufficient to change the organization of pedogenesis, is evident in the pH data between 30 and 35 cm, with descending pH values characterizing the upper soils, and ascending values characterizing the lower ones. This suggested perturbation event occurred sometime between 3805 and 5574 BC (5755 and 7524 calendrical years before the present [bp]). The lack of clear pedogenic patterns in the soil samples from below the discontinuity suggests that the lower soil (paleosol) was truncated (eroded) prior to the accumulation of the material above. In conjunction with the patterns discussed for the upper soil, the event responsible for this discontinuity would have constituted a severe perturbation event commensurate with land clearing. The change from Mesolithic, hunter-gathering cultures, to the Neolithic, agriculturist cultures, in this part of Central Europe occurred around 7500 years ago (5550 BC).

Two additional pedogenic events are suggested in the soils above the discontinuity. The 20 cm sample, and associated lower samples, show sorting of coarse and fine particle. This physiological pattern is consistent with a slight-to-moderate perturbation event affecting the soil surface. The specific event is not evident, but may represent either a limited cultural impact or a shift in climatic patterns. The associated OCRDATE is 665 BC

[1] The 2002 excavations was undertaken by the Czech-American Research Project (CARPro), Dr. Maximilian O. Baldia, Director and Principal Investigators, in cooperation with Dr. Miroslav Šmíd, Head, Prostějov Office of the Moravian Cultural Heritage Office, and the College of DuPage Experiential Archaeology Curriculum – Czech-American Archaeological Field School, Dr. John Staeck, Director.

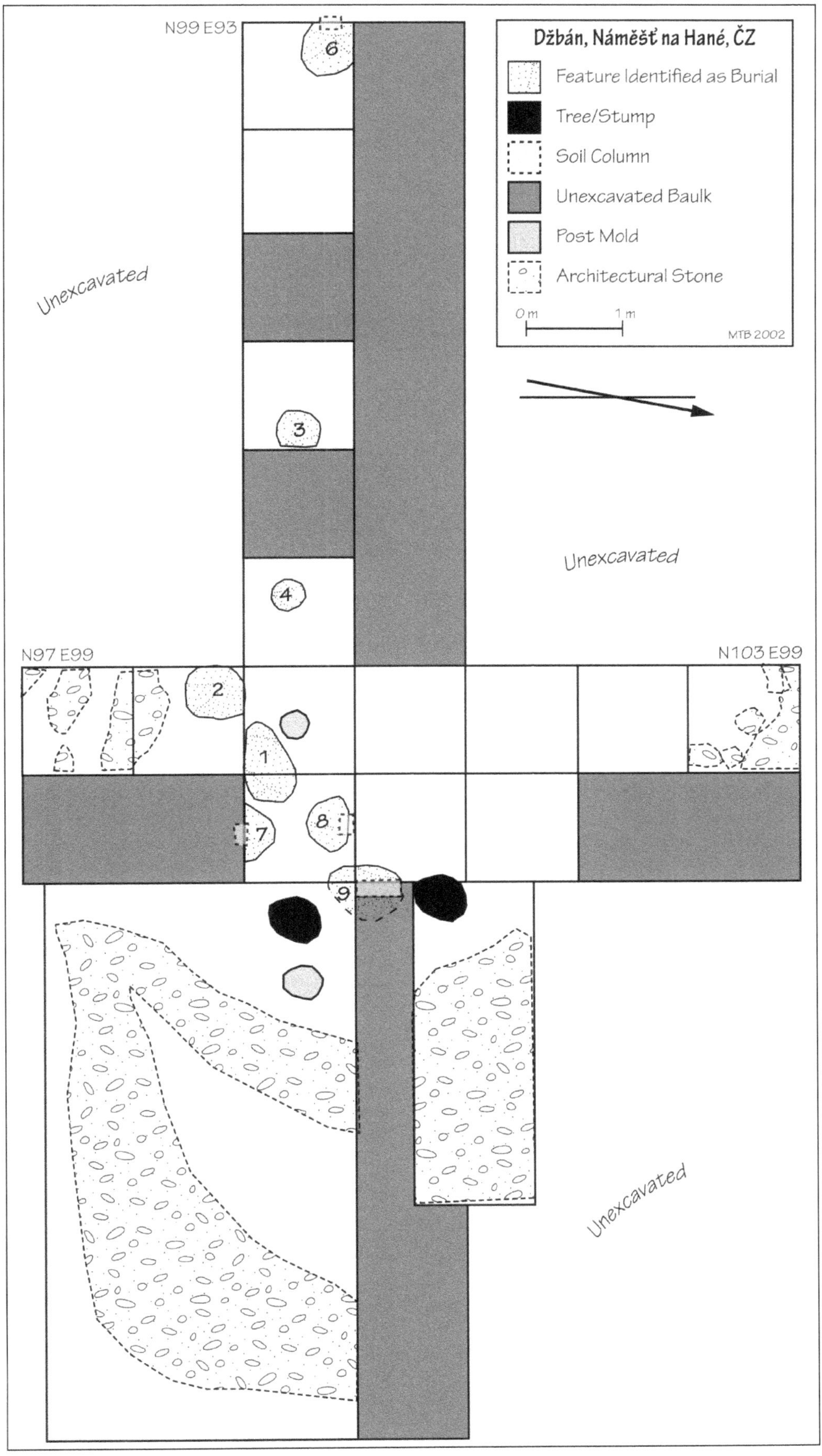

Figure 18.1. Plan of Džbán Mound 1, Náměšť na Hané, Olomouc, Czech Republic with excavation units (Drawing: M. Boulanger)

Table 18.1. OCR data for Džbán I N29 E91 – Off Mound Control

Soil Depth	pH	% Organic Carbon (LOI)	OCR Date	Very Coarse	Coarse	Medium	Fine	Very Fine	Coarse Silt	Fine Silt	Sample Id	% Oxidizable Carbon (WB)	OCR Ratio	Mn
5	4.4	12.037	282	1.069	.924	1.074	.873	.907	25.849	69.303	6104	5.72	2.10	52.33
10	4.3	6.558	753	2.807	1.767	1.594	.823	.746	15.658	76.605	6103	2.66	2.47	29.6
15	4.1	4.516	1369	2.733	2.100	1.729	1.513	.822	4.506	86.597	6102	1.83	2.47	12.37
20	3.3	2.320	2615	2.378	2.223	2.316	1.425	.843	21.602	69.213	6101	0.69	3.36	2.29
25	3.2	1.875	4048	6.975	3.046	1.727	1.975	.811	16.825	68.640	6100	0.50	3.75	1.145
30	3.2	1.998	5755	7.391	6.627	1.477	.984	.802	6.334	76.385	6105	0.43	4.65	3.49
35	3.0	2.058	7524	7.691	8.059	12.363	5.334	.616	33.793	32.143	6099	0.33	6.24	4.30
40	3.2	1.762	9385	5.000	3.897	1.528	.728	.924	21.233	66.692	6106	0.39	4.52	8.57
45	3.2	1.833	11593	4.206	4.854	2.512	.750	.507	16.128	71.048	6107	0.34	5.39	11.59

(2615 BP). The second pedogenic event is shown by a reversal in value for manganese associated with coarse and fine particle sorting. Manganese values decrease with depth in actively pedogenic soils down to the limits of biodegradation of carbon. This physiological pattern is consistent with a major perperturbation event and the associated OCRDATE of 3805 BC (5755 bp) would be indicative of the uninterrupted time span of the currently ongoing pedogenesis.

As the mound is a constructed monument, a high degree of stratigraphic complexity is expected and observed in the soil profiles. Generally, the soils consist of a 5 cm thick A/E surface horizon overlying a thin 2 cm thick E-horizon. An emergent stone line is contained within the Bs-horizon, which extends from 7 to 15 cm below the surface and overlies an incipient Bt-horizon. This upper solum rests on, and is forming in, non-pedogenic mound fills that extend from 15 to 37 cm below surface. The mound fill rests on a roughly 7 cm thick prepared floor. The sub-mound floor was created by adding pea-sized and smaller coarse material (river run sands) to the now buried paleo-surface (2Ab). A 2 cm thick, relatively unaffected portion of the 2Ab-horizons extends below the prepared floor. A 2Bb-horizon extends from 46 to 55 cm below surface and sits upon a well developed 2Btb-horizon that extends to the base of excavation at 60 cm.

One moderate-to-strong and one weekly defined pedogenic perturbation are evident in the OCR data from the upper portions of the mound (Table 18.2). The moderate-to-strongly defined perturbation is evident in the change in trending of the manganese along with textural sorting in samples from 15, 18, and 21 cm below surface. This recent perturbation to the mound dates to roughly AD 650 (1300 bp). The weekly defined perturbation, mostly erased by the latter and stronger one discussed above, is indicated by a decrease in pH and the OCR Ratio value. This weakly defined perturbation dates to 417 BC (2367 bp).

The sub-mound paleosol includes the prepared floor in the upper 2Ab horizon. A strong discontinuity, signifying the construction of the mound, exists between the upper mound fill (30 cm and above), and the lower paleosol (33 cm and below), and is evidenced by the pH, % organic carbon, and manganese values, as well as a scalar shift in textures. The construction of Mound 1 occurred after 3286 BC (5236 bp).

A moderate perturbation is expressed in the data from samples obtained from 48 and 51 cm below the surface, as marked by an increase in % organic carbon and pH values, along with textural sorting. A second, although weaker, perturbation is seen in the data from 54 and 57 cm. The OCRDATES for these two events are 5586 BC (7536 bp) and 7857 BC (9807 bp), respectively.

Additional information on diagenic process is evident along the toe-of-slope portions of the mound. A major discontinuity between 25 and 30 cm is evident in the OCR Ratio value and a scalar shift in textures (Table 18.3). The pedogenic soils above the discontinuity have been in place and undergoing pedogenic processes since 3173 BC (5123 bp). The non-pedogenic mound sediments rest on top of a paleosol forming the second major discontinuity between 48 and 52 cm. Again, the OCR Ratio value and scalar shift in textures are in evidence for this discontinuity. The absence of any evidence of pedogenesis in the colluvial sediments suggest that the paleosol was buried during mound construction after 3421 BC (5371 bp), but before 3173 BC (5123 bp).

Stages of construction and potential "ornamentation" of the monument can also be discerned. The initial cremation burial was dug into the paleosol and encapsulated in a clay matrix prior to the construction of the mound proper. The upper soil profile is similar to that of the nearby modal column presented earlier (Table 18.4). A discontinuity between samples at 40 and 45 cm marks the separation between the initial burial

Table 18.2. OCR data for Džbán Mound 1 N99 E99 – Apex of Mound 1

Soil Depth	pH	% Organic Carbon (LOI)	OCR Date	Very Coarse	Coarse	Medium	Fine	Very Fine	Coarse Silt	Fine Silt	Sample Id	% Oxidizable Carbon (WB)	OCR Ratio	Mn
3	2.9	9.629	170	14.304	3.566	1.841	1.429	2.850	14.802	61.208	6051	4.12	2.34	57.54
6	2.9	8.329	300	22.777	3.185	1.790	1.456	1.776	19.505	49.511	6052	4.07	2.05	30.36
9	2.9	5.246	515	33.856	4.038	1.692	1.437	18.597	7.338	33.041	6053	1.79	2.93	6.16
12	2.9	3.449	861	32.565	6.054	1.955	1.354	1.593	24.353	32.125	6054	1.00	3.45	5.41
15	2.9	3.065	1300	23.593	6.926	3.890	1.576	5.814	26.359	31.843	6055	0.815	3.76	5.60
18	2.9	2.663	1721	18.348	8.184	3.829	3.552	1.445	11.070	53.573	6056	0.88	3.03	9.27
21	2.8	2.434	2367	11.671	7.748	4.572	4.556	7.160	7.434	56.860	6057	0.63	3.86	10.89
24	2.8	2.071	3100	14.074	6.992	3.879	4.314	16.717	5.958	48.067	6058	0.48	4.31	13.14
27	2.8	2.112	3947	11.975	7.152	6.225	7.412	14.353	8.163	44.720	6059	0.42	5.03	15.26
30	2.8	1.994	5098	16.748	8.758	5.965	7.856	14.365	5.150	41.158	6060	0.28	7.12	19.45
33	3.1	2.011	5236	22.793	9.667	7.487	4.460	7.193	6.950	41.450	6061	0.53	3.79	47.81
36	3.2	2.244	5282	12.633	6.592	8.390	4.696	4.813	4.813	58.062	6062	0.54	4.16	50.92
39	3.3	2.256	5361	7.950	6.621	8.918	3.179	5.594	7.717	60.021	6063	0.43	5.25	55.67
42	3.4	2.206	6417	6.848	6.730	3.307	4.232	5.072	18.439	55.372	6064	0.375	5.88	48.74
45	3.5	2.056	6732	5.988	4.614	2.563	1.524	5.646	26.744	52.921	6065	0.38	5.41	44.935
48	3.7	2.402	7536	7.234	12.227	3.547	.436	3.099	22.386	51.071	6066	0.46	5.22	49.23
51	3.9	2.063	8647	8.691	6.033	3.030	1.038	10.719	14.360	56.129	6067	0.32	6.45	28.35
54	4.0	2.128	9807	10.596	6.335	7.717	14.319	14.992	10.971	35.070	6068	0.25	8.51	25.55
57	4.0	1.538	11644	8.159	8.610	6.422	17.365	9.990	9.731	39.723	6069	0.14	10.99	19.44

Table 18.3. OCR data for Džbán Mound 1, N99 E92 – Western Toe of Mound 1

Soil Depth	pH	% Organic Carbon (LOI)	OCR Date	Very Coarse	Coarse	Medium	Fine	Very Fine	Coarse Silt	Fine Silt	Sample Id	% Oxidizable Carbon (WB)	OCR Ratio	Mn
5	3.0	11.436	359	3.852	1.456	1.449	1.838	1.437	13.352	76.617	6110	5.17	2.21	71.01
10	2.9	4.365	1067	4.558	2.785	2.160	1.050	1.061	4.295	84.090	6111	1.74	2.51	15.64
15	2.9	2.889	2235	4.396	2.522	1.638	.975	.832	4.597	85.040	6112	0.90	3.21	5.96
20	2.8	2.359	3569	3.941	2.487	1.734	1.042	.736	6.440	83.618	6113	0.73	3.23	4.87
25	2.9	2.149	5123	9.128	1.944	1.237	1.228	1.092	10.471	74.901	6109	0.525	4.09	5.03
30	2.9	2.216	5283	14.134	3.778	1.613	.842	.754	6.647	72.233	6114	0.84	2.64	6.19
35	2.9	2.512	5277	13.127	5.514	2.126	1.054	.865	7.882	69.431	6115	0.90	2.79	6.87
40	2.9	2.570	5255	26.359	4.204	2.072	.870	.633	16.737	49.124	6116	0.84	3.06	7.71
45	2.9	2.249	5430	25.633	4.065	1.950	.941	1.327	6.835	59.250	6108	0.38	5.92	8.93
48	2.9	2.224	5276	26.502	5.141	2.529	1.103	.967	6.486	57.273	6118	0.685	3.25	11.45
52	2.9	2.303	5371	9.893	4.978	2.204	1.592	1.212	3.416	76.705	6119	0.61	3.78	15.3
55	2.9	2.073	5708	11.231	4.913	2.443	1.845	1.213	3.543	74.813	6120	0.39	5.32	16.81
60	2.9	2.098	6205	7.804	4.665	3.368	1.690	1.392	9.305	71.775	6117	0.29	7.23	20.69
65	3.0	2.281	6446	19.889	5.146	2.822	2.951	1.100	2.966	65.127	6121	0.45	5.07	14.76
70	2.9	2.350	7627	20.754	4.300	1.950	2.093	1.566	3.092	66.246	6122	0.56	4.20	17.03
75	2.9	2.231	9145	31.360	5.972	3.572	2.646	1.492	3.009	51.949	6123	0.35	6.37	20.73
80	2.9	2.118	10653	29.345	7.647	4.637	3.732	1.894	3.457	49.288	6124	0.34	6.23	23.93

Table 18.4. OCR data for Džbán Mound 1, N101 E101 – Central Burial

Soil Depth	pH	% Organic Carbon (LOI)	OCR Date	Very Coarse	Coarse	Medium	Fine	Very Fine	Coarse Silt	Fine Silt	Sample Id	% Oxidizable Carbon (WB)	OCR Ratio	Mn
5	3.6	6.492	277	22.772	5.436	2.754	1.247	1.125	5.490	61.176	6130	2.97	2.19	125.79
10	3.9	4.914	665	26.185	4.424	2.360	1.208	.962	4.356	60.505	6131	2.215	2.22	54.24
15	4.2	3.646	1499	16.783	5.810	2.281	1.643	1.079	5.340	67.063	6132	1.44	2.53	32.42
20	4.3	3.972	1872	17.591	10.044	2.738	1.367	1.172	5.116	61.973	6133	1.48	2.68	30.51
25	4.4	3.413	2984	12.098	7.783	3.523	1.533	1.376	4.875	68.813	6134	1.20	2.84	20.10
30	4.5	2.653	4411	16.777	7.211	2.334	1.634	1.281	4.715	66.048	6135	0.81	3.28	25.09
35	4.7	2.435	5010	11.465	9.339	2.586	3.303	1.686	4.858	66.762	6136	0.69	3.53	17.67
40	4.8	2.409	5146	11.462	7.634	2.274	2.243	2.009	5.210	69.168	6137	0.64	3.76	17.69
45	4.8	2.633	5223	21.300	10.499	4.857	2.558	2.391	5.348	53.047	6138	0.77	3.42	12.39
50	5.1	2.896	5267	16.441	14.563	8.605	4.888	4.431	4.315	46.756	6139	0.56	5.17	10.39
55	5.2	3.404	5206	17.690	12.085	6.624	4.115	4.559	6.334	48.593	6140	0.96	3.55	7.32
60	5.1	3.347	5808	11.030	13.000	9.986	5.996	9.039	5.270	45.680	6141	0.84	3.98	13.69
65	5.3	4.149	6285	21.680	17.889	12.542	6.912	5.243	3.702	32.033	6142	0.93	4.46	7.20
70	5.3	2.672	6619	38.114	21.071	5.693	1.028	.063	6.632	27.399	6143	0.91	2.94	8.41

covering (internal burial chamber) and the mound proper (external monumental architecture). A second discontinuity between 55 and 60 cm marks the separation between the burial and the paleosol underneath. The discontinuity separating the initial burial chamber and the overlaying mound is strongly expressed by a scalar change in textures and % organic carbon values. The timing of this event would fall between 3196 and 3256 BC (5146 and 5206 bp). The paleosol consists of a truncated, incipient 2Btb horizon down to the extent of excavation at 75 cm.

In addition to containing an intact burial, central to the mound (Fig. 18.1, No. 9), the soils within this column display significantly higher pH values than all other sample areas of Mound 1. The range in pH values is 3.6 to 5.4, while the range for all other sample areas on the mound is 2.7 to 4.0. As the high pH values are evident throughout the soil profile, the unique characteristic of these soils cannot be attributed to the burial, or the internal mounded burial chamber. The data suggest that the soils immediately above, and perhaps surrounding the burial, were amended differentially from the rest of the mound. Limestone, available within a kilometer of the site, may have been used hygienically to treat the burial, or symbolically to mark the burial. The color of the mound fill would have been either yellow or red (E, Bs, and Bt horizon soils), or black (A-horizon soils). Carbonate amendments as a decorative treatment to monumental structure would have contrasted greatly with the native soils used on the rest of the mound.[2]

DISCUSSION

The soil's environment includes other constituents that either directly affects the soil's trajectory of evolvement, or indirectly affects them by altering soil metabolic processes. These may include severe forest fires, down drafts associated with strong storms, animal activity, and local or regional seismic activity, among others.

Shifts in climatic patterns affect the soil's trajectory of evolvement by changing the patterns and rates of metabolic processes. Such shifts may also induce effects from both natural and human perturbations. For example, a change from cool and moist to warm and dry conditions can foster cataclysmic forest fires, while a change from warm and dry to cold and moist conditions can foster an increase in blow downs and solifluction (mass wasting). A change to colder and wetter conditions may lead to the subsequent increase in wood cutting for fuel and building material, and thereby changing the floral and faunal community and their effect on the soil.

A composite of the physiological alterations evidenced in the 10 soil columns obtained from the Džbán Mound Cluster are compared to known and suggested environmental and cultural events (Table 18.5). These correlations provide numerous hypotheses for testing at other sites locally and within the region.

Inseparable from any discussion of the direct effects of climate change to soils is the stress placed on human

[2] The idea of decorating the surfaces of monumental earthen architecture with different colored soils was presented by Dr. John Chapman during the 1st International Conference on Soils and Archaeology, Szazhalombatta, Hungary. Chapman, 2001.

Table 18.5. Summary of identified and dated soil perturbation events at Džbán Mound 1, Náměšť na Hané, Olomouc, Czech Republic and proposed with climatic and cultural events.

Location	OCR_{DATE}	Mean	Hypothesized Related Events
N99 E98	653 bp or 1297 AD		Climate Change – Medieval Warm to the Little Ice Age Wolf Min
N101 E101	665 bp or 1285 AD		Climate Change – Medieval Warm to the Little Ice Age Wolf Min
N99 E96	1049 bp or 901 AD		Climate Change – Medieval Minimum – 2
N110 E99	679 and 1177 bp or 1271 and 773 AD		Climate Change – Medieval Minimum – 1 Wolf Min
N99 E99	1300 bp or 650 AD		Climate Change ? 640 AD Volcano ?
N98 E100	1411 bp or 539 AD		Climate – Possible Comet, the 536 Event
N101 E101	1872 bp or 78 AD		Climate – Vesuvius 79 AD
N99 E99	2367 bp or 417 BC		Climate Change – Grecian Minimum
N29 E91	2615 bp or 665 BC		Climate Change – Late Homeric Minimum
N99 E99	2840 bp or 890 BC		Climate Change – Early Homeric Minimum
N101 E101	4411 bp or 2461 BC		Climatic / Cultural Change
N99 E92	5123 and 5255 bp or 3173 and 3305 BC	3239 BC	Initial Colluvium from Mound
N101 E101	5146 and 5206 bp or 3196 and 3256 BC	3226 BC	Mound Construction
N99 E99	5098 and 5236 bp or 3148 and 3286 BC	3217 BC	Mound Construction
N99 E100	5188 bp or 3238 BC		Mound Construction
Burial 9	5206 bp or 3256 BC		Central Grave
N101 E101	6285 bp or 4335 BC		Climatic Change Transition from Lengyel II to III
N99 E100	7369 bp or 5419 BC		Established Neolithic Agricultural Strategy
N101 E101	7480 bp or 5530 BC		Start of the Neolithic Agricultural Strategy
N99 E99	7536 bp of 5586 BC		Mesolithic-Neolithic Transition Advent of Agricultural Strategy?
N99 E92	7627 bp or 5677 BC		Mesolithic Forest management

populations that results from shifts in the climatic patterns. Such stress can induce changes in human exploitation patterns of the environment. While soil perturbations, evidenced through a physiological analysis, may be used as a proxy for climatic change, such proxies cannot be viewed out of context of cultural systems that similarly adjust to changing environmental conditions.

References

CORNWALL, I.W. 1958 – *Soils for the Archaeologist*. Phoenix House, London, 1958.

FRINK, D.S. 1992 – The Chemical Variability of Carbonized Organic Matter Through Time. *Archaeology of Eastern North America* 20, 1992:67-79.

FRINK, D.S. 1994 – The Oxidizable Carbon Ratio (OCR): A Proposed Solution to Some of the Problems Encountered with Radiocarbon Data. *North American Archaeologist* 15/1, 1994:17-29.

FRINK, D.S. 1995 – Application of the OCR Dating Procedure, and its Implications for Pedogenic Research. In M.E. Collins (ed.), *Pedological Perspectives in Archaeological Research*. Soil Science Society of America, Special Publication 44, 1995:95-106.

FRINK, D.S. 2001 – Temporal Values in a Universe of Perturbations: Application of the OCR Carbon Dating Procedure in Archaeological Site Formational Analyses and Pedogenic Evaluations. In G. Fuleky (ed.), *Proceedings of the 1st International Conference on Soils and Archaeology. Szazhalombatta, Hungary, 30 May – 3 June 2001*, Szent Istvan University Godollo Matrica Museum, Szazhalombatta, 2001:5-12.

FRINK, D.S. 2007 – *Explorations into a Dynamic Process-Oriented Soil Science*. Unpublished Ph. D. Dissertation, School of Geographical Sciences, Arizona State University, Tucson, 2007.

FRINK, D.S. 2011 – Explorations into a Dynamic Process-Oriented Soil Science. Waltham, MA, Elsevier Press, 2011.

HOLLIDAY, V.T. (ed.) 1992 – Soils in Archaeology: Landscape Evolution and Human Occupation. Smithsonian Institution Press, Washington, 1992.

JENNY, H. 1941 – Factors of Soil Formation: A System of Quantitative Pedology. New York, McGraw-Hill, 1941.

JOHNSON, D.L. 1993 – Dynamic Denudation Evolution of Tropical, Subtropical and Temperate Landscapes with Three-Tiered Soils: Toward a General Theory of Landscape Evolution. *Quaternary International* 17, 1993:67-78.

JOHNSON, D.L., E.A. KELLER and T. ROCKWELL 1990 – Dynamic Pedogenesis: New Views on Some Key Soil Concepts, and a Model for Interpreting Quaternary Soils. *Quaternary Research* 33, 1990: 306-319.

JOHNSON, D.L., and D. WATSON-STEGNER 1987 – Evolution Model of Pedogenesis. *Soil Science Society of America Journal* 143, 1987:349-366.

JOHNSON, D.L., and D. WATSON-STEGNER 1990 – The Soil-Evolution Model as a Framework for Evaluating Pedoperturbation in Archaeological Site Formation. In N.P. Lasca and J. Donahue (eds.), *Archaeological geology of North America.* Geological Society of America, Boulder, 1990:541-560.

JOHNSON, D.L., D. WATSON-STEGNER, D. JOHNSON and R. SCHAETZL 1987 – Proisotropic and Proanisotropic Processes of Pedoperturbation. *Soil Science Society of America Journal* 143, 1987:278-292.

LIMBREY, S. 1975 – *Soil Science in Archaeology*, Academic Press, London, 1975.

PHILLIPS, J.D. 1995 – Self-Organization and Landscape Evolution. *Progress in Physical Geography* 19, 1995:309-321.

PHILLIPS, J.D. 1997 – Simplexity and the Reinvention of Equifinality. *Geographical Analysis* 29/1, 1997:1-15.

PHILLIPS, J.D. 1998 – On the Relations Between Complex Systems and Factorial Model of Soil Formation (with Discussion). *Geoderma* 86, 1998:1-21.

PHILLIPS, J.D. 1999 – Divergence, Convergence, and Self-Organization in Landscapes. *Annals of the Association of American Geographers* 89/3, 1999:466-488.

PHILLIPS, J.D. 2000 – Contingency and Generalization in Pedology, as Exemplified by Texture-Contrast Soils. *Geoderma* 102, 2000:347-370.

RETALLACK, G.J. 2001 – Soils of the Past: An Introduction to Paleopedology. Blackwell Science, Oxford.

VOGEL, G. 2002 – *A Handbook of Soil Description for Archeologists.* Arkansas Archeological Survey Technical Paper 11 Fayetteville, Arkansas, 2002.

WOOD, W.R., and D. JOHNSON 1978 – A Survey of Disturbance Processes in Archaeological Site Formation. In M. Schiffer (ed.), *Advances in Archaeological Method and Theory* 1, 1978:315-381.

Chapter 19
NEOLITHIC SETTLEMENT IN THE CENTRAL EUROPEAN MOUNTAINS

Paweł VALDE-NOWAK
Institute of History, Dept. of Archaeology, Gdańsk University, Gdańsk.
Institute of Archaeology and Ethnology, Polish Academy of Sciences, Kraków

Abstract: *Intensively settled old centers of the Early Neolithic are well known in Central Europe, but they do not reflect all aspects of an early agricultural way of life. Although received wisdom holds that mountainous landscapes are avoided due to relatively low soil fertility, a shortened vegetation growth period, strong climatic variation, and other geographical conditions unfavorable for agriculture. Neolithic finds in the highlands, in most cases, consist of ground stone tools. Such stray finds confirm the impression of "empty" mountain zones, underlining the disparity even more between such regions and the archaeologically rich territories of the lowlands. However, the results from a series of surveys, conducted in different mountainous zones in Germany and Poland, indicate that numerous Neolithic stray finds have a characteristic material context, and suggest the existence of seasonal pastoral campsites represented in the highlands. The observable increase of such finds in the highlands during the Neolithic is a result of economic advances and demographic pressure, and not a response to climate change.*

INTRODUCTION

Maps of the distribution of Central European Neolithic settlements display the well-known concentration of sites in regions with the highest potential for primitive agriculture. Their location is usually characterized by loess or soils of similar fertility. Mountainous landscapes are avoided because of relatively low soil fertility, a reduced vegetation growth period, strong climatic variation, and other geographical conditions unfavorable for agriculture. This is thought to be corroborated by the distribution of various cultures.

In general, the mapping of Neolithic sites for Central Europe (Petrasch 2001:15 Fig. 1) is only detailed for a single region, i.e., the Saxonian part of Germany (Müller 2001:64 Fig. 8). This void leads us to postulate that the Neolithic settlement was concentrated in the fertile lowlands, avoiding the highlands. With some exceptions discussed later, this is valid for the Danubian cultures, i.e., the Linienbandkeramik (LBK) or Linear Pottery and Lengyel cultures, but it is inaccurate for the northern Neolithic sphere represented by Funnel Beaker (TRB), Globular Amphora, and later the Corded Ware cultures. This means that the tendency to occupy territories of low agricultural potential is more recognizable in the Young or Later and the Late Neolithic, although it applies to the whole period as well (Table 19.1).

The Early Neolithic LBK receives considerable attention across all of Central Europe, due to its central role in introducing agriculture and the resulting fundamental changes affecting Europe (Baldia this volume). However, while its formation and spread on the loess soils has received much attention, the apparent diversity of its settlement organization and economic strategies is still the subject of debate. Similarly, we are only at the verge of understanding the Middle and Late Neolithic economic and social implications of an increasing cultural diversity and its accompanying settlement expansion into previously uninhabited areas and less favorable lowland soil types. This diversification of economic strategies is also ***reflected*** in earthworks and settlement expansion into the North Alpine wetlands. However, the potential significance of the highlands is usually neglected. Consequently, the possibility of Neolithic land use in these areas is either not considered at all or flatly denied. This attitude is encouraged both by a negative assessment of the rare Neolithic finds from the highlands and by a lack of interest in ethnographic evidence suggesting transhumance as the pattern of land use most likely to be practiced in comparable situations. Since it is principally the lack of evidence that restricts our understanding of highland use during the Neolithic, the main results of a recent series of Polish-German archaeological projects in the highlands of southern Poland and Germany are presented here.

A MOUNTAINOUS "PERIPHERY"?

The most popular concept in center-periphery models presumes that the circulation of prestige goods determines the existence of a center and periphery system. Therefore, the lack of prestige symbols in prehistoric communities does not allow recognition of the center and periphery model. This conclusion is best reflected by recent socio-economic analyses of communities from the Late Neolithic-Early Bronze Age (Kristiansen 1987, Rowlands 1987, Kümmel 1998). However, little is known of social differentiation in the Neolithic. Even the nature of the luxury-prestige artifacts used in this period has yet to be determined. We expect that items such as bracelets made of *Spondylus* shells (found especially in Early Neolithic graves) or the first occurrence of copper artifacts, most probably played this role (van de Velde 1990, Nieszery 1995: 205-209, Valde-Nowak 2001a). However, we know little about the social

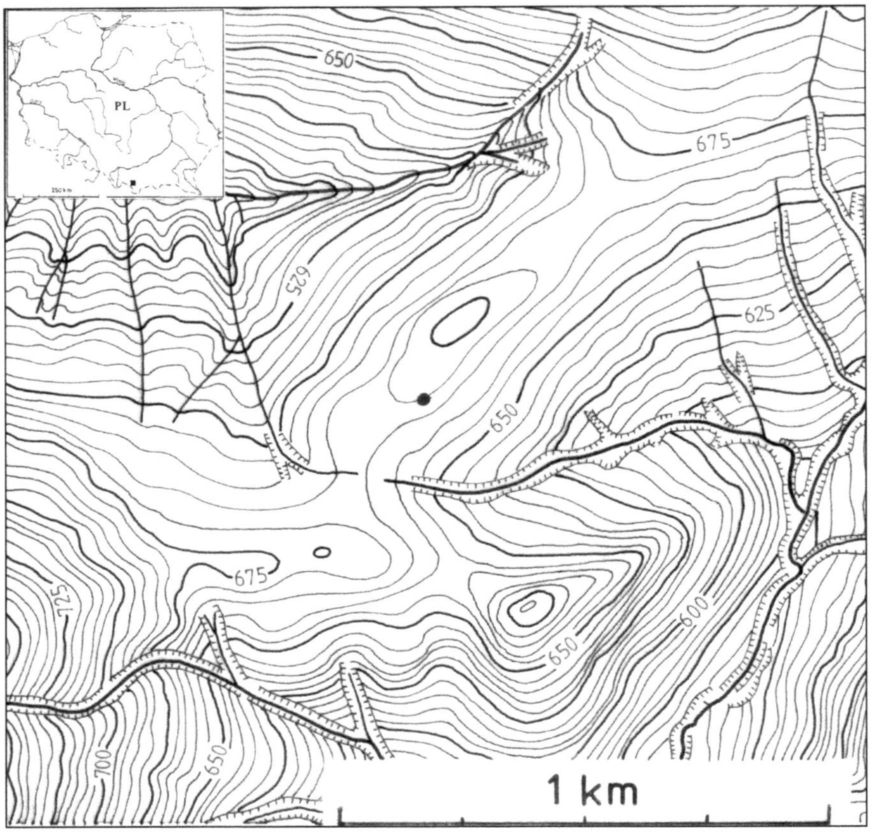

Figure 19.1. Zawoja, Middle Beskidy Mountains (Polish West Carpathians); topography of the Final Neolithic (Corded Ware Culture) site (after K. Tunia 1978, supplemented by the author)

Table 19.1. Simplified chronology of the Neolithic in Central Europe

Period	Culture	BC
Final Neolithic	Corded Ware	2800-2300
Late Neolithic	Baden, Globular Amphora, Cham	3500-2800
Later (Young) Neolithic	Michelsberg, Funnel Beaker	4300-3500
Middle Neolithic	Stroke Ornamented Pottery, Lengyel, Rössen	5000-4000
Early Neolithic	Linienbandkeramik (LBK)	5500-5000

(After Lüning 2000:7 and Milisauskas 2002:145, Furholt 2003, Włodarczak 2008, with modifications)

significance of other artifact forms, such as ground stone axes. These may have had only a utilitarian role or functioned as status symbols as well. The crucial issue of hammer-axes in the Neolithic may resolve problems with the center-periphery model in general and play a key role in understanding the significance of the highland settlements. Thus, if the distribution of Neolithic cultures as determined by the Polish-German field study is taken into account, it is easy to see that the highlands of central Europe, especially the highlands of southern Poland and Germany, are a prime example of a periphery.

"STRAY FINDS" AND LAND USE

The Neolithic "stray finds" of the highlands in most cases involve macrolithic ground tools, such as axes and hammer-axes. No typical Neolithic materials (e.g., ceramics, graves, pits, etc.) are known from these regions. Therefore, the large number of these supposedly "stray finds" are often excluded from maps illustrating the range of settlement. This confirms the impression of "empty mountains" and creates the impression of a disparity between them and archaeologically rich territories of the lowlands. The reason for excluding the highland finds from these site distributions is based on a historical dispute (Reinecke 1930, Mildenberger 1959, Nowothnig 1959). The dispute hinges on the interpretation of the ground stone axes, which are the most prominent finds.

One side of the dispute focuses on the medieval mythology and the notion that the axes were petrified "thunderbolts." It is argued that medieval and early modern belief in the axes' magic properties would have caused them to be intensively traded. Therefore, they do not reflect site locations of economic significance and have to be excluded from the Neolithic site distribution analysis (Reinecke 1930 and Mildenberger 1959:83, 1969:9-10). Based in many cases upon a rather poor knowledge of their archaeological context, the thunderbolt hypothesis is used to question any occupation of the highlands by Neolithic people (by way of contrast, see Nowothnig 1958, 1959 and Raddatz 1972a, 1972b).

The other side of the dispute emphasizes early exchange and cultural contact across the highlands. It draws attention to the importance of river valleys and passes as ways of communication between neighboring groups. It is asserted that stone axes were lost by travelers during their journeys along routes predetermined by the environment and the topography. However, the mechanisms of exchange (or cultural transmission) and the material evidence of such processes remain unspecified. Instead, the focus of interest is on settlement structure in the lowland areas or the origin of the artifacts. The mountainous regions between their point of origin and destination receives insufficient attention (Coblenz 1953:122, Simon and Hauswald 1995:9, Reinecke 1930:10).

A third argument generally recognizes Neolithic artifacts as evidence of human activity, and focuses on hunting and the search for wood, honey, fruit, stone, and ore deposits. Thus, for example, the Black Forest has been declared "Neolithic man's favorite hunting ground," a source of wood (Lais 1937:56 ff.), and a storehouse for blood jasper and cornelian (Paret 1925). Similar suggestions have been made with respect to other low mountain ranges such as the Thuringian Forest in Germany. It is proclaimed to be an early source of organic and mineral raw materials (Müller 1982/83:275, cf. Nowothnig 1953:15, 1958:115, 1959:55, Wein 1978:90). Recently, evidence of Neolithic quarrying for hematite has been established in the Black Forest, and other conspicuously colored minerals may also have attracted attention (Goldenberg *et al.* 1997). Neolithic peoples would thus have visited mountainous regions in search of raw materials otherwise unobtainable. However, rather than being a fundamental feature of the economy, these activities are seen as having supplemented a lowland-based way of life determined by agriculture and farming. Here too, Neolithic finds from mountainous areas are seen as the result of occasional losses.

A fourth argument hypothesizes a short-term presence of Neolithic people in the low mountain region. Stone axes are seen as deliberate deposits in unsettled areas associated either with drinking water or other conspicuous topographic features. S. Winghart (1987) expressed this view most directly with regard to both the Bavarian Forest and the Black Forest. However, as with the thunderbolt hypothesis, his study is characterized by a rather incomplete account of the archaeological context of the stone axes (Valde-Nowak 1995:158). Similarly, Winghart's (1987: 106-126) conception of the stone axes as weapons rather than tools remains problematic, because it depends solely on the assumption of a ritual cause for their deposition. Weapons, it seems, would be worth depositing while tools would not. The interpretation made by Winghart of the presence of Neolithic groups in terms of ritual rather than economics entails a specific interpretation of stone axes that remains unproven.

While none of these approaches should be entirely rejected, it is clear that they fail to account for the archaeological evidence from highland regions as a whole. For example, the secondary use of stone axes as ritual objects is a well-established fact (Reitinger 1976). To argue, however, that this practice utterly precludes the possibility of Neolithic land use is to ignore their archaeological context. Thus, as early as the 1930s, attention was drawn towards the existence of Neolithic finds other than stone axes, providing evidence against the ritual object ("thunderbolt") hypotheses (Lais 1937:45, Nowothnig 1959:51, Raddatz 1972a:341, 1972b:18).

A growing body of "stray finds" has been collected and there is now a large series of Neolithic chert and flint artifacts from the highlands (Valde-Nowak 1995a, 1999). On the basis of this evidence the majority of the axes should also be taken as an indicator of prehistoric economic activity. With regard to the Late Neolithic date of most stone axes (e.g., Winghart 1987) it should be noted that during this period there are hardly any deposits of axes known from lowland areas (Vencl 1975:12, 68). Furthermore, the apparent chronological clustering of axes in mountainous areas cannot be explained by the assumption of (early) historic trade from the lowlands where there is evidence of settlement throughout the Neolithic period (Raddatz 1972b:18, Gohlisch and Reisch 1999, Valde-Nowak 1995a, 1999, Valde-Nowak and Weißmüller 1994). On the other hand, with regard to the assumption of prehistoric exchange across the highlands, attention must be drawn to the actual topographic situation of most Neolithic finds (Valde-Nowak 1995a, 2002, Valde-Nowak and Weißmüller 1994, Winghart 1987:109-127, 131-134). Many of these stone axes, as well as flint artifacts, have been discovered away from rivers and natural passes. Instead they often occur in the dead end of valleys or plateaus and other regions that cannot be regarded as obvious transfer routes or passes, as in the case of Freiamt, Germany.

Studies analyzing both highlands and the surrounding lowlands provide a better understanding of the gradual expansion of Neolithic land use. Research on settlement structure and economy in neighboring zones of different topographic and climatic conditions is provided through surveys in the *Vogelsberg*, in Hessen, (Rehbaum-Keller 1984), the Rhine Valley and the *Rheinische Schiefer-gebirge* (Frank 1998), all in Germany, as well as in parts of the Ore Mountains (*Erzgebirge*) (Christl 1989, Beneš *et al.* 1992), and the lower parts of the Bohemian Forest (*Böhmer Wald*), both on the Czech-German border (Baštová 1986). However, much of this work suffers from a lack of clearly formulated concepts related to the process of expansion and its socio-economic underpinnings (cf. also Dauber 1934, Lais 1937, Nowothnig 1959, Raddatz 1972b, Biel 1974, Müller 1982/83, Löhr 1985, Weißmüller 1986, Schlichtherle 1988, Dehn 1999).

THE JOINT HIGHLAND PROJECT

A detailed multivariate verification of Neolithic finds in the different highland areas was recently concluded (Valde-Nowak 2002). This *Joint Highland Project* includes the Polish Carpathians and selected regions of the German highlands.[1]

The first synthesis of the Polish Carpathian research described material from the TRB (Valde-Nowak 1988). It demonstrated an expanded settlement range to the south. Following the Carpathian work, Polish studies turned to still ongoing research in selected regions of the German highlands. First a simple archaeological analysis was carried out in the Bavarian Forest, Black Forest, and the *Swebian Alb* (Suavian Alps) using criteria such as find type (axe, hammer-axe, etc.), find circumstances or context (i.e. on the surface, in soil, in a river, etc.), chronology of the material, and so on.

In all analyzed highlands areas, the number of axes exceeds the number of hammer axes. The Later (Young), Late, and Final Neolithic forms exceed the number of Early and Middle Neolithic finds. Some of the discoveries were made in areas covered by vegetation, which is disadvantageous for discovery. Results suggest that:

- The complex of finds from each highland region represents similar tendencies (chronology, etc.), a strong argument for their authenticity.
- A high number of undiscovered traces of Neolithic activity remain in the large vegetated zone, dominant in all Central European highlands.

The second phase in the analysis is connected with settlement and geographic analysis. Topographic and hypsometric location, soil class of the area of an accidental discovery, distance to water, and direction to cardinal points were recorded. Based on these results a topographic location model for "stray finds" has been delineated. Frequent location in places between river sources and the edge of the upland near local saddles, highland passes, and a generally high-altitude watershed zone represent the typical situation where highland Neolithic materials have been found.

CENTRAL EUROPEAN SURVEYS AND RESULTS

The next step in the analysis of the "stray" or more precisely isolated finds was a systematic survey in the Upper Harz Mountains (Valde-Nowak *et al.* 2004), Bavarian Forest, and Black Forest, all in Germany, as well as the Sudety and Carpathians Mountains in Poland. In all cases, surveys were undertaken in areas where an axe or hammer-axe had accidentally been found in the past (e.g., during agricultural work). The aim is to ascertain if "stray finds" have a material archaeological context.

A small valley (Zellertal) and a part of the larger Black Regen River valley in the Bavarian Forest were surveyed (Valde-Nowak and Weißmüller 1994). After two years of surveying, 28 sites with Neolithic stone implements, most probably with Cham culture affinities (blade and flake tools, splintered pieces, etc.), were collected. The most important discovery occurred in Dürfeld, where an axe together with a flint blade were found a few meters from each other (Weißmüller 1995).

In Freiamt-Ottoschwanden, in the Black Forest highlands, where five stone axe had been found previously, a two season survey of 58 agricultural fields yielded 166 mostly Neolithic stone implements (Valde-Nowak and Kienlin 2002, Kienlin and Valde-Nowak 2004). It is known that a boy found one of the stone axe on his way to school and gave it to his teacher in Freiamt-Langacker in the 1940s. We interviewed the discoverer and other informants, obtaining detailed information concerning the topography of the old find. In the process the archaeological context of the axe and additional old discoveries were revealed. Twelve Neolithic flint artifacts (but no pottery) occurred close to the place were the axe was reportedly found in the field.

A Polish-German joint project took place in the Sudetey Mountains. During 1996-1998 the systematic survey documented over 50 Neolithic sites. These sites were often close to known isolated finds, formerly interpreted as stray finds thought to be without context (Valde-Nowak 1999).

Similar surveys were conducted in the Beskidy Mountains of the Polish Carpathians. Among the resulting discoveries was the irregular outline of a 1.8 x 1.3 m feature with potsherds and possible traces of two posts, located 544 m above sea level, which can be interpreted as an eroded Baden culture grave (Valde-Novak 2008). This contrasts with older "stray finds", which were usually only sporadically documented. Therefore, interviews with the still living discoverers were conducted to recover contextual data. One good example comes from a high local pass in the village of Zawoja (Fig. 19.1). About 30 years ago, a Late Neolithic hammer-axe was found (Tunia 1978). The interview led to the precise location of where the isolated Corded Ware culture axe and excavation provided contextual data, including small Neolithic flint artifacts. A similar example from the Cergowa Mountains (Valde-Nowak 2001b) concerns a TRB hammer-axe that was found in the upper reaches of a local pass during agricultural work (Fig. 19.2). A year later archaeologists and the original discoverer visited the site and found a small collection of Neolithic flint artifacts (Fig. 19.3).

[1] The main difference between the Polish Carpathians and the German highlands as well as the Sudeten Mountains (Sudety or Sudetes) on the Polish/Czech border is that the Polish Carpathians have extensive foothills with highland characteristics.

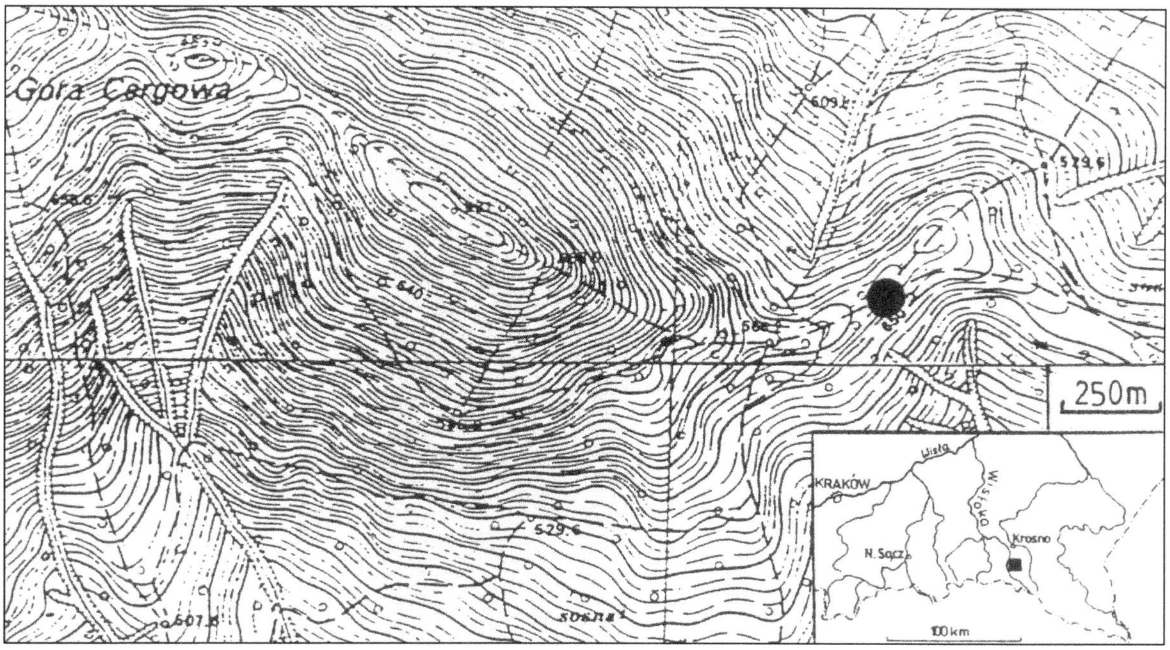

Figure 19.2. Lubatowa, Site 8; topography of the Later Neolithic (Funnel Beaker Culture) campsite in the Cergowa Góra Mountains (Lower Beskids, Polish West Carpathians) (after J. Gancarski, unpublished, supplemented by the author)

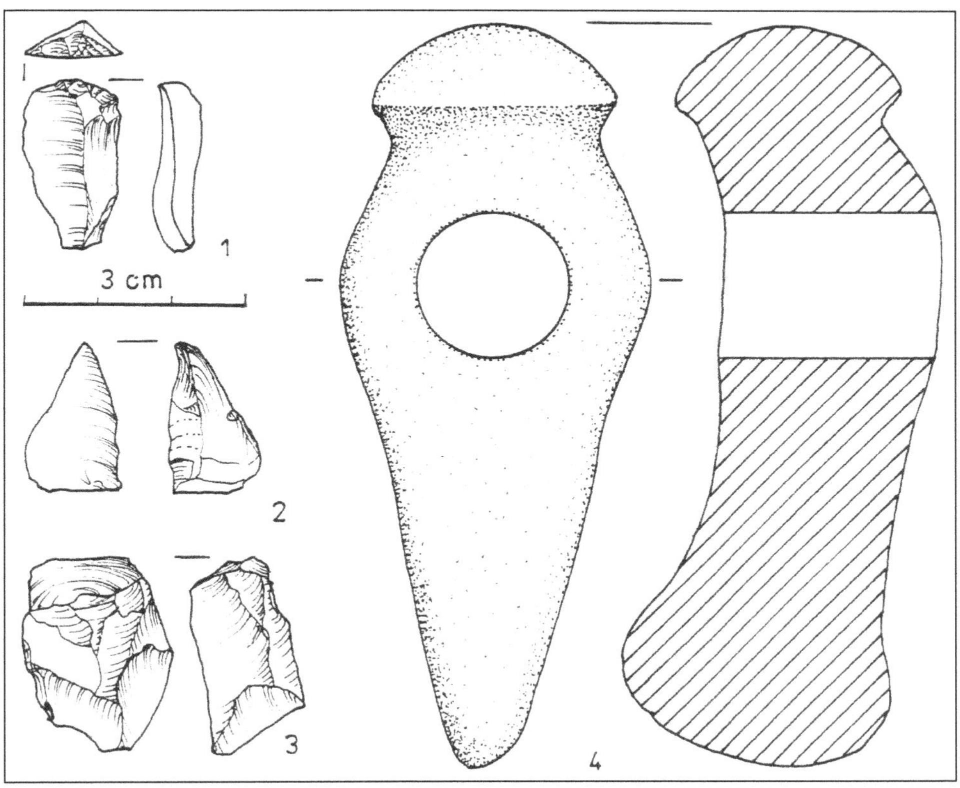

Figure 19.3. Lubatowa, Site 8; typical stone tool inventory from the Later Neolithic (Funnel Beaker Culture) campsites in the Polish West Carpathians (survey of J. Gancarski, unpublished; material located in the Sub-Carpathians Museum in Krosno)

Based on the above experiences the results of archaeological surveys and verification of archival data, generally speaking suggest that:

- We can assume that numerous enigmatic "stray finds" of various Neolithic axes have a characteristic archaeological context;

- We may consider the existence of an unknown form of Neolithic activity in the European highlands, i.e., seasonal campsites represented not only by various axes, but also by collections of small flint artifacts.

"STRAY FINDS" AND ALPINE ICEMAN

The Alpine Neolithic seems to differ from that in the Central European highlands. Archaeological research in the Alpine Neolithic has long included an abundant series of "stray finds" and evidence of stable settlement in exposed sites evidenced by ceramics and immovable objects (Leitner 1985, Lunz 1986). In spite of this fact, data from the mountains receives little attention in reports and later syntheses, including cartographic ones (Pauli 1980, Uslar 1995). However, a change in the Alpine research approach and formulation of a new methodology was announced by the International Colloquium PAESE'97 in Zürich (Della Casa 1999). It was brought about by an increasingly extensive Polish-German collaboration in studying the highland zones of Europe, as well as the sensational discovery of Ötzi the Iceman. The Iceman was found in the Tyrolean Ötztal Alps, on the border between Italy and Austria (Egg and Spindler 1993, Bagolini *et al.* 1995, Gostner *et al.* 2004). He was trapped in the Similaun Glacier during a period of sudden climate change near the end of the 4th millennium (Baldia this volume). The Neolithic "mummy" was released from the glacier in 1991 due to global warming.

Aside from the extraordinary elevation, the topography of the Iceman site is characteristic of many highland finds, it being close to a local pass. The only real difference is the exceptionally good preservation of the body, clothing, and equipment. Had the iceman died in a highland zone, we would have found an axe (in this case made of copper) and some flint artifacts, but no pottery or anything else. After 5000 years, everything made of organic material would have been destroyed. This reduced inventory would principally be the same as that of the Central European highland discoveries: an axe or hammer-axe and a poor series of flint artifacts.

SUMMER MOUNTAIN PASTURAGE, PALAEOBOTANY, AND ETHNOARCHEOLOGY

It is well known that mountainous territories are rich in peat bogs. Many pollen diagrams confirm the existence of Neolithic populations in various phases of this period, although mostly in the younger ones. This coincides with the tendency to intensify occupation during the Young and Late Neolithic. However, palaeobotanical data from mountainous regions cannot be compared with the anthropogenic effects documented in pollen diagrams from the lowlands (Hicks 1998).

Pollen profiles from the highlands do not present evidence of large-scale deforestation, burning, or crop cultivation, although human impact is recognizable. Some indicators are characteristic of pastoral activity. Many palaeobotanists agree that the impact can be explained as an effect of cattle herding, using leaf fodder in forests with only rare and very limited cleared areas (e.g., Knipping 1989; comp. Behre 1981, Austad 1988, Frenzel 1997).

In this context, ethnographic data should be kept in mind. In 1995, the author documented cattle herdsmen living in the Polish Carpathians (Silesian Beskids) who still practiced leaf foddering. Previously, experiments with leaf fodder were undertaken in Lejre in Denmark and analyses of dung layers from Swiss Neolithic sites (Baldia this volume) show clear evidence for the use of a broad spectrum of tree species (Haas and Rasmussen 1993, Rasmussen 1989, 1993).

Based on the above and as a result of these Polish-German collaborations, a new explanation for the use of specific highland sites is being proposed (Valde-Nowak 2002, Valde-Nowak and Kienlin 2002, Kienlin and Valde-Nowak 2004), suggesting pastoral activity close to transhumance (Chang and Koster 1986, Cribb 1991, Bar-Yosef and Khazanov 1992, Bartosiewicz and Greenfield 1999). This implies that outside of the traditional centers of Neolithic settlement, the landscape of non-loess areas dominated by dense forest during the Neolithic were used as fodder storage for cattle. In the summer, a herd was fed far from the stable settlement site to protect the foliage fodder resources around the villages. Thus, fodder from the vicinity of the home settlement was being saved for the winter. It can be presumed that various aspects of fodder storage and feeding were practiced in both the stable lowland settlements and the seasonal campsites in the hills. These included leaf and twig collection for winter (coppicing, shredding, and pollarding) and an economy of forest management in the neighborhood of permanent settlements.

TROUBLESOME DIFFERENCES IN ARCHEOLOGICAL INTERPRETATIONS OF HIGHLAND FINDS

There are differences in the archaeological interpretation in the Polish and German studies that are a consequence of the historical development of archaeology, including differing experiences and archaeological resources, in the two countries. Without going into details, two differences of the Polish-German highland projects need to be addressed.

The first problem relates to the unanimously accepted assumption in Western and Northern European countries that Neolithic populations used young shoots and leaves of deciduous trees as fodder for herds, mostly cattle. Terms such as pollarding, coppicing, or "forest management" have their own important positions in archaeological and paleobotanical hypotheses. These interpretations are exceptional in Polish Neolithic research (Wiślański 1979), which has focused on the role of grassland communities in the loess uplands as the basis of economic and settlement models for some time (Kruk

1980, 1981). As a result, attempts at extrapolating the loess land model to the Carpathians (Machnik 1998, 2001) have been completely erroneous, resulting in an archaeological controversy.

According to Machnik (2001), the TRB population supposedly engages in heavy deforestation in a large part of the region and survives to the start of the Corded Ware Culture. Through mass thinning of the forest, this activity by the TRB produces optimal conditions for herding. However, Carpathian palynological data do not suggest extensive expanses of deforested terrain or thinning through fire (Wacnik *et al.* 2001: 219). Indirect archaeological evidence of extensive deforestation should be preserved in burial mounds, but these and other TRB sites are not numerous and represent brief occupations (Zych 2004). Furthermore, the sites are only distributed in some regions and do not necessarily occur in the same area as the Corded Ware sites. Therefore, it is difficult to determine the relationship and economic activity of the two cultures.

The core of the problem, as elaborated by the Polish-German projects, is that the activities of the Neolithic populations (mainly in Late and End Neolithic) were centered on terrain with primeval forest, not deforested areas, as some would like to imagine. Paleobotanical diagrams, suggesting specific and limited deforestation through pasturage, confirm this. Therefore, conclusions from studies of highland sites point to a completely different behavior of Neolithic communities than those accepted for the European lowlands.

Another equally important problem is the long and advanced tradition of Polish Neolithic flint industry research, which includes household inventories, quarries, and processing sites. This research is missing in Germany, even though some very valuable exceptions exist (Reisch 1974, Arora 1986, Hassmann 2000, Tillman 2001). As noted, the verification of numerous Neolithic highland sites is based on finding small, non-ceramic inventories composed of flint artifacts, usually the byproducts of tool manufacture and other activities, which leave few traces. Thus, they are extremely difficult to date, especially since they are usually surface collections without a wider context. In the case of Poland the problem is less pronounced, because typological implications and manufacturing techniques of lithic material can be matched with existing Polish typological criteria. In German archaeology patterns in the lithic artifacts are less clear, making dating attempts of Neolithic flint material from the highland sites more difficult. Undeniably these two problems still produce barriers, which in turn will polarize Polish and German opinions in collaborative research of Neolithic highland settlement.

FINAL REMARKS: HIGHLAND OCCUPATION AND CLIMATE

The material spectrum of Neolithic activity reflects a mobile and most probably seasonal way of life. Early evidence for this observation came from the Polish Carpathians, where paleobotanical research had a significant impact on the development of archaeological investigations. Paleobotanists outdistanced archaeologists, predicting Neolithic occupation of the Polish Carpathians from paleobotanical data over 20 years before archaeologists reached the same conclusion (Ralska-Jasiewiczowa 1968; compare to Wacnik *et al.* 2001). However, the situation is complicated.

During the LBK the climate should have been most favorable for locating villages in the highlands (Baldia this volume), yet there is only limited evidence for this. For example, a group of villages situated on the southern foot of the Tatra Mountains (at an elevation of 700 masl) in the upper Poprad River basin, including Poprad-Matejovce (Novotny 1983), Strane pod Tatami (Sojak 2002), and Ganovce "Za Stodolami" (Sojak 1999, 2000) have been excavated (Fig. 19.4). Similar examples occur in the northern Carpathians, where an actual excavated long house of the LBK was located at Łoniowa (Site 18) on the top of the Wiśnicz Foothills (Fig. 19.5) (Excavations by the author, unpublished; see also Valde-Nowak 1998). In Germany, highland sites are known from the Usinger Becken (Baldia this volume; Wotzka *et al.* 2001) and Eschlipp in the Frankonian Jura (unpublished, Christian Züchner, personal communication). In addition, paleobotanical evidence of such Early Neolithic highland occupations was discussed some years ago by Kalis and Zimmermann (1988) from an archaeological point of view.

All above discussed LBK examples from the highlands represent large villages, with long houses, numerous pieces of pottery, and flint artifacts. The material culture of these highland sites do not differ from contemporary sites in the lowlands. However, in the later Neolithic, this situation changes. Generally climate conditions were less favorable for settlement in the highlands, yet in all Central European highlands the strong tendency to pastoral exploitation is well documented. Thus, highland

Figure 19.4. Poprad-Matejovce, Northern Slovakia; the area of one of the highest located villages of the Early Neolithic (LBK) in Europe. The Tatra Mountains are visible in the background. On the right is Dr. Marian Soják, the excavator of the neighboring LBK settlement in Stráne pod Tatrami (Photo: P. Valde-Nowak)

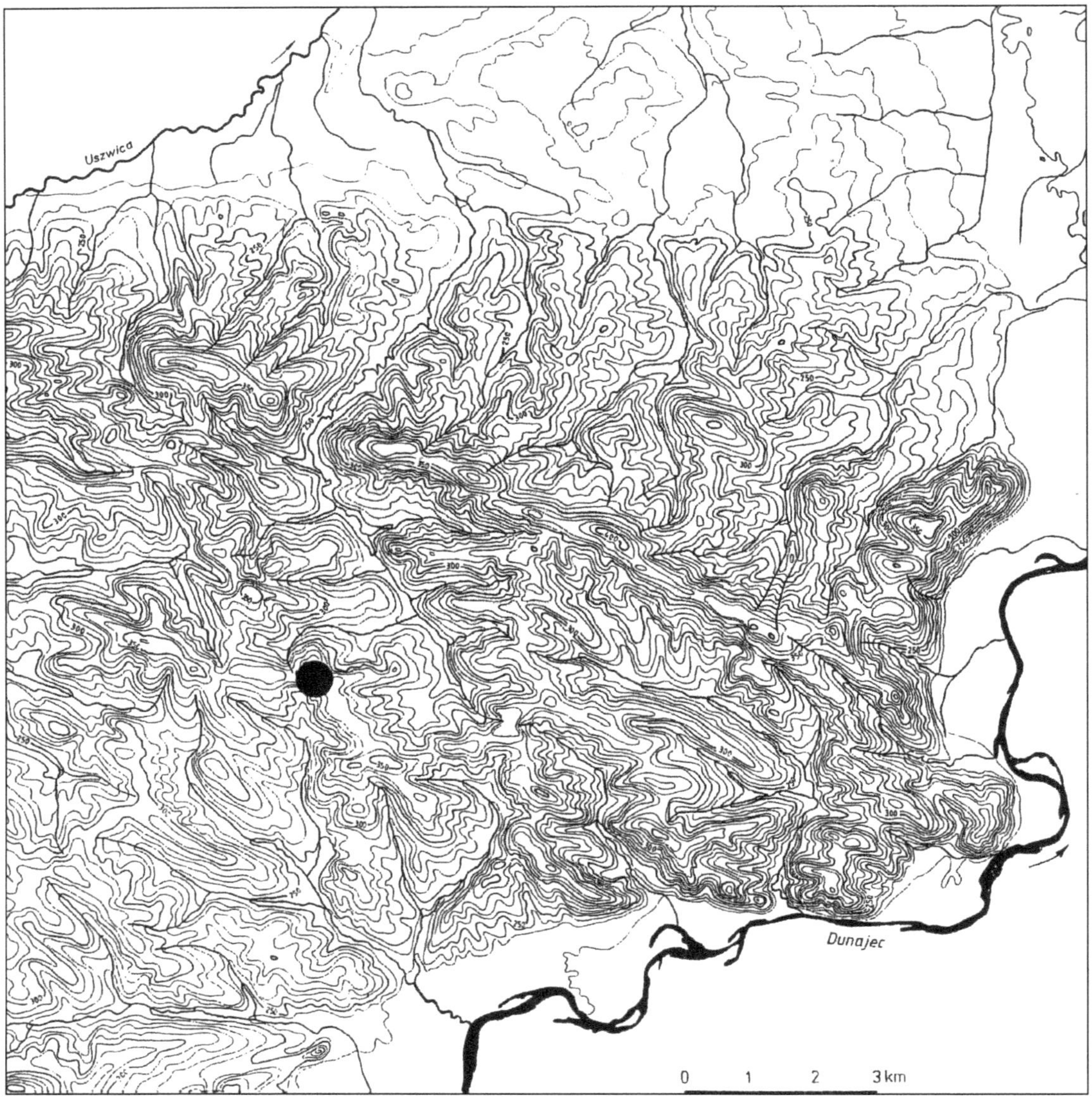

Figure 19.5. Łoniowa, Site 18; topography of the Early Neolithic (LBK) settlement situated on the top of Wiśnicz (Wiśnickie) Foothills (Polish West Carpathians) (after P. Valde-Nowak 2001b)

sites of the much later TRB, Cham, or Corded Ware cultures are very poor in ceramic artifacts. They are mostly represented by axes and a few flint pieces, and they do not provide any evidence of dwellings.

Although the opinion that Neolithic people infiltrated the highlands primarily as pastoralists is relatively unique and only one of several explanations to account for highland use, it is the most tangible one. It finds support in paleobotany and direct archaeological information. Similar suggestions concerning the possibility of existing Neolithic seasonal herding in the highlands have already been proposed (Kalis and Zimmermann 1988, Kruk 1980:118, 331, 1981, Kruk and Milisauskas 1999:250-257, Lüning 2000:39, 41, 148-150), but these were based solely on paleobotanical and not archeological information.

If the highlands were the domain of stable forms of human pastoral activity during the Neolithic (i.e., seasonal herding), then the term periphery for this type of settled area, especially in the Later, Late and End Neolithic, may be inadequate. Instead, these territories can be viewed as centers or epicenters of a different kind, characterized by behavior known only from these regions. Taking into account the less favorable environmental and climatic characteristics for agricultural activity in the beginning of the Subboreal phase of the Holocene, we should consider the increase of archaeological finds in the highlands to be a result of economic advances and demographic pressure and not response to climate changes.

References

ARORA, S.K. 1986 – Metallzeitliche Flintindustrie. Formenkundliche Aspekte einiger metallzeitlicher Steingeräte. In W. Hilgers (ed.), *Das Rheinische Landesmuseum Bonn.* Berichte aus der Arbeit des

Museums 1986/3-4. Rheinland-Verlag, Pulheim, 1986.

AUSTAD, I. 1988 – Tree pollarding in Western Norway. In H.H. Birks, H.J.B. Birks, P.E. Kalland and D. Moe (eds.), *The Cultural Landscape: Past, Present and Future*, Cambridge University Press, Cambridge, 1988:11-29.

BAGOLINI, B., L. DAL RI, A. LIPPERT and H. NORTDURFTER 1995 – Man im Eis: Die Fundbergung 1992 am Tisenjoch, Gem. Schnals, Südtirol. In K. Spindler, E. Rastbichler-Zissernig, E.H. Wilfing, D. zur Nedden and H. Northdurfter (eds.), *Der Man im Eis: Neue Funde und Ergebnisse*, Springer, Wien and New York, 1995, 3-23.

BARTOSIEWICZ, L., and H.J. GREENFIELD (eds.) 1999 – *Transhumant Pastoralism in Southern Europe*, Archaeolinqua, Budapest, 1999.

BAR-YOSEF, O., and A. KHAZANOV 1992 – Introduction. In O. Bar-Yosef and A. Khazanov (eds.), *Pastoralism in the Levant. Archaeological Materials in Anthropological Perspectives.* Monographs in World Archaeology 10, Prehistory Press, Madison, Wisconsin, 1992.

BAŠTOVA, D. 1986 – Vyvoj Pravekeho Osidleni Povodi Strely. *Archeologicke Rozhledy* 36, 1986:156-72.

BEHRE, K.H. 1981 – The Interpretation of Anthropogenic Indicators in Pollen Diagram. *Pollen and Spores* 23/1, 1981:225-245.

BENEŠ, J., M. KUNA, J. PEŠKE and M. ZVELEBIL 1992 – Rekonstrukce Stare Kulturni Krajiny v Severni Časti Čech: Československo-Britsky Projekt po Prvni Sezone Vyskumu. *Archeologicke Rozhledy* 44, 1992:337-342.

BIEL, J. 1974 – Zur neolithischen Besiedlung der Schwäbische Alb. *Fundberichte aus Baden-Württemberg* 1, 1974:53-64.

CHANG, C., and H.A. KOSTER 1986 – Beyond Bones: Toward an Archaeology of Pastoralism. *Advances in Archaeological Method and Theory* 9, 1986:97-148.

CHRISTL, A. 1989 – Höhengrenzen der urgeschichtlichen Besiedlung im Erzgebirge und dessen Umland, dargestellt an einem Ausschnitt. *Archeologicke Rozhledy* 39. 1989:386-405.

COBLENZ, W. 1953 – Vorgeschichtliche Besiedlungsnachweise im Erzgebirge. *Germania* 31, 1953:122.

CRIBB, R. 1991 – *Nomads in Archaeology*. Cambridge University Press, Cambridge.

DAUBER, A. 1934 – Die Steinbeile des nördlichen Schwarzwaldes. *Mannus* 26, 1934:237-243.

DEHN, R. 1999 – Ur- und Frühgeschichte. Von den Anfängen bis zum Beginn der Römerzeit im Landkreis Emmendingen. *Kreisbeschreibungen des Landes Baden-Württemberg*, Theiss, Stuttgart, 1999.

DELLA CASA, Ph. (ed.) 1999 – *Prehistoric Alpine Environment, Society, and Economy*, Habelt, Bonn, 1999.

EGG, M., and K. SPINDLER 1995 – Die Gletschermumie vom Ende der Steinzeit aus den Ötztaler Alpen: Vorbericht. *Jahrbuch des Römisch-Germanischen Zentralmuseums Mainz*. 39, 1995:3-113.

FRANK, T. 1998 – *Die neolithische Besiedlung zwischen der Köln-Bonner Rheinebene und dem Bergischen Hochfläche*, Deutsche Gesellschaft für Ur- und Frühgeschichte, Habelt, Bonn.

FRENZEL, B. 1997 – Vegetationsgeschichtliche und paläoökologische Probleme des Schwarzwaldes. *Veröffentlichungen des Alemanischen Instituts* 64, 1997:11-40.

FURHOLT, M. 2003 – *Die absolutchronologische Datierung der Schnurkeramik in Mitteleuropa und Südskandinavien.* Universitätsforschungen zur prähistorischen Archäologie 101. Habelt, Bonn, 2003.

GOHLISCH, T.H., and L. REISCH 1999 – Ein Survey im nordwestlichen Eulengebirge. In P. Valde-Nowak (ed.), *Die Anfänge der Besiedlung in den Sudeten*, Institute of Archaeology and Ethnology Polish Academy of Sciences, Kraków, 1999:1909-140.

GOLDENBERG, G., M. KAISER, and A. MAASS 1997 – Neolithischer Hämatitbergbau bei Sulzburg, Kreis Breisgau Hochschwarzwald. *Archäologische Ausgrabungen in Baden Württemberg* 1997:33-35.

GOSTNER, P., E. VIGL, and U. REINSTADLER 2004 – Der Mann aus dem Eis: Eine paläoradiologisch-forensische Studie zehn Jahre nach der Auffindung der Mumie. *Germania* 82, 2004:63-107.

HASSMANN, H. 2000 – *Die Steinartefakte der befestigten neolithischen Siedlung von Büdelsdorf, Kreis Rendsburg-Eckerförde*, Habelt, Bonn.

HAAS, J.N., and P. RASMUSSEN 1993 – Zur Geschichte der Schneitel- und Laubfutterwirtschaft in der Schweiz: Eine alte Landschaftspraxis kurz vor dem Aussterben. *Dissertationes Botaniceae* 196, J. Cramer, Berlin and Stuttgart, 1993.

HICKS, S. 1998 – Fields, Boreal Forests and Forest Clearings as Recorded by Modern Pollen Deposition. In M.-J. Galliard, B.E. Berglund, B. Frenzel and U. Huckriede (eds.), *Quantification of Land Surfaces Cleared of Forests During the Holocene: Modern Pollen/Vegetation/Landscape Relationship as an Aid the Interpretation of Fossil Pollen Data.* Paläoklimaforschung 27, Fischer, Stuttgart, Jena, Lübeck and Ulm. 1998:53-66.

KALIS, A and A. ZIMMERMANN 1988 – An Integrative Model for the Use of Different Landscapes in Linearbandkeramik Times. In J. Bintliff, D.A. Davidson and E. Grant (eds.), *Conceptual Issues in Environmental Archaeology*, Edinburgh University Press, Edinburgh, 1988:145-152.

KIENLIN, T.L., and P. VALDE-NOWAK 2004 – Neolithic Transhumance in the Black Forest Mountains, SW Germany. *Journal of Field Archaeology* 29/1-2, 2004:29-44.

KNIPPING, M. 1989 – Zur spät- und postglazialen Vegetationsgeschichte des Oberpfälzer Waldes.

Dissertationes Botanicae 140, J. Cramer, Berlin and Stuttgart, 1989.

KRISTIANSEN, K. 1987 – Centre and Periphery in Bronze Age Scandinavia. In M.J. Rowlands, M. Larsen and K. Kristiansen (eds.), *Centre and Periphery in the Ancient World*, Cambridge University Press, Cambridge, 1987:385-403.

KRUK, J. 1980 – *Gospodarka w Polsce południowo-wschodniej w IV-II tys. p.n.e.* Ossolineum, Wrocław, Warszawa, Kraków and Gdańsk, 1980.

KRUK, J. 1981 – Uwagi o rolnictwie neolitycznym w dorzeczu górnej Wisły' 21 *Acta Archaeologica Carpathica,* 1981:213-24.

KRUK, J., and S. MILISAUSKAS 1999 – *Rise and Fall of Neolithic Communities*, Institute of Archaeology and Ethnology Polish Academy of Sciences, Kraków, 1999.

KÜMMEL, C. 1998 – Marxistische Perspektiven in der gegenwärtigen englischsprachigen Archäologie. In M.K.H. Eggert and U. Veit (eds.), *Theorie in der Archäologie: Zur englischsprachigen Diskussion.* Waxmann, Münster, New York, München and Berlin, 1998:115-181.

LAIS, R. 1937 – Die Steinzeit im Schwarzwald' 13 *Badische Fundberichte,* 1937:29-66.

LEITNER, W. 1985 – *Die Urzeit. Geschichte des Landes Tirol.* Athensia, Bozen.

LÖHR, H. 1984 – Sammeln oder suchen? Anmerkungen zur archäologischen Feldbegehung. *Archäologische Informationen* 8/2, 1984:102-110.

LUNZ, R. 1986 – *Vor- und Frühgeschichte Südtirols: Die Steinzeit.* Manfrini, Trento, 1986.

LÜNING, J. 2000 – *Steinzeitliche Bauern in Deutschland: Die Landwirtschaft im Neolithikum.* Habelt, Bonn, 2000.

MACHNIK, J. 1997 – Uwagi o Najstarszym Osadnictwie Pasterskiej Ludności Kultury Ceramiki Sznurowej (III Tysiąclecie Przed Chr.) w Strefie Karpackiej. In J. Gancarski (ed.), *Dzieje Podkarpacia*, Podkarpackie Towarzystwo Historyczne 2. Krosno, 1997:99-120.

MACHNIK, J. 2001 – Kultura Ceramiki Sznurowej w Strefie Karpackiej (Stan I Perspektywy Badawcze). In J. Gancarski (ed.), *Neolit i początki epoki brązu w Karpatach Polskich*, Muzeum Podkarpackie w Krośnie, Krosno, 2001.

MILDENBERGER, G. 1959 – Zur Frage der neolithischen Besiedlung der Mittelgebirge. *Jahresschrift für mitteldeutsche Vorgeschichte* 43, 1959:76-86.

MILDENBERGER, G. 1969 – Verschleppte Bodenfunde: Ein Beitrag zur Fundkritik. *Bonner Jahrbuch des Rheinischen Landesmuseums in Bonn (im Landschaftsverband Rheinland) und des Vereins von Altertumsfreunden im Rheinlande* 169, 1969:1-28.

MILISAUSKAS, S. (ed.) 2002 – *European Prehistory. A Survey*, Kluwer Academic/Plenum Publishers, New York, Boston, Dordrecht, London and Moscow, 2002.

MÜLLER, D.W. 1983 – Gebirgsfunde aus dem Gothaer Land als Zeugen der Erschließung des Thüringer Waldes in Urgeschichtlicher Zeit, 22 *Acta Archaeologica Carpathica,* 1983:267-276, 1995.

MÜLLER, J. 2001 – *Soziochronologische Studien zum Jung- und Spätneolithikum im Mittelelbe-Saale-Gebiet (4100-2700 v. Chr.). Eine soziohistorische Interpretation prähistorischen Quellen,* VML, Rahden, Westfalen.

NIESZERY, N. 1995 – *Linearbandkeramische Gräberfelder in Bayern,* Internationale Archäologie16, VML, Espelkamp, 1995.

NOVOTNY, B. 1983 – Die Ausgrabungen in Poprad-Matejovce: Ein Beitrag zur Neolithisierung der Gebirgsregionen. *Acta Archaeologica Carpathica* 22. 1983:225-235.

NOWOTHNIG, W. 1953 – Zur Ur- und Frühgeschichte des Oberharzes. *Die Kunde*, N.F. 4/1-2, 1953:9-17.

NOWOTHNIG, G. 1958 – Die Vor- und Frühgeschichte des Oberharzes im Lichte neuerer Bodenfunde. *Jahreschrift für Mitteldeutsche Vorgeschichte* 41/42, 1958:107-225.

NOWOTHNIG, G. 1959 – Die Steinbeifunde im Oberharz: Zeugen einer steinzeitlichen oder mittelalterlichen Begehung des Gebirges. *Die Kunde* N.F. 10, 1959:51-61.

PARET, O. 1925 – Steinbeile im württembergischen Schwarzwald. In H. Mötefindt (ed.), *Studien zur vorgeschichtlichen Archäologie: Alfred Götze zu seinem 60 Geburtstag.* Kabitzsch, Leipzig, 1925.

PAULI, L. 1980 – *Die Alpen in Frühzeit und Mittelalter:. Die archäologische Entdeckung einer Kulturlandschaft.* Theiss München.

PETRASCH, J. 2001 – "Seid fruchtbar und mehret euch und füllet die Erde und machet sie euch untertan": Überlegungen zur demographischen Situation der bandkeramischen Landnahme. *Archäologisches Korrespondenzblatt* 31, 2001:13-25.

RADDATZ, K. 1972a – Zur Besiedlung des Solling in der Steinzeit: Zugleich ein Beitrag zur Frage der verschleppten Bodenfunde im südniedersächsisches Bergland. *Nachrichten aus Niedersachsens Urgeschichte* 41, 1972:1-21.

RADDATZ, K. 1972b – Probleme einer archäologischen Landesaufnahme im niedersächsischen Mittelgebirgsgebiet. *Neue Ausgrabungen und Forschungen in Niedersachsen* 7, 1972:341-380.

RALSKA-JASIEWICZOWA, M. 1968 – Ślady Osadnictwa Prehistorycznego w Diagramach Pyłkowych z Terenu Polski. *Folia Quaternaria* 29, 1968:163-82.

RASMUSSEN, P. 1989 – Leaf-foddering of Livestock in the Neolithic: Archaeobotanical Evidence from Weier, Switzerland. *Journal of Danish Archaeology* 8, 1989:51-71.

RASMUSSEN, P. 1993 – Analysis of Goat/Sheep Feces from Egolzwil 3, Switzerland: Evidence for Branch and Twig Foddering of Livestock in the Neolithic. *Journal of Archaeological Science* 20, 1993:479-502.

REHBAUM-KELLER, A. 1984 – Archäologisch-ökologische Studien zur vorgeschichtlichen Besiedlung von Wetterau und Vogelsberg. Unpublished Ph. D. Thesis, University of Gießen. Gießen, 1984.

REINECKE, P. 1930 – Die Grenzen vor- und frühgeschichtlicher Besiedlung Nordostbayerns. *Der Bayerische Vorgeschichtsfreund* 9, 1930:1-16.

REISCH, L. 1974 – *Der vorgeschichtliche Hornsteinabbau bei Lengfeld, Landkreis Kehlheim und die Interpretation "grobgerätiger" Silexindustrie in Bayern*, Materialhefte zur bayerischen Vorgeschichte 29, Michael Lassleben, Kallmünz, Opf., Germany, 1974.

REITINGER, J. 1976 – "Donnerkeile" aus Oberösterreich und Salzburg. In H. Mitscha-Märheim, H. Friesinger and H. Kerchler (eds.), Festschrift für Richard Pittioni zum siebzigsten Geburtstag II, Industriearchäologie und Metalltechnologie, Berger und Söhne Wien, 1976:511-546.

ROWLANDS, M. 1987 – Centre and periphery: a review of a concept. In M.J. Rowlands, M. Larsen and K. Kristiansen (eds.), *Centre and Periphery in the Ancient World*, Cambridge University Press, Cambridge. 1987:1-12.

SCHLICHTHERLE, H. 1987 – Das Jung- und Endneolithikum in Baden Württemberg. Zum Stand der Forschung aus siedlungsarchäologischer Sicht. In D. Planck (ed.), *Archäologie in Württemberg, Ergebnisse und Perspektive archäologischer Forschung von der Altsteinzeit bis zur Neuzeit.* Theiss, Stuttgart. 1987:91-110.

SIMON, K., and K. HAUSWALD 1995 – Kulmer Steig vor dem Mittelalter. Zu den ältesten sächsisch-böhmischen Verkehrswegen über das Osterzgebirge. *Arbeits- und Forschungsberichte zur sächsischen Bodendenkmalpflege* 37, 1995:9-98.

SOJÁK, M. 1999 – Záchranný výkop w Gánovciach. Studijne Zwesti. *Archeologický Ústav SAV* 33, 1999:35-51.

SOJÁK, M. 2000 – Neolitické osídlenie Spiša. *Slovenská Archeológia* 48/2, 2000:185-314.

SOJÁK, M. 2002 – Neolitické sídlisko v Stráňach pod Tatrami, okr. Kežmarok. In I. Cheben and I. Kuzma (eds.), *Otázky neolitu a eneolitu našich krajín – 2001*, Archeologický Ústav Slovenskej Akadémie Vied, Nitra, 2002:313-341.

TUNIA, K. 1978 – Neolityczne znaleziska z Beskidu Średniego. *Acta Archaeologica Carpathica* 18, 1978:135-137.

TILLMANN, A. 2001 – Zu den schnurkeramischen "Dolchgräbern in Südbayern". In T.H. Gohlisch and L. Reisch, *Die Stellung der endneolithischen Chamer Kultur in ihrem räumlichen und zeitlichen Kontext.* Kolloquien des Institutes für Ur- und Frühgeschichte Erlangen 1, Büchenbach, Erlangen, 2001:163-171.

TILLMANN, A. 1991 – Prähistorische Fundkarten der Alpen. In R. von Uslar, *Römisch-Germanische Forschungen* 48. Römisch-Germanisches Zentral-museum, DAI, Mainz, 1991:1-378.

VALDE-NOWAK, P. 1988 – *Etapy i Strefy Zasiedlenia Karpat Polskich w Neolicie i na Początku Epoki Brązu*. Ossolineum, Wrocław, Warszawa, Kraków, Gdańsk and Łódź, 2001.

VALDE-NOWAK, P. 1995 – *Osadnictwo Wczesnorolnicze Średniogórza Niemieckiego*, Instytut Archeologii i Etnologii PAN, Kraków.

VALDE-NOWAK, P. 2001a – Flammförmige Messer der Schnurkeramikkultur. In S. Kadrow (ed.), *A Turning of Ages*. Instytut Archeologii i Etnologii PAN, Kraków, 2001.

VALDE-NOWAK, P. 2001b – Etapy i Strefy w Badaniach nad Neolitem w Polskich Karpatach. In J. Gancarski (ed.), *Neolit i Początki Epoki Brązu w Karpatach Polskich.* Muzeum Podkarpackie w Krośnie, Krosno, 2001.

VALDE-NOWAK, P. 2002 – *Siedlungsarchäologische Untersuchungen zur neolithischen Nutzung der mitteleuropäischen Gebirgslandschaften*, Leidorf, Rahden, Westfalen, 2002.

VALDE-NOWAK, P. 2005 – Highland Settlement in Middle Europe During Neolithic Times: Polish-German Contribution. *Archaeologia Polona* 43, 2005:157-174.

VALDE-NOWAK, P. 2008 – An isolated Grave of the Baden Culture in the Beskidy Mountains. In M. Furholt, M. Szmyt and A. Zastawny (eds.), *The Baden Complex and the Outside World.* Proceedings of the 12th Annual Meeting of the EAA in Cracow, 19-24th September 2006. Studien zur Archäologie in Ostmitteleuropa (SAOISPEŚ) 4, Habelt, Bonn, 2008:139-145.

VALDE-NOWAK, P. (ed.) 1999 – *Die Anfänge der Besiedlung in den Sudeten.* Instytut Archeologii i Etnologii PAN, Kraków, 1999.

VALDE-NOWAK, P., and Kienlin, T. 2002 – Neolithische Transhumanz in Mittelgebirgen: Ein Survey in westlichen Schwarzwald. *Praehistorische Zeitschrift* 77/1, 2002:29-75.

VALDE-NOWAK, P.L. KLAPPAUF, and F.-A. LINKE 2004 – Neolithische Besiedlung der Gebirgsland-schaften: Fallstudie Harz. *Nachrichten aus der niedersächsischen Urgeschichte* 73, 2004:43-47.

VALDE-NOWAK, P., and W. WEISSMÜLLER 1994 – Eine archäologische Prospektion im inneren Bayerischen Wald: Zum Problem der neolithischen Nutzung der Mittelgebirge. *Archäologisches Korrespondenzblatt* 24, 1994:157-166.

VELDE, P. van de 1990 – Bandkeramik Social Inequality: A Case Study' 68 *Germania,* 1990:19-38.

VENCL, S. 1975 – Hromadne Nalezy Broušene Industrie z Čech. *Pamatky Archeologicke* 66/1, 1975:12-73.

WACNIK, A., K. SZCZEPANEK and K. HARMATA 2001 – Ślady Działalności Człowieka Neolitu i Brązu Obserwowane w Diagramach Pyłkowych z Okolic Przełęczy Dukielskiej i Terenów Przyległych. In J. Gancarski (ed.), *Neolit i Początki Epoki Brązu w Karpatach Polskich*, Muzeum Podkarpackie w Krośnie, Krosno, 2001:207-221.

WINGHART, S. 1987 – Vorgeschichtliche Deponate im ostbayerischen Grenzgebirge und im Schwarzwald: Zu Horten und Einzelfunde in Mittelgebirgslandschaften. *Bericht der Römisch Germanischen-Kommission* 67, 1987:89-201.

WEIN, G. 1978 – Vor- und Frühgeschichte. In G. Mauer (ed.), *Der Kreis Freudenstadt, Stuttgart-Aalen*, Theiss, München.

WEISSMÜLLER, W. 1986 – *Postmesolithische Funde aus Höhlen und Abris am Beispiel des südlichen Riesrandgebiets*, British Archaeological Research International Series 179, Oxford, 1986.

WEISSMÜLLER, W. 1995 – Ein archäologischer Survey im inneren Bayerischen Wald zur neolithischen Nutzung der Mittelgebirge. *Archäologisches Jahr in Bayern*, 1995:55-58.

WIŚLAŃSKI, T. 1979 – Kształtowanie się Miejscowych Kultur Rolniczo-Hodowlanych. Plemiona Kultury Pucharów Lejkowatych. In W. Hensel and T. Wiślański (eds.), *Prahistoria ziem Polskich* 2, Ossolineum, Wrocław, Warszawa, Kraków and Gdańsk. 1979:165-260.

WŁODARCZAK, P. 2008 – Corded Ware and Baden Cultures: Outline of Chronological and Genetic Relations Based on Finds from Western Little Poland. In M. Furholt, M. Szmyt and A. Zastawny (eds.) in cooperation with E. Schalk, *The Baden Complex and the Outside World.* Proceedings of the 12th Annual Meeting of the EAA in Cracow, 19-24th September 2006. Studien zur Archäologie in Ostmitteleuropa, Habelt, Bonn, 2008: 247-262.

WOTZKA, H.-P., E. LAUFER, M. POSSELT, and B. STAROSSEK 2001 – "Periphere": Plätze der späten Bandkeramik im Usinger Becken (Östlicher Hintertaunus, Hessen). *Berichte der Kommission für Archäologische Landesforschung in Hessen* 6, 2001:53-75.

ZYCH, R. 2004 – *Osadnictwo Kultury Pucharów Lejkowatych w Kotlinie Sandomierskiej i w Karpatach Polskich.* Unpublished Ph. D. Thesis, University of Rzeszów, Rzeszów.

Chapter 20
SEPARATING NATURAL AND ANTHROPOGENIC INFLUENCES ON PAST ECOSYSTEMS: THE TESTATE AMOEBAE AND QUANTITATIVE PALEOENVIRONMENTAL RECONSTRUCTION

Edward A.D. MITCHELL
Laboratory of Soil Biology, University of Neuchâtel, Neuchâtel, Switzerland

Abstract: *Testate amoebae (protozoa) are ubiquitous in almost all terrestrial environments. They respond to changes in ecological conditions such as hydrology, water chemistry, or nutrient status. The tests (shells) they produce are preserved in peat and lake sediments, thus allowing paleoenvironmental reconstruction. Testate amoebae are rapidly becoming a precious tool for multi-proxy in paleoecological studies aiming to sort apart natural from anthropogenic influences on ecosystems. Testate amoebae are small (mostly 20-200 μm), abundant (10^3-10^4 ind. g^{-1} peat), diverse (about 1900 species), and respond fast (i.e. within weeks to months) to environmental changes. Using modern ecological data sets, quantitative transfer functions can be built and allow quantitative paleoenvironmental reconstruction of environmental variables. An example from the Swiss Jura Mountains is used to illustrate how paleohydrological reconstruction based on testate amoebae and past human land-use inferred from pollen can be combined to reach a synthetic understanding of the history of a site and the region surrounding it. Providing an archaeological site is located close to a lake or a peatland, testate amoebae should be considered as a potential tool for archaeologists interested in understanding how environmental conditions might have changed during the period of interest, and to what extent humans might be responsible for the observed changes.*

INTRODUCTION

Testate amoebae are unicellular eukaryotic organisms. They are ubiquitous in almost all terrestrial environments such as lakes, rivers, mosses, and soils, but also occur in estuarine environments (Meisterfeld 2002a, Meisterfeld 2002b). They are small (mostly 20-200 μm – roughly the same size as a pollen grain), abundant (e.g. 10^3-10^4 ind. g^1 dry weight peat) (Gilbert and Mitchell 2006), and diverse (about 1900 species described so far). Testate amoebae build a test (shell) either from proteinaceous, calcareous, or siliceous material they secrete off by gluing together particles they find in their environment. This test, which allows identification to the species level, is usually well preserved in peat and lake sediments (Warner 1990). Agglutinated tests can be made of organic or mineral debris, or may contain elements of the tests of prey organisms, among which are other testate amoebae. In addition to this diversity of building material testate amoebae cover a relatively broad range of sizes (over one order of magnitude) and a surprisingly high diversity of morphologies: Most tests have one aperture (referred to as pseudostome), but one family, Amphitrematidae, has two. Test with a single aperture range from the shape of a round or flattened bottle (*Nebela*, many *Difflugia* species) to that of a flying saucer (*Arcella*). Some taxa, especially those adapted to dry conditions (e.g. *Plagiopyxis*) have a vestibule protecting the main chamber from desiccation. In addition, many species, especially in very wet environments produce spines (e.g. many *Euglypha* species, *Placocista spinosa*) or horns (e.g. *Centropyxis aculeata*, *Difflugia leidyi*). Some examples are illustrated in Figure 1.

Recent studies have shown that testate amoebae are polyphyletic (i.e. they do not constitute a homogeneous taxonomic unit but are instead composed of a heterogeneous assemblage of several taxonomically unrelated groups of organisms) (Bhattacharya, *et al.* 1995, Nikolaev, *et al.* 2005, Wylezich, *et al.* 2002). However, because of the similarities in size, feeding habit, general biology, and the presence of a test (shell) produced by the organism, it remains practical to study them as a single functional group of organisms from an ecological standpoint and this reasoning also applies to their use in paleoecology. Testate amoebae are also named testaceans, shelled amoebae, rhizopods, Arcellaceans. This multiplicity of names is rather unfortunate as some of these are either too broad (e.g. rhizopods include both naked and shelled organisms) or are confusing (e.g. Arcellaceans suggest an affinity with the Arcellinida, which is not the case for one of the major groups of testate am.

RESPONSE OF TESTATE AMOEBAE TO ECOLOGICAL GRADIENTS, BIOINDICATION, AND TRANSFER FUNCTIONS

Owing to their sensitivity to multiple natural and anthropogenic influences on the environment, testate amoebae are very useful for a broad range of research questions in neo- and paleo-ecology, from the long-term (e.g. Holocene) developmental history of a site and climatic fluctuations to the assessment of human influence on ecosystems (Mitchell, *et al.* 2008).

Testate amoebae respond to environmental gradients such as hydrology, water chemistry, or nutrient status. The tests produced by testate amoebae can be recovered from peat and lake sediments using a simple sieving method

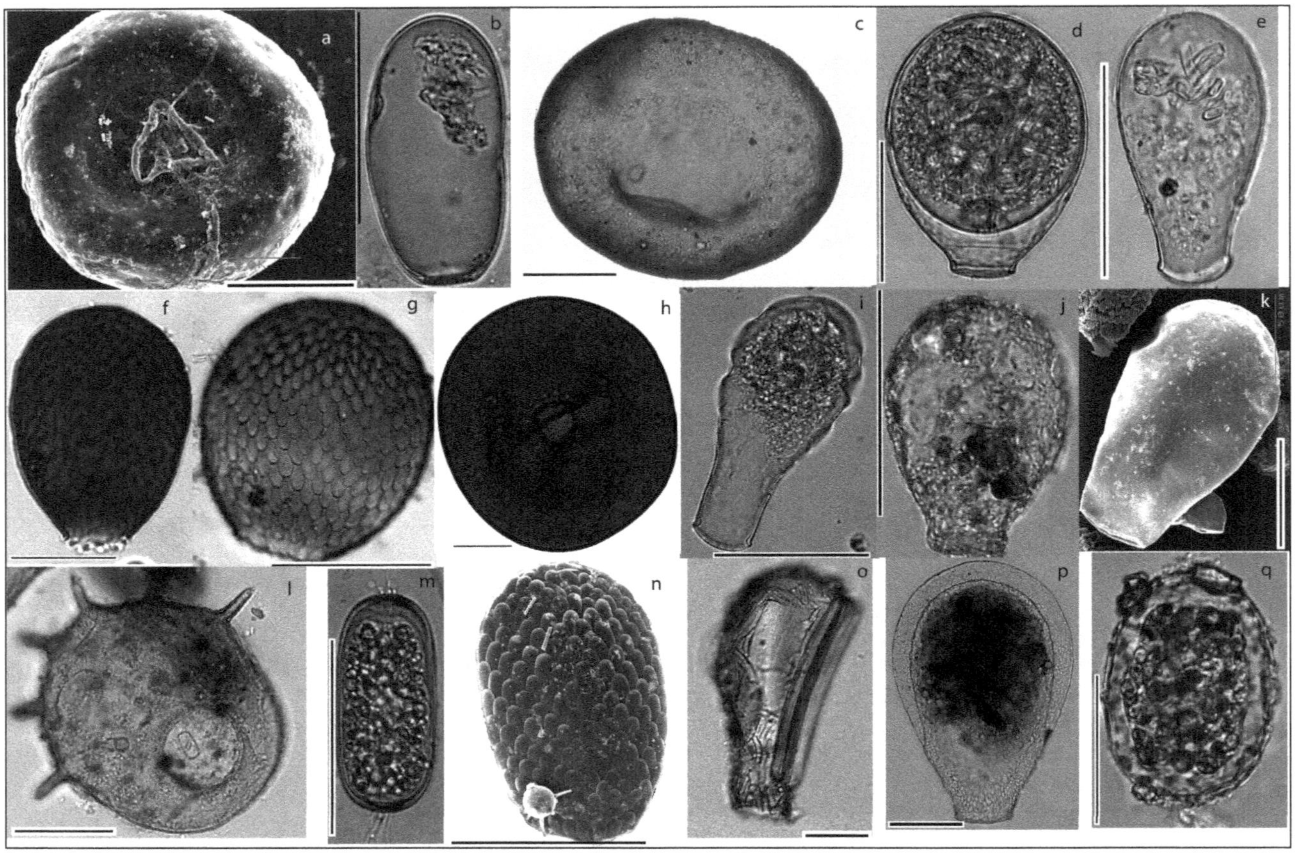

Figure 20.1. Examples of testate amoebae: a) *Trigonopyxis arcula*, b) *Hyalosphenia subflava*, c) *Bullinularia indica*, d) *Nebela tincta*, e) *Nebela militaris*, f) *Assulina muscorum*, g) *Assulina seminulum*, h) *Arcella arenaria*, i) *Hyalosphenia elegans*, j) *Physochila (Nebela) griseola*, k) *Hyalosphenia papilio*, l) *Centropyxis aculaeta*, m) *Archerella (Amphitrema) flavum*, n) *Placocista spinosa*, o) *Difflugia bacillifera*, p) *Nebela carinata*, q) *Amphitrema wrightianum*. Scale bars indicate approximately 50 µm except for A. muscorum: 20 µm

(Hendon and Charman 1997, Warner 1990). Sub-fossil testate amoebae communities can then be used with the help of transfer functions derived from the study of modern communities (surface samples), to infer quantitatively past environmental conditions (Charman 2001). If sedimentation or peat accumulation rates are high, as can be the case in *Sphagnum* peatlands or lakes, a high resolution of environmental reconstruction can be achieved (Lamentowicz, *et al*. 2010).

In order to understand how testate amoebae respond to their environment is it important to keep in mind that they are aquatic micro-organisms. They probably evolved in aquatic freshwater environment and later colonized terrestrial ecosystems where they became adapted to living in increasingly dry and variables moisture conditions. But even the species living in relatively dry environments such as mineral soils or mosses growing on vertical surfaces such as walls or the bark of trees will only thrive in the, sometimes very thin and temporally absent, water film. Most species seem to be able to encyst under unfavourable conditions. This feature was certainly a key adaptation to the colonization of land, comparable for plants to the invention of resistant structures such as spores and pollen grains. The size and morphology of species can to some extent be used to predict the potential ability of a species to live in low moisture micro-sites: a small amoeba will be able to remain active under drier conditions (i.e. a thinner capillary water film) than a large species. The responses of testate amoebae to their environment may be due to direct (e.g. desiccation) or indirect (e.g. abundance of prey organisms) environmental influences. To paleoecologists and archaeologists, the exact nature of this relationship is of secondary importance as long as a statistically significant correlation can be established. To ecologists, however such correlations are useful indications of possible causal relationships that need to be studied in detail.

Testate amoebae being very small respond to micro-environmental gradients that may not be visible to the naked eye of the human observer. For example, in a homogeneous 40 x 60 cm surface of a single peat moss species (*Sphagnum magellanicum*), Mitchell *et al*. (2000a) found important heterogeneities in the testate amoebae communities and these were correlated to the (almost flat; 6 cm height difference between the extremes) microtopography. In this homogeneous surface, two samples up to about 12 cm apart had a significant probability of containing a similar testate amoebae community. When the distance between samples was increased to about 17 cm, communities were no longer significantly similar. In the presence of a steep environmental gradient, changes in community structure are likely to be found over an even smaller distance. In some cases the kind of sensitivity testate amoebae offer can represent an advantage. In other cases it may make the interpretation more difficult. Paleoecologists therefore need to consider the size of the organisms they are using as bioindicators and the spatial scale of their samples when interpreting their results.

Under optimal conditions, testate amoebae have generation times of a few days (Schönborn 1986). When conditions change, different species are likely to respond differently. Some will thrive under the new conditions while others may have a lower rate of reproduction, encyst until conditions improve, or die out. The result is that within weeks to months the structure of the community will change. When ecological conditions change in a cyclic way, for example over the course of one year with the pattern of seasons, the community structure will follow a similar pattern from year to year. But if, to this cyclic pattern, a clear trend in one or more environmental conditions exists, then the community structure will rapidly change. Thus testate amoebae have the potential to respond very fast to environmental change. Indeed, at the community level, their response is much faster than that of vascular plants or even bryophytes, which are more commonly used for paleoenvironmental reconstruction (through the analysis of plant macrofossil or pollen and spores) than testate amoebae. This potential advantage of testate amoebae over plants is indeed confirmed by paleoecological studies (Blundell and Barber 2005). The point here is however not to set one method against another but to illustrate the strengths and weaknesses of each. Pollen and spores, unlike testate amoebae provide information not only on the local coring site but also on the vegetation of the studied ecosystem and even the surroundings, which may be the actual sites of interest for archaeologists. The combination of different methods, the so-called multi-proxy approach clearly offers the greatest potential for accurate interpretation of past environmental changes including the influence of human activities.

Given their sensitivity to natural environmental gradients, it is not surprising that testate amoebae also respond to various anthropogenic influences on the environment such as air, water and soil pollution, the conversion of natural ecosystems to agricultural ecosystems (Foissner 1997) and agricultural fertilization (Aescht and Foissner 1994, Mitchell 2004). In fact the list seems to be limited only by what potential influences have been studied to this date. Anthropogenic influences typically cause a lowering of either diversity or abundance or both (Gilbert, *et al.* 1998a, Gilbert, *et al.* 1998b, Mitchell 2004, Mitchell, *et al.* 2003). Studies have for example shown that the diversity of testate amoebae decreases with increasing atmospheric NO_2 concentrations (Nguyen-Viet, *et al.* 2004), soil pollution (road dust and heavy metal) (Balik 1991, Kandeler, *et al.* 1992), or water pollution (Kumar and Patterson 2000, Reinhardt, *et al.* 1998). Just as they respond to perturbations of their environment, testate amoebae can also be used to monitor ecosystem recovery either as biomonitors or using a paleoecological approach but over a short period of time (Jauhiainen 2002, Laggoun-Défarge, *et al.* 2004, Vickery and Charman 2004, Wanner and Dunger 2001).

ECOLOGICAL AND PALEOECOLOGICAL APPLICATION OF TESTATE AMOEBAE ANALYSIS

Testate amoebae are rapidly becoming a precious tool for multi-proxy paleoecological studies in lakes and peatlands as attested by the apparently exponential increase in the number of studies using this tool.

Lakes

In lakes, testate amoebae have long been shown to be reliable indicators of the trophy level (or productivity) of lakes (Beyens, *et al.* 1991, Burbidge and Schröder-Adams 1998, Schönborn 1965, Schönborn, *et al.* 1965). The response of testate amoebae to the trophy level of lakes has for example been used to reconstruct the history of Lake Winnipeg and to assess how the biologic productivity changed after the retreat of Lake Agassiz (Burbidge and Schröder-Adams 1998).

Lake testate amoebae also respond clearly to the alkalinity-acidity gradient (Beyens, *et al.* 1991). In a Canadian lake, testate amoebae were used to reconstruct the long-term changes in lake acidity (Kumar and Patterson 2000). This study revealed that lake acidity had decreased, most likely due to natural causes, long before any significant pollution from mining activities had started to take place. Similarly, testate amoebae were used to reconstruct the change in acidity of an English lake over 11000 years and separate natural from anthropogenic influences (Ellison 1995).

Paleoecological studies in Canada suggest that lake testate amoebae responded well to climate change: as conditions improved in the early Holocene, the community composition changed and the overall diversity increased (McCarthy, *et al.* 1995). However, more research is still needed to determine if testate amoebae respond to climate directly or if their apparent response to climate is not due to indirect effects through changes in the lake chemistry or prey organisms (Wall, *et al.* 2010). Such indirect relationships would not preclude the use of testate amoebae as climate indicators, but knowing the causal relationship would improve the interpretability of the signals.

Peatlands

Peat preserve the remains of plants (macrofossils) and some animals (e.g. beetles) and microorganisms (e.g. testate amoebae and fungi) and these remains are used to reconstruct the developmental history of a site as well as changes in the climate or in the vegetation over the broader landscape, including human-induced changes.

Testate amoebae are increasingly used in paleoecological studies of peatlands, usually in a multi-proxy approach in combination with pollen and spores and macrofossils. Similar to plant macrofossils, they provide information on the communities living at the exact coring location. However, testate amoebae have several important advantages over plant macrofossils: First the diversity is usually higher (10-30 species in a given sample) than that of plants (vascular and bryophytes combined). Second they are very numerous (about 10^3 to 10^4 ind. g-1 peat dry weight). Third most species appear to be cosmopolitan and therefore interpretation are not biased by biogeography or patterns of post-glacial species dispersal that may or may not be documented (Mitchell, *et al.* 2000b); one can therefore assume that if environmental conditions are suitable for a species to occur, it should be present. It should however be noted that although most testate amoebae indeed seem to be cosmopolitan this question is still debated and some species at least seem to have a limited distribution (Foissner 1999). Finally, as mentioned above, their generation time is faster than that of plants and their response to environmental changes can therefore be expected to be faster.

The initial abiotic and biotic characteristics of peatlands are initially determined by the geomorphological setting and the climate. In later developmental phases, autogenic factors such as vegetation succession play an increasingly important role, together with allogenic factors such as global climatic change, or direct and indirect human influences (Charman 2002). Testate amoebae respond primarily to hydrology and are therefore precious proxy indicators of paleohydrology. Raised bogs, which depend primarily on precipitation for their water supply are very sensitive to climatic changes and paleomoisture reconstructions derived from testate amoebae correlate well with paleoclimate (Blackford 2000, Chambers and Charman 2004). In such cases testate amoebae can thus be used as indirect indicators of paleoclimate (Charman, *et al.* 2004).

In addition to natural processes affecting peatlands, human activities exert an increasing influence over the hydrology of peatlands as well as the local climate, and the global climate (Chapman, *et al.* 2003). Total peatland losses add up to or about 16% of their former extent. An estimated 50% of the area of active peatland in the temperate and boreal zones has been lost to agriculture, 30% to forestry (mostly since the middle of the 20th century), 10% to peat extraction and the remainder to urbanization, erosion, water reservoirs, and other uses (Joosten and Clarke 2002). Although the most dramatic impacts are clearly recent, the exploitation of peatlands by man started long ago. Peat has probably been used for millennia as fuel in the Scottish islands and Ireland to be followed later by industrial extraction of fuel peat in the 19th century (Lappalainen 1996). In Southern Sweden, the cultivation of peat soils dates back to the early Iron Age (Egelmark 2000) and in Finland, many of the more fertile peatlands (fens) had been used for agriculture long before that time (Heikkilä and Lindholm 2000). More extensive drainage of bogs and fens for agricultural use in Europe began in the 17th century. Finland, for example, has lost 60% of its former extensive active peatland area to forestry since the 1950s. In addition to the wealth of paleoecological information that lies in peatlands, they may also represent a significant archaeological resource, especially in regions where they represented a significant proportion of the landscape and were thus less likely to be avoided or ignored by resident human populations (Buckland 1993, Buckland and Dinnin 1992).

Following centuries of destruction, in many regions attempts are now made to restore damaged peatlands. To be effective, such initiatives require a good understanding of the functioning of these ecosystems. Testate amoebae can be useful in two ways: First as paleoecological indicators. Knowing the history of a site and how it reacted to direct or indirect human impacts will help managers make appropriate decisions. Paleoecology is a powerful, but still under-used tool for the management of natural or human-impacted peatlands (Buttler, *et al.* 1996, Lavoie, *et al.* 2001). Second as early indicators of change (biomonitors), they can provide useful information on the micro-environmental conditions existing at the peatland surface (Laggoun-Défarge, *et al.* 2004, Vickery and Charman 2004).

A CASE STUDY: THE PRAZ-RODET BOG, SWISS JURA MOUNTAINS

An example from the Swiss Jura Mountains, the Praz-Rodet Bog is used to illustrate how paleohydrological reconstruction based on testate amoebae and past human land-use inferred from pollen can be combined to reach a synthetic understanding of the history of a site and the region surrounding it (Mitchell, *et al.* 2001). The Jura Mountains are constituted of limestone and therefore the hydrology is strongly influenced by karst. The water never stays long at the surface and rivers and lakes are quite rare in the region. Instead, most of the water drains in subsurface networks of caves. For this reason it is a somewhat unusual place for peatlands to form, but nevertheless, many peatlands do exist. Their development is made possible by the existence of either impermeable bedrock or glacial sediment. The presence of karst drainage limits their lateral extension. Through time, but especially in the last 100 years, many peatlands have been partially or totally destroyed through peat cutting and conversion to agricultural land. These practices have now ceased and peatlands are recognized as valuable components of the natural history heritage. For this reason it is also important to know more about their development and the extent to which human activities have modified them.

One of the goals of the Praz-Rodet study was to assess to what extent the present pine tree cover (*Pinus rotundata*) of the bog was natural or not. Were these trees planted by man, or were they growing there naturally in which case they may represent a relict from glacial times? In a management perspective the answer to this question could determine if the trees should be left on the mire or removed to restore a more natural condition. Using a multi-proxy approach combining pollen and spores plant macrofossils and testate amoebae analyses, we were able to reconstruct the developmental history of the site over the entire Holocene. Testate amoebae allowed inferring quantitatively past water table depth and pH. This information was compared to the plant macrofossil data providing information on the local vegetation changes and to the pollen and spores data providing information on the local and regional vegetation as well as on the intensity of human exploitation of the surrounding land for forestry and agriculture.

The paleoenvironmental data showed that the development of the Praz-Rodet bog followed the standard terrestrialization sequence, starting from a small lake, which evolved into a fen and then a *Sphagnum*-dominated bog. The first evidence of human activity occurred around 6850 cal yr. bp as attested by the occurrence of *Plantago lanceolata*-type pollen, a grazing indicator. This date is in line with the early Neolithic period in the Jura Mountains (SPM-II 1995). An interesting wet shift took place after 2800 cal yr. bp. This is attested by testate amoebae indicators (*Amphitrema wrightianum, A. flavum,* and *Difflugia globulosa*) and pollen indicators (*Scheuchzeria* and *Drosera*). This wet shift occured at a time when human impact on the surround landscape was increasing as attested by *Plantago lanceolata*-type and Gramineae pollen. *Pinus rotundata* macrofossil evidence at 1500 cal yr bp suggested that this species was a natural component of the mire vegetation and had not been introduced there recently.

Humans appear to have had little or no direct impact on the bog until the middle of the 18th century when the pollen data shows evidence of forest clearance and creation of extensive pasturelands. After this time, the bog surface became drier and *Pinus rotundata, Calluna vulgaris* expanded on the bog. An independent dendroecological study in comparable peat bogs of the region showed that the growth and establishment of pine trees was strongly stimulated by the clearance of the forest growing on the mineral soil surrounding the bog (Freléchoux, *et al.* 2000). The clearance of the forest growing around the bog removed a windbreak that protected the bog from the desiccating effect of the wind in summer. Without this protection, the evapotranspiration increased on the bog and this allowed ericaceous shrubs and pine trees to grow better. Then through positive feedback effects, this dynamic fed itself until nearly the entire bog surface was covered by a pine forest. The coring site was chosen in the centre of the mire. Therefore conditions at the coring site remained relatively wet for some time after the clearance of the surrounding forest. But as the pine forest expanded from the periphery towards the centre, the water table declined owing to the evapotranspiration and eventually even the centre of the mire became drier as shown by changes in testate amoebae communities and, with a lag of 15-60 years, the bryophytes.

In Northwest Europe and North America, forest clearance has often induced paludification through raising ground-water tables (Warner, *et al.* 1989). The case of Praz Rodet is opposite from this situation because (1) the hydrology of the mire is independent from its surroundings as far as groundwater is concerned because of the karstic character of the landscape; and (2) summer winds are much hotter and drier and desiccate more strongly than in NW Europe due to (a) a more continental climate and (b) a more southerly position close to the southern distribution limit of raised bogs in Europe. Indeed, the development of the Praz-Rodet bog is unusual, but similar cases may exist elsewhere in analogous mesoclimatical and geomorphological conditions. The paleoecological history of this site provides an interesting example of how an ecosystem can be profoundly affected by human activities even without any apparent direct impact.

CONCLUSIONS

Providing an archaeological site is located close to a lake or a peatland, testate amoebae should be considered as a potential tool for archaeologists interested in understanding how environmental conditions might have changed during the period of interest, and to what extent humans might be responsible for the observed changes. The list of possible applications of testate amoeba analysis is increasing constantly as new research is done on the ecology of these organisms and their responses to various environmental gradients or perturbations. With more and more researchers becoming interested in these organisms there are good reasons to be optimistic with respect to their usefulness in paleoenvironmental and archaeological research.

Acknowledgements

Edward Mitchell was supported by EU project RECIPE. RECIPE was partly supported by the European Commission (n° EVK2-2002-00269) and partly, for the Swiss partners EPFL-ECOS and WSL-AR, by the OFES (Swiss Federal Office for Education and Science), Switzerland.

References

AESCHT, E., and W. FOISSNER 1994 – Effects of Organically Enriched Magnesite Fertilizers on the Testate Amebas of a Spruce Forest. *European Journal of Soil Biology* 30, 1994:79-92.

BALIK, V. 1991 – The Effect of the Road Traffic Pollution on the Communities of Testate Amoebae (Rhizopoda, Testacea) in Warsaw (Poland). *Acta Protozoologica* 30, 1991:5-11.

BEYENS, L., D. CHARDEZ and D. de BAERE 1991 – Ecology of Aquatic Testate Amoebae in Coastal Lowlands of Devon Island (Canadian High Arctic). *Archiv für Protistenkunde* 140, 1991:23-33.

BHATTACHARYA, D, T. HELMCHEN and M. MELKONIAN 1995 – Molecular evolutionary analyses of nuclear-encoded small subunit ribosomal RNA identify an independent rhizopod lineage containing the Euglyphina and the Chlorarachniophyta 42/1, J. Eukaryot. Microbiology 1995:65-69.

BLACKFORD, J. 2000 – Palaeoclimatic Records from Peat Bogs. *Trends in Ecology and Evolution* 15, 2000:193-198.

BLUNDELL, A., and K. BARBER 2005 – A 2800-year Palaeoclimatic Record from Tore Hill Moss, Strathspey, Scotland: The need for a Multi-Proxy Approach to Peat-Based Climate Reconstructions. *Quaternary Science Review* 24, 2005:1261-1277.

BUCKLAND, P.C. 1993 – Peatland Archaeology: A Conservation Resource on the Edge of Extinction. *Biodiversity and Conservation* 2, 1993:513-527.

BURBIDGE, S.M., and C.J. SCHRÖDER-ADAMS 1998 – Thecamoebians in Lake Winnipeg: a tool for Holocene Paleolimnology. *Journal of Paleolimnology* 19, 1998:309-28.

BUTTLER, A., B.G. WARNER, P. GROSVERNIER and Y. MATTHEY 1996 – Vertical Patterns of Testate Amoebae (Protozoa: Rhizopoda) and Peat Forming Vegetation on Cutover Bogs in the Jura, Switzerland. *New Phytologist* 134, 1996:371-382.

CHAMBERS, F.M., and D.J. CHARMAN 2004 – Peatland Dynamics and Holocene Environmental Change: A Review and Editorial. *The Holocene* 14, 2004:1-6.

CHAPMAN, S., A. BUTTLER, A.-J. FRANCEZ, F. LAGGOUN-DÉFARGE, H.M. VASANDER, S. COMBE, J. GROSVERNIER, P.H. HARMS, D. EPRON, D. GILBERT and E.A.D. MITCHELL 2003 – Commercial Exploitation of Peatlands and Maintenance of Biodiversity: A Conflict Between Economy and Ecology. *Frontiers in Ecology and the Environment* 1, 2003:525-532.

CHARMAN, D.J. 2001 – Biostratigraphic and Palaeoenvironmental Applications of Testate Amoebae. *Quaternary Science Reviews* 20, 2001:1753-1764.

CHARMAN, D.J., A.D. BROWN, D. HENDON and E. KAROFELD 2004 – Testing the Relationship Between Holocene Peatland Palaeoclimate Reconstructions and Instrumental Data at two European Sites. *Quaternary Science Reviews* 23, 2004:137-143.

ELLISON, R.L. 1995 – Paleolimnological Analysis of Ullswater using Testate Amebas. *Journal of Paleolimnology* 13, 1995:51-63.

FOISSNER, W. 1997 – Protozoa as Bioindicators in Agroecosystems, with Emphasis on Farming Practices, Biocides, and Biodiversity. *Agriculture Ecosystems and Environment* 62, 1997:93-103.

FOISSNER, W. 1999 – Protist Diversity: Estimates of the Near-Imponderable. *Protist* 150, 1999:363-368.

FRELÉCHOUX, F., A. BUTTLER, F.H. SCHWEINGRUBER and J.M. GOBAT 2000 – Stand Structure, Invasion, and Growth Dynamics of Bog Pine (*Pinus uncinata* var. *rotundata*) in Relation to Peat Cutting and Drainage in the Jura Mountains, Switzerland. *Canadian Journal of Forest Research-Revue Canadienne De Recherche Forestiere* 30, 2000:1114-1126.

GILBERT, D, C. AMBLARD, G. BOURDIER and A.J. FRANCEZ 1998a – "The microbial loop at the surface of a peatland: Structure, function, and impact of nutrient input" 35(1) Microb. Ecol. 83-93.

GILBERT, D., C. AMBLARD, G. BOURDIER and A.J. FRANCEZ 1998b – Short-Term Effect of Nitrogen Enrichment on the Microbial Communities of a Peatland. *Hydrobiology* 374, 1998:111-119.

GILBERT, D., and E.A.D. MITCHELL 2006 – Microbial Diversity in *Sphagnum* Peatlands. In I.P. Martini, A. Matinez Cortizas and W. Chesworth (eds.), *Peatlands: Basin Evolution and Depository of records on Global Environmental and Climatic Changes*. Elsevier, Amsterdam, 2006:287-318.

HENDON, D., and D.J. CHARMAN 1997 – The preparation of testate amoebae (Protozoa: Rhizopoda) samples from peat. *The Holocene* 7, 1997:199-205.

JAUHIAINEN, S. 2002 – Testacean Amoebae In Different Types of Mire Following Drainage and Subsequent Restoration. *European Journal of Protistology* 38, 2002:59-72.

KANDELER, E., G. LUFTENEGGER and S. SCHWARZ 1992 – Soil Microbial Processes and Testacea (Protozoa) as Indicators of Heavy-Metal Pollution. *Zeitschrift für Pflanzenernahrung und Bodenkunde* 155, 1992:319-322.

KUMAR, A. and R.T. PATTERSON 2000 – Arcellaceans (Thecamoebians): New Tools for Monitoring Long- and Short-Term Changes in Lake Bottom Acidity. *Environmental Geology* 39, 2000:689-697.

LAMENTOWICZ, M., W.O. van der KNAAP, Ł. LAMENTOWICZ, J.F.N. van LEEUWEN, E.A.D MITCHELL, T. GOSLAR and C. KAMENIK 2010 – "A near-annual palaeohydrological study based on testate amoebae from a sub-alpine mire: surface wetness and the role of climate during the instrumental period" 25(2) J. Quat. Sci. 190-202.

LAVOIE, C., C. ZIMMERMANN and S. PELLERIN 2001 – Peatland restoration in southern Quebec (Canada): A paleoecological perspective. *Ecoscience* 8, 2001:247-258.

McCARTHY, F.M.G., E.S. COLLINS, J.H. McANDREWS, H.A. KERR, D.B. SCOTT and F.S. MEDIOLI 1995 – A Comparison of Postglacial Arcellacean (Thecamoebian) and Pollen Succession in Atlantic Canada, Illustrating the Potential of Arcellaceans for Paleoclimatic Reconstruction. *Journal of Paleontology* 69, 1995: 980-993.

MEISTERFELD, R. 2002a – Order Arcellinida Kent, 1880. In J.J. Lee, G.F. Leedale and P. Bradbury (eds.), *The Illustrated Guide to the Protozoa* 2, Society of Protozoologists, Lawrence, Kansas, 2000:827-860.

MEISTERFELD, R. 2002b – Testate Amoebae with Filopodia. In J.J. Lee, G.F. Leedale and P. Bradbury (eds.), *The Illustrated Guide to the Protozoa* 2, Society of Protozoologists, Lawrence, Kansas, 2000:1054-1084.

MEISTERFELD, R., M. HOLZMANN and J. PAWLOWSKI 2001 – Morplogical and Molecular Characterization of a New Terrestrial Allogromiid Species: *Edaphoallogromia australica* gen. et spec. nov. (Foraminifera) from Northern Queensland (Australia). *Protist* 152, 2001:185-192.

MITCHELL, E.A.D. 2004 – Response of Testate Amoebae (Protozoa) to N and P Fertilization in an Arctic wet Sedge Tundra. *Arctic Antarctic and Alpine Research* 36, 2004:77-82.

MITCHELL, E.A.D., D. BORCARD, A.J. BUTTLER, P. GROSVERNIER, D. GILBERT and J.M. GOBAT 2000a – Horizontal Distribution Patterns of Testate Amoebae (Protozoa) in a *Sphagnum magellanicum* Carpet. *Microbial Ecology* 39, 2000:290-300.

MITCHELL, E.A.D, A. BUTTLER, P. GROSVERNIER, H. RYDIN, C. ALBINSSON, A.L. GREENUP, M.M.P.D. HEIJMANS, M.R. HOOSBEEK and T. SAARINEN 2000b – "Relationships among testate amoebae (Protozoa), vegetation and water chemistry in five *Sphagnum*-dominated peatlands in Europe" 145(1) New Phytol. 95-106.

MITCHELL, E.A.D., D.J. CHARMAN and B.G. WARNER 2008 – Testate Amoebae Analysis in Ecological and Paleoecological Studies of Wetlands: Past, Present and Future. *Biodiversity and Conservation* 17, 2008:2115-2137.

MITCHELL, E.A.D., D. GILBERT, A. BUTTLER, P. GROSVERNIER, C. AMBLARD and J.-M. GOBAT 2003 – Structure of Microbial Communities in *Sphagnum* Peatlands and Effect of Atmospheric Carbon Dioxide Enrichment. *Microbial Ecology* 46, 2003:187-199.

MITCHELL, E.A.D., W.O. van der KNAAP, J.F.N. van LEEUWEN, A. BUTTLER, B.G. WARNER and J.M. GOBAT 2001 – The Palaeoecological History of the Praz-Rodet Bog (Swiss Jura) Based on Pollen, Plant Macrofossils and Testate Amoebae (Protozoa). *The Holocene* 11, 2001:65-80.

NGUYEN-VIET, H., D. GILBERT, N. BERNARD, E.A.D. MITCHELL and P.-M. BADOT 2004 – Relationship Between Atmospheric Pollution Characterized by NO_2 Concentrations and Testate Amoebae Abundance and Diversity. *Acta Protozoologica* 43, 2004:233-239. (Electronic copy: http://www.nencki.gov.pl/pdf/ap/ap753.pdf).

NIKOLAEV, S.I., E.A.D. MITCHELL, N.B. PETROV, C. BERNEY, J. FAHRNI and J. PAWLOWSKI 2005 – The Testate Lobose Amoebae (ord Petrov er Arcellinida Kent, 1880) Finally Find Their Home Within Amoebozoa. *Protist* 156/2, 2005:191-202.

REINHARDT, E.G., A.P. DALBY, A. KUMAR and R.T. PATTERSON 1998 – Arcellaceans as Pollution Indicators in mine Tailing Contaminated Lakes near Cobalt, Ontario, Canada. *Micropaleontology* 44, 1998:131-148.

SCHÖNBORN, W. 1965 – Die Sedimentbewohnenden Testaceen einiger Masurischer Seen. *Acta Protozoologica* 3, 1965:297-309.

SCHÖNBORN, W. 1986 – Population Dynamics and Production Biology of Testate Amoebae (Rhizopoda, Testacea) in raw Humus of two Coniferous Forest Soils. *Archiv für Protistenkunde* 132, 1986:325-342.

SCHÖNBORN, W., D. FLOSSNER and G. PROFT 1965 – Die Limnologische Characterisirung des Profundals einiger Seen mit Hilfe von Testaceen-Geimenschaften. *Limnologica* 3, 1965:371-380.

WANNER, M and W. DUNGER 2001 – Biological Activity of Soils from Reclaimed Open-Cast Coal Mining Areas in Upper Lusatia Using Testate Amoebae (Protists) as Indicators. *Ecological Engineering* 17, 2001:323-330.

WARNER, B.G. 1990 – Testate Amoebae (Protozoa). In B.G. Warner (ed.), *Methods in Quaternary Ecology*, Geoscience Canada, St. John's, Newfoundland, 1990:65-74.

WARNER, B.G., H.J. KUBIW and K.I. HANF 1989 – An Anthropogenic Cause for Quaking Mire Formation in Southwestern Ontario. *Nature* 340, 1989:380-384.

WALL, A, M. MAGNY, L. MILLET, B. VANNIERE and D. GILBERT 2010 – "Environmental and climatic changes during the Lateglacial-Holocene transition as reflected by variations in thecamoebian assemblages at Lake Lautrey (Jura Mountains, eastern France)" 25 Quat. Res. in press.

WYLEZICH, C., R. MEISTERFELD, S. MEISTERFELD and M. SCHLEGEL 2002 – Phylogenetic Analyses of Small Subunit Ribosomal RNA Coding Regions Reveal a Monophyletic Lineage of Euglyphid Testate Amoebae (Order Euglyphida). *Journal of Eukaryotic Microbiology* 49, 2002:108-118.

Chapter 21
ENVIRONMENTAL AND CULTURAL CHANGE IN THE ALPS: SEEKING CONTINUITY IN THE BRONZE AGE LAKE-DWELLING TRADITION

Francesco MENOTTI
Institute of Prehistory and Archaeological Science, University of Basel, Basel, Switzerland

Abstract: *Ever since their discovery, the famous prehistoric Alpine lake dwellings have produced new insights into past human societies. Following an abrupt change in climatic conditions, the Middle Bronze Age northern Alpine lake-dwellers abandoned their settlements at the end of the sixteenth century BC to the beginning of the twelfth century BC. Paleo-lacustrine environments have been reconstructed and lake level fluctuations simulated. This, together with archaeological evidence, demonstrates that the Middle Bronze Age lake-dwellers temporarily adapted to a drier habitat until the lakeshores became exposed again for re-settlement.*

INTRODUCTION

The vertical timbers or piles of prehistoric houses occur near lakes in several areas of Europe (Menotti and Prankenaite 2008). Since the middle of the nineteenth century the Alpine region's lake-dwelling sites have received the most attention. This is mainly due to the large amount and quality of archaeological evidence found within them. However, other types of settlement on dry land have received less attention in this region. The situation has left a vacuum of knowledge during periodic settlement interruptions, which occur during the whole lake-dwelling phenomenon. One of the best examples of such an absence is the well-studied Middle Bronze Age (MBA) hiatus (fifteenth-twelfth centuries BC) (Menotti 1999a, 2001, 2004, 2009).

Programs of rescue archaeology, resulting from road construction, have recently re-focused attention to the areas away from the lakes and onto 'dry' places, where traces of MBA settlements are found. The large number of these sites can counterbalance the preponderance of data from the lake-dwelling environs. They also afford us an opportunity to study what actually happened during the 'MBA hiatus'.

Evidence supports the theory of a marked inland shift of lake-dwelling communities at the end of the Early Bronze Age (EBA). Although it would be premature to conclude that the exodus from the lakeshores was homogeneous and sudden everywhere, this large-scale abandonment of the lake environs was due to a major rise in lake levels. It occurred between the fifteenth and the twelfth centuries BC (Gross-Klee and Ritzmann 1990, Menotti 1999a, 1999b, 2001). The inland areas of the Alpine region were obviously populated at this time, but the settlements of terrestrial type are mostly located fairly far from the lakes. The recent discovery of MBA 'terrestrial' sites in the surrounding area of the lakes raise the question of whether these are examples of habitations built by former lake-dwellers.

A few MBA sites of possible lacustrine origin have been found near the northwestern side of the Bodensee (Lake Constance) in Germany (Schlichtherle 1995). Further inland in the Singen region more relevant MBA settlements were discovered (Dickmann 1989, 1990, Aufdermauer and Dickmann 1995). Remarkable evidence of such sites was found on the Swiss lakeshores in the Kreuzlingen area near Konstanz city (Rigert 1998, 1999, 2001). A similar situation is present in the Lake Zürich region where a few MBA settlements have been located in the surrounding hills of Küsnacht, at Erlenbach, and Dietikon near the river Limmat, plus two more on Lake Greifen and Lake Pfäffikon (Bauer 1992, Fischer 1997). The gentle hills surrounding the northern shores of Lake Zug yielded similar sites, such as Cham-Oberwil Hof, Cham-Oberwil Hinterbüel and Steinhausen Schlossberg (Hochuli 1995, Gnepf 1995).

The increasing number of MBA settlements built on 'dry' land near the formerly settled lake areas throws new light on the marked discontinuity in lakeside settlements, which is such a salient feature of this period. It is clear that there was a moderately widespread pattern of population displacement in this region. Lakeside settlements shifted to new locations as a result of significant changes in the previously preferred environments (Fig. 21.1).

CLIMATE AND LAKES: THE LAKE LEVEL FLUCTUATION THEORY

Long-term as well as short-term variations in climatic conditions have always influenced the hydrological balance of the Alpine lakes, causing fluctuations in water level that affected communities living in the proximity of the lakes (Magny 2004, Menotti 2009). One of the most striking settlement gaps in the whole lake-dwelling chronology is the MBA hiatus, which seems to have had a very rapid onset. This hypothesis can be confirmed by examining three EBA lacustrine sites situated on two of

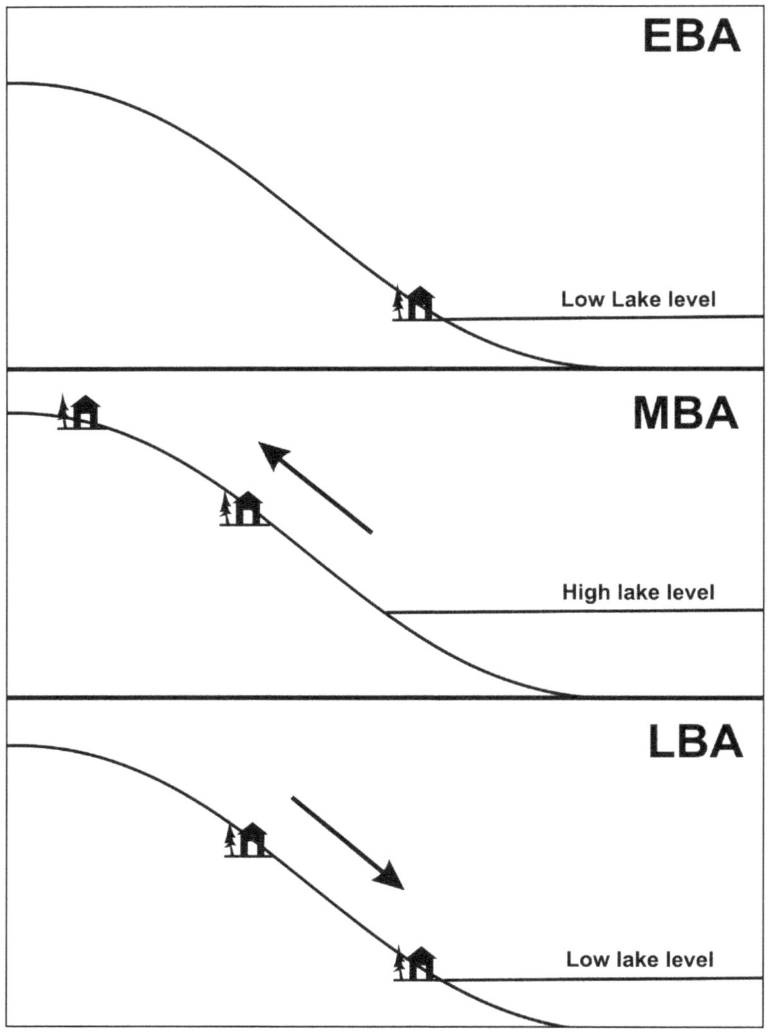

Figure 21.1. The MBA lake-dwellings inland shift in the northern Alpine region

the main northern Alpine lakes, namely Lake Constance and Lake Zürich. In fact, archaeological as well as environmental analyses show that Bodman-Schachen 1 (Lake Constance), Arbon-Bleiche 2 (Lake Constance), and ZH-Mozartstrasse (Lake Zürich) were all abandoned within five years of each other. This occurred in the last decade of the sixteenth century BC (Menotti 1999a, 1999c, 2001). However, if there was indeed an abrupt change in climate, is it possible that all the various lakes in the Alpine region were affected in the same way?

The answer is obviously no, although some similarities can be noticed. For instance, Lake Constance and Lake Zürich are part of the same micro-climatic area (the northern Alpine *Vorland*), sharing a similar environment in terms of land morphology, geology, and vegetation. There are nevertheless some differences concerning their hydrologic balance, which is mainly due to the size of the lakes and their hydrologic catchments. Lake Constance lies at an altitude of 396 m above sea level (masl), and it is surrounded by gentle hills, which extend as far as Lake Zürich in the southwest, and the Federsee (Lake Feder) in the northeast. The only mountainous part of its catchment is situated on the southeast of the lake on the east bank of the Rhine (Rhein) River (Wessels 1995). Sedimentological analyses show that it is very unlikely that the level of Lake Constance has ever exceeded the 400 masl contour line in the past seven millennia. The only time in which the level reached the maximum (400 masl) lake level was in the MBA, around the fourteenth century BC (Schlichtherle 1995).

Geologically Lake Zürich and Lake Constance are very similar. The only substantial difference is size. Lake Zürich is six times smaller than Lake Constance (Schüepp 1979). The Zürich basin (406 masl) is divided into two parts by the narrow peninsula of Hurden, which extends across the lake almost to the city of Rapperswil. Therfore, the smaller southeastern part of Lake Zürich, also called Obersee, is almost independent. The northern surroundings of the lake correspond to the southern areas of Lake Constance, whereas the eastern and southern surroundings are at the foot of the northern Alpine chain. A mixture of gentle hills and fairly fertile plains are found west of Zürich bay. Because of its inlet-outlet balance, the maximum and minimum levels of Lake Zürich can only be respectively 407.5 m and 401.5 masl. (Schindler 1971, 1981).

Taking into account all these environmental factors and adding archaeological data, GIS analyses have allowed us to graphically simulate the impact of the fluctuating lake levels on the surrounding environment of the three aforementioned EBA lacustrine sites (Bodman-Schachen 1, Arbon-Bleiche 2 and ZH-Mozartstrasse). The result is that changes in lake level would have had profound effects on the existing settlements here. They also provide useful insights as to how MBA settlements can be located after the displacement (Menotti 2001, 2002, 2004).

ADAPTING TO A DIFFERENT ENVIRONMENT

It is difficult to isolate a single reason why prehistoric people chose to occupy the lakeshores, marshes, and other wetland environments. It could have been the easy accessibility to food resources (fish, lakeside plants), defense, land division, transportation, or even all of the above. Whatever the reasons were, they had to adapt to the environment, not only in terms of subsistence, but also in the way their habitations had to be constructed. For instance, on top of the unconsolidated organic lake sediments (peat, marl, and lake-mud), vertical timber (piles) were driven directly into the ground rather than placed in dug-out post holes (Menotti and Prankenaite 2008). This is in stark contrast to the morainic clay, which dominated environments away from the lakeside. The sediment demands a very different approach to construction. Therefore, the whole architectural tradition, as well as economic habits, had to be reformulated to new conditions during the MBA inland shift.

One of the strong advantages of human beings is their flexibility in coping with unexpected happenings. This allows inventive responses to social or environmental variability, even in the event of residential displacement. Sedentary communities can be forced to abandon their settlements for a multitude of reasons and resettle in a completely different environment. One likely outcome of this forced migrating process is the radical change of the society and its coping mechanisms. The process of adaptation to a new area may, for example, cause social transformations. As a result, these changes are reflected in the material culture of the group, especially if the new environment is different from the previous one. Another important aspect of population displacement is acculturation when resulting from contact with other groups that are already occupying the area.

Migratory groups not only modify their culture to adapt to a new environment, but they also change the environment itself. The loss of tillable land, for example, forces the displaced groups to look for new cultivable areas or pastures for their cattle. A possible low availability of space might results in massive deforestation, which would transform the landscape substantially. Evidence of this cultural adaptation as well as culture-driven environmental change is preserved in the archaeological record, and the study of the latter eventually allows archaeologists to put the jigsaw-puzzle together and reconstruct the apparently 'lost' chronological continuity of societies' development.

THE "TRANSITIONAL" MIDDLE BRONZE AGE LAND SETTLEMENTS

In this section we will take into account some of the most relevant MBA 'inland' settlements, which are possibly of lacustrine origin. Unfortunately, the state of preservation of most of these sites is fairly poor and wooden structures are barely present. Nevertheless, the living floors, postholes, and the house layouts demonstrate typological similarities to earlier lacustrine settlements, as well as terrestrial ones. Continuity in style of house construction can be traced from the EBA to the Late Bronze Age (LBA). While the lacustrine human occupation followed a very irregular pattern, land settlement occupation was more continuous. Inland sites were occupied without any major interruption throughout the entire Bronze Age. Nevertheless, archaeological evidence shows an increase in land settlements in the immediate vicinity of lakes in the northern Alpine region during the MBA and in particular from the fifteenth to the twelfth century BC. Recent studies have revealed that these sites might be particularly important for the study of the MBA lacustrine hiatus.

On the German side of Lake Constance and more precisely in the Bodman-Schachen area, a Middle Bronze Age settlement has been found in a locality called *Breite* a few hundred meters from the lake shore (Fig. 21.2-B) (Schlichtherle 1995). Other relevant MBA sites have been discovered further inland at Rielasingen-Worblingen, Hilzingen-Duchlingen, Hilzingen-Unter-Schoren and Mühlhausen-Ehingen near Singen (Fig. 21.2-A) (Dickmann 1989, 1990, Aufdermauer and Dickmann 1995).

On the Swiss shores of the Lake near the city of Konstanz, an archaeological survey, preceding the construction of the Motorway N7 Schwaderloh-Kreuzlingen, has revealed a number of MBA sites. They are Tägerwilen (ARA-Strasse), Kreuzlingen (Töbeli), Tägerwilen (Hochstross), Tägerwilen (Spuelacker), Tägerwilen (Im Ribi), Kreuzlingen (Ribi Brunegg), Tägerwilen (Ribi-Girsbergtunnel), Kreuzlingen (Schlossbühl), Kreuzlingen (Bernrain) and Kreuzlingen (Wildenwis) (Fig. 20.2-C) (Rigert 1998, 1999, 2001).

The hills surrounding Lake Zürich also have a number of MBA settlements, which are particularly relevant to the study of the MBA occupational gap. These are situated at Küsnacht, Erlenbach and Dietikon near the river Limmat (Fig. 21.2-H). Two more sites, namely Fällenden-Wigartenstrasse and Hotzenweid/Steinacker are also to be found respectively on Lake Greifen and Lake Pfäffikon (Fig. 21.2-D-G) (Bauer 1992, Fischer 1997).

Another lake where several MBA site with possible lacustrine origin have come to light is Lake Zug (fifteen km south of Lake Zürich). The sites of Cham-Oberwil

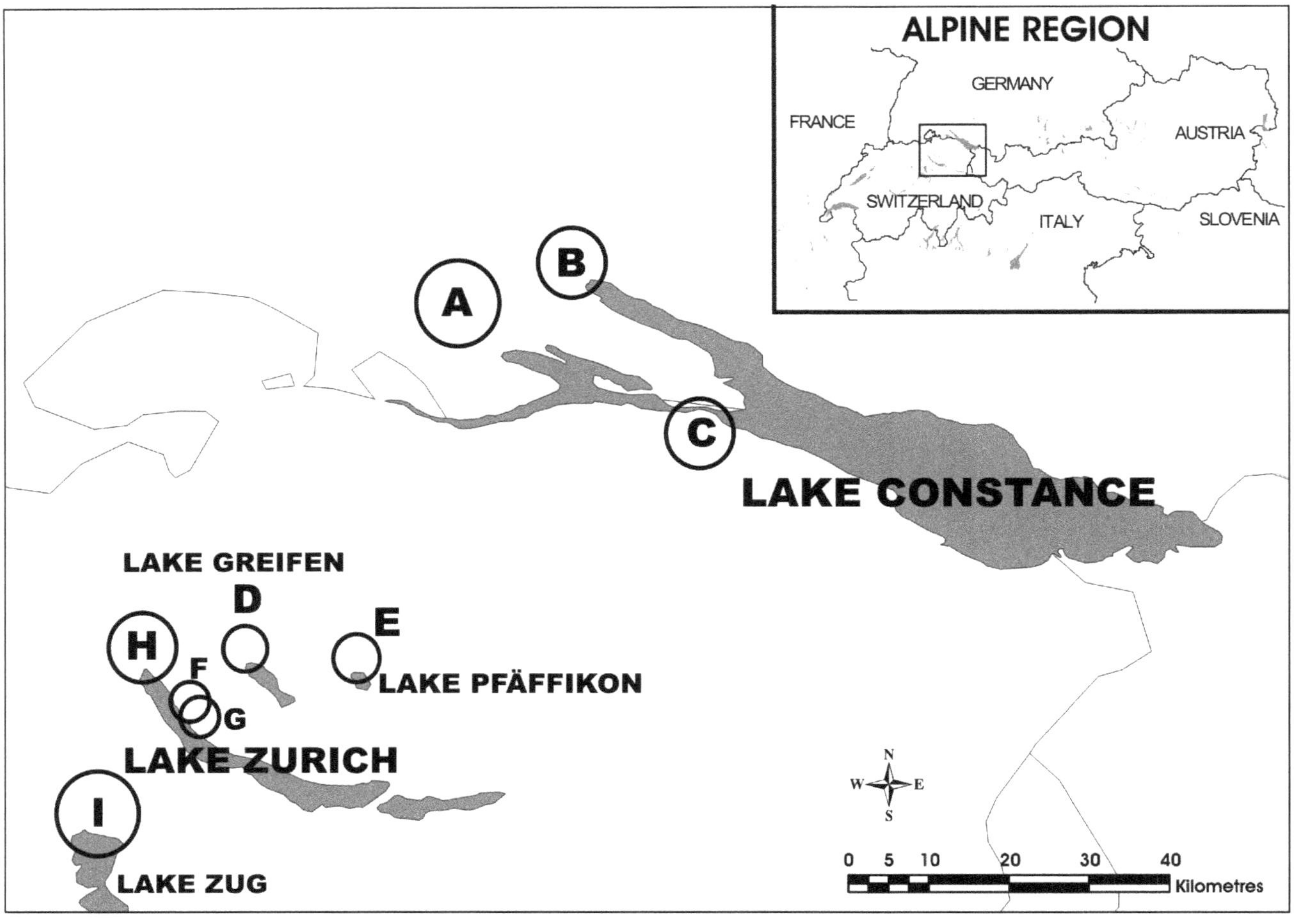

Figure 21.2. Location of the northern Alpine region Middle Bronze Age 'transitional' settlements with lacustrine tradition

Hof, Cham-Oberwil Hinterbüel and Steinhausen Schlossberg are particularly relevant to the MBA lacustrine hiatus. They were occupied throughout the fifteen to twelfth century BC gap. Some continued into the LBA (eleventh century BC). Not all of them were occupied before the fifteenth century BC (Fig. 21.2-I) (Hochuli 1995, Gnepf *et al.* 1996).

The trend of moving inland in search of a drier environment is also noted outside the Alpine region, towards the big southern German plains north of Lake Constance. The best example is the MBA 'land' settlement of *Uttenweiler-Offingen 'Bussen'* on the hills surrounding the Federsee basin. The large quantity of diagnostic pottery along with ^{14}C dates, have placed the site in the MBA, exactly during the MBA lake-dwelling hiatus (Krumland 1998, Schlichtherle and Strobel 2001).

RECONSTRUCTING CONTINUITY

Because of their location, style of house construction, and artifact typology, it is argued that all the MBA sites mentioned in the previous section might represent settlements built by displaced lacustrine populations perhaps already acculturated with terrestrial communities. Recent research on pottery typology and house construction techniques has allowed us to pinpoint cultural transformations and reconstruct continuity.

Among the large variety of decorative patterns in the pottery belonging to the lake-dwelling tradition in the northern Alpine region, there are two that are particularly common. These are embossed and zigzag patterns. They are found throughout the whole Bronze Age (Conscience 1998). With minor variations, these two ways of decorating pottery are present on the majority of the EBA and LBA lacustrine ceramics. Some of the best-known EBA sites exhibiting these patterns are Arbon-Bleiche 2 and Bodman-Schachen 1 (Lake Constance), ZH-Mozartstrasse and Meilen-Schellen (Lake Zürich), several sites in Zürich bay, as well as settlements on Lake Zug (Köninger 1996, Gross 1987, Hochuli 1994, Ruoff 1996, Menotti 2001). The same decorative patterns occur in lacustrine LBA sites occupied after the MBA hiatus (Ruoff 1996, Hochuli 1998, Schöbel 1995).

How could the two above-mentioned ways of decorating pottery have survived during the entire lake-dwelling occupational hiatus (fifteenth to twelfth centuries BC) and reappear in the lake villages more than three centuries later? The MBA 'land' settlements discussed earlier in this chapter provide the possible missing link in the chain of this tradition. A thorough comparative analysis of the pottery assemblages found on these sites has revealed striking similarities to the EBA and LBA pottery (Rigert 2001, Menotti 2003). For instance, at Lake Zürich, the 'land' settlement of Erlenbach (500 m inland from the lake) shows embossed and zigzag ceramic decorations similar to the EBA and LBA lake-dwelling of ZH-Mozartstrasse (Fischer 1997). Similarly, Cham-Oberwil situated in the vicinity of Lake Zug, retains notable similarities to the EBA site of Zug-St. Andreas and the LBA site of Zug-Sumpf (Gnepf 1995, Gnepf *et al.* 1996, Seifert 1992, 1996). Significant evidence comes from the Kreuzlingen area on the Swiss shore of Lake Constance. The majority of sites found there, and in particular Tägerwilen-Hochstross, yielded the two characteristic patterns of ceramic decoration. They are easily linked to both the EBA lake villages of Arbon-Bleiche 2 and Bodman-Schachen 1 on the one hand, and the LBA lacustrine settlements of Konstanz-Rauenegg and Unteruhldingen on the other (all of them situated on Lake Constance) (Rigert 2001).

Decorative patterns on pottery are not the only tool to trace the former lake-dwellers' moves during the MBA. Another effective analytic method concerns house construction, house layouts, and site locations. As already stated, the house remains found on these MBA 'land' settlements are limited and in most cases not very well preserved. Nevertheless, they still retain some fundamental features of construction, which allow us to detect differences as well as similarities to both EBA and LBA lacustrine villages. There are some construction techniques within the Bronze Age that are specifically characteristic of either inland or lacustrine settlements. For example, ground-joints, perforated base-plates, wooden floors, and posts simply driven into the ground are building techniques present in both the EBA and LBA lake-dwellings (Hochuli 1994, Gross 1987). On the other hand, post holes with sustaining piles of stones are only considered a feature of the MBA terrestrial settlements (Regeth 1998, Gollnisch and Seifert 1998). Finally, cobbled floors are found in both MBA 'land' sites and LBA lake-dwellings (Dickmann 1991, Fisher 1997, Seifert 1996).

A striking characteristic of the MBA 'land' settlements found in the vicinity of the lakes is that the majority of them exhibit more or less of all these features. Location, therefore, is an important variable in the equation. Typical MBA terrestrial settlements in the Alpine region are usually situated much further inland from the lakes than those so-called 'transitional land settlements', which are always located in the vicinity of lacustrine areas.

These 'transitional' settlements are not only crucial in reconstructing the MBA lake-dwelling chronology, but they also help us understand cultural and economic adaptability to a new environment resulting from climate change. Furthermore, they also show how past traditions will always be part of people's future cultural make up. In fact, despite the acculturation of these former lacustrine settlers after the displacement, they kept vital connections to the lakes, especially in terms of subsistence and food production. This is evidenced by a fifteenth/fourteenth centuries BC MBA wooden structure, resembling an enormous fish trap. It was found at Zug-Chollerpark on Lake Zug (Hochuli and Röder 2001). The former lake-dwellers changed the location of their settlements, but their everyday economic activities such as fishing, communication, and transport remained related to the lakes. Maybe it was indeed these never-lost links to the wetlands that made those groups resettle the lakes once the transgressive waters were no longer a threat.

CONCLUSIONS

There have been various occupational interruptions throughout the entire lake-dwelling phenomenon in the Alpine region. However, the MBA hiatus, which occurred in the northern part of the Alps between the fifteenth and the twelfth century BC, has always intrigued a larger number of scholars than any other period. Over the years, a number of attempts to shed light on this mysterious period have been made.

The latest results show that the northern Alpine MBA lake communities did not stray far from their native area, once they abandoned their lake-dwelling communities. In response to the climatologically induced rise in lake level, they sought a drier and safer environment in the immediate surroundings of the lakes. However, they kept liaisons with their former stomping grounds, perpetuating their traditions as much as possible. At the beginning of the LBA, when climate change reduced the water-levels and the shores became once again safe to be resettled, they returned to their ancestors' land.

Acknowledgements

I would like to express my sincere gratitude to the British Academy, the Lockey Bequest – University of Oxford, the Meyerstein Award – School of Archaeology, Oxford. Special thanks are also due to all the staff of the Institute of Archaeology, Oxford and in particular to Professor Andrew Sherratt, who provided me with encouragement and useful suggestions. Finally, I would like to thank Maximilian O. Baldia for asking me to contribute this chapter.

References

AUFDERMAUER, J., and B. DICKMANN 1995 – Mittelbronzezeitliche und frühbronzezeitliche Siedlungsbefunde aus Mühlhausen-Ehingen, Kreis Konstanz. *Archäologische Ausgrabungen in Baden-Württemberg*, 1994:65-69.

CONSCIENCE, A.C. 1998 – Wädenswil-Vorderau: Eine Ufersiedlung am Übergang vom 17. zum 16. Jahrhundert v. Chr. im Rahmen der Frühbronzezeit am Zürichsee. Unpublished Master Thesis, University of Zürich, Zürich, 1998.

DICKMANN, B. 1989 – Eine Siedlung der ausgehenden Frühbronzezeit bei Hilzingen, Kreis Konstanz. *Archdologische Ausgrabungen in Baden-Württemberg*, 1989:53-58.

DICKMANN, B. 1990 – Neue Forschungsergebnisse zur Jungsteinzeit im Hegau und Hornstaad am Bodensee. In Schweizerisches Landmuseum (ed.), *Die erste Bauern* 2. Schweizerisches Landesmuseum Zürich 1990:157-169.

DICKMANN, B. 1991 – Sondage in den mittelbronzezeitlichen Siedlungen von Hilzingen, Rielasingen-Worblingen und Hilzingen-Duchtlingen, Kreis Konstanz. *Archäologische Ausgrabungen in Baden-Württemberg*, 1990:56-62.

FISCHER, C. 1997 – *Innovation und Tradition in der Mittel-und Spätbronzezeit: Gräber und Siedlungen in Neftenbach, Fällanden, Dietikon, Pfäffikon und Erlenbach.* Monographien der Kantonsarchäologie Zürich 28, Direktion der öffentlichen Bauten des Kantons Zürich, Hochbauamt, Kantonsarchäologie, Fotorotar AG/Neue Medien, Zürich, 1997.

GNEPF, U. 1995 – Acht neue prähistorische Fundstelle aus dem Kanton Zug. *Tugium* 11, 1995:60-73.

GNEPF, U., P. MOSER and J. WEISS 1996 – Morastige Wege und stattliche Häuser im mittelbronzezeitlichen Cham. *Archäologie der Schweiz* 19/2, 1996:64-67.

GROSS, E. 1987 – Die Lage des Grabungsplatzes. In Berichte der Zürcher Denkmalpflege (ed.), *Zürich "Mozartstrasse": Neolithische und bronzezeitliche Ufersiedlungen.* (Monographien 4, Band 1), Orell Füssli, Zürich, 1987:13-18.

GROSS-KLEE, E., and C. RITZMANN 1990 – Die neolithische und bronzezeitliche Siedlungen im Züricher Seefeld. In Schweizerisches Landesmuseum (ed.), *Die erste Bauern* (Band 1). Schweizerisches Landesmuseum, Zürich, 1990:161-176.

HOCHULI, S. 1994 – *Arbon-Bleiche, die neolithischen und bronzezeitlichen Seeufersiedlungen: Ausgrabungen 1985-1991.* Amt für Archäologie des Kantons Thurgau, Frauenfeld, 1994.

HOCHULI, S. 1995 – Die frühe und mittlere Bronzezeit im Kanton Zug. *Tugium* 11, 1995:74-96.

HOCHULI, S. 1998 – Archäologie im Zugersee. *Nachrichtenblatt Arbeitskreis Unterwasserarchäologie* 4, 1998:16-24.

HOCHULI, S., and B. RÖDER 2001 – Bronzezeitliches Strandgut mit rätselhaften Holzobjecten aus Steinhausen ZG. *Archäologie der Schweiz.* 24/1, 2001:2-13.

KÖNINGER, J. 1996 – *Bodman-Schachen 1: die frühbronzezeitlichen Ufersiedlungen (Tauchsondagen 1982-1984 und 1986).* Freiburg: UB Freiburg, 1996.

KRUMLAND, J. 1998 – *Die bronzezeitliche Siedlungskeramik zwischen Elsass und Böhmen. Studien zur Formenkunde und Rekonstruktion der Besiedlungsgeschichte in Nord- und Südwürttemberg.* International Archaeology 49. VLM, Rahden, Westfalen, 1998.

MAGNY, M. 2004 – The Contribution of Palaeoclimatology to Lake-Dwellings. In F. Menotti (ed.), *Living on the Lake in Prehistoric Europe.* Routledge, London, 2004:207-217.

MENOTTI, F. 1999a – The Abandonment of the ZH-Mozartstrasse Early Bronze Age Lake- Settlement: GIS Computer Simulations of the Lake-Level Fluctuation Hypothesis. *Oxford Journal of Archaeology* 18/2, 1999:143-155.

MENOTTI, F. 1999b – The abandonment of the Early Bronze Age lake-settlement of Bodman-Schachen 1: A CAD and GIS approach to the lake-level fluctuation hypothesis. In J.A. Barceló, I. Briz and A. Vila (eds.), *New techniques for old times: Computer Applications and Quantitative Methods in Archaeology.* BAR International Series 757, Oxford, 1999:265-269.

MENOTTI, F. 1999c – Die Aufgabe der frühbronzezeitlichen Uferrandsiedlung von Bodman-Schachen 1 *Plattform* 7/8, 1999:58-65.

MENOTTI, F. 2001 – *The 'Missing Period': Middle Bronze Age Lake-Dwelling Occupational Hiatus in the Northern Alpine Region.* BAR International Series 968, Oxford.

MENOTTI, F. 2002 – Climatic Change, Flooding and Occupational Hiatus in the Lake-Dwelling Central European Bronze Age. In J. Grattan and R. Torrence (eds.), *Natural Disasters and Cultural Change.* Routledge, London, 2002:235-249.

MENOTTI, F. 2004 – Displacement, Readaptation and Cultural Continuity: A Lake-Dwelling Perspective. In F. Menotti (ed.), *Living on the Lake in Prehistoric Europe.* Routledge, London, 2004:207-217.

MENOTTI, F. 2009 – Climate Variations in the Circum-Alpine Region and their Influence on Neolithic-Bronze Age Lacustrine Communities Displacement and\or Cultural Adaptation. *Documenta Praehistorica* XXXVI, 2009:61-66.

MENOTTI, F., and E. PRANKENAITE 2008 – Lake-Dwelling Building Techniques in Prehistory: Driving Wooden Piles into Lacustrine Sediments. *EuroRAE* 5, 2008:3-7.

REGETH, J. 1998 – Gebäude in Graubünden. In Schweizerische Gesellschaft für Ur- und Frühgeschichte (ed.), *Die Schweiz vom Paläolithikum bis zum frühen Mittelalter: SPM Bronzezeit.* Schweizerische Gesellschaft für Ur- und Frühgeschichte, Basel, 1998:206-211.

RIGERT, E. 1998 – Fundbericht 1997 (Bronzezeit). *Jahrbuch der Schweizerischen Gesellschaft für Ur - und Frühgeschichte* 81, 1998:266-279.

RIGERT, E. 1999 – Fundberichte 1998. *Jahrbuch der Schweizerischen Gesellschaft für Ur -und Frühgeschichte* 82, 1999:250-290.

RIGERT, E. 2001 – *A7-Ausfahrt Archäologie: Prospektion und Grabungen im Abschnitt Schwaderloh-Landesgrenze.* Amt für Archäologie des Kantons Thurgau, Frauenfeld.

RUOFF, U. 1996 – Meilen-Schellen et Zürich-Mozartstrasse, Deux Sites Lacustres au Bord du Lac de Zürich. In C. Mordant and O. Gaiffe (eds.), *Cultures et Sociétés du Bronze Ancien en Europe.* Comité des Travaux Historiques et Scientifiques, Paris, 1996:199-210.

SCHINDLER, C. 1971 – Geologie von Zürich und ihre Beziehung zu Seespiegelschwankungen. *Vierteljahrsschrift der Naturforschenden Gesellschaft in Zürich.* 116/2, 1971:285-315.

SCHINDLER, C. 1981 – Geologische Unterlagen zur Beurteilung archäologischer Probleme in den Seeufergebieten. *Helvetica Archaeologica* 45/48, 1981:71-88.

SCHLICHTHERLE, H. 1995 – Eine Mineralbodensiedlung der Mittelbronzezeit in Bodman, Gde. Bodman-Ludwigshafen, Kreis Konstanz. *Ausgrabungen in Baden-Württemberg.* 1994:61-65.

SCHLICHTHERLE, H., and M. STROBEL 2001 – Ufersiedlungen – Höhensiedlungen: Extremfälle unbekannter Siedlungsmuster der Früh- und Mittelbronzezeit im südwestdeutschen Alpenvorland. In B. Eberschweiler, J. Köninger, H. Schlichtherle and C. Strahm (eds.), *Frübronzezeit und frühen Mittelbronzezeit im nördlichen Alpenvorland.* Offsetdruck Bernauer, Freiburg, 2001:79-92.

SCHÖBEL, G. 1995 – Tauchuntersuchungen in den Siedlungen der Spätbronzezeit am Bodensee. In Landesdenkmalamt Baden-Württemberg (ed.), *Archäologie unter Wasser* 1, Theiss, Stuttgart, 1995:51-57.

SCHÜEPP, M. 1979 – Meteorologische und hydrologische Aspekte und Verhältnisse. In Neue Zürcher Zeitung (ed.), *Der Zürichsee und seine Nachbarseen.* Office du Livre, Fribourg, 1979:35-60.

SEIFERT, M. 1992 – Die Keramik der spätbronzezeitlichen Siedlungsstelle Zug-Sumpf. *Tugium* 8, 1992:64-91.

SEIFERT, M. 1996 – *Die spätbronzezeitlichen Siedlungen von Zug-Sumpf: Die Dorfgeschichte.* Kantonsarchäologie, Zug, 1996.

WESSELS, M. 1995 – Bodensee-Sedimente als Abbild von Umweltänderungen im Spät-und Postglazial. *Göttinger Arbeiten zur Geologie und Paläontologie* 66, 1995:1-105.

Chapter 22
SOCIETY AND ECOLOGY DURING THE MIDDLE BRONZE AGE OF SOUTHERN SCANDINAVIA

Lars LARSSON
Institutionen för Arkeologi och Antikens Historia, Lund University, Lund, Sweden

Abstract: *The most marked prehistoric human impact of southern Scandinavia occurs in the middle part of the Bronze Age. It coincides with a complex climate oscillation dated between ca. 1200-900 BC. This landscape transformation is evaluated from an interregional perspective and compared to the preceding and succeeding periods. Prior to climatic oscillations a large number of barrows were built, containing many bronze grave goods, which legitimized the ritual and social significance of chiefly lineages. During the climate oscillation, monuments and marked symbols are lacking. Thereafter, a new legitimization crisis occurs as new burial mounds are built.*

INTRODUCTION

As a reaction against the view that the individual is a passive component in the cultural system, more importance is attached to societal relationships, and the physical environment is seen as something that humans exploit and shape. This view of humankind and our physical environment has become much more multifaceted. Studies of the attitude of prehistoric societies to the landscape mean that new dimensions are integrated into the research. The prehistoric landscape is seen as not being used exclusively as an economic resource. The shaping of the landscape also has an important social and mental dimension. The perception of the landscape is changed in parallel to the physical change brought about by human impact. Therefore, this chapter attempts to provide a new understanding of the largest landscape transformation in Southern Scandinavia. Different explanations, well-known as well as new, are evaluated from an interregional perspective.

The most marked human influence on the landscape during the entire prehistory of southern Scandinavia is dated to the middle part of the Bronze Age around 1000 BC. This change coincided with a climatically complex period with a cool and wet phase followed by warm and dry conditions. Around 800 BC the climate is again discernibly cooler and wetter. To set the framework within the ongoing archaeological discussion, other innovation periods can be recognized before and after 1000 BC. Before the climate oscillations, a large number of barrows were built. They contain burials with a numerous bronze grave goods of high-quality local or regional bronze artifact manufacture as Southern Scandinavia is established as an independent region within Bronze Age Europe. This is interpreted as reflecting the formation of a new elite and enforcing the ritual and social significance of a chiefly lineage. Following the climate oscillations and human impact on the landscape, i.e. around 800 BC, a new legitimization crisis appears with an increased intensity of hoarding that might reflect an increase in ritual. In between these two periods, around 1000 BC, is a phase that lacks marked symbols in the material culture as well as the formation of monuments. But how are we to understand this transformation of the landscape? Different explanations, well-known as well as new, are evaluated from an interregional perspective.

THEORETICAL DEVELOPMENTS AND SWEDISH CULTURAL LANDSCAPE RESEARCH

In the 1980s, the multidisciplinary Ystad Project entitled *The cultural landscape during 6000 years in southern Sweden* aimed at studying the long-term prehistoric cultural landscape changes since the initial adoption of agriculture in the region (Berglund 1991b, Larsson *et al.* 1992). Collaboration between representatives of the natural sciences and the humanities underlay the project, which began in 1982.

The initiators of the project were greatly influenced by the processual approach in archaeology. There was great interest in paleoecological studies, since these were expected to lead to a better understanding of social organization. Strangely enough, this interest was chiefly embraced by archaeologists involved in Stone Age research. Yet, paleoecological research should really be of greater importance to archaeologists concentrating on later periods, when the relationship between landscape and society should be much more apparent and complex.

Even if most of the research was conducted during the late 1980s, not much of the post-processual thinking influenced the interpretation of the archaeological data of the project. Most of the relationship between society and the environment were looked upon as different modes of adaptation to changes in natural resources.

The replacement of the word adaptation by response might look like a change of minor importance or even unnecessary, but it was a change of major significance. It emphasizes social organization, and the physical environment becomes something that humans exploit and

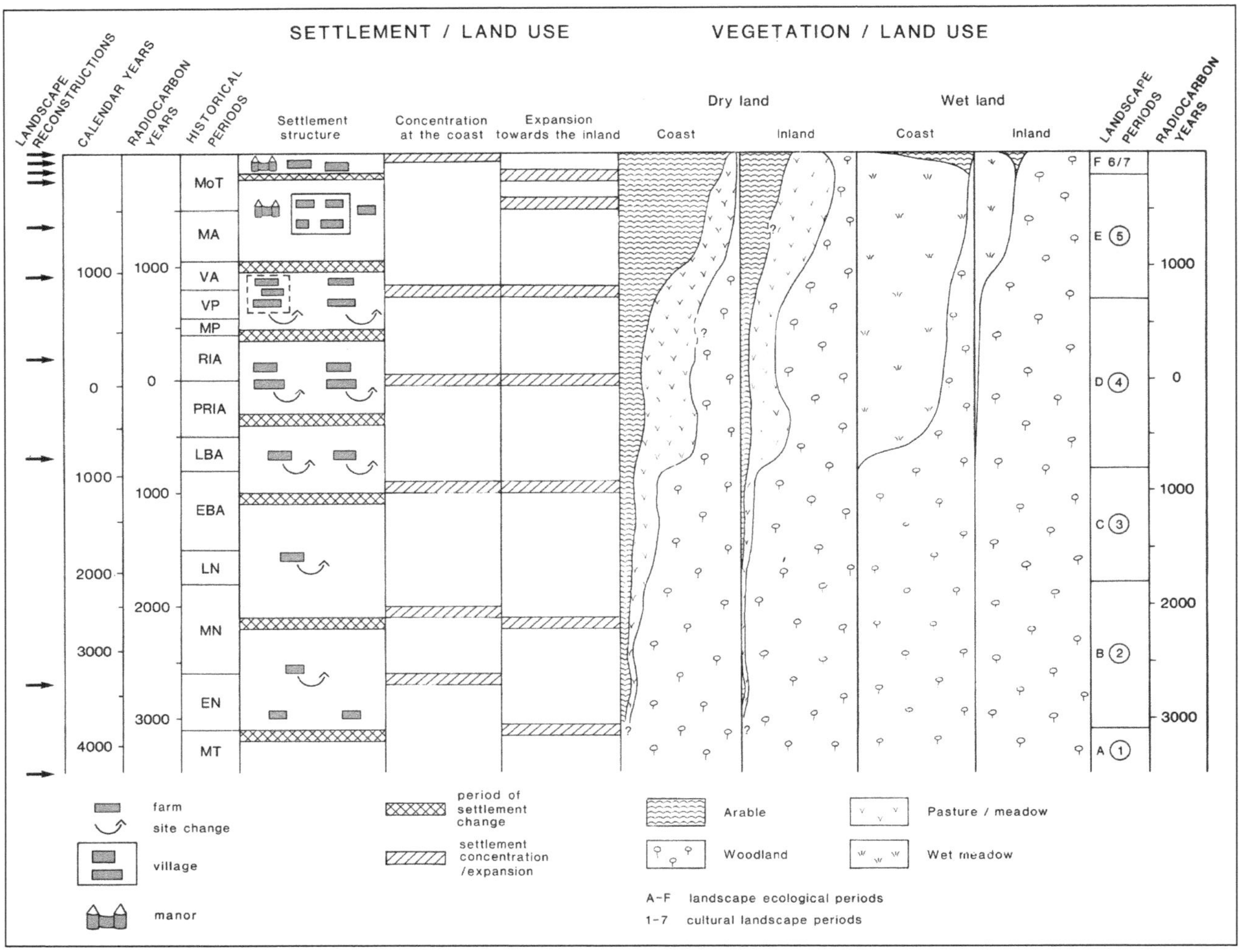

Figure 22.1. A synthesis of long-term changes in the landscape based on research within the Ystad Project (after Berglund *et al.* 1991)

manipulate (Hodder 1982). Post-processual theory sometimes even goes so far that the significance of natural changes are played down at the expense of cultural phenomena, and in some research camps the significant role of the environment is partially denied (Shanks and Tilley 1987, Thomas 1991). As a result, the interest in paleoecological research declined considerably until recently. Now there are clear signs to suggest that the British and American ideas can once again view natural conditions as being of interest to cultural scholars as well. These notions are of considerable significance to Swedish archaeological research.

Although the archaeologist can once again consider natural conditions as a subject of interest, this does not mean a return to the earlier situation. The new view of humankind and our physical environment has become much more multifaceted. As a result the interaction of humankind and our physical environment is perceived as being more complex, in a way that I hope will be evident in what follows below.

The theoretical as well as the methodological basis for the Ystad Project was founded on the interpretation of pollen diagrams from southern Sweden, published in the late 1960s. The result led to the recognition of at least four different prehistoric phases of increased human impact on the vegetation (Berglund 1969). The first phase coincided with the introduction of farming into Sweden. The second was primarily dated to the latest part of the Neolithic (ca. 2000 BC).

The second phase of human impact on the environment was in good agreement with the archaeological evidence, which suggested a geographic expansion of society, especially during the Late Neolithic. Among the lines of evidence was the appearance of the first monuments, such as the megalithic gallery graves erected in the interior of southern Sweden. Another was the fact that at the same time the material culture becomes well represented in both coastal and inland areas of the research area.

THE GREAT CHANGE AROUND 1000 BC

In the 1980s the project engaged in new paleoecological studies. Their results provide a somewhat different picture (Fig. 22.1). The human influence on the landscape

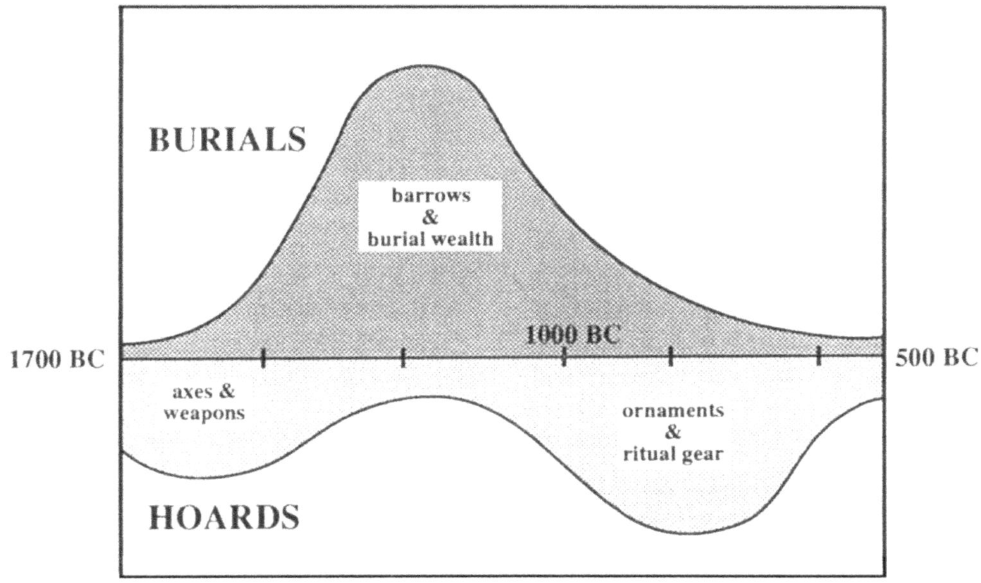

Figure 22.2. Variation of monument construction and wealth depositions during the Bronze Age (after Kristiansen 1991)

was shown to increase during the late part of the Neolithic, but no distinct expansion phase was detected. Instead, a marked change can be identified and provisionally dated to the middle part of the Bronze Age.

This Bronze Age expansion includes the coast as well as the interior. The observable increase of cultivated plants may be explained either as an expansion of arable land, mainly at the expense of semi-natural vegetation, or as intensification of cultivation. The effects are a major reduction of wet meadows and fens in the coastal wetland zones. Inland the same type of landscape was much less effected.

This arable land expansion was initially dated to about 1000 BC based upon the rate of deposition of numerous layers of peat and turf. Unfortunately, radiocarbon assays could not date these layers precisely, and despite its importance, it has not been the target of intensified studies in Sweden since discovery in the early 1990s.

In Zealand, eastern Denmark, studies of Holocene vegetation sequences indicate the same distinctly marked change as in southern Sweden at ca. 1000 BC (Odgaard and Rasmussen 2000). The special Danish study was based on samples from several small lakes. Pollen assemblages dated to around AD 1800 were compared with the prehistoric sequence at each sampling point since the introduction of agriculture. The vegetation around 1000 BC shows a strikingly different composition. From this time onward there is an almost continuous process of change, lasting throughout the rest of prehistory. Furthermore, around 1000 BC, there is the same variation from woodland to heath-land in the landscape between the different sampling points, as was the case around AD 1800. An increase in sedimentation rate can also be registered at about 1000 BC (Løvberg and Odgaard 2000).

All this suggests that until 1000 BC, the vegetation cover was generally less influenced by human impact. Yet, perhaps more importantly, the landscape displayed a more mosaic structure and pattern of change prior to 1000 BC. Furthermore, the transition was not just on a large scale; the vegetation also became more uniform. One is forced to conclude that this change seems to mark the introduction of a more modern way of utilizing the landscape, combined with a marked intensification of human impact.

The intensified human impact and introduction of a more modern landscape utilization coincide with a climatically complex period of cool and wet conditions that were later followed by a warm and dry phase (van Geel *et al.* 1996). Shortly thereafter, at about 800 BC, a period of cooler and wetter conditions occurs, known in Scandinavia as the Sub-Boreal/Sub-Atlantic transition.

Before exploring the prehistoric social structure during this important period of change, it needs to be stressed that this period has not yet been satisfactorily dated. Although the discerned increase in human impact indicates that the change was rapid, the ^{14}C samples provide a range of 1100-700 BC. This is partly because this period coincides with a ^{14}C reservoir effect, which causes a major problem in understanding the complex changes in social interaction that occurred during the most important change in human impact on the landscape in Swedish prehistory.

THE ARCHAEOLOGICAL FRAMEWORK

To set the framework within the archaeological discussion, this period of great change is preceded and followed by periods of innovation (Fig. 22.2). The first dates ca. 1500-1200 BC, the second ca. 800-600 BC.

The first period of innovation falls into the middle part of the Early Bronze Age, when Southern Scandinavia was established as an independent region within Europe. A large number of barrows are built at that time and the primary inhumation burial within these mounds contain several bronze grave goods that reflect the high-quality of local or regional bronze artifact manufacture (Kristiansen 1998).

During the same period, a marked change in the design of houses occurs. Houses with a two-aisled structure change to a three-aisled type – a layout that becomes totally dominant for long-houses throughout the rest of prehistory. This new house-type, along with the barrows, high skill in bronze manufacture, and intensive contacts with other parts of Europe reflects social changes of major importance. This development is sometimes interpreted as reflecting the formation of a new elite and the concomitant enforcement of the ritual and social significance of the chiefly lineage (Kristiansen 1998).

In the second innovation period (ca. 800-600 BC) the number of bronze artifacts increases (Kristiansen 1998). The technical skill of bronze casting reaches its maximum, as exemplified by the bronze lurs – an elaborately designed form of trumpet (Fig. 22.3). However, only a small number of the bronze artifacts are found in the graves in which cremations in urns become completely dominant. Instead of placing the artifacts into the graves, most bronze objects are now hoards found as wetland depositions.

This period is in several respects similar to the expansion of the first period (1500-1200 BC). Contacts with continental Europe once again increase, as indicated, among other things, by objects such as swords and vessels of chased (hammered or forged) bronze plates. The largest houses of the Bronze Age are dated to this phase.

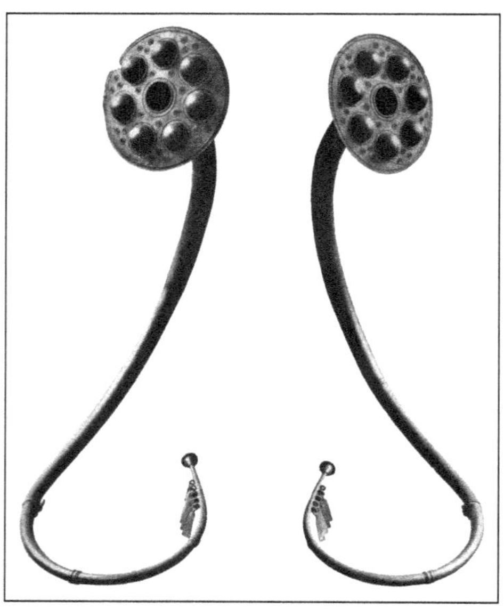

Figure 22.3. Bronze Age lurs
(after Broholm *et al.* 1949)

CULTURAL DYNAMICS: HUMAN RESPONSE TO THE GREAT CHANGE

The foundation for the Bronze Age social organization is laid around 1500 BC and appears to have been consolidated by 1000 BC. After about 900 BC, a new legitimization crisis seems to be initiated, explained by some archaeologists in terms of a degradation of land. Increased intensity of hoarding might reflect an increase in ritual (Kristiansen 1998). Thus, compared with what we know about the deposition of wealth in graves and hoards around 1000 BC, the transition between the Early and the Late Bronze Age around 900 BC (Fig. 22.1) seems to be the period during which the smallest amount of bronze objects were invested in graves as well as hoards. Furthermore, if one examines the variety of types and decoration, the bronze artifact production seems to lack the creativity evidenced both before and after this period, which also lacks marked symbols in the material culture and the formation of monuments.

From an archaeological point of view, therefore, the time around 1000 BC seems somewhat dull. However, from an environmental and climate perspective, the pollen diagrams suggest that it was one of the most dynamic phases in the whole of prehistory. How are we to understand this transformation of the landscape?

A small number of bronze types appear during the time in question. These include socketed axes and sickles – forms that can be directly linked to domestic activities, such as land clearing and harvesting of grasses of grains. This contrasts with most bronzes that can be related to different expressions of power.

A demographic expansion cannot explain the major transformation in vegetation. According to the study of the coastal as well as the inland settlement of the research area, no firm indication exists of any major agricultural expansion into virgin or low-intensity-use areas (Olausson 1992). So what remains?

The rise of cultivation is interpreted as being based upon innovations, such as manuring or the introduction of new crops, e.g., hulled barley. However, the importance of manuring may be overestimated (Lagerås and Regnell 1999) and as a whole, it is questionable whether such a major increase of human impact on the environment was caused by these technological developments. At the same time, no changes in human influence on the landscape are detectable in Jutland, western Denmark, to the west.

Something took place in eastern Denmark and southern Sweden (including the northern part of the state of Götaland). An archaeological explanation would be to view the change as caused by social interactions, but with manifestations of another character than that evident during the Early and Late Bronze Age, respectively (Skoglund 1999a). The houses became smaller, probably as a result of new social systems based on a nuclear family organization. The importance of the nuclear family in society facilitated the colonization of new land.

Large barrows and rich grave goods likely witness the rise of competing chiefly elites. During the time in question, no new barrows are built and the number of grave goods is small, partly caused by a decreasing metal supply (Kristiansen 1998). Even if the chiefdoms were consolidated, there may still have been competition and conflicts at both regional and local scales.

As monument building appears to go out of fashion and production of bronze artifacts decreases, demand for other expressions of power might be sought at the level of interaction between society and nature, which could be very difficult to recover archaeologically. The vegetation change at 1000 BC was more affected by increasing pastures than by arable fields. The importance of pastureland can also be noted for the inner part of southern Sweden. Here, large cairns were still being erected during the time in question. However, these were not located close to traditional settlements and arable land, but in areas some distance away, where pasture seems to be of major importance (Larsson 1994).

One explanation could be an interest in increasing the number of livestock or crops to create a surplus. In fact, at about 1000 BC the climate becomes somewhat wetter. This might have improved pasturing (Schmidt and Gruhle 2003:295). According to the osteological remains, stocks of sheep and pig increase in numbers. Yet, at large settlements with special political power, cattle numbers increase (Kristiansen 1998). In continental Europe the embryo of what later is regarded as the Celtic expression appears at this time. In Celtic society, power based on the number of cattle played a very important social role. Therefore, competition among elite groups might have been expressed in keeping large stocks of cattle used for exchange, great feasts, and bull-baiting.

The middle part of the Bronze Age is the period when stalling of cattle is presumed to have been introduced in southern Scandinavia, because of the form of the house structure. One part appears to be the dwelling section and the other, almost an equal part, seems to function as a stable or briar (Årlin 1998). Besides the interpretation of climatic change as a cause for stalling animals, a change from collective to individual ownership of cattle has been suggested (Pearson 1984, Barker 1999). Cattle might have been of greater ideological importance and a more important part of the cosmology (Olausson 1999). It had become acceptable to quarter the cattle in the same living quarters as humans. It may be that just part of the herd was stalled, such as the milk-producing cows or special animals of high value, needing protection and supervision. The importance of cattle must be viewed not only in a quantitative perspective, but also in a qualitative one. Special animals like large bulls or those with special markings might have been highly desired, especially if cattle increased in value for ritual activities.

Lurs are regarded as perhaps the most exquisite bronzes ever made in northern Europe. They had an important role in the society (Broholm *et al.* 1949, Larsson 2001) (Fig. 22.3). Rock art illustrates the playing of lurs combined with ritual activities. I happened to take part in a meeting where replicas of bronze lurs were played. It took place at a local zoo with cattle. The reaction of the bull to the sound of the lur was amazing. He recognized the sound as the presence of a competitor, for whom he started searching in a very agitated mood. It is precisely during the middle part of the Bronze Age that bronze lurs – today an internationally known signature of the Scandinavian Bronze Age – started to be made. Supposedly they are used in ritual performances – and bull-baiting could be one of these. Thus, there might be a link between vegetation change and an emphasis on bronze lur production.

That new aspects of social life might have caused such widespread changes in the vegetation pattern may be exemplified by a study related to changes from the Early to the Later Bronze Age (Skoglund 1999b). During the Bronze Age, an increasing weed content in domestic areas has been noted. Hulled barley becomes dominant, as flax and gold of pleasure are introduced. Instead of marking a change in agricultural technology, it is interpreted as an increasing interest in cooking weeds instead of grinding domesticated grains to produce flour for baked goods. Houses for just one family become more common, and cooking, as an activity closely connected with the household hearth, was one of several aspects of social change. This is viewed as one of several processes that were introduced as a package in contact with the late Urnfield culture of continental Europe.

CONCLUSION

During the transition from the Early to the Later Bronze Age (ca. 1000 BC) southern Sweden exhibits major ritual, economic, technological, and socio-cultural changes. These coincide with a climatically complex period that starts with cool, wet conditions, followed by a warm and dry phase. This is succeeded at ca. 800 BC by the Scandinavian Sub-Boreal/Sub-Atlantic paleoclimate transition, which provides noticeably cooler and wetter conditions. During this period, staring around 1000 BC, the landscape is altered through human impact on the environment. However, to fully understand this period of great changes, there is a need for further analysis, including the precise role of human influence on the landscape and its relationship to climate oscillations. Let us hope that our common interest in the relationship between society and nature might serve as the platform for a multidisciplinary and international project in the near future.

References

ÅRLIN, C. 1999 – Under Samma tak: Om 'Husstallets' Uppkomst och Betydelse under Bronsåldern ur ett Sydskandinaviskt Perspektiv. In M. Olausson (ed.), *Spiralens öga.* Tjugo Artiklar Kring Aktuell Bronsåldersforskning. Riksantikvarieämbetet. Avdelningen för Arkeologiska Undersökningar Skrifter 25, Riksantikvarieämbetet, Stockholm, 1999:291-207.

BARKER, G. 1999 – Cattle-Keeping in Ancient Europe: To live Together or Apart. In C. Fabech and J. Ringtved (eds.), *Settlement and Landscape.* Proceedings of a Conference in Århus, Denmark, May 4–7 1998. Jutland Archaeological Society, Aarhus, 1999:273-280.

BERGLUND, B.E. 1969 – Vegetation and Human Influence in South Scandinavia During Prehistoric Time. In B.E. Berglund (ed.), *Impact of Man on the Scandinavian Landscape During the Late Post-Glacial.* Oikos 12 (Supplement), Munksgaard, Copenhagen, 1969:9-28.

BERGLUND, B.E. 1991a – The Project: Background, Aims, and Organization. In B.E. Berglund (ed.), *The Cultural Landscape During 6000 Years in Southern Sweden: The Ystad Project.* Ecological Bulletin 41, Munksgaard, Copenhagen, 1991:13-17.

BERGLUND, B.E. (ed.) 1991b – *The Cultural Landscape During 6000 Years in Southern Sweden: The Ystad Project.* Ecological Bulletin 41, Munksgaard, Copenhagen, 1991.

BERGLUND, B.E., L. LARSSON, N. LEWAN, G.A. OLSSON and S. SKANSJÖ 1991 – Ecological and Social Factors Behind the Landscape Changes. In B.E. Berglund (ed.), *The Cultural Landscape During 6000 Years in Southern Sweden: The Ystad Project.* Ecological Bulletin 41, Munksgaard, Copenhagen, 1991: 425-435.

BROHOLM, H.C., W.P. LARSEN and G. SKJERNE 1949 – *The Lures of the Bronze Age.* Gyldendal, København, 1949.

HODDER, I. 1982 – *Symbols in Action: Ethnoarchaeology Studies of Material Culture.* Cambridge University Press, Cambridge, 1982.

KRISTIANSEN, K. 1991 – Magt som Historisk Process. In L. Larsson and E. Ryberg (ed.), *Arkeologi och Makt.* University of Lund, Institute of Archaeology Report Series 40. Institute of Archaeology, Lund, 1991.

KRISTIANSEN, K. 1998 – Europe Before History. Cambridge University Press, Cambridge, 1998.

LAGERÅS, P., and M. REGNELL 1999 – Agrar Förändring Under Sydsvensk Bronsålder. In M. Olausson (ed.), *Spiralens öga.* Tjugo artiklar kring aktuell bronsåldersforskning. Riksantikvarieämbetet, Avdelningen för arkeologiska undersökningar skrifter 25, Riksantikvarieämbetet, Stockholm 1999:263-276.

LARSSON, L. 1994 – *Säterdrift i Sydskandinaviskt Neolitikum? Odlingslandskap* och *Fångstmark. En Vänbok till Klas-Göran Selinge,* Riksantikvarieämbetet, Stockholm, 1994.

LARSSON, L. 2001 – Society and Nature: Forests, Trees and Lures. *Archaeologia Polona* 39, 2001:37-54.

LARSSON, L., J. CALLMER and, B. STJERNQUIST (eds.) 1992 – *The Archaeology of the Cultural Landscape: Field Work and Research in a South Swedish Rural Region.* Acta Archaeologica Lundensia 40/19, Almqvist and Wiksell International, Stockholm, 1992.

LØVBERG, T., and B.V. ODGAARD 2000 – Long-Term Development and Human Impact on the Vegetation Around Lake Birkerød, Eastern Denmark, as Indicated by Pollen and Mineral Magnetic Analysis. In P. Sandgren (ed.), *Environment Changes in Fennoscandia During the Late Quaternary.* Lundqua Report 37, Department of Quaternary Geology, Lund University, Lund, 2000.

ODGAARD, B.V., and P. RASMUSSEN 2000 – Origin and Temporal Development of Macro-Scale Land-Cover Patterns in the Cultural Landscape of Denmark. In P. Sandgren (ed.), *Environment Changes in Fennoscandia During the Late Quaternary.* Lundqua Report 37, Department of Quaternary Geology, Lund University, Lund, 2000.

OLAUSSON, D. 1992 – The Archaeology of the Bronze Age Cultural Landscape: Research Goals, Methods, and Results. In L. Larsson, J. Callmer and B. Stjernquist (eds.), *The Archaeology of the Cultural Landscape. Fieldwork and Research in a South Swedish Rural Region.* Acta Archaeologica Lundensia 4/19, Almqvist and Wiksell International, Stockholm, 1992:251-282.

OLAUSSON, M. 1999 – Herding and Stalling in Bronze Age Sweden. In C. Fabech and J. Ringtved (eds.), *Settlement and Landscape.* Proceedings of a conference in Århus., Denmark, May 4–7 1998, Jutland Archaeological Society, Aarhus, 1999:319-328.

PEARSEN, M.P. 1984 – Economic and Ideological Change: Cyclical Growth in the Pre-State Societies of Jutland. In D. Miller and C. Tilley (eds.), *Ideology, Power and Prehistory.* Cambridge University Press, Cambridge, 1984:69-92.

SCHMIDT, B and W. GRUHLE 2003 – Niederschlalags-schwankungen in Westeuropa während der letzten 8000 Jahre: Versuch einer Rekonstruktion. *Archäologisches Korrespondenzblatt* 33, 2003:281-300.

SHANKS, M., and C. TILLEY 1987 – *Re-constructing Archaeology: Theory and Practice.* Cambridge University Press, Cambridge, 1987.

SKOGLUND, P. 1999a – De Enskilda Hushållens Betydelse för Landskapsutvecklingen Under Bronsålder. In M. Olausson (ed.), *Spiralens öga.* Tjugo artiklar kring aktuell bronsåldersforskning. Riksantikvarieämbetet, Avdelningen för Arkeologiska Undersökningar Skrifter 25, Riksantikvarieämbetet, Stockholm, 1999:277-289.

SKOGLUND, P. 1999b – Diet, Cooking and Cosmology: Interpreting the Evidence from Bronze Age Plant Macrofossils. *Current Swedish Archaeology* 7, 1999:149-160.

THOMAS, J. 1991 – *Rethinking the Neolithic.* Cambridge: Cambridge University Press, 1991.

Van GEEL, B., J. BUURMAN and H.T. WATERBOLK 1996 – Archaeological and Palaeoecological Indications of an Abrupt Climate Change in the Netherlands, and Evidence for Climatological Teleconnections Around 2650 BP. *Journal of Quaternary Science* 11/6, 1996:451-460.

Chapter 23
SUMMARY AND CONCLUSIONS

Maximilian O. BALDIA, Timothy K. PERTTULA and Douglas S. FRINK

PART I: PALEOCLIMATE AND SOCIO-CULTURAL CHANGE IN THE NEW WORLD

Chapter 2: Joel Gunn, William J. Folan, and Joseph M. Herbert, *Dangerous Regions: A Source of Cascading Cultural Changes.* Paleoclimate, environmental and archaeological data from the Maya lowlands of the Yucatan Peninsula and the U.S. Coastal Plain of North and South Carolina are compared. The shifts across critical subsistence level boundaries appear in many instances to lead to cascading cultural changes as populations and ideas are flushed out of more environmentally sensitive regions into surrounding less sensitive ones. As global temperatures decline or elevate from the optimum, negative impacts on the complexity of the social organization are observed. However, the precise mechanisms of the impacts vary significantly.

Chapter 3: Timothy K. Perttula, *Risky Business: Caddo Farmers Living at the Edge of the Eastern Woodlands.* The successes and failures of the Native American Caddo farmers during times of rapidly fluctuating climatic conditions (ca. AD 1400-1680) are analyzed. The farmers lived at the western edge of the eastern Woodlands in Texas, Arkansas, Louisiana, and Oklahoma during repeated periods of drought-like conditions. This includes the major cool and dry period (ca. AD 1430-1470). By reorganizing the structure, location, and dispersion of settlements, the Caddo lessened competition for plant and animal resources, thereby alleviating, or at least minimizing, stress during such periods. A similar development is suggested for Europe (Chapters 16-18).

Chapter 4: Dean Snow: *Environmental Change, Population Movements, and the Archaeological Record.* North American archaeology provides examples on how archaeology can solve large-scale demographic problems using modern dating techniques and geographic information systems. The cases illustrate that large-scale processes involve many small-scale migrations and demonstrate the timing of migration episodes with climatic shifts. They also illustrate the common emergence of matrilineal social organizations amongst migrating communities of horticulturalists, providing basic processes that characterize the adaptive radiation of such societies as a response to long-term environmental changes.

PART II: PALEOCLIMATE AND SOCIO-CULTURAL CHANGE IN THE OLD WORLD: AFRICA

Chapter 5: Ralf Vogelsang and Birgit Keding, *Climate, Culture, and Change: From Hunters to Herders in Northeastern and Southwestern Africa.* The focus is on the process of economic change during the middle and late Holocene in Northeastern and Southwestern Africa. Both areas show similar economic sequences with a transition from foraging to pastoralism. However, differences in the transition process are observed. There was a complete change of the economic system in the Sudanese Sahara. In Southwestern Africa, the socio-cultural response differs, because only some of the pastoral life ways were adopted. Therefore, variability in the climate and environment only partially explain the differences.

Chapter 5: David K. Wright, *Fits and Starts: Why Did Domesticated Animals 'Trickle' Before They 'Splashed' Into Sub-Saharan Africa?* Environmental and climate data is used to evaluate two theories that aim to account for the relatively slow development pastoralism south of the Lake Turkana Basin between 2000 and 1000 BC. One suggests the spread of domesticated animals south of the equator between ca. 2000 – 1000 BC was hindered by arid conditions. The situation changed with the onset of pluvial conditions after 1000 BC. The second theory argues that episodes of disease hindered the spread of domesticated cattle into eastern Africa. Based on data from Tsavo, Kenya both theories apply suggesting a more nuanced relationship between culture and climate change.

Chapter 7: Munyaradzi Manyanga, *Socio-Cultural Responses to a Changing Environment: The Shashe-Limpopo Valley Since ca. AD 900.* The research incorporates southern Africa's first known state system, Mapungubwe, which developed between the 11th and the 12th century AD. Past archaeological research and environmental reconstruction are re-examined along with data from extensive surveys. Despite numerous fluctuations between semi-arid and arid conditions, the area appears to have been continuously occupied since the Early Stone Age. This is contrary to previous assumptions, which suggested that climate change interrupted human occupation of the region. Observations of living cultures are used to explain the ability of hunter-gatherers to state-level agriculturalists to respond to climate change.

PART III: PALEOCLIMATE AND SOCIO-CULTURAL CHANGE IN THE OLD WORLD: EUROPE

Chapter 8: Olena V. Smyntyna, *Early Prehistoric Migration as Socio-Cultural Response to a Changing World.* Migration is often invoked as the typical human response to climate change. Traditionally, Late Palaeolithic and Mesolithic groups are said to have changed their territory in time and space. In most cases, these movements supposedly occur due to the necessity for a secure food supply. This oversimplified interpretation is reconsidered in light of migration theory and ethnic contact. It is suggested that a complicated system of different kinds of migrations between the Black Sea, the Middle and Lower Danube, east Central and Northern Europe existed even before the introduction of Near Eastern domesticates. The theoretical discussion of migration, which focuses among other things on human agency, is just beginning to be enhanced by new insights resulting from Black Sea sea-level changes resulting from climatic oscillations.

Chapter 9: David Calado, Francisco Nocete, Dimas Martín-Socas, Maria Dolores Càmalich, José Miguel Nieto, *The Early Megaliths of SW Atlantic Europe and the Inference of the Socio-economic Organization of their Builders (8th to 6th millenniums BC).* Contrasting with other parts of Europe, there may be relatively little impact on the environment due to climate change in the research area. Therefore, the surprisingly early pattern of dense settlements with standing stones may be attributable to a relatively stable environment. This led to early sedentary complex hunter-gatherer or food producer communities who did not use Near Eastern domesticates. These communities were probably responsible for a heavy economic impact on the environment.

Chapter 10: Pál Sümegi, *Preneolithization: Reconstructing the Environmental Background to Life Way Changes in the Late Mesolithic of the Carpathian Basin.* The interaction between climate change, environment, and humanity is used to form the hypothesis that the Mesolithic/Neolithic transition was autochthonous. The data for this hypothesis comes from new bog cores in the Carpathian Basin. Accordingly, Early Holocene climate changes triggered significant social and technical changes in the Mesolithic. During the Late Mesolithic, the climate oscillations were less severe and human induced environmental changes become evident. Some Neolithic techniques and methods develop between 7000-6000 BC, i.e. before the use of Near Eastern cultigens. The resulting independent Pre-Neolithic phase continues in the foothills of the Carpathians during the Neolithic Körös culture. It is concluded that climate change played a significant role in the Hungarian Mesolithic, but not during the Mesolithic/Neolithic transition and environmental change is largely attributed to human impact.

Chapter 11: Pál Sümegi, Róbert Kertész, Imola Juhász, Gábor Tímár, Sándor Gulyás, *The Mesolithic/Neolithic Transition in the Carpathian Basin: Was there an Ecological Trap During the Neolithic?* New geoarchaeological and environmental data confirm that agricultural expansion came to a temporary halt in the Carpathian Basin due to a complex mosaic environment. However, autochthonous hunter-gatherer pre-adaptation and interaction with agriculturists eventually led to the Neolithic transition culminating in the Bandkeramik culture, which brought agriculture to the rest of Central and Eastern Europe. Climate change is not considered to play a major role in this process. Contradictory evidence is considered in Chapter 16, this volume.

Chapter 12: Imola Juhász, *New Data Concerning the Detection and Nature of Human Impact on the Mohos Lakes, North-East Hungary.* Palynological analysis from the cores of the Kis-Mohos Marsh yield new data about the vegetation changes at the Pleistocene/Holocene boundary of the Kelemér region. This provides evidence of the human landscape utilization and transformation. The palynological data are correlated with the process of pre-neolithization, the deforestation activities in the Neolithic, as well as the environmental conditions in the Copper and Bronze Age, the Keltic (Celtic) period, and the Migration period (including the immigrating Hungarians). Environmental change is largely attributed to human impact.

Chapter 13: Pál Sümegi, Imola Juhász, Gábor Timár, Sándor Molnár, Katalin Herbich, Mariann Imre, Gabriella Szegvári, Sándor Gulyás, *Late Neolithic Man and Environment in the Carpathian Basin: A Preliminary Geoarcheological Report from Csőszhalom at Polgár.* The site of Csőszhalom (ca. 5000-4600 BC) at Polgár, Hungary, is established following the demise of the Bandkeramik culture. The unusual site consists of a tell surrounded by concentric ditches and an adjacent unenclosed village. The Paleogeography of the area is determined through sedimentological and micromorphological analysis. Pedological and paleobotanical analyses suggest that human activity is concentrated near the tell and the settlement. Minimal activity can be identified in the alluvial plain. Pollen and pedological analyses lead to the conclusion that the highest flood protected areas were used for cultivation. Macrofossil research indicates the extension of the croplands to the seasonally flooded areas. The research indicates that Late Neolithic populations, living in a mosaic environment, totally altered the original the vegetation cover of the loess levees, while the flora composition of marshes and forests were only marginally modified. Climate change is not considered a major factor.

Chapter 14: Sándor Gulyás and Pál Sümegi, *Freshwater Mussels and Life in the Late Neolithic Tell of Hódmezővásárhely-Gorzsa, southeastern Hungary.* The research takes advantage of comparative archeological data from excavations and multi-disciplinary research from various of the Old and New World to analyze the shell-fish diet in the Late Neolithic tell near Hódmezővásárhely-Gorzsa, Southeastern Hungary. The Hungarian shellfish are analyzed archeozoologically, morphometrically and paleoecologically. It is concluded that shellfish was a secondary food source and its intensified exploitation periods may correspond with new settlement stages rather than climate change.

Chapter 15: Imola Juhász, *Imprints of the Anthropogenic Influences in a Peat Bog from Transdanubia, Hungary*. The Holocene vegetation history of the large Pötréte swamp in the Hahót Basin of the Zala Hills, in southwest Hungary is reconstructed. Palynological data is correlated with archaeological information to determine the anthropological influences on the vegetation dynamics. The chronology of the Holocene mesophile forest taxa and the anthropogenic influence from the Copper Age to the Migration Period are delineated. The changes in vegetation are suggested to be primarily due to human agency rather than climate change.

Chapter 16: Maximilian O. Baldia, *Breaking Unnatural Barriers: Comparative Archaeology, Climate, and Culture Change in Central and Northern Europe (6000-2700 BC)*. The chapter evaluates the punctuated expansion of agriculture from the Carpathian Basin to Scandinavia and its relationship to paleoclimate changes, using high-resolution proxy climate records. A variety of the socio-cultural responses to different kinds and degrees of paleoclimate change are observable. Socio-cultural adjustments, including technological innovations, are developed in response to these changes. Paradoxically, these adjustments and innovations ultimately impact the environment and possibly even the climate itself starting around 3000 BC.

Chapter 17: Matthew Boulanger, *Cultural Geography in the Context of Climatic and Environmental Change: Neolithic and Eneolithic Transitions of the Upper Morava Valley*. GIS, GPS, aerial photography, soil types, geologic and river system formation are used in a spatial, temporal, and environmental analysis of changing settlement patterns during the Neolithic/Copper Age in the Eastern Czech Republic. These data, when compared to various paleoclimate proxies, suggest that society in the Eastern Czech Republic behaved in an organic fashion, developing with, and responding to, a dynamic environment.

Chapter 18: Douglas Frink, *Identification and Interpretation of Taphonomic Processes Affecting Monumental Earthen Architecture: A Pedo-Physiological Approach*. Application of the pedo-physiological method to soil analysis during archaeological excavations allows correlation between the autonomous dynamic systems of climate, soils, and past human behavior. The method enables rigorous hypothesis building and testing. Close interval sampling along a vertical soil column is used to determine the archaeological and temporal context of artifacts and turbational features. Data from recent excavations of the Neolithic-Copper Age Funnel Beaker/Baden-Boleráz culture of the Džbán Long-barrow Group, Moravia (Czech Republic) is correlated with soil turbation events and monumental construction. Climate change and the decay process of monumental architecture are delineated through the analysis of pedogenic processes.

Chapter 19: Paweł Valde-Nowak, *Neolithic Settlement in the Central-European Highlands*. The highlands of Central Europe are thought to have been avoided in the Neolithic, due to relatively low soil fertility, a reduced vegetation growth period, strong climatic variation, and other geographical conditions unfavorable for agriculture. This is negated by paleobotanical data and stray finds evaluated through surveys from Poland and Germany. Pointing to the array of preserved organic material of the Austro-Italian iceman, which were fortuitously preserved by an alpine glacier, it is argued that the stray finds of the central European highlands represent an unrecognized complexity. Therefore, the highlands are far from peripheral to Neolithic activity and may have been used for pastoral pursuits. Furthermore, the favorable climate between 5500-5000 BC allowed highland utilization by the Linienbandkeramik farmers. This coupled with the less favorable climatic conditions for agricultural activity and an increasing number of stray highlands finds during the later parts of the Neolithic are seen as a result of economic advances and demographic pressure. Therefore, the increasing highland utilization is not seen as a response to climate changes. This conclusion differs markedly from those reached for the same period in Chapters 16 to 18.

Chapter 20: Edward Mitchell, *Separating Natural and Anthropogenic Influences on Past Ecosystems: The Testate Amoebae and Quantitative Paleoenvironmental Reconstruction*. Testate amoebae (protozoa) respond to ecological change, such as hydrology, water chemistry, or nutrient status. Their response to environmental change within months or even weeks makes them a high-resolution climate proxy. Thus, one can create transfer functions for quantitative paleoenvironmental reconstruction, using modern ecological data sets. An example from the Praz-Rodet Bog in the Swiss Jura Mountains is used to illustrate the method. Together with palynological data, it demonstrates among other things, that agricultural activity first impacted the area of the former lake around 4900 BC.

Chapter 21: Francesco Menotti, *Environmental and Cultural Change in the Alps: Seeking Continuity in the Bronze Age Lake-Dwelling Tradition*. Ever since their discovery in the 19th century, the famous prehistoric Alpine lake dwellings have produced advanced insights into past human societies. The Alpine region also has precise dendro-chronological control with high resolution climate records. These data are used to analyse the Middle Bronze Age abandonment of the northern Alpine lakeshores by its lake dwellers. The reconstructed palaeo-lacustrine environments and simulated lake level fluctuations demonstrate that the abandonment was the socio-cultural response to abrupt environmental change. It is concluded that, after having moved inland at the end of the sixteenth century BC, renewed climate change at the beginning of the twelfth century BC brought the lake dwellers back to the lakeshores.

Chapter 22: Lars Larsson, *Society and Ecology During the Middle Bronze Age of Southern Scandinavia.* During Bronze Age, which starts there around 1700 BC, large barrows with inhumation graves appear. They contain numerous bronze grave goods. Around 1000 BC the strongest human impact on the environment known in southern Scandinavian prehistory is recorded. This ca. 360 year-long phase heralds a more modern landscape utilization, as well as changes in house construction and religious rituals in southern Scandinavia. The period coincides with complex climate changes. A renewed switch in climate around 800 BC coincides with further socio-cultural changes, including a possible increase in ritual.

CONCLUSIONS

Overall, the specific concepts of culture, environment and climate are vague. In fact, environment and climate are often used interchangeably, even though they are different overlapping systems. At minimum, environment should be seen as the physical surroundings defined by the local and regional geography, geology, topography, hydrology, as well as the spectrum of plants and animals within those surroundings. This distinguishes it from climate, i.e. the meteorological conditions, which lead to long-term weather patterns (decadal or millennial) in a region large enough to contain a recognizable socio-cultural system. Future research should improve on these distinctions.

As illustrated in the summary, Comparative Archaeology is a global approach which can profitably explore the variety of human adjustment to climate change and point to the similarities and differences in the trajectory of human behavior. The multidisciplinary and nonlinear approaches presented, combine data from the hard sciences with analysis of archaeological information to determine socio-cultural adjustment strategies of humans in the New and Old Worlds.

Comparison of the contributions shows that perspectives and methods vary. They include a wide variety of approaches derived from ethnography, cultural geo-graphy, palyno-logy, pedo-physiology, paleoclimatology, etc. Among the variety of methods from the hard sciences is the analysis of the testate amoebae, which allows the reconstruction of paleohydrology and past human land use.

Differences in the perspectives expressed in various contributions are invaluable, because they force the research community to examine the data from different angles. The differences begin with the definition of culture. Contributors with training in Cultural Anthropology generally search for similarities and differences in how cultures interact with the separate but overlapping systems of the environment and climate through time and space. Contributors with a training in the culture-historical tradition search for geographically definable boundaries of specific ethnic populations in which environment and climate limit ethnic groups to specific regions. Thus, migration is generally proposed as the human response to climate change. In general, those who favor human agency as the prime mover of culture change minimize the impact of climatic change, irrespective of their anthropological or culture-historical perspective.

From a global perspective, the results from the Americas and Africa may indicate that shifts over subsistence level borders result in cultural adjustments which in some instances appear first in some of the environmentally more sensitive regions. This contrasts with perspectives from temperate Europe, where innovations are largely thought to originate in the less marginal zones, such as the Danube region. Even the Central European highlands are thought not to be entirely marginal.

While long-term climatic events have traditionally been considered to be most significant, several contributions profitably address the significance of short term climatic events, such as precipitation, evaporation, temperature, floods etc. on the regional and even local level. Cultural adjustments to such changes are documented in the form of precisely dated water wells, log ways (corduroy roads) to village translocation. Changes in house architecture and settlement pattern related to climatic fluctuations best documented through the so-called lake-dwellings of the Neolithic and Bronze Age. Other technological changes range from the location of storage pits in Africa to the development of wooden inter-village roads in temperate Europe. The development of rudimentary roads and wheeled vehicles during strong climatic fluctuations suggests that long-term social security was promoted by increasing communication efficiency.

Over time more complex societies develop and man-made barriers seem to become more prominent than those related to climate. Simultaneously, human impact on the environment becomes ever more pronounced. After 3000 BC the traces of copper mining and smelting begin to leave an increasing footprint as far away as Greenland (Chapter 16, this volume). Thus, the refinements in socioeconomic strategies, aimed at coping with a changing world begin to create a feedback loop that seems to amplify environmental and climatic changes. The loop's continuing effect is underlined by the recent opening of the fabled Northwest Passage through the Arctic ice sheet. This long searched for shortcut from Europe to Asia may soon be exploited as a commercial shipping lane. However, this pursuit for ever more efficient communication and exchange may exacerbate Arctic ice sheet melting. Melting of the polar icecaps and accelerating sea-level rise is usually attributed to man-made CO_2 emissions by industrialized nations. Man-made or not, global warming transcends modern climatic and cultural barriers.

In contemplating this conclusion it becomes clear that the subject matter of this volume has invited an impersonal tone. However, climate and socio-cultural change plays out on a very personal level, even if people do not take notice of climatic change in everyday life. For instance, when the senior author visited Austria's highest peak with his small children in 1973, he observed that the Pasterze Glacier had shrunk considerably since his own childhood, but he did not think of long-term global warming. Similarly, in late October 2012 most New Yorkers around Battery Park may have thought about upcoming Halloween and perhaps even of the impending tropical storm, but not about the significance of the ca. 40 cm sea-level rise observed there since the mid eighteen-hundreds. Still, this rise together with the high tide during hurricane Sandy's landfall changed their lives. Lower Manhattan and other heavily populated coastal regions were flooded. The elderly and infirm suffered the most. Living a short distance inland, the senior author personally experienced only minor episodes of heavy rain, but high winds toppled even trees, which had withstood storms for two hundred years or more. All forms of modern communication shut down for days. Electricity was not restored to the home for nearly twelve days and food spoiled in the modern storage pits (refrigerator and freezer). Traffic came to a halt in the region for days, gasoline was rationed, and five weeks after the storm public transportation is barely reaching a semblance of normalcy. The personal cost to those who lost loved ones, houses, or their livelihood is not fathomable by outsiders. Yet, the storm is only the latest manifestation of the accelerating climatic oscillations. In the previous year tropical storm Irene and an unseasonable early snow storm caused similar upheaval. Although, the unexpected snow storm was followed by the virtual lack of snow and mild temperatures last winter, heavy snowfalls and very cold temperatures marked the preceding two winters. Simultaneously, summer temperatures keep reaching new highs. This must serve as a reminder that climatic change exacts a heavy toll, which is always born by individual human beings, whether they lived in long past cultures or exist in the technologically most advanced societies of the present.

www.ingramcontent.com/pod-product-compliance
Ingram Content Group UK Ltd.
Pitfield, Milton Keynes, MK11 3LW, UK
UKHW061845041026
15591UKWH00015B/919

9 781407 310640